This is the first book devoted to Bose–Einstein condensation (BEC) as an interdisciplinary subject, covering atomic and molecular physics, laser physics, low temperature physics, nuclear physics and astrophysics. It contains 18 authoritative review articles on experimental and theoretical research on BEC and associated phenomena.

Bose–Einstein condensation is a phase transition in which a macroscopic number of particles all go into the same quantum state. It has been known for some time that this phenomenon gives rise to superfluidity in liquid helium. However, recent research deals with BEC in a number of other systems, such as excitons in semiconductors, spin-polarized hydrogen, laser-cooled atoms, high-temperature superconductors, and subatomic matter. The search for BEC in these new physical systems is an exciting field of research, and this unique book gives an in-depth report on recent progress, as well as suggesting promising research for the future.

This book will be of interest to graduate students and research workers in all fields of physics, but especially those in condensed matter, low temperature, atomic and laser physics.

Bose–Einstein Condensation

BOSE–EINSTEIN CONDENSATION

Edited by

A. GRIFFIN
University of Toronto, Toronto, Canada

D. W. SNOKE
University of Pittsburgh, Pittsburgh, USA

S. STRINGARI
University of Trento, Povo, Italy

Published by the Press Syndicate of the University of Cambridge
The Pitt Building, Trumpington Street, Cambridge CB2 1RP
40 West 20th Street, New York, NY 10011-4211, USA
10 Stamford Road, Oakleigh, Melbourne 3166, Australia

First published 1995
Reprinted 1996
First paperback edition 1996

A catalogue record of this book is available from the British Library

Library of Congress cataloguing in publication data

Bose–Einstein condensation / edited by A. Griffin, D.W. Snoke, S. Stringari.
p. cm.
Includes index.
ISBN 0-521-46473-0
1. Bose–Einstein condensation. I. Griffin, Allan. II. Snoke, D. W.
III. Stringari, S.
QC175.47.B65B67 1995
530.4′2–dc20 94-27795 CIP

ISBN 0 521 46473 0 hardback
ISBN 0 521 58990 8 paperback

Transferred to digital printing 2002

TAG

Contents

Preface *page* xi
Preface to paperback edition xiii

1 Introduction: Unifying Themes of Bose–Einstein Condensation *D. W. Snoke and G. Baym* 1

Part one: Review Papers 13
2 Some Comments on Bose–Einstein Condensation *P. Nozières* 15
3 Bose–Einstein Condensation and Superfluidity *K. Huang* 31
4 Bose–Einstein Condensation in Liquid Helium *P. Sokol* 51
5 Sum Rules and Bose–Einstein Condensation *S. Stringari* 86
6 Dilute-Degenerate Gases *F. Laloë* 99
7 Prospects for Bose–Einstein Condensation in Magnetically Trapped Atomic Hydrogen *T. J. Greytak* 131
8 Spin-Polarized Hydrogen: Prospects for Bose–Einstein Condensation and Two-Dimensional Superfluidity *I. F. Silvera* 160
9 Laser Cooling and Trapping of Neutral Atoms *Y. Castin, J. Dalibard and C. Cohen-Tannoudji* 173
10 Kinetics of Bose–Einstein Condensate Formation in an Interacting Bose Gas *Yu. Kagan* 202
11 Condensate Formation in a Bose Gas *H. T. C. Stoof* 226
12 Macroscopic Coherent States of Excitons in Semiconductors *L. V. Keldysh* 246
13 Bose–Einstein Condensation of a Nearly Ideal Gas: Excitons in Cu_2O *J. P. Wolfe, J. L. Lin and D. W. Snoke* 281
14 Bose–Einstein Condensation of Excitonic Particles in Semiconductors *A. Mysyrowicz* 330

15 Crossover from BCS Theory to Bose–Einstein Condensation *M. Randeria* 355
16 Bose–Einstein Condensation of Bipolarons in High-T_c Superconductors *J. Ranninger* 393
17 The Bosonization Method in Nuclear Physics *F. Iachello* 418
18 Kaon Condensation in Dense Matter *G. E. Brown* 438
19 Broken Gauge Symmetry in a Bose Condensate *A. J. Leggett* 452

Part two: Brief Reports 463
20 BEC in Ultra-cold Cesium: Collisional Constraints *E. Tiesinga, A. J. Moerdijk, B. J. Verhaar, and H. T. C. Stoof* 465
21 BEC and the Relaxation Explosion in Magnetically Trapped Atomic Hydrogen *T. W. Hijmans, Yu. Kagan, G. V. Shlyapnikov and J. T. M. Walraven* 472
22 Quest for Kosterlitz–Thouless Transition in Two-Dimensional Atomic Hydrogen *A. Matsubara, T. Arai, S. Hotta, J. S. Korhonen, T. Mizusaki and A. Hirai* 478
23 BEC of Biexcitons in CuCl *M. Hasuo and N. Nagasawa* 487
24 The Influence of Polariton Effects on BEC of Biexcitons *A. L. Ivanov and H. Haug* 496
25 Light-Induced BEC of Excitons and Biexcitons *A. I. Bobrysheva and S. A. Moskalenko* 507
26 Evolution of a Nonequilibrium Polariton Condensate *I. V. Beloussov and Yu. M. Shvera* 513
27 Excitonic Superfluidity in Cu_2O *E. Fortin, E. Benson and A. Mysyrowicz* 519
28 On the Bose–Einstein Condensation of Excitons: Finite-lifetime Composite Bosons *S. G. Tikhodeev* 524
29 Charged Bosons in Quantum Heterostructures *L. D. Shvartsman and J. E. Golub* 532
30 Evidence for Bipolaronic Bose-liquid and BEC in High-T_c Oxides *A. S. Alexandrov* 541
31 The Dynamic Structure Function of Bose Liquids in the Deep Inelastic Regime *A. Belić* 550
32 Possibilities for BEC of Positronium *P. M. Platzman and A. P. Mills, Jr* 558
33 Bose–Einstein Condensation and Spin Waves *R. Friedberg, T. D. Lee and H. C. Ren* 565
34 Universal Behaviour within the Nozières–Schmitt-Rink Theory *F. Pistolesi and G. C. Strinati* 569

35 Bound States and Superfluidity in Strongly Coupled Fermion Systems *G. Röpke* 574
36 Onset of Superfluidity in Nuclear Matter *A. Hellmich, G. Röpke, A. Schnell, and H. Stein* 584

Appendix. BEC 93 Participant List 595
Index 597

Preface

In recent years, the phenomenon of Bose-Einstein condensation (BEC) has become an increasingly active area of research, both experimentally and theoretically. Long associated with the study of superfluid ^{4}He and ^{3}He-^{4}He mixtures, current research increasingly deals with other condensed matter systems, including spin-polarized hydrogen, excitons, laser-cooled atoms, high-temperature superconductors, and subatomic matter.

The present volume contains a series of authoritative review articles on current BEC research and related phenomena. The editors invited the leading experts to review their research field in such a way that their articles would introduce non-experts to current research and at the same time highlight some of the most promising topics for study in the next decade. These articles contain new material which is not available elsewhere. This is the first book devoted to BEC as an inter-disciplinary subject in its own right. It covers research in atomic and molecular physics, laser physics, low temperature physics, subatomic physics and astrophysics.

In the opening chapter, Snoke and Baym introduce the various review articles by discussing them in the context of the dominant themes in current BEC studies. These themes include broken gauge invariance, phase coherence in equilibrium and non-equilibrium situations, time scales for the formation of a Bose condensate, Bose particles with a finite lifetime, and BEC vs. BCS superfluidity in fermionic systems with attractive interactions.

The interdisciplinary character of modern BEC research has led to its being largely ignored by the regular low temperature conferences. The genesis of the present volume was the suggestion in 1992 by David Snoke that an international workshop devoted to current BEC research was long overdue. The idea then developed that the main focus of this first BEC workshop should be review-type talks by carefully chosen international experts and that these talks would be the basis of an authoritative book on BEC research, suitable for someone just getting interested in this topic.

The *International Workshop on Bose–Einstein Condensation* (BEC 93) was duly organized and held in Levico Terme, a small resort town near Trento in northern Italy, 31 May to 4 June, 1993. It was a resounding success from every point of view. Over 70 participants from over 20 countries (listed in an appendix at the end of this book) represented all the major research groups currently working on BEC, and included a good mix of experimentalists and theorists. Everyone had a sense that this first BEC workshop was an historic occasion, helped by the fact that the participants included many of the "founding fathers" of BEC studies, such as K. Huang, S.A. Moskalenko and L.V. Keldysh. The unifying themes of BEC research were stressed by having talks on quite different sub-fields given each day.

We decided that the invited review articles could be submitted well after the workshop, so that they would reflect the excitement and the interdisciplinary insights which arose from the workshop. The present volume contains a few additional invited review articles covering topics which were not emphasized at BEC 93. At the workshop, over thirty poster contributions were also presented. The second part of this book contains brief research reports and mini-reviews based on some of these contributions. The refereeing was done by the editors and the authors of the invited papers. Many of the contributed papers have since been submitted to research journals and for this reason are not included here.

There is only one review article devoted to superfluid ^{4}He. It was felt that since a recent book has given an extensive review of superfluid ^{4}He (A. Griffin, *Excitations in a Bose-Condensed Liquid*, Cambridge University Press, 1993), it was not necessary to emphasize this subject in the present volume.

The organizing committee of BEC 93 at Levico Terme consisted of the editors, with the assistance of F. Lalöe and G. Baym. In addition, A. Mysyrowicz, S. Tikhodeev and D. van der Marel were very supportive at the very early stages of organization. The major source of external funding of BEC 93 was from the European Centre for Theoretical Studies in Nuclear Physics and Related Areas (ECT*), which has recently been set up at the University of Trento. In addition, BEC 93 was financially assisted by grants from CNRS (France) and the Department of Physics at the University of Toronto, as well as funds from NSERC of Canada.

A crucial role was played by the workshop secretary, Jennifer Tam (Toronto).

The local organization at the University of Trento and Levico Terme was carried out by Sandro Stringari, and we thank especially F. Dalfovo and S. Giorgini for their help. The Grand Hotel Bellavista in Levico Terme played an important role in producing a successful workshop, providing the right atmosphere as well as fine Italian cuisine. Late into each evening, heated discussions on topics such as timescales and broken symmetry could be heard coming from the hotel bar.

Preparation of the book in the Cambridge University Press LaTeX format was carried out by David Snoke and Karl Schroeder (Toronto). Most of the contributors made this job much easier for us by sending in their articles in TeX or LaTeX. We would also like to thank Simon Capelin of Cambridge University Press for his enthusiastic support right from the beginning.

We hope this unique volume will introduce many readers to the challenging problems involved in the quest for BEC and set the stage for future progress.

Allan Griffin, Toronto
David Snoke, Pittsburgh
Sandro Stringari, Trento

Preface to paperback edition

With the achievement of Bose–Einstein condensation (BEC) in a dilute, ultracold gas of alkali atoms (M.H. Anderson, J.R. Ensher, M.R. Matthews, C.E. Wieman and E.A. Cornell, *Science*, **269**, 198 (1995)), the study of BEC and its consequences has become a topic of intense worldwide interest. For recent research, see the BEC homepage (http://amo.phy.gasou.edu/bec.html).

The review articles in this book provide an excellent source of background information about the properties of Bose-condensed systems. For the many new researchers who have become interested in studying BEC in trapped atomic gases, this book also puts the new discoveries into the wider context of the search for BEC in other physical systems. We hope this paperback edition will be a useful reference for this exciting new area of research.

A. Griffin, D. Snoke, S. Stringari

[illegible] provided by the workshop secretary, Jennifer [illegible]

[illegible] The local organization [illegible] and we thank [illegible] for their help [illegible] and important [illegible] as well as [illegible] such as [illegible] could be heard coming from the hotel [illegible]

[illegible] of the book [illegible] Cambridge University Press [illegible] was carried out by [illegible] the composition [illegible] by [illegible] We would also like to thank [illegible] Cambridge University Press for [illegible]

[illegible] in the [illegible] in the [illegible]

[illegible]

[illegible] C.E. [illegible] and E.A. [illegible] study of [illegible] its consequences [illegible] For recent references see [illegible]

[illegible] in this book [illegible] this paperback edition will be [illegible] for the [illegible]

1

Introduction: Unifying Themes of Bose–Einstein Condensation

D. W. Snoke†

Mechanics and Materials Technology Center
The Aerospace Corporation
El Segundo, CA 90009-2957, USA

Gordon Baym

Department of Physics
University of Illinois at Urbana-Champaign
1110 W. Green St, Urbana, IL 61801, USA

After many years as a phenomenon with only one experimental example, superfluid ^{4}He, the concept of Bose–Einstein condensation (BEC) has in recent years emerged in an array of exciting new experimental and theoretical systems; indeed, the study of BEC has become a field in its own right, no longer primarily a subfield of liquid helium studies. As the articles in this book make apparent, BEC is a common phenomenon occuring in physics on all scales, from condensed matter to nuclear, elementary particle, and astrophysics, with ideas flowing across boundaries between fields. The systems range from gases, liquids, and solids, including semiconductors and metals, to atomic nuclei, elementary particles and matter in neutron stars and supernova explosions. Generally, the bosonic degrees of freedom are composite, originating from underlying fermionic degrees of freedom. Table 1 lists bosonic systems presently under study.

The articles in this book bring out several unifying themes as well as common problems in the study of BEC which transcend specific systems. In this introduction, we give an overview of some of these major themes.

(i) Broken gauge symmetry. What is Bose–Einstein condensation? What is the essential underlying physics? In the past forty years the concept of *broken gauge symmetry* (also described as *off-diagonal long-range order* (ODLRO) or *long-range phase coherence*), which is one of

† Present address: Dept of Physics and Astronomy, University of Pittsburgh, Pittsburgh, PA 15260, USA.

Table 1. *Bosons under study*

Particle	Composed of	In	Coherence seen in
Cooper pair	e^-, e^-	metals	superconductivity
Cooper pair	h^+, h^+	copper oxides	high-T_c superconductivity
exciton	e^-, h^+	semiconductors	luminescence and drag-free transport in Cu_2O
biexciton	$2(e^-, h^+)$	semiconductors	luminescence and optical phase coherence in CuCl
positronium	e^-, e^+	crystal vacancies	(proposed)
hydrogen	e^-, p^+	magnetic traps	(in progress)
^{4}He	$^4He^{2+}$, $2e^-$	He-II	superfluidity
^{3}He pairs	$2(^3He^{2+}, 2e^-)$	^{3}He-A,B phases	superfluidity
cesium	$^{133}Cs^{55+}$, $55e^-$	laser traps	(in progress)
interacting bosons	nn or pp	nuclei	excitations
nucleonic pairing	nn or pp	nuclei neutron stars	moments of inertia superfluidity and pulsar glitches
chiral condensates	$\langle \bar{q}q \rangle$	vacuum	elementary particle structure
meson condensates	pion condensate $= \langle \bar{u}d \rangle$, etc. kaon condensate $= \langle \bar{u}s \rangle$	neutron star matter	neutron stars, supernovae (proposed)
Higgs boson	$\langle \bar{t}t \rangle$ condensate (proposed)	vacuum	elementary particle masses

the more remarkable implications of quantum mechanics, has come to the fore as the fundamental principle for understanding BEC. Nozières, Huang, Leggett, Stringari, Stoof, and several other authors in this book discuss the implications of the theory of broken gauge symmetry. In the opening review, Chapter 2, Nozières outlines the essential physics which determines whether a Bose condensate will win over competing phases in a wide variety of systems. Leggett, in the concluding review, Chapter 19, examines the validity of the concept of broken symmetry at a very fundamental level.

The concept of broken gauge symmetry, with the accompanying macroscopic wavefunction describing the condensate, was first introduced in explaining superconductivity and superfluidity. Since then it has also been widely applied in nuclear and elementary particle physics. Bosonic condensates, e.g., with non-vanishing expectation values of quark operators, $\langle \bar{u}u \rangle$, $\langle \bar{d}d \rangle$, $\langle \bar{s}s \rangle$, are believed to be fundamental features of the vacuum and the structure of elementary particles such as the nucleon, underlying the spontaneous breaking of the chiral symmetry of the strong interactions [1]. In fact, the Higgs boson, which gives rise to the breaking of the SU(2)×U(1) symmetry of the weak interactions, and is invoked to generate the masses of elementary particles, has been suggested as arising from a condensation of the top and antitop quarks, with a non-vanishing expectation value $\langle \bar{t}t \rangle$ in the vacuum [2]. Mesonic condensates, e.g., pion and kaon condensates, which are nuclear analogs of exciton condensates, may be an important feature in dense neutron star matter, as Brown discusses in his article. In nuclear physics, BCS-type pairing is used to explain the reduced moments of inertia of heavier nuclei. Both the neutron and proton components of nuclear and neutron star matter, more generally, undergo pairing to become superfluid.

In high energy physics, the normal state of the vacuum contains a condensate, and thus one can ask the opposite of the question asked in condensed matter physics, namely, can one produce *non*-condensed states? A program to do so, in collisions of ultrarelativistic heavy-ions, has been underway at CERN and Brookhaven. In the normal vacuum, quark condensates, $\langle \bar{q}q \rangle$, are non-zero; in a collision, on timescales of about 10^{-23} s and distance scales of order 10^{-12} cm, it is believed possible to generate regions with $\langle \bar{q}q \rangle = 0$, in which chiral symmetry becomes unbroken and the quarks deconfined [3]. Somewhat later the quarks form hadrons, mostly π mesons, and the state returns to $\langle \bar{q}q \rangle \neq 0$. The dense pion gas might itself Bose condense, although most likely its entropy is too high.

(ii) The telltale signal for BEC. What is the proof that a system is Bose–Einstein condensed? Superfluid helium has been studied for decades, and while we know that it is Bose–Einstein condensed, what is the "smoking gun?" Almost thirty years ago, Hohenberg and Platzman [4] proposed that neutron scattering experiments should yield the classic signal, a delta function for the occupation number of the particles with zero momentum associated with long-range phase coherence. As Sokol discusses in his article in this book, however, such a clean signal is not likely to appear any time soon in the experimental liquid helium data. The reason is that liquid helium is a strongly interacting system, so that the condensate is strongly depleted; furthermore, the short lifetime of the final state of the recoiling atom broadens the neutron scattering data considerably. The neutron scattering data have been shown to be consistent with many-body numerical calculations of helium which predict about 9% of the atoms with zero momentum at $T = 0$ [5, 6]. Nevertheless, the strongest proof of the existence of a condensate in liquid helium probably comes from analysis of the critical exponents of the superfluid phase transition. (They are characteristic of a two-component order parameter.) Obtaining accurate data on critical exponents in other experimental systems may turn out to be much more difficult. In a recent book, Griffin [6] has given a detailed review of the dynamical properties of liquid ^{4}He based on the role of an underlying Bose condensate.

What is the general relation between BEC and superfluidity? Do the superfluid properties of ^{4}He prove that it involves a Bose–Einstein condensate? The classic work of Hohenberg and Martin, as well as Bogoliubov, in the 1960's showed how the two-fluid description of superfluid ^{4}He was a direct consequence of the existence of Bose broken symmetry (see chapter 6 of Ref. [6]). That a condensate is not *necessary* is seen in superfluid two-dimensional Kosterlitz–Thouless systems [7], which, rather than having an infinitely long-ranged phase coherence, have a particle correlation function $\langle\psi^\dagger(r)\psi(0)\rangle$ with a power-law dependence on r. Lasers illustrate that a condensate is not *sufficient.* Huang presents a model of condensation in the presence of a highly disordered substrate to suggest that, more generally, Bose–Einstein condensation is neither necessary nor sufficient for superfluidity of boson systems.

(iii) Bose–Einstein vs. Fermi–Dirac degrees of freedom. Bosonic degrees of freedom are in general composite; indeed the only fundamental bosons in nature appear to be the photon, the three massive vector bosons mediating the weak interaction, the gluon, which mediates the strong interaction, and the graviton. We must thus face the problem of

understanding how bosonic degrees of freedom emerge from underlying fermionic degrees of freedom. Fermionic systems, for example liquid ^{3}He and electrons in metals, become superfluid; to what extent can these systems be regarded as Bose–Einstein condensates? As discussed by Randeria in this book, the same underlying theory based on the formation of Cooper pairs can yield both BEC and BCS in different limits, BEC in the limit of particle–particle correlation length very short compared to the average interparticle spacing, and BCS in the opposite limit. One cannot in general always make a clean distinction between a superfluid system of Cooper pairs and a Bose–Einstein condensate. Superfluidity and superconductivity both stem from the same underlying physics. In either case, the bosonic nature of the pairs is critical for understanding the "super" behavior. Ranninger and others in this volume suggest possible mechanisms for understanding the new cuprate high-T_c superconductors as a BEC of small Cooper pairs.

In condensed-matter physics, the internal fermion degrees of freedom of the nuclei only play the role of determining the statistics of the nucleus, since nuclear energy scales are so vastly larger than atomic scales. The nuclear domain itself provides important examples of the transition between Bose and Fermi degrees of freedom. Examples include the Interacting Boson Model of nuclei, reviewed here by Iachello, the description of mesons in terms of their underlying quark structure by Rho [8] and the deconfinement transition between bosonic mesons and fermionic quarks, as expected in ultrarelativistic heavy-ion collisions and in the early universe [3]. In this book Röpke and Hellmich *et al.* treat the BEC–BCS crossover in nuclear matter, analogous to that discussed by Randeria in superconductors.

(iv) BEC of a weakly interacting gas. In superfluid helium, many features associated with BEC are masked by the strongly interacting character of the liquid [6]. Many of the novel systems currently studied in the laboratory are *weakly interacting* boson gases, whose momentum spectra and other properties can be much more amenable to theoretical analysis. The theory of a weakly interacting gas has been developed at length over the years, starting with London and Bogoliubov, and aspects of the theory are reviewed by Huang, Stoof and Laloë in this volume; Greytak and Silvera review experimental attempts to observe BEC in spin-polarized hydrogen; Castin, Dalibard and Cohen-Tannoudji review the search for BEC of laser-trapped atoms, in particular cesium; and Wolfe, Lin and Snoke as well as Mysyrowicz review experimental work on BEC of excitons, in particular in Cu_2O. Platzman and Mills report on

a proposal for observation of BEC of positronium. All these systems are expected to remain weakly interacting gases at the BEC phase transition, and one expects that their behaviors in the quantum degenerate regime can be well understood [9].

The question of the "smoking gun" for BEC, discussed above, has become more important because of the recent progress on excitons in Cu_2O. Work over the last ten years has shown that excitons can exceed the critical density for condensation, that they exhibit an extremely narrow energy distribution with full width at half maximum much less than $k_B T$, and that at high densities they move without friction through the crystal over macroscopic distances. This system is clearly a highly quantum-degenerate, weakly interacting boson gas. Determining the extent to which this system exhibits off-diagonal long-range order will require ingenuious experimental tests. One has the advantage, compared with ^{4}He, that measurement of exciton recombination luminescence provides fairly direct information about the particle momentum distribution.

An important property of a weakly interacting Bose gas is the phenomenon of "Bose narrowing," i.e., the energy distribution of a weakly interacting gas should decrease in average energy at a given temperature as density increases, opposite to the increase with density in the average energy of a Fermi gas at a given temperature. So far, this effect, which occurs at densities greater than about one-tenth of the critical density for BEC, has been demonstrated only for excitons. Laser Doppler measurements of the energy distribution of cold atoms may see Bose narrowing in the near future, giving the first indication of their quantum degeneracy.

Another prediction for the weakly interacting gas is *spatial* condensation at the center of a three-dimensional potential well. As discussed in the articles by Greytak and by Wolfe *et al.*, this kind of trap is feasible for both atoms and excitons. If BEC occurs in such a trap, the appearance of an extremely narrow peak in the *spatial distribution* of the gas would provide a dramatic "smoking gun."

Finally, weakly interacting gases can allow study of Bose–Einstein statistics far from equilibrium. The process of stimulated emission, familiar for photons, occurs for all scattering processes for bosons, since scattering rates contain a factor $(1 + N_f)$ for each final scattered Bose particle, where N_f is the occupation number of the final state. This enhancement of the density of final states, or *stimulated scattering*, is opposite to the suppression of scattering of fermions by the factor $(1 - N_f)$ from the Pauli exclusion principle; occupied states "attract" other bosons just as fermions "repel" like fermions. This effect can lead to an increase by orders of magnitude of the total scattering rate even in the normal state, which should be observable in time-resolved studies of the momentum spectra or spatial distribution of weakly interacting systems created at or above the critical density but far from thermal equilibrium. The case of short-lived biexcitons in the semiconductor CuCl illustrates this effect. Although the biexciton lifetime is too short for a spontaneous appearance of BEC, when a condensate is placed "by hand" via a laser tuned to the ground state energy or very near to it, the biexcitons at low energy "attract" other biexcitons to the same region of momentum space [10]. Mysyrowicz reviews this work in CuCl here and discusses evidence for the same effect in transport measurements in Cu_2O.

(v) The timescale for formation of BEC. One important feature in all the weakly interacting boson systems discussed here is that the particles have, in general, a *finite lifetime* to remain in the system. Spin-aligned atomic hydrogen can flip its spin (resulting in the formation of a molecule), and atoms can evaporate from a trap. Excitons and positronium are composed of particle–antiparticle pairs and can thus decay; since these are not conserved, what are the experimental consequences of the spontaneously broken gauge symmetry being only approximate? To what extent can one describe systems with finite lifetime as Bose condensed? For example, as discussed in several articles in this book, excitons in Cu_2O appear to move without friction through the crystal when they reach supercritical densities consistent with BEC. Should excitons ever exhibit superfluidity? Kohn and Sherrington, in an oft-cited paper [11], argued that excitons cannot be superfluid in a rigorous sense, because they are particle–antiparticle pairs. Hanamura and Haug [12] responded, however, that Kohn and Sherrington's argument depends on an *equilibrium* picture, while an excitonic condensate is in fact an inherently non-equilibrium ("quasiequilibrium") state. Keldysh briefly addresses this issue in this book. Generally, the problem of seeing evidence of superfluidity on timescales comparable to the particle lifetime is challenging for both theory and experiment. Can one see analogs of persistent currents, or of the Meissner effect in superconductors, or reduced moment of inertia in He-II, and if so, what is the appropriate superfluid mass density?

The controversial question of the timescale for onset of Bose–Einstein condensation is reviewed at length in articles by Kagan and Stoof. The general issue is the growth and evolution in space and time of the boson correlation function, $\langle\psi^\dagger(r)\psi(0)\rangle$. Several timescales are involved: Suppose that a boson gas has density above the critical density for condensation, but is in a metastable state with no condensate present. How long does the quasiparticle distribution of the system take to develop a peak at zero kinetic energy of width less than $k_B T$? How long does it take to develop a *quasicondensate*, i.e., a condensate whose magnitude is locally that of the equilibrium condensate, but with phase fluctuations arising from a random macroscopic distribution of topological defects? Then how long does it take for the condensate to achieve long-range phase coherence, leading to the characteristic delta function at $p = 0$ in the momentum distribution? Finally, when does superfluidity appear?

(vi) Two-dimensional degenerate boson systems. Formation of a quasicondensate is also the basis of the explanation of the Kosterlitz–Thouless (KT) transition of a boson gas in two dimensions, which leads to superfluidity even though a true condensate cannot appear. The KT transition has been seen for (two-dimensional) helium films [7]; Silvera reviews work on observation of a KT transition of spin-polarized hydrogen adsorbed on a surface, and Matsubara *et al.* briefly report on recent experimental work toward this goal. Semiconductor quantum-well structures provide another natural way to examine physics in two dimensions. One such system contains as excitations the "dumbbell exciton" in which the electron and hole in a two-dimensional well are separated by a potential barrier[13] or by an applied electric field [14]. Early indications [14] of a KT transition in this type of structure were later realized to be obscured by trapping in random potentials created by surface irregularities (similar to the scenario considered by Huang in this book) which actually produced a Fermi–Dirac distribution of excitons [15]. The possibility nevertheless still remains for observation of the Kosterlitz–Thouless transition with better-grown quantum-well structures.

Various kinds of two-dimensional semiconductor structures have been proposed which contain charged bosonic excitations that could in principle become superfluid, conceivably at room temperature, and which would be superconductors. "Magnetoexcitons," which have received much attention in recent years [16], move perpendicular to applied magnetic and electric fields in a two-dimensional well, and could undergo Bose–Einstein condensation [17, 18]. In this volume Shvartsman and Golub speculate that the possibility exists for engineering the band structure of GaAs to yield a "bihole," a boson in two dimensions with charge $+2e$. Since band-structure engineering of semiconductor structures has reached such an advanced state, one can imagine great advances in the near future in the manufacture of new kinds of superconductors based on BEC.

(vii) BEC and lasing. While excitons in Cu_2O have lifetimes long compared to interaction times, in most semiconductors the interaction time is comparable to the lifetime of the excitons due to coupling to the photon field. In the limit of very short exciton lifetime, the excitonic BEC essentially becomes indistinguishable from a laser. The observation of a narrow photon emission peak from a semiconductor does not therefore immediately indicate the presence of an excitonic BEC, since in many cases it may equally well be described as superradiant emission due to lasing [12]. As discussed by Keldysh in this volume, lasing and excitonic (or biexcitonic) BEC can be seen as two limits of the same theory. Lasing occurs in the case of strong electron–photon coupling (recombination rate large compared to the interparticle scattering rate) while the excitonic condensate occurs in the case of weak electron–photon coupling (recombination rate slow compared to interparticle scattering rate). A laser, in fact, can be seen as a Bose–Einstein condensate in which the long-range order involves the photon states, while in the excitonic condensate the long-range order involves the electronic states [19]. Many of the same issues of the timescale for onset of condensation, discussed above, have already been discussed in the context of the onset of lasing [20]. The polariton condensate, reviewed by Keldysh (see also Beloussov and Shvera) in this volume, represents a middle case between lasing and excitonic BEC, since the polariton has a mixed character with both exciton-like and photon-like behavior.

The overlap between BEC and lasing has been illustrated recently in the context of understanding the origin of dark matter in the early universe, where it has been argued that decay of a heavy species of neutrino into a light fermion and a boson could give rise either to a Bose–Einstein condensate [21] or to a "neutrino laser" [22].

(viii) BEC of small clusters of particles. The usual theory of BEC assumes the thermodynamic limit of an infinite system. The concept of bosonization, including BEC and superfluidity, has become valuable in understanding the properties of nuclei, however, in which pairs of nucleons bind via the strong force to produce effective bosonic degrees of freedom. Iachello reviews the theory of bosonization; Röpke as well as Hellmich *et al.* discuss the crossover from BEC to BCS in the theory of nuclear matter. These concepts work quite well in small nuclei, which are systems far from the thermodynamic limit; this is perhaps not so surprising since numerical models of helium using of order 64 atoms have been shown to reproduce the bulk superfluid properties of helium [23]. The theory of BEC of small clusters of bosons [24] may find extension in

the future to atomic and molecular systems, since the experimental study of atomic clusters has progressed tremendously in recent years. What is the minimum size of a system that we may still call a Bose–Einstein condensate?

(ix) BEC in random potentials. The appearance of a Bose condensate in disordered systems is a very active area of research in condensed-matter physics (the so-called "dirty boson" problem.) The key issues are highlighted in the review article by Nozières, and some specific aspects are treated in the articles by Huang and Stringari, but this important topic is not reviewed in depth in this book. Experimental systems to which this theory may apply include excitons in quantum wells with rough surfaces, and helium in porous media such as Vycor, the superconductor–insulator transition in disordered films, and vortices in type-II superconductors.

As one can see from this short survey, the abundant themes in BEC go beyond specific system properties, and present a unifying basis to the study of the phenomenon in different systems. We hope that this volume will serve as a stimulus to deeper understanding of these issues.

Acknowledgements. We appreciate the useful comments of A. Griffin. This work has been supported in part by the Aerospace Sponsored Research Program (D.S.) and by National Science Foundation Grant DMR91-22385 (G.B.)

References

[1] J. Gasser and H. Leutwyler, Phys. Reports **87**, 77 (1982).
[2] C. T. Hill, Phys. Lett. B **266**, 419 (1991) and references therein.
[3] See, e.g., *International conferences on quark matter and ultra-relativistic nucleus–nucleus collisions,* Nucl. Phys. A **544** (1992); in press (1994).
[4] P. C. Hohenberg and P. M. Platzman, Phys. Rev. **152**, 198 (1966).
[5] D.M. Ceperley and E.L. Pollock, Phys. Rev. Lett. **56**, 351 (1986).
[6] A. Griffin, *Excitations in a Bose-Condensed Liquid* (Cambridge University Press, Cambridge, 1993).
[7] For a review, see D.R. Nelson, in *Phase Transitions and Critical Phenomena* **7**, C. Domb and J.L. Lebowitz, eds. (Academic, New York, 1983).
[8] M. Rho, "Cheshire Cat Hadrons," Phys. Reports, in press (1994).
[9] See, e.g., V.N. Popov, *Functional Integrals and Collective Excitations,* (Cambridge University Press, Cambridge, 1987), ch. 6.

[10] N. Peghambarian, L.L. Chase, and A. Mysyrowicz, Phys. Rev. B **27**, 2325 (1983).
[11] W. Kohn and D. Sherrington, Rev. Mod. Phys. **42**, 1 (1970).
[12] E. Hanamura and H. Haug, Sol. State Comm. **15**, 1567 (1974); Phys. Reports **33**, 209 (1977).
[13] F.M. Peeters and J.E. Golub, Phys. Rev. B **43**, 5159 (1991).
[14] T. Fukuzawa *et al.*, Phys. Rev. Lett. **64**, 3066 (1990).
[15] J.A. Kash *et al.*, Phys. Rev. Lett. **66**, 2247 (1991).
[16] D.S. Chemla J.B. Stark, and W.H. Knox, *Ultrafast Phenomena VIII*, (Springer, Berlin, 1993), p. 21.
[17] I.V. Lerner and Yu. E. Lozovik, Sov. Phys. JETP **53**, 763 (1981).
[18] D. Paquet *et al.*, Phys. Rev. B **32**, 5208 (1985).
[19] The analogy between a laser and superfluid ^{4}He is nicely reviewed by P.C. Martin, in *Low Temperature Physics— LT9*, J.E. Daunt, D.O. Edwards, F.J. Milford, and M. Yaqub, eds. (Plenum, New York, 1965), p. 9.
[20] See, e.g., H. Haken, Rev. Mod. Phys. **47**, 67 (1975).
[21] J. Madsen, Phys. Rev. Lett. **69**, 571 (1992).
[22] N. Kaiser, R. A. Maleney, and G. D. Starkman, Phys. Rev. Letters **71**, 1128 (1993).
[23] Finite size effects are discussed by E.L. Pollock, Phys. Rev. B **46**, 3535 (1992).
[24] Ph. Sindzingre, M. Klein and D. M. Ceperley, Phys. Rev. Lett. **63**, 1601 (1989).

Part one

Review Papers

2

Some Comments on Bose–Einstein Condensation

P. Nozières

Institut Laue-Langevin
B.P. 156, 38042 Grenoble Cedex 9
France

Abstract

The paper reviews some important features of Bose–Einstein condensation such as the nature of the condensate, exchange interactions, phase locking. The competition with other physical effects (crystallization, dissociation, localization) is briefly discussed.

Bose–Einstein condensation was first described for an ideal gas of free bosons, with mass m, density N. The chemical potential μ is obtained from the conditions

$$n_k = \frac{1}{e^{\beta(\varepsilon_k - \mu)} - 1}, \quad \sum_k n_k = N,$$

where $\varepsilon_k = \hbar^2 k^2/2m$. In a dimension $d > 2$, μ reaches 0 at some temperature T_c such that

$$\int d\varepsilon \nu(\varepsilon) \frac{1}{e^{\beta_c \varepsilon} - 1} = N, \tag{1}$$

where $\nu(\varepsilon) \sim \varepsilon^{\frac{d}{2}-1}$ is the one particle density of states. Below T_c a macroscopic number of particles N_0 accumulates in the lowest state, μ being locked to 0 (Fig.1). When $d \leq 2$ the integral in (1) diverges: Bose–Einstein condensation does not occur (put another way, T_c is 0).

These conclusions hold for a uniform infinite system. It was pointed out by Bagnato and Kleppner [1] that the condensation can be restored at low dimensions if the system is put in a confining potential

$$V \sim r^\eta.$$

Then the region accessible to a particle with energy ε has a radius $L \sim \varepsilon^{1/\eta}$

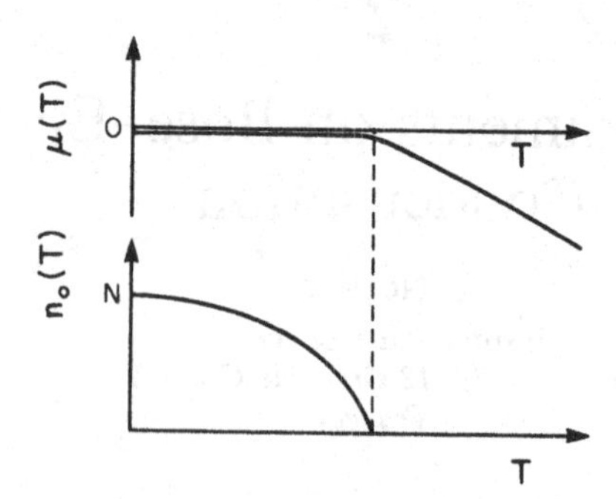

Fig. 1. The chemical potential μ and condensate population n_0 of an ideal Bose gas as a function of temperature.

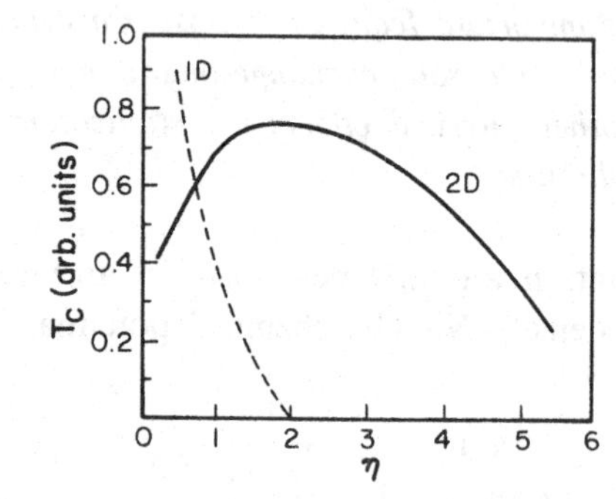

Fig. 2. A sketch of the critical temperature of a confined ideal Bose gas as a function of the exponent η in the confining potential for $d = 1$ and $d = 2$ (after Ref. 1).

and the corresponding density of states behaves as

$$v(\varepsilon) \sim L^d \varepsilon^{\frac{d}{2}-1} \sim \varepsilon^\alpha, \text{ with } \alpha = \frac{d}{\eta} + \frac{d}{2} - 1.$$

Playing with the exponent η one can make the integral in (1) convergent and T_c finite. The result is sketched in Fig. 2.

1 Fragmentation of the Condensate

The above standard wisdom actually leaves one central question unanswered: do the condensate particles accumulate in a single state? Why don't they share between several states that are degenerate, or at least so close in energy that it makes no difference in the thermodynamic limit?

The answer is non-trivial: *it is the exchange interaction energy that makes condensate fragmentation costly.* Genuine Bose–Einstein condensation is not an ideal gas effect: it implies interacting particles!

Consider the simplest type of scalar interaction for structureless particles

$$V = \frac{1}{2} \sum_{k,k',q} V_q b_k^* b_{k'}^* b_{k'-q} b_{k+q}.$$

The fully condensed ground state has all particles in the lowest state

$$|\psi_0\rangle = \frac{1}{\sqrt{N!}} (b_0^*)^N |\text{vac}\rangle. \tag{2}$$

The corresponding interaction energy is

$$E_0 = \frac{1}{2} V_0 \langle b_0^* b_0^* b_0 b_0 \rangle = \frac{1}{2} V_0 N(N-1) \simeq \frac{1}{2} V_0 N^2.$$

Note that V_0 is necessarily > 0 (i.e., *repulsive*): otherwise the system would collapse spontaneously to infinite densities N.

We compare that state with one in which the condensate is shared between states 1 and 2, with occupancies N_1 and $N_2 = N - N_1$

$$|\psi_0\rangle = \frac{1}{\sqrt{N_1! N_2!}} (b_1^*)^{N_1} (b_2^*)^{N_2} |\text{vac}\rangle. \tag{3}$$

The kinetic energies are both close to 0 and negligible. The interaction energy involves all possible contractions of operators in the expectation value of V. It contains a Hartree term (between all species) and a Fock *exchange* term. The latter only occurs between different species, since there is only one type of contraction in the expectation value of $b_1^* b_1^* b_1 b_1$. Hence

$$\begin{aligned} E &= \frac{V_0}{2}[N_1(N_1-1) + N_2(N_2-1) + \underbrace{2N_1N_2}_{\text{Hartree}}] + \underbrace{V_{1-2}N_1N_2}_{\text{Fock}} \\ &\simeq \tfrac{1}{2} V_0 N^2 + V_0 N_1 N_2. \end{aligned} \tag{4}$$

Since $V_0 > 0$ fragmentation of the condensate costs a macroscopic *extensive* exchange energy. *Genuine Bose–Einstein condensation is due to exchange, a feature which is often overlooked.*

It may be objected that the use of Hartree–Fock wave functions is unwarranted. Indeed, short range correlations modify the wave function at short distances between particles, thereby leading to a *depletion* of the condensate ($N_0 < N$ at $T = 0$). The ground state is complicated, leading to all the subtleties of Bose liquid theory. The argument against fragmentation nevertheless remains true, as it relies on the comparison of two situations with the same amount of depletion and virtual excitation, with or without fragmentation. The only difference is in the Fock "intracondensate" exchange, everything else remaining the same. This exchange energy is reduced by depletion, but it remains extensive.

The crucial role of particle interactions in Bose liquids is well known. It is responsible for the *finite compressibility* of the liquid, and therefore for the existence of sound modes at long wave length instead of free particle excitations ($\varepsilon_k \simeq k$ instead of k^2). It leads to a finite coherence length

$$\frac{\hbar^2}{m\xi^2} = NV_0.$$

As a result the effect of confinement in one and two-dimensions disappears [2]. In two-dimensions a transition remains which is of the Kosterlitz–Thouless type (unbinding of vortices) – yet another consequence of interactions. The effect of interactions is equally important for the *kinetics* of condensation in a thermally quenched normal gas. While for a free gas the energy narrowing of the N_0 condensed particles is very slow (the corresponding width decays as $1/t$), exchange effects make the relaxation time atomic once the condensate has been nucleated.

2 Internal Degrees of Freedom

Assume now that the bosons have an internal quantum number α. This may be nuclear spin for polarized atomic hydrogen, or the electric dipole moment for optically active excitons, or orbital and spin total momentum for a triplet pair of ^{3}He atoms. *Is the condensate a pure state or a mixture?* We may describe it by a density matrix

$$\rho_{\alpha\beta} = \langle b^*_{0\alpha} b_{0\beta} \rangle,$$

which can be diagonalized in the appropriate basis

$$\rho_{\alpha\beta} = n_\alpha \delta_{\alpha\beta}.$$

Is there one or several non-zero n_α?

In order to answer this question we write the interaction within the condensate as

$$H_{\alpha'\beta'}^{\alpha\beta}\, b^*_{0\alpha} b^*_{0\alpha'} b_{0\beta'} b_{0\beta},$$

from which we infer the condensate energy

$$E_0 = \sum_{\alpha} H_{\substack{\alpha\alpha \\ \alpha\alpha}} n_\alpha(n_\alpha - 1) + \sum_{\alpha \neq \alpha'} n_\alpha n_{\alpha'} \left(H_{\substack{\alpha\alpha \\ \alpha'\alpha'}} + H_{\substack{\alpha\alpha' \\ \alpha'\alpha}} \right) \qquad (5)$$

$$\simeq \sum_{\alpha,\alpha'} A_{\alpha\alpha'} n_\alpha n_{\alpha'}.$$

Here we neglect the depletion via virtual excitation which should also depend on the internal state α. The selection argument relies only on intracondensate interactions, which is certainly an approximation. We note first that the energy E_0 must be positive for any choice of $n_\alpha > 0$: otherwise the density would collapse. The nature of the minimum then depends on the matrix A. Consider, for instance, a two-fold situation

$$A = \begin{pmatrix} \varepsilon & \eta \\ \eta & \varepsilon \end{pmatrix} \Rightarrow E_0 = \varepsilon[n_1 + n_2]^2 + 2(\eta - \epsilon)n_1 n_2.$$

The ground state is a pure state if $\eta > \epsilon$ (n_1 or $n_2 = 0$). It is, on the contrary, a mixture if $\eta < \epsilon$ ($n_1 = n_2 = 1/2$). Note that in the Hartree–Fock approximation the selection occurs to first order in V. It may be that the degeneracy is not lifted at that order: then one must go beyond Hartree–Fock.

As an example consider polarized atomic hydrogen, $H_\downarrow$. In a magnetic field the atoms undergo a nuclear Zeeman splitting, $\pm\mu B$. The lowest energy state is undoubtedly a pure state in which all particles are in the lowest nuclear state. But one can work at given magnetization $M_z = N_\uparrow - N_\downarrow$. If the condensate atoms are in a pure spin 1/2 state, a transverse magnetization appears, the same for every atom, as shown in Fig. 3. Internal coherence of the condensate implies *nuclear ferromagnetism* (of course this magnetization precesses at the Larmor frequency). Conversely, if one allows for a mixture of $\downarrow$ and $\uparrow$ atoms the transverse M can be killed.

In practice the most obvious interactions are

(i) A scalar interaction

$$V\rho_1\rho_2 \rightarrow H_{\substack{\alpha\beta \\ \alpha'\beta'}} = V\delta_{\alpha\beta}\delta_{\alpha'\beta'}.$$

The corresponding interaction energy

$$E_0 = \frac{1}{2}Vn_\alpha n_{\alpha'} = \frac{1}{2}VN^2$$

does not depend on the condensate structure.

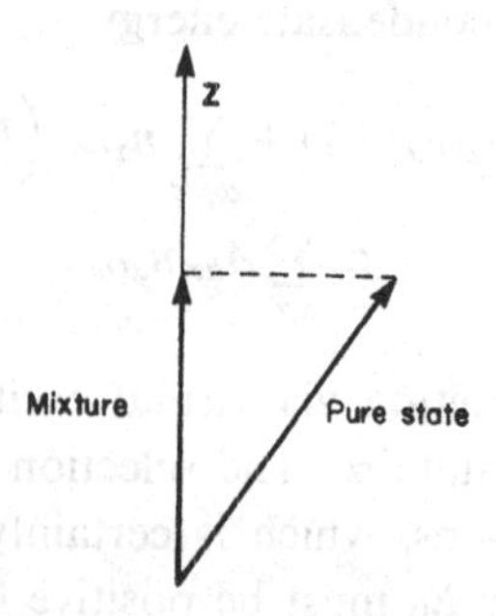

Fig. 3. The onset of transverse ferromagnetism for a pure state $H_\downarrow$ condensate.

(ii) An exchange interaction $J\mathbf{S}_1 \cdot \mathbf{S}_2$ which is equally ineffective in lifting the degeneracy (orbital symmetry implies a triplet spin state for a pair of bosons, such that $\mathbf{S}_1 \cdot \mathbf{S}_2 = 1/4$). An anisotropic exchange, $J_z \neq J_\perp$, would probably do the job, but where does it come from?

3 Condensate Fluctuations and Phase Locking

We return to structureless bosons. Ensuring a non-fragmented condensate we gain the *Fock exchange energy*, first order in V. Actually one can do better allowing for coherent fluctuations of the condensate population N_0. Instead of state (2) let us try a superposition

$$|\psi_0\rangle = e^{\phi b_0^*}|\text{vac}\rangle; \tag{6}$$

ϕ is a complex order parameter, the phase θ of which is locked to a specific value. Locking θ means fluctuations in N. Indeed, N and θ are canonically conjugate: state (2) is obtained by an integration over θ

$$|\psi_0(N)\rangle \sim \int_0^{2\pi} d\theta e^{-iN\theta}|\psi_0(\theta)\rangle.$$

It is easily verified that $\langle N\rangle = |\phi|^2$: the energy is the same in (2) and (6).

The situation is different when we consider interaction terms such as

$$b_0^* b_0^* b_k b_{-k} + \text{c.c.}$$

which allow virtual excitations of two particles out of the condensate. In order to take such a hybridization into account, we try a variational wave function

$$|\psi_0\rangle = e^{[\phi b_0^* + \lambda_k b_k^* b_{-k}^*]}|\text{vac}\rangle; \tag{7}$$

λ_k is obtained minimizing the energy. Equation (7) is nothing but the well-known Bogoliubov approximation for Bose liquids. When calculating the expectation value of the energy, we recover the usual Hartree and Fock exchange contributions, which here have corrections of order

$$V_0|\lambda_k|^2 \ , \ V_k|\lambda_k|^2.$$

They are the same whether N or θ is locked. The novelty is the so-called Bogoliubov term, which appears only when the phase is locked

$$V_k\langle b_0^* b_0^* \rangle \langle b_k b_{-k} \rangle \sim \phi \lambda_k.$$

Such a linear coupling between ϕ and λ_k always lowers the energy, by an extensive amount of order V^2: hybridization is bound to occur. It follows that *phase locking is implied by Bose–Einstein condensation.* Note that the Bogoliubov energy favours pure state condensates: it is enough to lift the degeneracy.

Phase locking is a genuine *symmetry breaking*, distinct from Bose–Einstein condensation. It is responsible for superfluidity (superfluid currents are due to a gradient in the locked phase θ). Its physical consequences are well known : quantization of circulation in multiples of h/m, vortices, Josephson effect through a weak link, etc.

As an unconventional example let us consider Bose condensation of excitons in Cu_2O, extensively discussed by Wolfe *et al.* and Mysyrowicz [3]. These excitons exist in two brands, *para* (lowest in energy) and *ortho* (some 12 meV higher). They are seen through phonon assisted fluorescence (momentum k is not conserved), from which one infers the exciton energy distribution $n(\varepsilon)$. Most of the data are concerned with orthoexcitons, less "forbidden" than the para. As the density N grows, quantum deviations to $n(\varepsilon)$ are clearly visible, but somehow one never reaches Bose–Einstein condensation. One "surfs" on the transition line without being able to cross it – why?

A tentative explanation might go as follows. Being lower, the paraexcitons should be more numerous and consequently they should Bose condense first. But they are harder to see: their condensation may escape observation. Assume now that the ortho density grows through the transition. Then a Josephson process is possible

$$O + O \rightarrow P + P.$$

Direct conversion of a single exciton is forbidden by symmetry, but there is no such problem for a pair. That Josephson current is a.c., at a frequency ω corresponding to the 12 meV splitting – hence a strong dissipation that precludes further cooling. One thereby would explain why any onset of ortho condensation would immediately stop the process.

The above example may be wrong, but it points to an essential fact: phase locking is a crucial feature of condensed systems.

4 The Enemies of Bose–Einstein Condensation

Condensation of bosons, or of fermion pairs through the appropriate generalization, is a fragile effect that may easily be superseded by other instabilities. We now survey some of the main competitors.

4.1 Crystallization

The energy per particle of a liquid phase is roughly an interaction V_m averaged over a unit cell (the distance between atoms is free to vary, except for some short range correlations). This may be compared with a solid phase in which the particles are localized at lattice sites. The latter energy contains a kinetic part due to localization of the particles, and a potential energy V_a corresponding to the lattice spacing a

$$\frac{E_0}{N} \sim \frac{\hbar^2}{ma^2} + V_a.$$

Because the repulsive potential $V(r)$ decreases with r, we have $V_a < V_m$. It follows that the crystalline state will be favoured for strong coupling (it optimizes the potential energy), while the Bose condensed liquid will win at weak coupling (it minimizes the kinetic energy). An obvious example is ^{4}He, which is a superfluid at low pressure and which freezes at higher densities.

Note that the kinetic energy that is relevant to the above competition is that associated to *coherent hopping* of the particle from place to place, from which one may construct Bloch states (remember that phase coherence is an essential ingredient). Such a coherence may be very hard to achieve. Consider for instance *bipolarons*, i.e. two fermions trapped in the same cloud of phonons that they have generated. These bipolarons are candidates for high temperature superconductivity. What matters here is elastic *coherent* hopping of the bipolaron, leaving the phonon bath in its ground state. The corresponding amplitude W involves the overlap of phonon ground states Φ_i and Φ_j when the particle is at sites i or j,

$$W = W_0\langle\Phi_i|\Phi_j\rangle.$$

This overlap is usually very small and the kinetic energy is negligible as compared to potential energy. Charge density waves (crystallization) win!

The above competition is nicely formulated in the old lattice gas model of Matsubara and Matsuda [4]. Consider N hard core bosons on $2N$ sites. Due to the single occupancy constraint, each site has two states, empty or filled. The bosons have commutation rules on different sites, which is just the algebra of spins. The hard core boson gas is therefore isomorphous to a spin $1/2$ system. The hopping amplitude between nearest neighbours is equivalent to a transverse (xy) magnetic coupling:

$$b_i^* b_j^* \rightarrow S_i^+ S_j^- \rightarrow \text{energy } J_\perp.$$

A nearest neighbour repulsion is equivalent to an Ising coupling

$$n_i n_j \rightarrow S_{iz} S_{jz} \rightarrow \text{energy } J_z.$$

Hence there are two possibilities:

(i) $J_z > J_\perp$: the equivalent magnetic system is an Ising antiferromagnet, corresponding to a crystal in which one sublattice is empty and the other occupied.

(ii) $J_z < J_\perp$: the magnetic ground state is an (xy) transverse ferromagnet. The transverse magnetization is the superfluid order parameter, its azimuth being the phase θ. The competition between kinetic energy and interactions is clear. One can enrich the model by adding further couplings, and it is easy to construct a phase diagram using a mean-field approximation [5].

An interesting issue is the possible coexistence of Bose–Einstein condensation and crystallization. Put another way do *supersolids* exist? The question is relevant for vacancies in solid ^{4}He. It has been argued [6] that zero point vacancies could persist at zero temperature (if the tunneling bandwidth is larger than the creation energy), and indeed there is experimental evidence for an unusually large concentration of such vacancies [7]. If that holds vacancies might become superfluid, leading to unusual Herring–Nabarro plastic flow of solid ^{4}He. In practice this is unlikely since vacancy coherent hopping is subject to a strong polaronic reduction.

4.2 Dissociation of Composite Bosons

We now turn to the more frequent case of bosons made out of two fermions, appropriately bound, as they occur in superconductors or excitons. Let $c_{k\sigma}$ be the fermion destruction operator. The condensing entity is a bound pair characterized by

$$b_q^* = \sum_k \varphi_k c^*_{k+q,\uparrow} c^*_{-k,\downarrow};$$

q is the total momentum of the pair, φ_k its internal wave function. The Hartree ground state for the pairs (equivalent to (6)) is just

$$\begin{aligned} e^{[\phi b_0^*]}|\text{vac}\rangle &= e^{[\phi\varphi_k c^*_{k\uparrow} c^*_{-k\downarrow}]}|\text{vac}\rangle \\ &= \prod_k [u_k + v_k c^*_{k\uparrow} c^*_{-k\downarrow}]|\text{vac}\rangle, \end{aligned} \tag{8}$$

in which we have set

$$\frac{u_k}{v_k} = \phi\varphi_k \ , \ 2\sum_k v_k^2 = N.$$

(8) is nothing but the usual BCS wave function, viewed here as pair Bose–Einstein condensation. v_k is a variational parameter which allows an interpolation between low and high densities. (Note that (8) ignores the effect of *screening* and its interplay with *binding*: the real problem of Mott transitions is not tackled.)

The superfluid state (8) exists only if the particles are subject to an attraction. Two limiting cases are simple:

(i) In the dilute "molecular" limit the attraction must be strong enough to create a bound state of two fermions ("dilute" means a particle separation large as compared to the bound state radius). The bound pairs behave as "point" bosons that condense in the usual way. v_k is $\ll 1$ and it reflects the internal state of the underlying fermions. The gap Δ is the molecular binding energy, while the critical temperature T_c is controlled by *centre of mass motion* of the bound pairs. For free bosons $T_c \sim N^{\frac{2}{3}}$, while for a Hubbard lattice gas $T_c \approx 1/U$ (one must virtually break the pair in order to make it hop).

(ii) In the opposite dense limit, the pairs strongly overlap and the exclusion principle becomes the dominant feature. $|v_k|^2$ cannot exceed 1: it extends to higher k in order to accommodate the N particles – in practice to the Fermi momentum k_F. In the limit of high densities we recover the usual BCS result, with an exponentially small gap Δ. The fermions are normal except for a narrow region of width Δ near

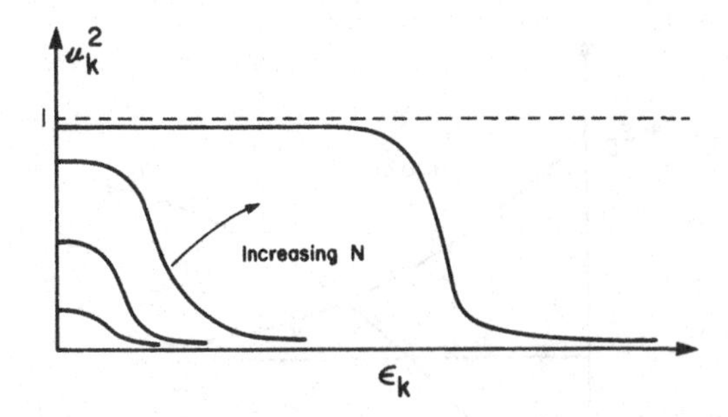

Fig. 4. The evolution of the BCS parameter v_k^2 as a function of energy ε_k for increasing densities. Note the saturation due to the exclusion principle.

k_F. Binding is now a cooperative effect, which occurs only due to Bose–Einstein condensation (in contrast to the dilute limit in which the two effects were distinct). The critical temperature T_c is controlled by pair breaking: it is of order Δ.

In between these two limits the evolution is smooth, as shown on Fig. 4: contrary to what has been sometimes claimed, superconductivity is indeed Bose–Einstein condensation of pairs. The transition occurs when the pair binding energy ε_B is comparable to the Fermi energy ε_F. The behaviour of the gap Δ and chemical potential μ as a function of N is sketched in Fig. 5.

While the condensation of dilute "point" pairs is a robust phenomenon, the BCS state is fragile when $\Delta \ll \varepsilon_F$. If the interaction near the Fermi surface has repulsive parts, Bose condensation may well disappear, leading to a standard *normal* Fermi liquid. In the case of dense excitons, any anisotropy will kill condensation: the Fermi surfaces of electrons and holes need not be the same and pairing cannot occur. We conclude that overlap and resulting *dissociation of composite bosons is antagonizing to Bose–Einstein condensation.*

4.3 Disorder and Localization

Superfluidity implies a long range phase coherence, while localization means no coherent extended state: clearly they are antagonist. In practice, two very different localization mechanisms must be considered:

(i) *Mott localization* is due to interaction between particles. When the number of particles N is commensurate to the number of sites N_L,

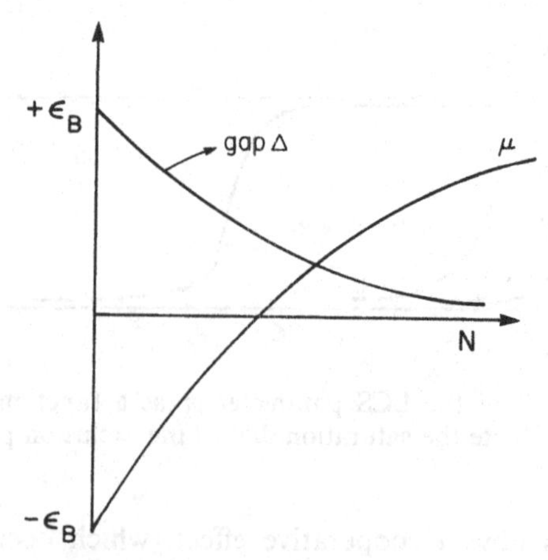

Fig. 5. The evolution of the gap Δ and chemical potential μ as a function of density for attracting bosons.

the ground state becomes insulating beyond a critical interaction U_c. Such a localization is the ultimate stage of crystallization, above and beyond appearance of a periodic order (a real solid is both *periodic* and *localized*). The hard core Bose lattice mentioned earlier is one such example. Mott transition occurs equally well for a full lattice, $N = N_L$, past a critical U_c. (In that case it is pure, unspoiled by charge density waves.)

(ii) *Anderson localization* is due to disorder. A random one-body potential creates destructive interference on the wave function. At low enough kinetic energy the long range phase coherence is suppressed. For moderate disorder a *mobility edge* ε^* appears on either side of the electron band, separating extended states in the middle and localized states near the edges. If the disorder is strong enough the edges may close in: no extended state subsists and localization is complete.

Strictly speaking, the effect does not depend on statistics – except that something must prevent accumulation of particles in the lowest localized states. For fermions the exclusion principle does it. For bosons one must rely on repulsive interactions that forbid large local densities. At first sight one expects a transition from a localized state at *small* coupling – the so-called *Bose glass* – to a regular superfluid at large coupling, when interactions force the chemical potential μ into the region of extended states. But one might also argue that strong

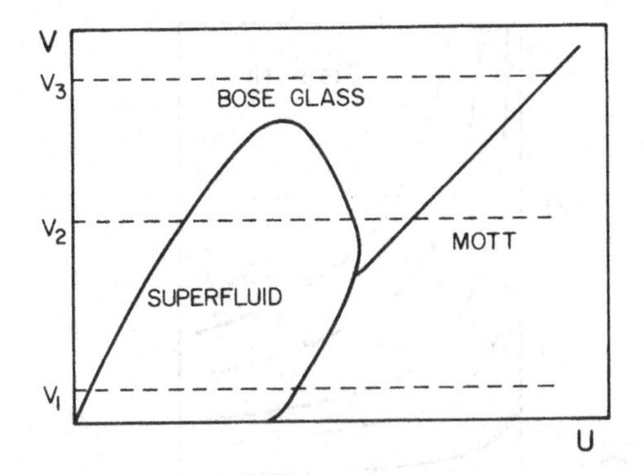

Fig. 6. A sketch of the phase diagram of a disordered two-dimensional Bose lattice gas with $N = N_L$ (after Ref. 9).

local repulsion produces constraints that favour localization. Such a contradiction gives a feeling of how primitive the present theoretical understanding is. The subject is indeed "hot" and a number of attempts have been put forward [8]. Generally speaking, they fall in two categories:

(i) Either they put emphasis on localization of *individual* particles, referred to the mobility edge.
(ii) Or they consider the rigidity of the superfluid phase. Breakdown of superfluidity corresponds to a vanishing rigidity – a truly collective point of view.

Efforts to characterize the critical behaviour are somewhat premature, since even the qualitative nature of the effect is not well understood.

Extensive numerical work has been performed on lattices, where Anderson localization is mixed with the Mott transition. They confirm the above qualitative picture. Results of Trivedi [9] for a two-dimensional lattice with $N = N_L$ are shown in Fig. 6: the Bose glass is found at strong disorder – yet another enemy of Bose condensation.

The problem of disordered Bose liquids is not a theorist's amusement: it has important experimental applications. Liquid ^{4}He in a porous medium (e.g. Vycor) is one example. A Cooper pair in disordered alloys is another candidate. A particularly fascinating problem is that of *disordered thin films* in which disorder is monitored by film thickness d. One can make d much smaller than the film superconducting coherence length ξ, in which case the system is two-dimensional. When

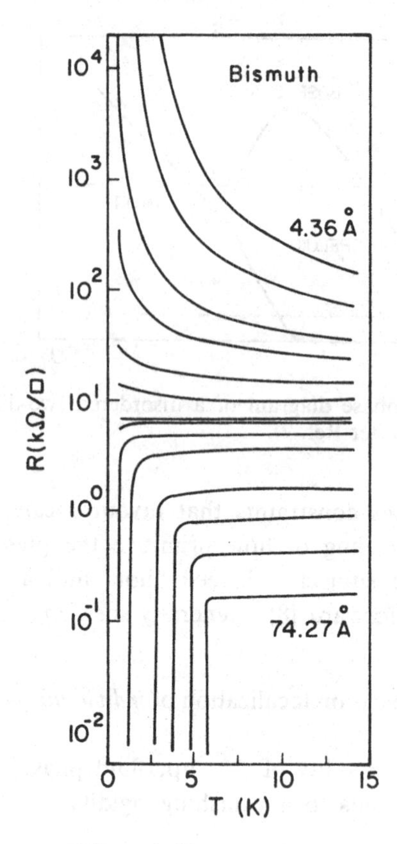

Fig. 7. The resistance of Bismuth films as a function of temperature for varying thickness d (after Ref. 10).

the square resistance is comparable to the quantum scale $h/4e^2$ a sharp transition occurs at $T = 0$ as a function of d, between a *superconducting* and an *insulating* state. This is shown dramatically in Fig. 7, which displays the resistance of Bismuth films as a function of temperature for varying d [10]: the direct passage from 0 to ∞ at zero temperature is spectacular.

The problem is important and, to a large extent, open. It is attacked either via renormalization methods or through numerical simulation. But we are far from a real understanding. Even simple questions such as the relative importance of disorder *scale* (vs ξ) and disorder *strength* (vs $\hbar^2/m\xi^2$) are not answered.

5 Conclusion

In this brief review we have emphasized a few simple questions and remarks:

(i) Bose–Einstein condensation is more *subtle* than is usually assumed. For instance the nature of the condensate (fragmentation, pure state or mixture) is a non-trivial issue.

(ii) Interactions between particles are absolutely crucial – one may say that genuine condensation is an effect of *exchange coupling*.

(iii) Bose–Einstein condensation is *fragile*. It competes with other instabilities and it may be destroyed by disorder, dissociation, etc.

One question remains: can a Bose liquid be *normal*, breaking no symmetry whatsoever (translation, gauge, etc.)? We argued that a hard core Bose liquid is isomorphic to a spin 1/2 system: a normal state would correspond to a *singlet* ground state of that equivalent magnetic system. Clearly this is what happens for composite bosons when they turn into a normal Fermi liquid (the ground state of which is a singlet). Is such a normal state possible for point bosons? It would correspond to what present wisdom calls an "RVB" state. Rather than playing with uncontrolled approximate calculations for fermions, it would be much better to start with point bosons. It has been known for a long time that interactions deplete the condensate. Can that depletion proceed to 100%? If it does, what is the physical nature of the resulting state? These questions are among the outstanding puzzles of Bose–Einstein condensation.

Acknowledgements. This lecture was not given at the Trento workshop, but at a meeting held in Paris at the Ecole Normale Supérieure on 9 June, 1993 in honour of Claude Cohen-Tannoudji on his sixtieth birthday. I would like to take this occasion to renew my congratulations to Claude: Bose–Einstein condensation is far from his usual worries, but I am acting as "le jongleur de Notre Dame". I am also grateful to Franck Laloë for many passionate discussions – and for his disagreements!

References

[1] V. Bagnato and D. Kleppner Phys. Rev. A **44**, 7439 (1991).
[2] S.I. Sevchenko, Sov. Phys. JETP **73**, 1009 (1991).

[3] J.P. Wolfe *et al.* and A. Mysyrowicz, this volume.
[4] T. Matsubara and H. Matsuda, Prog. Theoret. Phys. **16**, 569 (1956); **17**, 19 (1957).
[5] K.S. Liu and M.E. Fisher, J. Low Temp. Phys. **10**, 655 (1973).
[6] A.F. Andreev and I.M. Lifshitz, Sov. Phys. JETP **29**, 1107 (1969).
[7] G.A. Lengua and J.M. Goodkind, J. Low Temp. Phys. **79**, 251 (1990).
[8] R. Ma, B.I. Halperin and P.A. Lee, Phys. Rev. B **34**, 3136 (1986); M.P.A. Fisher, P.B. Weichman, G. Grinstein and D.S. Fisher, Phys. Rev. B **40**, 546 (1989). An excellent review by T.V. Ramakrishnan is in press.
[9] N. Trivedi, in *Computer Simulation Studies in Condensed Matter Physics V*, (Springer, Berlin, 1993); W. Krauth, N. Trivedi and D. M. Ceperley, Phys. Rev. Lett. **67**, 2307 (1991).
[10] D.B. Haviland, Y. Liu and A.M. Goldman, Phys. Rev. Lett. **62**, 2180 (1989).

3

Bose–Einstein Condensation and Superfluidity

Kerson Huang
Department of Physics
and
Center for Theoretical Physics, Laboratory for Nuclear Science
Massachusetts Institute of Technology
Cambridge, MA 02139
USA

Abstract

We review generally accepted definitions of Bose–Einstein condensation and superfluidity, emphasizing that they are independent concepts. These ideas are illustrated in a dilute hard-sphere Bose gas, which is relevant to experiments on excitons and spin-aligned atomic hydrogen. We then discuss superfluid ^{4}*He in porous media, as simulated by different models in different regimes. At low coverage, we model it by a dilute hard-sphere Bose gas in random potentials, and show that superfluidity is destroyed through the pinning of the Bose condensate by the external potentials. At full coverage, we model the random medium by an ohmic network of random resistors, and argue that the superfluid transition is a percolation transition in d=3, with critical exponent* 1.7.

This book is devoted to the phenomenon of Bose–Einstein condensation [1, 2] and inevitably, its relevance to superfluidity [3]. To provide some background for other articles in this volume, I would like to summarize some commonly accepted views on these phenomena, and illustrate them in the context of a dilute hard-sphere Bose gas, a model in which we have some control over the approximations made. I will also describe some recent work on the effect of randomness on the Bose condensate, which shows that Bose–Einstein condensation does not automatically give rise to superfluidity.

1 Bose–Einstein Condensation

In an ideal Bose gas in three spatial dimensions (d=3), Bose–Einstein condensation occurs when there is a finite density of particles in the zero-momentum ($\mathbf{k}$=0) state:

$$n_0 \equiv \frac{1}{\Omega}\langle a_0^\dagger a_0 \rangle > 0 \tag{1}$$

where $a_\mathbf{k}$ is the annihilation operator for the single-particle state of momentum $\hbar\mathbf{k}$, Ω is the total volume and $\langle\ \rangle$ denotes thermal averaging. In an interacting system, a Bose condensate can be defined in the same manner, with thermal averaging taken with respect to the interacting Hamiltonian. The Bose–Einstein condensation is a first-order phase transition in an ideal gas, but not in interacting systems generally.

The condensate density appears in a correlation function sometimes called the "one-particle density matrix":

$$\langle \psi^\dagger(\mathbf{x})\psi(\mathbf{y})\rangle = \frac{1}{\Omega}\langle a_0^\dagger a_0\rangle + \int \frac{d^3k}{(2\pi)^3}\, e^{i\mathbf{k}\cdot(\mathbf{x}-\mathbf{y})}\langle a_k^\dagger a_k\rangle \tag{2}$$

where $\psi(\mathbf{x})$ is the boson field operator. The relation

$$\langle \psi^\dagger(\mathbf{x})\psi(\mathbf{y})\rangle \underset{|\mathbf{x}-\mathbf{y}|\to\infty}{\longrightarrow} n_0 \tag{3}$$

has been used to theoretically estimate [4], and experimentally determine [5] the condensate fraction in ^{4}He, with the result $n_0/n \approx 0.1$ in the limit $T = 0$.

It has taken nearly forty years from the first discussion of Bose–Einstein condensation to the realization that its true essence lies in the twin phenomena of broken gauge symmetry, and phase coherence. These are embodied in what is now generally accepted as the criterion for Bose–Einstein condensation:

$$\langle \psi^\dagger(\mathbf{x})\psi(\mathbf{y})\rangle \underset{|\mathbf{x}-\mathbf{y}|\to\infty}{\longrightarrow} \langle\psi(\mathbf{x})\rangle^*\langle\psi(\mathbf{y})\rangle \tag{4}$$

where $\langle\psi(\mathbf{x})\rangle \neq 0$ is called the "condensate wave function", which, being a complex number, has a magnitude and a phase. Broken symmetry refers to the fact that, when the condensate wave function is non-zero, the vacuum state must depend on its phase, even though the Hamiltonian is invariant under a constant phase change of $\psi(\mathbf{x})$ (global gauge invariance). Phase coherence refers to the fact that, in order to have non-vanishing thermal average, the phase of $\psi(\mathbf{x})$ must be spatially correlated over the entire system. The criterion (4) was first proposed by Penrose and Onsager [4], and its significance emerged gradually, through the work of many people, among them Goldstone [6], who advanced the idea of spontaneous symmetry breaking, Yang [7], who termed the phenomenon "off-diagonal long-range order (ODLRO)", and Anderson [8], who emphasized the phase coherence.

It is instructive to repeat Anderson's argument [8, 9] that Bose–Einstein condensation implies phase coherence. Suppose, for simplicity, that all N particles in the system are in the k=0 state. Imagine that the system is subdivided into K macroscopic boxes of the same size. Thermodynamic extensiveness requires that each box contain N/K particles in the k=0 state. Let A_i be the annihilation operator of k=0 particles in the ith box. We must have

$$\begin{aligned} a_0 &= K^{-1/2}\sum_i A_i \\ \langle a_0^\dagger a_0\rangle &= N \\ \langle A_i^\dagger A_i\rangle &= N/K \end{aligned} \tag{5}$$

which lead to the relation $\langle a_0^\dagger a_0\rangle = K^{-1}\sum_i\langle A_i^\dagger A_i\rangle + K^{-1}\sum_{i\neq j}\langle A_i^\dagger A_j\rangle$, or

$$N = \frac{N}{K} + \frac{1}{K}\sum_{i\neq j}\langle A_i^\dagger A_j\rangle \tag{6}$$

For large K, the second term on the right side must dominate, and therefore the phases of the A_i in different boxes must be correlated.

The correct order parameter for Bose–Einstein condensation is thus the condensate wave function. As defined, however, it is a microscopic property of the system, and we can rarely make headway with microscopic calculations. It is more practical to adopt the Ginsberg–Landau approach, and introduce a local order parameter $\phi(\mathbf{r})$ that is a coarse-grained version of $\langle\psi(\mathbf{r})\rangle$:

$$\phi(\mathbf{r}) = \sqrt{R(\mathbf{r})}e^{i\alpha(\mathbf{r})} \tag{7}$$

whose phase is related to the superfluid velocity through

$$\mathbf{v}_s = \frac{\hbar}{m}\nabla\alpha \tag{8}$$

where m is the boson mass. Since ϕ should be single-valued, the change of α over any closed path must be a multiple of 2π. This leads to the Onsager–Feynman quantization condition [10, 11]

$$\oint d\mathbf{s}\cdot\mathbf{v}_s = \frac{2\pi\hbar\kappa}{m} \quad (\kappa = 0, \pm1, \pm2, \cdots) \tag{9}$$

The Landau free energy is taken to be [12]

$$E[\phi] = \int d^dx\left[\frac{\hbar^2}{2m}|\nabla\phi|^2 + c_2|\phi|^2 + c_4|\phi|^4 + \cdots\right] \tag{10}$$

As usual, in d=3 we put $c_2 = \gamma(T - T_c)$, to model the Bose–Einstein condensation as a second-order phase transition. The dots in (10) represent terms that can be neglected near the transition point, when ϕ is small. The partition function is given by the functional integral $Z = \int(D\phi)(D\phi^*)e^{-E[\phi]/kT}$. Critical exponents associated with the transition can be calculated in the mean-field approximation, and, more elaborately, in a power series expansion in ϵ, with $d = 4 - \epsilon$ [13]. By regarding (10) as a Hamiltonian, we can derive an equation of motion:

$$-\frac{\hbar^2}{2m}\nabla^2\phi + c_2\phi + c_4|\phi|^2\phi = i\hbar\frac{\partial\phi}{\partial t} \tag{11}$$

which is referred to as "the non-linear Schrödinger equation" (NLSE), or Gross–Pitaevsky equation [14, 15]. It admits static vortex solutions [12], spawns vortices in flows past obstacles [16], and can be used to discuss the kinetics of Bose–Einstein condensation [17]. The Ginsberg–Landau theory needs retooling in d=2, because strong phase fluctuations of the local order parameter destroy the Bose condensate. To see this in a simple way, let us assume that $|\phi| > 0$, due to the potential-energy terms in (10). Ignoring amplitude fluctuations, we consider the correlation function

$$C(\mathbf{r}) \equiv \langle e^{i\alpha(\mathbf{r})}e^{-i\alpha(0)}\rangle \tag{12}$$

where the average is taken with Gaussian weight (i.e., with only the kinetic term in the Landau free energy). The calculation is elementary, and yields $C(\mathbf{r}) \propto e^{-G(\mathbf{r})}$, where $G(\mathbf{r})$ is the Green's function of the Laplace equation: $G(\mathbf{r}) \propto \ln r$ for d=2, and $G(\mathbf{r}) \propto 1/r$ for d=3. Thus, up to a multiplicative constant,

$$C(\mathbf{r}) \underset{r\to\infty}{\longrightarrow} \begin{cases} r^{-bT} & \text{(d=2)} \\ 1 & \text{(d=3)} \end{cases} \tag{13}$$

where b is a constant. We see that phase coherence is maintained to infinite distances in d=3, but not in d=2 at finite temperatures. A rigorous proof of the absence of Bose–Einstein condensation in d=2 at finite temperatures was given by Hohenberg [18]. Equivalent statements are that long-ranged order does not exist in d=2 (Mermin–Wagner theorem) [19], and that spontaneous symmetry breaking does not occur in d=2 (Coleman's theorem) [20].

Even though the correlation function in d=2 goes to zero asymptotically, it does so slowly, with power-law behavior. This means that phase coherence persists over a finite but large distance, and a local Bose condensate exists. The local Bose condensate supports vortex–antivortex pairs, whose unbinding leads to the Kosterlitz–Thouless transition [21]. Thus, despite the fact that there is no Bose condensate, there is a phase transition, associated with what Kosterlitz and Thouless termed "topological order."

In the Ginsberg–Landau picture, the Kosterlitz–Thouless transition in d=2 should not be associated with a sign change of c_2. It is caused by vorticity associated with phase fluctuations, and to support vortices we must must have $c_2 < 0$. It would be interesting to discover such a transition within the Ginsberg–Landau theory; but this has not been attempted. Instead, one takes the two-dimensional XY model [22] as a more expedient starting point. As we shall discuss later, this model exhibits superfluidity in the low-temperature phase, even though there is no Bose condensate.

The local Bose condensate in d=2 is relevant to current thinking in high-T_c superconductivity, where one speaks of Bose–Einstein condensation of "holons" in the two-dimensional copper oxide plane. What one means is a "local condensation", which presumably becomes global when a small amount of interplane coupling is turned on.

2 Superfluidity

The term superfluidity covers a group of experimental phenomena in liquid ^{4}He, including frictionless flow, persistent current, wave propagation on liquid surfaces, and other kinetic effects. Our understanding on this subject is considerably less than that for Bose–Einstein condensation.

These first attempts to define superfluidity were based on specific pictures of liquid ^{4}He. Landau [23] proposed a low-temperature model in which the excitations are quasiparticles, which become phonons in the long-wavelength limit. Feynman [24] justified the picture from a microscopic point of view, and argued that Bose–Einstein statistics rule out all excitations except density fluctuations (phonons) in the immediate neighborhood of the ground state.

Landau argued that a system with only phonon excitations will flow without friction, because it cannot absorb arbitrarily small amounts of energy-momentum transfer. The argument is as follows. At absolute zero the only way to interact with the system is to excite phonons. Suppose an external object transfers energy ΔE and momentum $\Delta \mathbf{P}$ by creating n_p phonons of momentum $\mathbf{p}$ and energy cp, where c is the sound velocity.

Then

$$\Delta E = \sum_{\mathbf{p}} cpn_p$$

$$\Delta \mathbf{P} = \sum_{\mathbf{p}} \mathbf{p}n_p. \tag{14}$$

Thus $|\Delta\mathbf{P}| \leq \sum_p pn_p$, or $\Delta E \geq c|\Delta\mathbf{P}|$. If the external object is moving with velocity $\mathbf{v}_e$, we must have $\Delta E = \mathbf{v}_e \cdot \Delta\mathbf{P}$. Therefore $v_e \geq c$.

Using the quasiparticle picture, Khalatnikov [25] derived an expression for the normal fluid density at finite temperatures, based on Gallilean invariance. Suppose the energy of a quasiparticle of momentum $\mathbf{p}$ is ω_p, and the average occupation number for quasiparticles is $n(\omega) = [\exp(\omega/kT) - 1]^{-1}$. In a two-fluid picture, with superfluid velocity $\mathbf{v}_s$ and normal fluid velocity $\mathbf{v}_n$, the quasiparticle energy in a reference frame moving with the superfluid is

$$\omega_p' = \omega_p - \mathbf{p} \cdot (\mathbf{v}_n - \mathbf{v}_s). \tag{15}$$

The current density in the co-moving frame is given by

$$\mathbf{j} = \int \frac{d^3p}{(2\pi)^3} \mathbf{p} n(\omega_p') \equiv \rho_n(\mathbf{v}_n - \mathbf{v}_s) \tag{16}$$

where the right side is a definition of the normal fluid mass density ρ_n. By expanding in powers of $(\mathbf{v}_n - \mathbf{v}_s)$ and identifying coefficients on both sides, one easily obtains

$$\rho_n = -\frac{1}{3} \int \frac{d^3p}{(2\pi)^3} p^2 \frac{\partial n(\omega_p)}{\partial \omega_p} \tag{17}$$

The superfluid mass density is $\rho_s = \rho - \rho_n$, where ρ is the total mass density. In the limit $T \to 0$, the occupation number vanishes like $n(\omega) \approx kT/\omega$, and thus $\rho_n \to 0$. This shows that the ground state is pure superfluid. This result is not valid for fermionic quasiparticles, because $n(\omega)$ does not vanish at absolute zero.

A more general definition of the superfluid density is proposed by Hohenberg and Martin [26]. They emphasized that, unlike the condensate density, the superfluid density is not an equilibrium quantity but a transport coefficient, and should be defined in terms of linear-response theory.

Suppose we impose on the system an infinitesimal velocity field $\delta\mathbf{v}(\mathbf{x}, t)$, which should be turned on and off adiabatically:

$$\delta\mathbf{v}(\mathbf{x}, \mathbf{t}) \underset{|t|\to\infty}{\longrightarrow} 0 \tag{18}$$

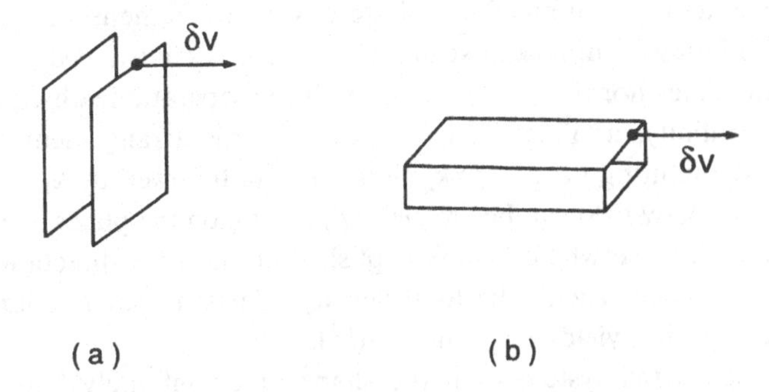

Fig. 1. (a) Longitudinal response: Place liquid between parallel plates that are moving at velocity $\delta\mathbf{v}$. The response gives the total density. (b) Transverse response: Place liquid in long pipe moving with velocity $\delta\mathbf{v}$. The response gives the normal fluid density.

The linear response, as measured by the changed in the momentum density $\mathbf{g}$ of the system, can be expressed in terms of the appropriate Fourier transforms in the form

$$\delta\langle g_i(\mathbf{k},\omega)\rangle = \chi_{ij}(\mathbf{k},\omega)\delta v_j(\mathbf{k},\omega) \tag{19}$$

which defines the susceptibility $\chi_{ij}(\mathbf{k},\omega)$. We are concerned only with the static susceptibility $\chi_{ij}(\mathbf{k}) \equiv \chi_{ij}(\mathbf{k},0)$. By rotational invariance, it can be decomposed into a longitudinal and a transverse part:

$$\chi_{ij}(\mathbf{k}) = \frac{k_i k_j}{k^2}A(k^2) + \left(\delta_{ij} - \frac{k_i k_j}{k^2}\right)B(k^2) \tag{20}$$

In the limit $k \to 0$, the longitudinal response is the total density (statement of the f-sum-rule), while the transverse response is defined as the normal fluid density. Thus, we have

$$\begin{aligned} \rho &= \lim_{k\to 0} A(k^2) \\ \rho_n &= \lim_{k\to 0} B(k^2) \\ \rho_s &= \lim_{k\to 0}[A(k^2) - B(k^2)]. \end{aligned} \tag{21}$$

An elementary explanation for these definitions due to Baym [27] goes as follows. Since $A(k^2)$ and $B(k^2)$ are rotationally invariant, it suffices to examine any component of χ_{ij}, say χ_{11}. That is, we consider the momentum response in the x direction, $\delta g_1(\mathbf{k})$, due to an imposed velocity field in the x direction, $\delta v_1(\mathbf{k})$.

Let the system be contained in a finite box, whose dimensions eventually tend to infinity. Suppose first that the system is sandwiched between two infinite plates normal to the x axis, with the separation subsequently tending to infinity, as illustrated in Fig.1(a). This arrangement corresponds to the limiting procedure $k_2 \to 0$, $k_3 \to 0$, followed by $k_1 \to 0$. In this case, a velocity field can be created by pushing on the plates, with the consequence that the whole system is pushed along the x direction. The response is associated with the total density. Carrying out this limiting procedure explicitly yields $\rho = \lim_{k \to 0} A(k^2)$.

Next suppose the system is in the shape of an infinitely long pipe, along the x axis, with the cross section of the pipe tending to infinity subsequently, as illustrated in Fig.1(b). This arrangement corresponds to the limiting procedure $k_1 \to 0$, followed by $k_2 \to 0$, $k_3 \to 0$. In this case, a velocity field can be created by dragging the walls of the pipe, and the part of the the system responding to the shear force is identified as the normal fluid. Carrying out this limiting procedure explicitly gives $\rho_n = \lim_{k \to 0} B(k^2)$.

These considerations apply equally well in d=2. In fact, using the definitions (21), Nelson and Kosterlitz [28] calculated the superfluid density in the two-dimensional XY model, and showed that it has a finite jump $\Delta\rho_s$ at the Kosterlitz–Thouless transition temperature T_c:

$$(\hbar/m)^2 k T_c \Delta\rho_s = 2/\pi \tag{22}$$

where m is the mass of a particle in the fluid. This result demonstrates that superfluidity can occur without Bose–Einstein condensation. The experimental verification of this jump in ^{4}He films [29] validates the linear-response definition of the superfluid density.

It is well to re-emphasize that superfluidity is a kinetic property of a system, and is generally more complicated than equilibrium properties. The arguments of Landau and Khalatnikov were formulated in the context of liquid helium in the quasiparticle picture, and one can find fault with them in a more general setting [30]. The linear-response definition, on the other hand, is more general; but it may not be unique. Indeed, one can question whether there exists a unique definition of superfluidity in general [31].

3 Hard-Sphere Bose Gas

A Bose gas in d=3 with hard-sphere interactions can be treated analytically in the limit of low temperatures and densities, the criteria for which

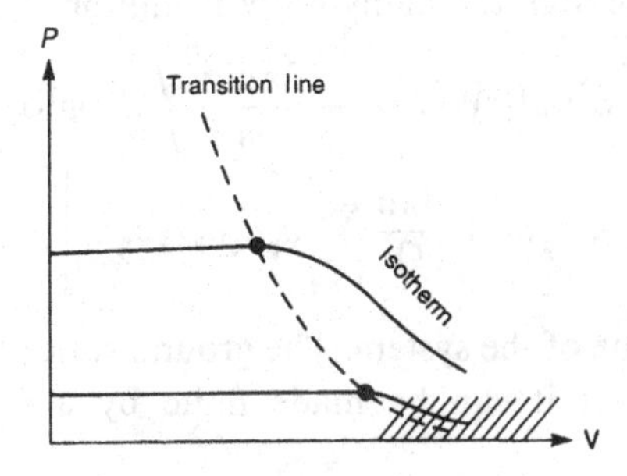

Fig. 2. Shaded region indicates schematically the low-density and low-temperature region in which our calculations are valid. It includes the transition line of the Bose–Einstein condensation for the ideal Bose gas.

are

$$\begin{aligned} a/\lambda &\ll 1 \\ na^3 &\ll 1 \end{aligned} \tag{23}$$

where a is the hard-sphere diameter, n the particle density, and

$$\lambda = \sqrt{2\pi\hbar^2/mkT} \tag{24}$$

is the thermal wavelength. The region satisfying these criteria, shown shaded in the P–V diagram in Fig. 2, includes the transition line of the ideal Bose gas. Experimental efforts are underway to observe superfluidity in such a dilute gas, in the form of spin-aligned atomic hydrogen [32] or excitons [33].

In the low-temperature regime, the hard-sphere diameter a enters as the s-wave scattering length of the interparticle potential, the detailed shape of which is irrelevant. On this basis, Lenz [34] first showed that, to lowest order, the system behaves like a medium with a refractive index, and the ground state energy per particle is shifted by an amount proportional to na. Fermi [35] showed that, for low-energy scattering, all potentials with the same scattering length a can be effectively replaced by a pseudopotential

$$V(\mathbf{r}) = \frac{4\pi a\hbar^2}{m}\delta^3(\mathbf{r}) \tag{25}$$

Using this, we can get Lenz's result in first-order perturbation theory.

With the pseudopotential, the many-body Hamiltonian reads

$$
\begin{aligned}
H &= -\frac{\hbar^2}{2m}\int d^3r\psi^\dagger(\mathbf{r})\nabla^2\psi(\mathbf{r}) + \frac{2\pi a\hbar^2}{m}\int d^3r\psi^\dagger(\mathbf{r})\psi^\dagger(\mathbf{r})\psi(\mathbf{r})\psi(\mathbf{r}) \\
&= \frac{\hbar^2}{2m}\left[\sum_{\mathbf{k}} k^2 a_{\mathbf{k}}^\dagger a_{\mathbf{k}} + \frac{4\pi a}{\Omega}\sum_{\mathbf{k},\mathbf{p},\mathbf{q}} a_{\mathbf{p}}^\dagger a_{\mathbf{q}}^\dagger a_{\mathbf{p}+\mathbf{k}} a_{\mathbf{q}-\mathbf{k}}\right],
\end{aligned} \tag{26}
$$

where Ω is the volume of the system. The ground state energy is divergent in higher orders; but it can be made finite by a simple subtraction procedure [36].

The Hamiltonian can be diagonalized using Bogoliubov's method [37], in which a_0 is replaced by a c-number $\sqrt{N_0}$. The parameter N_0, which labels the state we are considering, is the number of k=0 particles in the unperturbed state. We write it in the form

$$
N_0 = \xi N, \tag{27}
$$

where N is the total number of particles. The diagonalized form of the Hamiltonian to order $a^{5/2}$ is given by [38, 39]

$$
\begin{aligned}
H &= E_0 + \sum_{\mathbf{k}} \hbar\omega_{\mathbf{k}} b_{\mathbf{k}}^\dagger b_{\mathbf{k}} \\
\frac{E_0}{N} &= \frac{\hbar^2}{2m} 4\pi a n \left[1 + (1-\xi)^2 + \frac{128}{15}\sqrt{\frac{na^3}{\pi}}\xi^{5/2}\right] \\
\omega_{\mathbf{k}} &= \frac{\hbar k}{2m}\sqrt{k^2 + 16\pi n a\xi},
\end{aligned} \tag{28}
$$

where $n = N/\Omega$ is the particle density. The quasiparticle annihilation operator $b_{\mathbf{k}}$ is related to $a_{\mathbf{k}}, a_{\mathbf{k}}^\dagger$ through a Bogoliubov transformation:

$$
\begin{aligned}
a_{\mathbf{k}} &= \frac{1}{\sqrt{1-\alpha_{\mathbf{k}}^2}}(b_{\mathbf{k}} - \alpha_{\mathbf{k}} b_{\mathbf{k}}^\dagger) \\
a_{\mathbf{k}}^\dagger &= \frac{1}{\sqrt{1-\alpha_{\mathbf{k}}^2}}(b_{\mathbf{k}}^\dagger - \alpha_{\mathbf{k}} b_{\mathbf{k}})
\end{aligned} \tag{29}
$$

where

$$
\begin{aligned}
\alpha_{\mathbf{k}} &= 1 + x^2 - x\sqrt{x^2+2} \\
x &= \frac{\xi n}{8\pi a}k^2.
\end{aligned} \tag{30}
$$

The sound velocity deduced from the Bogoliubov quasiparticle spectrum is $c = (\hbar/m)\sqrt{4\pi a n\xi}$, which agrees with that calculated via the compressibility by differentiating E_0.

The condensate density is defined by $n_0 = n - \Omega^{-1} \sum_{\mathbf{k}\neq 0}\langle a_{\mathbf{k}}^{\dagger} a_{\mathbf{k}}\rangle$. In the ground state, which corresponds to $\xi = 1$, we have

$$\frac{n_0}{n} = 1 - \frac{8}{3}\sqrt{\frac{na^3}{\pi}}. \tag{31}$$

This shows that the interactions deplete the Bose condensate; but the superfluidity density at absolute zero is n, from Khalatnikov's formula (17).

The partition sum should extend over all values of ξ. For thermal properties near the transition point, the important states have $\xi \approx 0$, and in first approximation we keep ξ only to lowest order. This means that the quasiparticle spectrum is replaced by a shifted free-particle spectrum $(\hbar^2/2m)(k^2 + 8\pi a n\xi)$. The calculation of the free energy is elementary [39], and gives

$$F = F^{(0)} + \frac{N\hbar^2}{m} 4\pi a n \left(1 - \frac{1}{2}\bar{\xi}^2\right), \tag{32}$$

where $F^{(0)}$ is the free energy of the ideal Bose gas, and $\bar{\xi}$, the thermodynamic average of ξ, is the same as that in the ideal Bose gas:

$$\bar{\xi} = \begin{cases} 1 - (T/T_c)^{3/2} & (T < T_c) \\ 0 & (T > T_c), \end{cases} \tag{33}$$

where $kT_c = 2\pi\hbar^2 n/[m(2.612)^{3/2}]$. To this order, the transition temperature is the same as that of the ideal Bose gas. In a higher-order calculation, the transition temperature is lowered from that of the ideal gas by $O(\sqrt{na^3})$ [39, 40].

The isotherms are given in this approximation by

$$P = P^{(0)} + \begin{cases} (4\pi\hbar^2 a/m) v^{-2} & (v > v_c) \\ (2\pi\hbar^2 a/m)(v^{-2} + v_c^{-2}) & (v < v_c), \end{cases} \tag{34}$$

where P is the pressure, $P^{(0)}$ the pressure of the ideal Bose gas, $v = 1/n$, and $v_c = 2\pi\hbar^2/[mkT(2.612)^{3/2}]$. They are sketched in Fig. 3. The Bose–Einstein condensation now appears as a second-order phase transition. In higher-order calculations, it appears to be a first-order phase transition with a very narrow transition region, of order $\sqrt{na^3}$, which may not be significant. According to (34), the isothermal compressibility just above v_c is $\kappa_0 = mv_c^2/8\pi\hbar^2 a$. When the gas is compressed just across the transition point, it drops abruptly by a factor of 2. This is a signal for Bose–Einstein condensation that one hopes to observe in spin-aligned hydrogen.

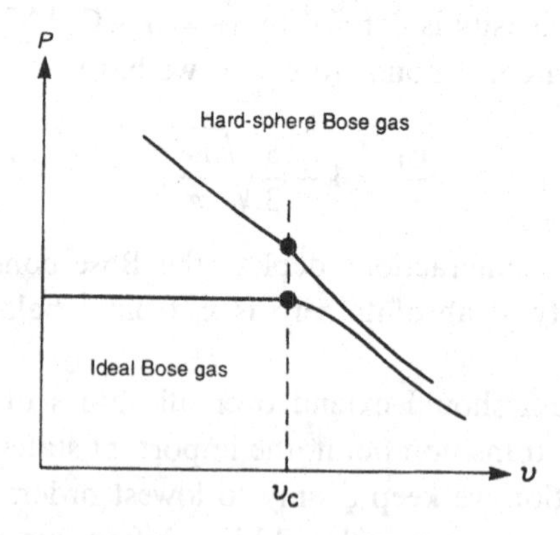

Fig. 3. Qualitative sketch of isotherm of hard-sphere Bose gas to lowest order of approximation, compared with that of ideal Bose gas at the same temperature. The isothermal compressibility changes discontinuously by a factor of 2 at the transition point.

A more realistic simulation of ^{4}He consists of adding an attractive tail to the hard-sphere potential [39, 41]. Such a model qualitatively reproduces the phase diagram of ^{4}He more faithfully. There is now a first-order gas–liquid phase transition, and Bose–Einstein condensation occurs within the liquid phase, dividing it into liquid I and liquid II, just like the λ transition in liquid ^{4}He.

4 Superfluidity in Random Media

We have seen that Bose–Einstein condensation is not necessary for superfluidity. We now show that it is not sufficient. The basic idea is very simple. Imagine a Bose gas in a randomly chosen potential well, which presumably has bound states. If there are no interactions, then at absolute zero all particles will condense into the lowest bound state. As long as the bound-state wave function has a non-vanishing spatial integral (k=0 projection), we have a Bose condensate; but it is pinned by the potential, which may be regarded as a kind of external wall. Thus, the condensate belongs to the normal fluid rather than the superfluid. This model is of course unrealistic. We need repulsive interparticle interactions to make the energy extensive, and the picture becomes more complicated. But the underlying idea remains, namely, localization effects in random media may destroy superfluidity. This idea has been developed further in tight-binding models, in which bosons can hop from site to site in

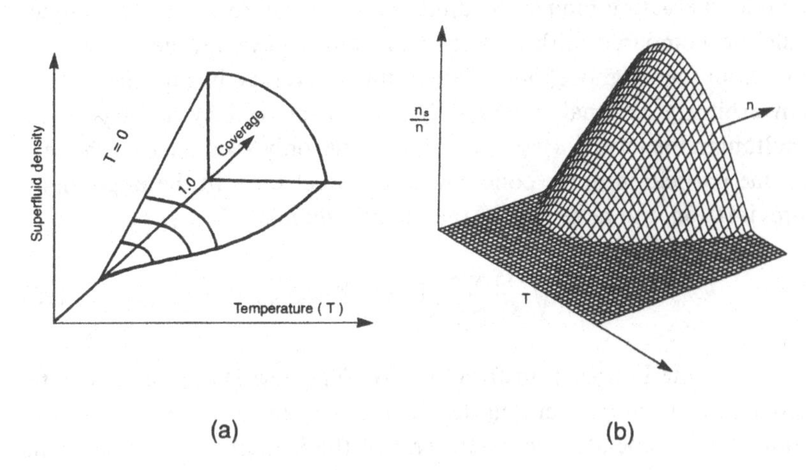

Fig. 4. (a) Transition surface for superfluid transition of ^{4}He in porous media. Superfluidity disappears at T=0 below a critical coverage. After *Physics Today*, July, 1989, p.22. (b) Transition surface for superfluid transition in a hard-sphere Bose gas in random external potentials in d=3. The total density is n, and the superfluid particle density is n_s.

a lattice, simulating quantum tunneling between localized states. The phase diagram for such a system has been discussed. In particular, a non-superfluid phase is proposed, with very specific properties dubbed the "Bose glass" [42, 43, 44].

The experimental impetus for these theoretical considerations come from observations on superfluid ^{4}He adsorbed in various porous media: Vycor, xerogel and aerogel [45].

The superfluid density was measured as a function of temperature and coverage, using a torsional pendulum. A qualitative sketch of the superfluid transition surface, synthesized from different experiments, is shown in Fig.4(a). We can see that, at absolute zero, the superfluid density as a function of coverage vanishes below a certain critical coverage. A system of bosons in random potentials is an attempt to simulate some aspects of these experiments. So far, signals for the "Bose glass" have not been detected experimentally [46].

I would like to describe a model somewhat different from the tight-binding kind – a dilute hard-sphere Bose gas in random potentials at low temperatures, in d=3 [47]. Since, unlike the much-studied case of fermion localization, interparticle interactions are necessary for stability, we cannot start with an ideal Bose gas. The next best thing is to start with a soluble interacting model, the dilute hard-sphere Bose gas. This model should be compared with experiments at low coverage near absolute zero, i.e., in the neighborhood of the critical coverage in Fig. 4(a).

An arbitrary external potential $U(\mathbf{r})$ is introduced by adding to the Hamiltonian the term $\int d^3r\psi^\dagger U\psi$, of which we only keep the part that excites one particle from the condensate, and vice versa. In the Bogoliubov approximation, the new term in the Hamiltonian is

$$H' = \sqrt{\frac{N_0}{\Omega}} \sum_{\mathbf{k}\neq 0} \left(U_{-\mathbf{k}} a_{\mathbf{k}}^\dagger + U_{\mathbf{k}} a_{\mathbf{k}} \right), \tag{35}$$

where $U_{\mathbf{k}}$ is the Fourier transform of $U(\mathbf{r})$. The new Hamiltonian can be diagonalized by supplementing the Bogoliubov transformation (29) with a c-number, $\mathbf{k}$-dependent term. In term of the field operator, this means that we write $\psi(\mathbf{r}) = v(\mathbf{r}) + \eta(\mathbf{r})$, where $v(\mathbf{r})$ is a c-number vacuum field, and $\eta(\mathbf{r})$ an operator. We then calculate the free-energy and relevant correlation functions, all to lowest order in the external potential, which is arbitrary but fixed.

We make quenched averages of the free energy and correlation functions over external potentials by putting

$$\frac{1}{\Omega}\langle U_{\mathbf{k}} U_{-\mathbf{k}'} \rangle_{av} = \delta_{\mathbf{k}\mathbf{k}'} R_0, \tag{36}$$

where R_0 is the single parameter that characterizes the random potentials. It has dimension $(energy)^2(length)^3$. If we picture $U(\mathbf{r})$ to consist of scattering centers of various positions and strengths, R_0 would represent the average of $(density)(strength)^2$ of the scatterers.

The validity of the approximations made are subject to the conditions (23) plus the fact that R_0 is treated only to first order. This means that the only effect of the random potentials is to deplete or enhance the condensate, while rescattering effects are neglected.

I will only quote the results at absolute zero. In the unperturbed ground state, the total density n is equal to that of the condensate $n_0 = N_0/\Omega$. When perturbations are turned on, the condensate is depleted, due to both the hard-sphere interactions and to the random potentials. The total density is thus made up of three terms:

$$n = n_0 + n_1 + n_R, \tag{37}$$

where n_1 arises from the hard-sphere interactions, and can be read off (31), while n_R arises from the random potentials:

$$n_1 = c_1 a^{3/2} n_0^{3/2}$$
$$n_R = c_2 R_0 \sqrt{\frac{n_0}{a}}, \tag{38}$$

where $c_1 = 8/(3\pi^{1/2})$, $c_2 = m^2/(8\pi^{3/2})$ (in units with $\hbar = 1$.) Substituting (38) into (37), we obtain a transcendental equation for n_0. The singularity $n_R \sim a^{-1/2}$ when $a \to 0$ signals the collapse of the system when the interparticle interactions are removed.

To physically understand n_R, note that it can be represented in the form

$$\frac{n_R}{n} \propto \frac{a n^{1/3}}{\tau E}, \tag{39}$$

where the factor $E = m^{-1} n^{2/3}$ supplies an energy scale natural for a Bose gas, τ is the collision time for the scattering of a particle out of the condensate:

$$\frac{1}{\tau} = R_0 \rho(E), \tag{40}$$

where $\rho(E)$ is the density of final states, taken to be one-phonon states of momentum k and energy ck, with $c \propto \sqrt{na}/m$:

$$\rho(E) = \int d^3k \; \delta(E - ck) \propto E^2/c^3. \tag{41}$$

For the approximations to be valid, both a and R_0 must be small. Thus n_1 must be small; but $n_R \propto R_0/\sqrt{a}$ can have arbitrary values. Solving for n_0 from (37) and (38), we can easily see that n_0 never vanishes, however large R_0 might be. Thus, at absolute zero, there will always be a Bose condensate.

The superfluid density is calculated by calculating the particle density n_n of the normal fluid using (21). At absolute zero we find $n_n = (4/3) n_R$, and thus the superfluid particle density is given by

$$n_s = n - \frac{4}{3} n_R. \tag{42}$$

The remarkable factor 4/3 means that the random potentials generate an amount of normal fluid 4/3 times what it took from the condensate. This implies that part of the condensate belonging to the normal fluid, being pinned by the random scatterers – a example of boson localization.

We can now see that superfluidity can be destroyed, even though there is a Bose condensate. From (37) and (42), we can deduce the relation

$$n_s = \frac{1}{3}[4(n_0 + n_1) - n], \tag{43}$$

where n_1 may be neglected. Thus, n_s vanishes when the condensate is depleted by random scattering to the value $n_0 = n/4$. The finite-temperature calculations yield a superfluid transition surface shown in Fig. 4(b), which is to be compared with the experimental one in Fig. 4(a). They have similar qualitative features, but there is a re-entrant effect in the theoretical graph which has not been seen experimentally. The detailed nature of the superfluid transition, as well as properties of the "Bose glass" phase, is beyond the capabilities of this calculation.

In quenched averaging, there is no interference between members of the ensemble over which we average. Our results thus indicate that localization happens in a good fraction of the potentials in the ensemble. The calculation with any single given potential, of course, cannot be carried out analytically.

The d=2 case cannot be treated in the same manner, because there is no Bose condensate at any finite temperature, due to strong phase fluctuations. Thus we cannot replace a_0 by a c-number of fixed phase. The way to proceed is to make a canonical transformation $\psi = \sqrt{\rho}\exp(i\theta)$. We can then replace $a_0^\dagger a_0$ by a c-number n_0 in the expansion $\rho = a_0^\dagger a_0 + \cdots$. A Bogoliubov-like transformation then diagonalizes the Hamiltonian. In this manner, it is found that, just as in the d=3 case, superfluidity can be destroyed by random potentials [48, 49].

Finally, let us turn to superfluid flow in a fully saturated porous medium. This corresponds to a cut at maximum coverage in the phase diagram of Fig. 4(a). Experimental interest centers on the critical exponent ζ at the superfluid transition, which is reported to be, respectively, 0.67, 0.89 and 0.80, for Vycor, xerogel and aerogel [45], as compared with $\zeta = 0.672$ for the bulk liquid. It is thought, therefore, that each medium represents a new universality class. I would like to suggest that actually there is only one universality class with $\zeta = 1.7$, as appropriate for percolation in d=3. The difference in the apparent values of ζ can be attributed to the fact that the experiments have not reached the true critical point [50].

The physics here is different from that in the low-density regime, and calls for a different model. We simulate the flow channels in the sponge-like medium as bonds in a cubic lattice, with channel radii chosen at random from a given distribution. The momentum density for superfluid

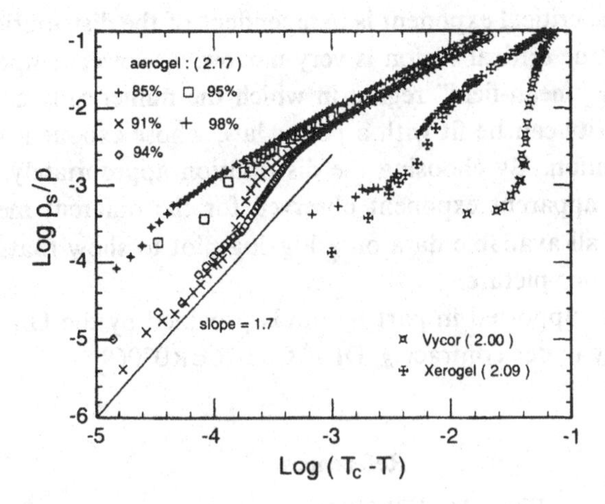

Fig. 5. Log–log plot of superfluid density vs. temperature for liquid ^{4}He in fully saturated media. The number in braces after the name of each medium gives the value of T_c used in the plot. Percentages correspond to porosity of aerogel samples [51].

flow is given by

$$\mathbf{j} = \rho_s \mathbf{v}_s + \rho_n \mathbf{v}_n \tag{44}$$

where the two terms refer to the superfluid and normal fluid respectively. Assuming that all channels are so small that the normal fluid is pinned, we put $\mathbf{v}_n = 0$. Thus, writing $\mathbf{v}_s = \nabla\Phi$, where Φ is proportional to the superfluid phase, we have

$$\mathbf{j} = \rho_s \nabla\Phi \tag{45}$$

This looks like Ohm's law with current density **j**, electrostatic potential Φ and electrical conductivity ρ_s. Thus, each bond can be replaced by a resistor, and the superfluid density of the medium corresponds to the conductivity of the network.

In a long channel filled with liquid ^{4}He, the superfluid transition temperature decreases with the radius according to a power law [52]. At sufficiently high temperatures, all bonds in the lattice are non-conducting, and the overall conductance is zero. As the temperature is lowered, some bonds begins to conduct; but the overall conductance remains zero, until there is a percolating cluster. Therefore, the true critical point corresponds to percolation in d=3.

This picture is verified by solving Kirchoff's equations numerically to obtain the conductivity of the network, for various distributions of the resistances. The critical exponent is independent of the distribution used. However, the true critical region is very narrow. At lower temperatures, there is a wide "mean-field" region in which the numerically calculated superfluid density can be fit with a power law, whose exponent depends on the distribution. By choosing the distribution appropriately, we can reproduce the apparent exponent observed for the different media. In Fig. 5 we plot all available data on a log–log plot to show that they do not contradict our picture.

This work is supported in part by funds provided by the US Department of Energy under contract # DE-AC02-76ER03069.

References

[1] S.N. Bose , Z. Phys. **26**, 178 (1924).

[2] A. Einstein, Sitzber. Kgl. Preuss. Akad. Wiss., (1924), p.261 ; (1925), p.3 . The story of Bose–Einstein statistics is told in A. Pais, *Subtle is the Lord, The Science and the Life of Albert Einstein* (Clarendon Press, Oxford, 1982), Ch. 23.

[3] L. Tisza, Nature, **141**, 913 (1938),

[4] O. Penrose and L. Onsager, Phys. Rev. **104**, 576 (1956).

[5] P.E. Sokol, this volume.

[6] J. Goldstone, N. Cim. **19**, 154 (1961).

[7] C.N. Yang, Rev. Mod. Phys. **34**, 4 (1962).

[8] P.W. Anderson, Rev. Mod. Phys. **38**, 298 (1966).

[9] P.W. Anderson, *Basic Notions of Condensed Matter Physics* (Benjamin-Cummings, Menlo Park,1984), pp.15ff.

[10] L. Onsager, Suppl. N. Cim. **6**, 249 (1949).

[11] R.P. Feynman, in *Progress in Low Temperature Physics*, vol.1, C.J. Gorter, ed. (North-Holland, Amsterdam, 1955), p.17.

[12] V.L. Ginsberg and L.P. Pitaevsky, Sov. Phys. JETP **34**, 858 (1958).

[13] K.G. Wilson and J. Kogut, Phys. Rep. **12C**, 75 (1974).

[14] E.P. Gross, N. Cim. **20**, 454 (1961); J. Math. Phys. **4**, 195 (1963).

[15] L.P. Pitaevsky, Sov. Phys. JETP **13**, 451 (1961).

[16] T. Frisch, Y. Pomeau, and S. Rica, Phys. Rev. Lett. **69**, 1644 (1992).

[17] See the contributions of Yu. Kagan and H.T.C. Stoof in this volume.

[18] P.C. Hohenberg, Phys. Rev. **158**, 383 (1967).

[19] N.D. Mermin and H. Wagner, Phys. Rev. Lett. **22**, 1133 (1966).

[20] S. Coleman, Comm. Math. Phys. **31**, 259 (1973).

[21] J.M. Kosterlitz and D.J. Thouless, J. Phys. C **6**, 1181 (1973).

[22] J.V. José, L.P. Kadanoff, S. Kirkpatrick, and D.R. Nelson, Phys. Rev. B **16**, 1217 (1977).
[23] L.D. Landau, J. Phys. USSR **5**, 71 (1941).
[24] R.P. Feynman, Phys. Rev., **94**, 262 (1954).
[25] I.M. Khalatnikov, *Introduction to the Theory of Superfluidity* (Benjamin, New York, 1965), p.13.
[26] P.C. Hohenberg and P.C. Martin, Ann. Physics (NY) **34**, 291 (1965).
[27] G. Baym, in *Mathematical Methods of Solid State and Superfluid Theory*, ed. R.C. Clark (Oliver and Boyd, Edinburgh, 1969), p.121; the argument is reproduced in D. Forster, *Hydrodynamics Fluctuations, Broken Symmetry, and Correlation Functions* (Benjamin, Reading, MA, 1975), pp.219 ff.
[28] D.R. Nelson and J.M. Kosterlitz, Phys. Rev. Lett. **39**, 1201 (1977); D.R. Nelson in *Phase Transitions and Critical Phenomena*, vol.7, C. Domb and J.L. Lebowitz, eds. (Academic, New York, 1983).
[29] I. Rudnick, Phys. Rev. Lett. **40**, 1454 (1978); D.J. Bishop and J. Reppy, Phys. Rev. Lett. **40**, 1727 (1978); E. Webster, G. Webster, and M. Chester, Phys. Rev. Lett. **42**, 243 (1978); J.A. Roth, G.J. Jelatis, and J.D. Maynard, Phys. Rev. Lett. **44**, 333 (1978).
[30] See the article by L. V. Keldysh, this volume.
[31] See the article by A. Leggett, this volume.
[32] See the articles by T. Greytak and I. Silvera, this volume.
[33] See the article by J. P. Wolfe, *et al.*, this volume.
[34] W. Lenz, Z. Phys. **56**, 778 (1929).
[35] E. Fermi, Riverca Sci. **7**, 13 (1936).
[36] K. Huang and C.N. Yang, Phys. Rev. **105**, 767 (1956).
[37] N.N. Bogoliubov, J. Phys. USSR, **2**, 23 (1947).
[38] T.D. Lee, K. Huang, and C.N. Yang, Phys. Rev. **106**, 1135 (1957).
[39] K. Huang, in *Studies in Statistical Mechanics*, vol.2, J. De Boer and G.E. Uhlenbeck, eds. (North-Holland, Amsterdam, 1964).
[40] K. Huang, C.N. Yang, and J.M. Luttinger, Phys. Rev. **105**, 776 (1957).
[41] K. Huang, Phys. Rev. **115**, 765 (1959); **119**, 1129 (1960).
[42] J.A. Hertz, L. Fleishman, and P.W. Anderson, Phys. Rev. Lett. **43**, 942 (1979).
[43] D.S. Fisher and M.P.A. Fisher, Phys. Rev. **61**, 1847 (1988).
[44] M.P.A. Fisher, P.B. Weichman, G. Grinstein, and D.S. Fisher, Phys. Rev. B **40**, 546 (1989).
[45] For a comprehensive review, see J.D. Reppy, J. Low Temp. Phys. **87**, 205 (1992).

[46] J.D. Reppy, private communication.
[47] K. Huang and H.F. Meng, Phys. Rev. Lett. **69**, 644 (1992).
[48] H.F. Meng, Superfluidity and Random Media, Ph.D. Thesis, Mathematics Department, MIT, 1993.
[49] H.F. Meng, Quantum Theory of Two-Dimensional Interacting Boson Systems, (MIT Mathematics Department preprint, 1993).
[50] K. Huang and H.F. Meng, Phys. Rev. B **48**, 6687 (1993).
[51] The data in Fig. 5 is from J. D. Reppy and H. M. W. Chan, private communication.
[52] R. Donnelly, R. Hills, and P. Roberts, Phys. Rev. Lett. **42**, 75 (1979).

4

Bose–Einstein Condensation in Liquid Helium

P. E. Sokol
Physics Department
Pennsylvania State University
University Park, PA 16802
USA

Abstract

Liquid helium is the prototypical example of a superfluid – a liquid that flows without viscosity and transfers heat without a temperature gradient. These properties are intimately related to the Bose condensation that occurs in this strongly interacting liquid. Bose condensation is most directly observed in the single particle atomic momentum distribution, where the Bose condensate appears as a delta function singularity. In this article, we discuss the experimental techniques used to observe the condensate and the current status of measurments of the Bose condensate in liquid helium.

1 Introduction

Liquid helium (^{4}He) has fascinated physicists ever since Kammerlingh-Onnes liquified the last of the so called permanent gases in 1908. However, evidence of what is without doubt the most fascinating property of this unique liquid, superfluidity, was not reported until almost 25 years later [1]. The superfluid phase, where heat is transferred without a thermal gradient and mass flows without a driving pressure, is a macroscopic manifestation of microscopic quantum effects governing the behavior of atoms [2]. Bose–Einstein condensation was first proposed by London [3] as the microscopic explanation for these fascinating phenomena.

Helium is unique among condensed atomic systems since the bulk properties of the liquid are dominated by quantum effects. The most important of these are the statistical effects that arise for identical particles. For bosons, such as ^{4}He atoms, there is no limitation on the number of particles that can occupy a single quantum state. Thus, in the absence of interactions, all the atoms in the liquid would occupy a single momentum state at zero temperature.

Liquid helium is also unique among the Bose systems considered in this volume. Other systems, such as spin polarized hydrogen and excitons, are expected to exhibit Bose condensation. (See the review articles by Greytak and Silvera, Castin *et al.,* Mysyrowicz and Wolfe *et al.* in this volume.) At present, however, liquid helium is the only system where the existence of an experimentally attainable Bose condensed phase is almost universally accepted.

Unlike many of the other systems considered in this volume, liquid helium is a strongly interacting system. The interaction between the atoms, which is dominated by the hard core repulsion, is strong enough that the properties of the liquid cannot be treated as a simple perturbation of the Ideal Bose Gas, as is the case for more weakly interacting systems. As we will see, these strong interactions have greatly complicated both the theoretical and experimental studies of the Bose-condensed phase of superfluid ^{4}He.

We begin this review with a discussion of the Ideal Bose Gas, a model system consisting of non-interacting particles. This model, while not applicable to the strongly interacting liquid phase of helium, is useful in that it provides a system where the effects of Bose condensation are clearly delineated. The microscopic properties of liquid helium, which we discuss next, are quite different from the ideal gas due to the strong interactions between atoms. Despite these differences, similarities between the ideal gas and the liquid systems remain. In particular, a Bose-condensed phase still appears in liquid ^{4}He at low temperatures, although the fraction of particles in the condensate is considerably reduced. Therefore, in this review, the analogies between an ideal Bose gas and superfluid ^{4}He are stressed, even though the former does not exhibit the long-range phase coherence associated with Bose-broken symmetry. (For a discussion of this, see especially the articles in this book by Nozières, Huang, Stringari and Stoof.)

The single-particle momentum distribution is one of the few quantities which directly reflects the appearance of a Bose condensate. Therefore, we next turn our attention to the experimental technique that provides information on the momentum distribution: Deep Inelastic Neutron Scattering (DINS). This technique, due to the information on $n(p)$ that it provides, provides one of the few opportunities to determine n_0 directly [4, 5, 6].

Early attempts to measure the momentum distribution, and to obtain direct information on the condensate using DINS, were limited by the fluxes and energies of neutrons available. Recent advances in neutron sources, which now have both higher fluxes and higher energies, have led to a new generation of experiments. These new experiments are reviewed, and the information that can be extracted, both directly and with the help of models, are described.

Finally, we turn our attention to the direct evidence for the condensate in liquid helium. The experimentally observed scattering shows distinct changes consistent with the appearance of a condensate. However, the sought after direct evidence, an observation of the δ-function singularity associated with the condensate, is not observed. Nevertheless, comparison to current theoretical predictions and to empirical expressions based on realistic models give a strong case for a finite value of n_0 and reasonable results for its temperature and density dependence.

2 The Ideal Bose Gas

Before examining the properties of liquid helium, it is instructive to first examine the Ideal Bose Gas (IBG), whose properties are entirely controlled by the Bose–Einstein statistics obeyed by the particles. All the complicating effects of interactions, which play a major role in determining the properties of liquid helium, are absent. The effects of Bose–Einstein condensation can, therefore, be studied in fairly direct fashion.

The IBG, in its purest form, consists of a collection of non-interacting particles in a box [7]. The eigenstates of a single particle in a box are plane wave states with a well-defined momentum p. In the case of the IBG, these s·ates are also the eigenstates of the many particle system. The statistical description of the particles in this box consists of specifying the number of particles occupying each of the single-particle eigenstates. This function is the single-particle momentum distribution, $n(p)$, which specifies the number of particles in a state with momentum p.

The particles in the gas can be characterized by their thermal DeBroglie wavelength, which is given by

$$\lambda_D = \frac{h}{\sqrt{3mk_B T}}, \tag{1}$$

where m is the mass of the particle and T is the temperature. When λ_D is much less than the interparticle spacing, quantum effects will be negligible and the particles will behave classically. Conversely, when

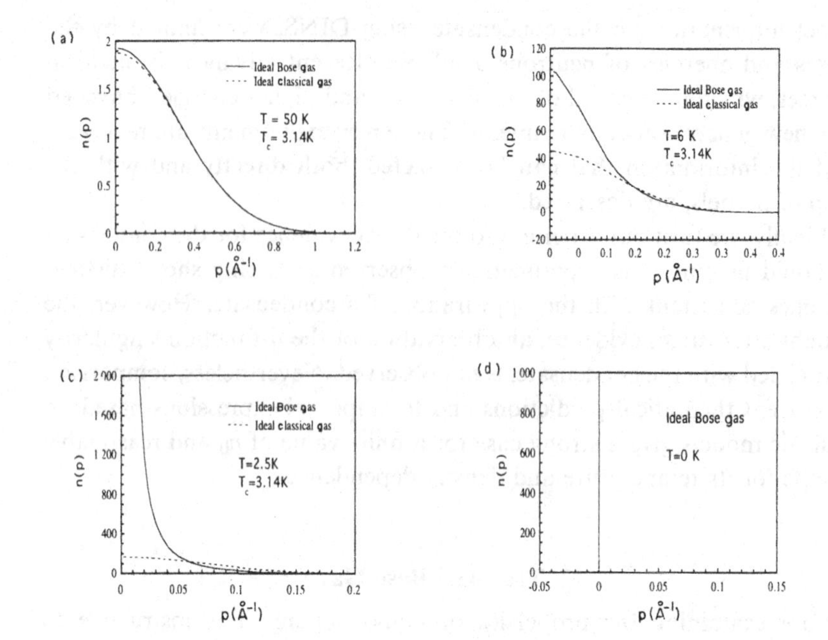

Fig. 1. Temperature dependence of the momentum distribution in the Ideal Bose Gas (solid line) and the classical result (dashed line). The Bose–Einstein condensation temperature is 3.14 K, corresponding to the density of liquid helium at SVP. (a) $T = 50$ K, (b) $T = 6$ K, (c) $T = 2.5$ K (a condensate is present but is not visible in this diagram due to the singular behavior of the uncondensed component) and (d) $T = 0$.

λ_D is large compared to the interparticle spacing, there will be a large overlap of the wavefunctions for the particles, and quantum effects will be important.

The properties of the ideal Bose gas at high temperatures, where λ_D is smaller than the atomic spacing, are dominated by the thermal motion of the particles. The momenta of the particles will be distributed according to the classical Maxwell–Boltzman distribution. This is simply a Gaussian with a width that is proportional to the temperature, as illustrated in Fig. 1(a).

Upon lowering of the temperature, λ_D of the particles begins to approach the interparticle spacing. The effects of quantum statistics become evident as $n(p)$ deviates from its classical Gaussian shape. In the case of bosons, where there is no restriction on the occupancy of a state, the quantum behavior is reflected by an increase in the occupation of states at small p. This reflects the particles taking advantage of the lack of restrictions on the occupancy of a state to reduce the energy of the system. Thus, $n(p)$ assumes a non-Gaussian shape and becomes more peaked at small p, as shown in Fig. 1(b). It is interesting to note that this behavior becomes evident at relatively large temperatures for the IBG. Even at twice the critical temperature, deviations from the classical Gaussian behavior are quite pronounced.

As the temperature is lowered further , the quantum statistical effects become the dominant factor in determining the properties of the IBG, and we eventually have a transition to the Bose–Einstein condensed phase. The Bose–Einstein transition temperature is

$$T_{BE} = (2\pi\hbar^2/1.897mk_B)\rho^{2/3}, \tag{2}$$

where ρ is the density of the gas. This transition temperature is simply determined by the condition that λ_D is equal to some appropriate measure of the mean interparticle spacing.

At temperatures higher than T_{BE}, the occupancy of any particular one-body momentum state remains finite. Below this temperature, a macroscopic (i.e. of order the total number of particles in the system, N) occupation of the $p = 0$ momentum state develops. This is the famous "Bose–Einstein condensation". It is reflected in $n(p)$ by the appearance of a term proportional to a Dirac delta function $\delta(p)$, with a coefficient which determines the (finite) fraction of particles residing in this Bose condensate.

At finite temperatures lower than T_{BE}, $n(p)$ consists of a δ function component with a weight determined by the condensate and a component resulting from atoms not in the condensate, as shown in Fig. 1(c). As the temperature is lowered, the magnitude of the condensate term increases and the non-condensed component decreases. At $T = 0$, the condensate component contains all of the intensity, representing the fact that every particle in the system is in the ground state, as shown in Fig. 1(d).

A common misconception about the IBG is that the non-condensed part of $n(p)$ is similar to a classical distribution, i.e. a broad Gaussian. Associated with this is the idea that the uncondensed $n(p)$ is similar to the distribution above T_λ. Neither of these ideas is correct. The momentum distribution below T_λ becomes singular, with the leading term going as $1/p^2$, as can be seen in Fig. 1(c). Above T_λ, $n(p)$ becomes sharply peaked, but not singular. Neither of these distributions is anything like the simple Gaussian form often depicted.

3 The Interacting Liquid

We now turn our attention to liquid ^{4}He. Unlike the IBG, helium atoms do interact with each other. Due to these interactions, the properties of liquid helium are significantly different from those of the IBG. However, one can still have a macroscopic number of atoms with zero momentum (the Bose condensate). The interactions do, however, significantly reduce the occupation of the ground state below its value in an ideal gas.

Liquid helium is, in one sense, a weakly interacting system. The attractive part of the helium potential is quite weak [8], which, when coupled with the light mass of ^{4}He, leads to a large zero-point motion that prevents the atoms from being strongly localized even in the condensed phases. Thus, the liquid phase is not even stable until very low temperatures and a solid phase, in which the atoms must be localized about lattice sites, does not exist unless substantial external pressure is applied [9]. This weak localization leads to a large overlap of the atomic wavefunctions which make quantum statistics an important factor in determining the properties of the condensed phases.

At the same time, liquid ^{4}He is also a very strongly interacting system [10]. The hard core repulsion between atoms is very large due to the closed shell electronic structure. The large zero-point motion would lead to a large overlap of the core wavefunctions. The hard core repulsion prevents this and leads to large repulsive interaction energies between the atoms.

The effects of the interactions in liquid helium can be highlighted by comparing the phase diagram of the IBG and liquid helium, as shown in Fig. 2. The IBG exhibits a single-phase transition from the high temperature gas phase to the Bose-condensed phase. This transition, which is marked by the appearance of a finite fraction of particles in the ground state, increases with increasing density. Liquid helium, on the other hand, has a considerably more complex phase diagram [11]. There is no low density phase at temperatures where Bose condensation might occur. At low temperatures, the low density gas is not stable and condenses to form a high density liquid. Thus, there is no region where Bose condensation of ^{4}He atoms is possible where the interactions might be considered weak.

Another difference between the liquid and the ideal gas is the dependence of the transition temperature on density. In the IBG the transition temperature increases with density, reflecting the fact that overlap of the wavefunctions occurs at higher temperatures as the density increases. In

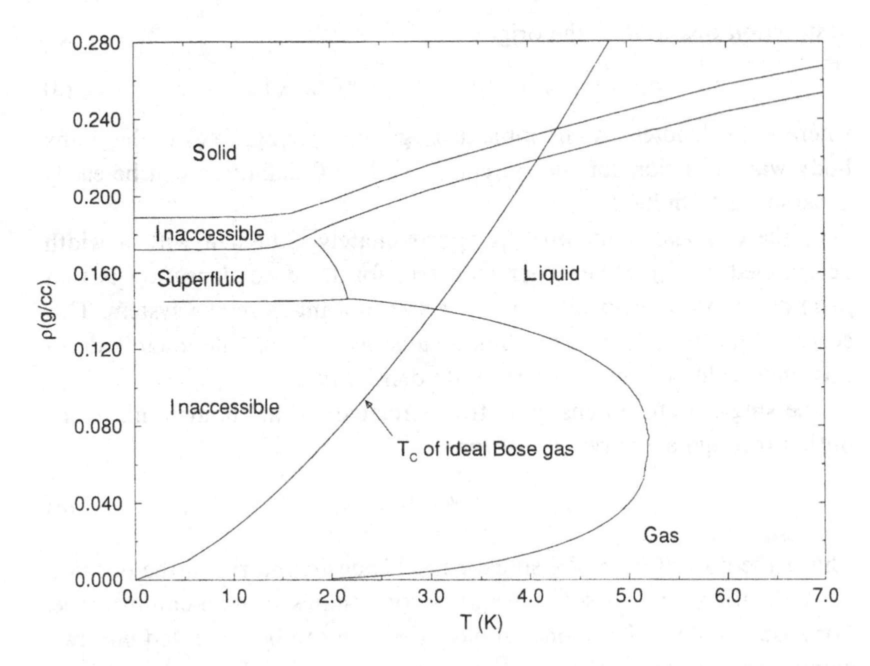

Fig. 2. The phase diagram of liquid helium in the temperature–density plane. The regions marked inaccessible are not stable in the bulk liquid. The Bose–Einstein transition temperature as a function of density is also shown.

liquid helium, on the other hand, the transition temperature *decreases* with increasing density (as shown by pressure studies), indicating that the strong interactions present in the liquid destroy the condensate. Thus, the IBG is not an appropriate model for liquid helium.

The eigenstates of the particles in the non-interacting gas are free particle states. Thus, p is a "good" quantum number and n_0 is just the density of particles in the ground state. The nature of Bose condensation in the liquid is more complicated since we are no longer dealing with free particle states and p is no longer a "good" quantum number. However, one still finds that there can be a finite fraction of atoms with zero momentum, and define this as the "condensate fraction".

A more general definition of Bose condensation that is applicable to interacting systems is required. The concept of off-diagonal, long-range order [12, 13, 14] (ODLRO) provides the necessary new definition. (See also the article by Huang in this volume.) This new phase is described in terms of the one-body density matrix which is defined as the average of the product of the field creation operator at point $\mathbf{r}$ and the field destruction operator at the origin,

$$\rho_1(\mathbf{r}) \equiv < \psi^*(\mathbf{r}, \mathbf{r}_2, ..., \mathbf{r}_N)\psi(\mathbf{r}, \mathbf{r}_2, ..., \mathbf{r}_N) >, \tag{3}$$

where $< >$ denotes an ensemble average and $\psi(\mathbf{r}, \mathbf{r}_2, ..., \mathbf{r}_N)$ is the many body wave function for the system. This $T = 0$ definition can be easily generalized to finite T.

In the classical limit, $\rho_1(r)$ is approximately Gaussian with a width determined by λ_D. The general criteria for Bose condensation is that $\rho_1(r)$ develops an eigenvalue of finite size in a macroscopic system. This corresponds to $\rho_1(r)$ having a finite value as $r \to \infty$. The magnitude of this finite value is just the condensate density n_0.

The single-particle density matrix is related to the momentum distribution through a Fourier transform:

$$n(p) = \int d\mathbf{r} e^{-i\mathbf{p}\cdot\mathbf{r}} \rho_1(r). \tag{4}$$

This is the projection of the single-particle density matrix, in the position representation, onto a set of single-particle states in momentum space. For a Bose system, the momentum distribution can be separated into two components, a sharp delta-function term representing the condensate and a smooth component corresponding to the atoms with finite momentum,

$$n(p) = n_0\delta(\mathbf{p}) + n'(p), \tag{5}$$

where the condensate density is determined by

$$\lim_{r\to\infty} \rho_1(r) = n_0 \tag{6}$$

and the non-condensate portion has the Fourier representation

$$n'(p) = \int [\rho_1(r) - \rho_1(\infty)] e^{-i\mathbf{p}\cdot\mathbf{r}} d\mathbf{r}. \tag{7}$$

Typical behavior for both $n(p)$ and $\rho_1(r)$ for an interacting Bose fluid shown in Fig. 3. The dynamical short-range correlation due to the hard core repulsion governs the small-r behavior of $\rho_1(r)$ and the large-p behavior of $n(p)$. The effects of statistics play little role in this small-r range and $\rho_1(r)$ would have a similar shape for Fermion or Boltzmann (classical) particles. The effects of statistical correlations are most apparent in $\rho_1(r)$ at large r and in $n(p)$ at small p. The condensate, which gives rise to a finite value of $\rho_1(r)$ at infinity, manifests itself as a δ-function at $p = 0$ in $n(p)$.

A simple physical interpretation of $\rho_1(r)$ exists in the case of bosons [15]. As can be seen from the definition of $\rho_1(r)$, it is the product of a

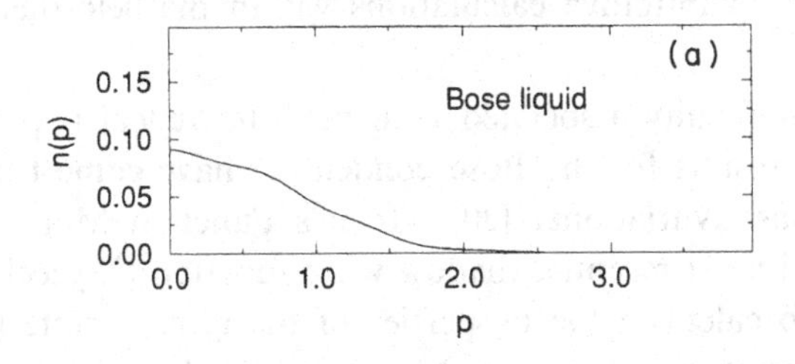

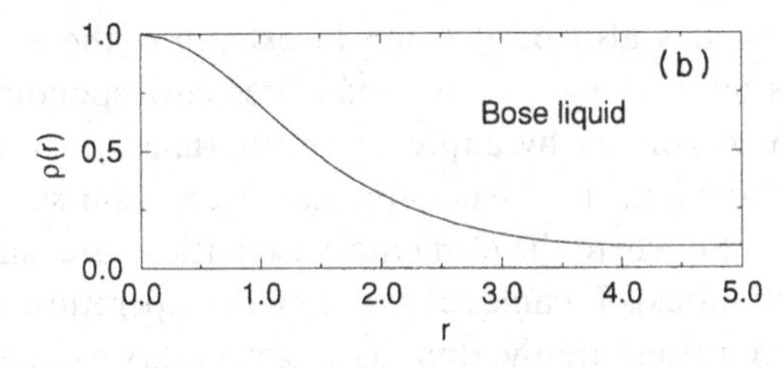

Fig. 3. Typical behavior of the momentum distribution (a) and single-particle density matrix (b) for a Bose-condensed system.

particle annihilation operator at $r = 0$ and a particle creation operator at r. Thus, $\rho_1(r)$ can be viewed as the overlap in the wavefunction of the system when a particle is removed at the origin and replaced a distance r away. In a classical system, this overlap would die away quickly, with a distance scale on the order of λ_D. However, when a condensate exists, this overlap persists out to infinite separations. Thus, there is a finite possibility of removing a particle at the origin and placing it back in the same quantum state, no matter what the separation. This interpretation emphasizes the fact that the Bose-condensed phase has long-range macroscopic coherence.

The dual nature of liquid helium, being both strongly and weakly interacting, has proved an extreme challenge for theoretical studies. Consequently, it has been difficult to develop a comprehensive microscopic theory of the dense liquid phase using perturbative techniques [16]. Some specific results can be obtained using field-theoretical Green's function techniques [17]. For example, this approach shows how the superfluid properties are related to the broken Bose symmetry associated with a condensate. One is able to predict that $n(p)$ should become singular at small p when a condensate is present, due to coupling of fluctuations in the condensate with phonons [18, 19]. It is still difficult, however, to carry out realistic, quantitative calculations within the field-theoretical approach.

In view of the difficulty associated with field-theoretical approaches, the most detailed results for the Bose condensate have come from numerical calculations. Variational [20], Green's Function Monte Carlo (GFMC) [21], and most recently, shadow wave function [22] techniques have been used to calculate the properties of the ground state (i.e., at $T = 0$). Perturbative techniques, such as Correlated Basis Functions, have also been extensively applied to ^{4}He [23].

Recently, progress has also been made in extending these results to finite temperatures by working with excited states corresponding to elementary excitations generated by applying appropriate operators to the ground state wavefunction. This technique has been quite successful at lower temperatures [20] where the elementary excitations are simple density fluctuations (phonons). Techniques for higher temperatures closer to the transition, where rotons are the dominant elementary excitations, are still being developed [24].

Another technique which is extremely powerful is the Path Integral Monte Carlo method [25], in which the single-particle density matrix at some low temperature is formulated in terms of path integrals over density matrices at higher temperatures. Since the single-particle density matrix at high temperature can be calculated quite accurately, this method allows calculation of the momentum distribution at any temperature, limited only by the computational power available.

These numerical techniques can be considered as, in a sense, first-principles calculations. They use as input only the mass of the ^{4}He atom, the measured inter-atomic interaction potential, and the density. There are no adjustable parameters that can be varied to obtain agreement with experiment. In view of this, the excellent agreement of these calculations with several measured properties for the ground state of the liquid [26], such as the total energy and static structure of the liquid, is quite impressive.

The most prominent feature in the ground state $n(p)$ obtained from these numerical techniques is the appearance of a condensate, a δ-function singularity at $p = 0$. The intensity of this δ-function, which is the condensate density n_0, is about 9% of the total intensity. All recent calculations of the ground state of helium have obtained essentially the same result. It is interesting to note that this is quite similar to the original estimates by Penrose and Onsager for a simple hard-sphere liquid [12].

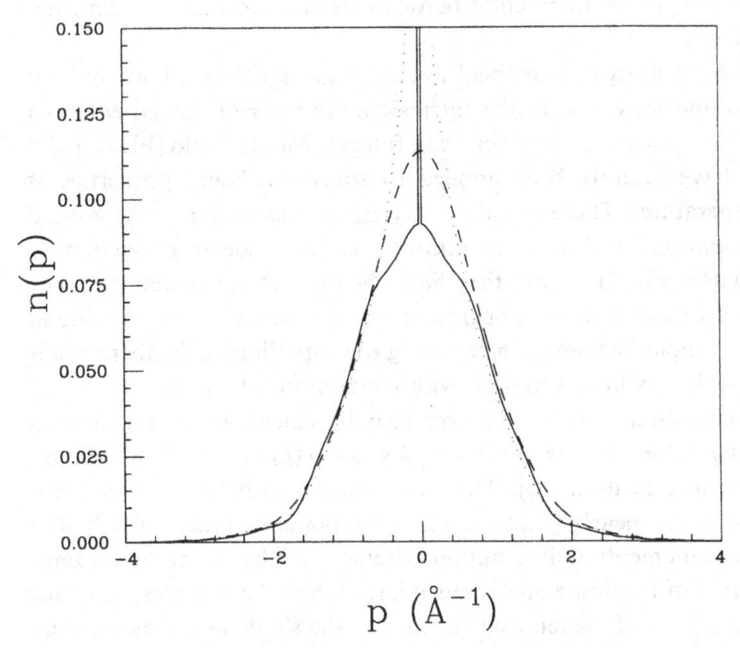

Fig. 4. Theoretical calculations of the momentum distribution. The ground state momentum distributions have been calculated using GFMC (solid) and variational (dotted) techniques. The momentum distribution in the normal liquid (dashed) has been calculated using PIMC techniques.

Fig. 4 shows two recent calculations of $n(p)$ in the ground state, based on GFMC [21] and variational [20] techniques.

The remainder of the momentum distribution, excluding the condensate δ-function, is known as the uncondensed momentum distribution and exhibits several interesting features. At small p, which corresponds to long-range interactions among the atoms, the effects of statistics are the dominant factor. For example, singular behavior [18, 19] due to the coupling of the condensate to long-wavelength collective excitations (phonons) appears. This singular behavior can be obtained exactly at small p, where the phonons are well defined, and goes as $1/p^2$ in the ballistic regime and $1/p$ in the hydrodynamic regime. The variational $n(p)$ in Fig. 4 shows this behavior explicitly at small p. This singular behavior is not present in the GFMC results, presumably due to the relatively small size of the samples used in the numerical calculations. Alternately, the large p behavior of $n(p)$ is determined primarily by the short-range repulsive interaction between atoms, and the Bose statistics play little role.

The great majority of numerical studies of the liquid have concentrated on the ground state due to the intrinsic limitations of the GFMC and variational approaches. However, Path Integral Monte Carlo (PIMC) [25] methods have recently been applied to study the liquid properties at finite temperatures. These calculations yield results similar to the ground state calculations at low temperatures, i.e. a condensate fraction of approximately 9%. However, they have the distinct advantage that they can provide results at finite temperature. The condensate fraction is found to exhibit a rapid increase of n_0 entering the superfluid as temperature is lowered, with very little variation with temperature thereafter.

The momentum distribution may also be calculated in the normal liquid using PIMC. For example, Fig. 4 shows $n(p)$ at 3.33 K, well above the superfluid transition [25]. The momentum distribution is broad and featureless with a nearly Gaussian form, the familiar classical result. The width of the momentum distribution is determined by the quantum zero-point motion of the liquid and is much larger than the width expected for classical particles. However, aside from this, the shape of the momentum distribution in the normal liquid shows little effect due to quantum statistics, in contrast to the IBG discussed in Section 2.

4 Deep Inelastic Neutron Scattering

Inelastic neutron scattering at large momentum transfer Q provides the most direct means of obtaining information on $n(p)$ and hence the condensate [10]. The scattering of neutrons by a condensed sample, such as liquid helium, is described by the double differential scattering cross section, which can be written as [10]

$$\frac{d^2\sigma}{d\omega d\Omega} = \frac{\sigma_c}{\sigma_c}\frac{k_i}{k_f} S_c(Q,\omega) + \frac{\sigma_i}{\sigma_T} S_i(Q,\omega) \ , \tag{8}$$

where σ_T, σ_i and σ_c are the total, incoherent and coherent scattering cross sections and k_i and k_f are the initial and final neutron wave vectors. The interaction of the neutron with the sample is described by the coherent and incoherent dynamic structure factors $S_c(Q,\omega)$ and $S_i(Q,\omega)$, where ω and Q are the energy and momentum transfer of the scattered neutron.

The dynamic structure factors contain information on spatial and temporal correlations between atoms in the sample [27]. The coherent dynamical structure factor involves correlations between different atoms and provides information on the structure and time-dependent behavior related to phonons and rotons. The incoherent cross section, in contrast, contains information only on the correlations of a single atom at two different times.

The type of information provided by neutron scattering measurements depends on the wavelength of the probing particle. Neutrons with wavelengths $2\pi/Q$ comparable to the interparticle separation provide information on both the collective and single-particle correlations. However, when the wavelength is much less than the average interparticle spacing, neutron scattering measurements only probe single-particle correlations. In this limit the wavelength of the neutron is too short to probe collective effects. The scattering is then determined totally by the incoherent structure factor and provides information only on the single-particle motion of atoms. Even in this single-particle limit, however, the scattering still probes the full many body properties of the sample through the interactions of the scattering atom with the other atoms in the sample.

The scattering simplifies even further if the energy transferred to an atom by a neutron is large compared with the interaction energies with neighboring atoms. This limit [4], known as the Impulse Approximation (IA), assumes that the neutron scatters from a single atom in a time so short that the neighboring particles cannot react to the perturbation caused by the neutron. The response of the He atom is then determined entirely by $n(p)$. It recoils from the collision in a free particle state of high momentum and energy and the scattering directly probes the initial many body wavefunction of the system.

The scattering in this high momentum and energy transfer limit is described by the Compton profile

$$J_{IA}(Y) = \frac{1}{4\rho\pi^2} \int_{|Y|}^{+\infty} pn(p)dp, \tag{9}$$

where ρ is the density. The scattering in the IA does not depend on the energy and momentum transfer separately, but only through the scaling variable

$$Y \equiv (M/Q)(\omega - \omega_r), \tag{10}$$

where ω and Q are the energy and momentum transfer of the scattered neutron, M is the mass and $\omega_r = Q^2/2M$ is the recoil energy of the scattering atom.

The scattering exhibits some characteristic features [28]; it is symmetric, centered at $Y = 0$ and depends on ω and Q only through the scaling variable Y. These conditions have often been taken as indicative of scattering in the IA limit. However, it is important to note that they are necessary, but not sufficient, conditions for the validity of the IA. It is possible for the scattering to exhibit these features even when the IA is not valid, such as in the case of a gas of hard spheres [29].

The scattering in the IA limit is directly related to $n(p)$ and, in the case of a liquid, measurements of $J(Y)$ can be used to determine $n(p)$ directly. For example, a Gaussian $n(p)$ implies a Gaussian $J(Y)$ with the same second moment. Furthermore, the condensate, which appears as a three-dimensional δ-function in $n(p)$, is a one-dimensional δ-function in $J(Y)$. Unfortunately, the extraction of information is hampered by the fact that prominent – and physically interesting – features in $n(p)$ may not be strongly reflected in $J(Y)$ [6].

To see this, consider the two recent calculations of the ground state $n(p)$ discussed earlier and shown in Fig. 5(a). Both calculations predict a condensate fraction, which appears as a delta function, with n_0=9.2 %. Both calculations also predict quite similar behavior at intermediate and large p. However, they differ markedly at small p. The variational calculation exhibits singular behavior due to coupling of long wavelength density fluctuations (phonons) to the condensate which are not present in the GFMC result, presumably due to finite size effects in the calculation.

While $n(p)$ for these two calculations is quite different, the corresponding $J_{IA}(Y)$, shown in Fig. 5(b), is remarkably similar. The singular behavior, which is the dominant feature in the variational $n(p)$ at small p, is quite small in $J(Y)$. When instrumental broadening is taken into account the (now small) differences between the predicted scattering for the two calculations all but disappear, as shown in Fig. 5(c). The predictions of the two quite different $n(p)$s are now nearly indistinguishable! In principle, the differences between the two different $n(p)$s are still present in Fig. 5(c). In practice, a measurement of the scattering would need fantastically good statistical accuracy to ever hope to observe these differences.

This insensitivity to the details at small p is a direct consequence of the IA. The Compton profile (9), which is proportional to the measured scattering, is the momentum distribution in the direction of the momentum transfer averaged over the longitudinal components. Since this is the integral of $pn(p)$ singular behavior at small p will be suppressed due to the factor of p in the integrand. Alternately, features at large p will be enhanced for exactly the same reason.

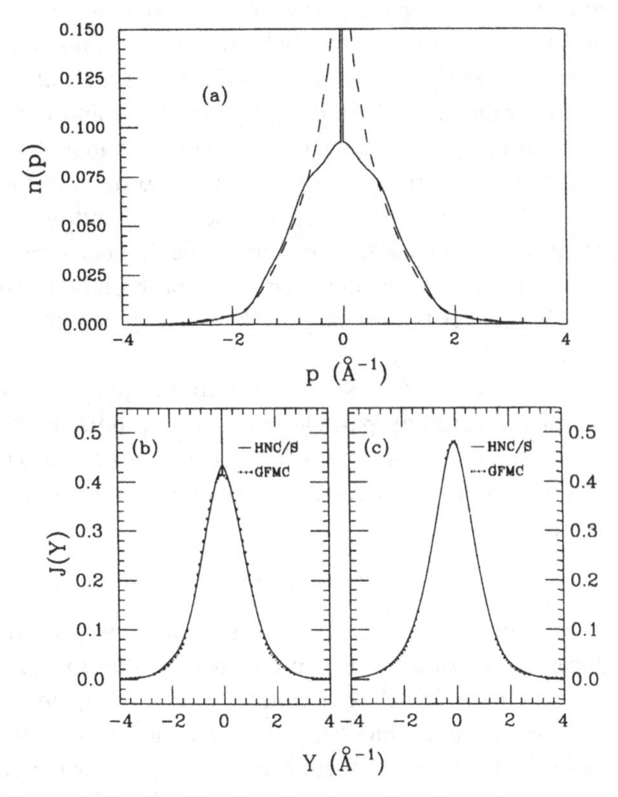

Fig. 5. (a) Two theoretical calculations of $n(p)$ in liquid helium at $T = 0$. The solid curve is the GFMC result, and the dashed curve is the HNC/S variational result. (b) The longitudinal momentum distributions described by $J(Y)$ corresponding to the two ground state $n(p)$ in (a). (c) The combined effects of the convolution of $J(Y)$ in (b) with the instrumental resolution function and Silver's final-state broadening.

Directly extracting $n(p)$ from $J(Y)$ suffers similar problems. Statistical noise, which is present in any measurement, will allow a variety of different scattering functions to be consistent with the data. This problem is particulaly severe at small p where the inversion procedure magnifies the errors. Therefore, in view of the difficulties in extracting $n(p)$ from the experimental results, we find it more appropriate to work directly with $J(Y)$. Theoretical predictions can be compared to the experimental data using the IA. Working with $J(Y)$, as opposed to $n(p)$, has the distinct advantage that the statistical errors on the data provide a direct measure of the 'goodness-of-fit' between theory and experiment.

The discussion above has implicitly assumed that we can measure the scattering in the IA limit, complicated only by the effects of instrumental resolution. However, formally, the IA is only valid in the limit of infinite ω and Q. In real experiments, where ω and Q are finite and collisions between atoms are always present, deviations from the IA must be taken into account [30]. The important questions are then the size and form of the deviations from the IA and the extent to which they limit our ability to infer $n(p)$ from scattering data. Obviously, the IA becomes more accurate with larger ω and Q (i.e. the shorter the probe/struck-particle interaction time), the weaker the interactions between the struck particle and its neighbors.

At high ω and Q, the most important deviations from the IA are due to collisions of the recoiling particle with its neighbors, known as final-state effects. Elementary arguments [4] lead to the prediction that the IA result will experience a Lorentzian broadening with a full width at half maximum (FWHM)

$$\Delta Y_{FWHM} \approx \rho\sigma_{\text{tot}}(Q), \tag{11}$$

where $\sigma_{\text{tot}}(Q)$ is the atom–atom scattering cross section. Heuristically, in terms of the uncertainty principle, this broadening is due to the finite 'lifetime' of the struck particle before it collides with its neighbors. This simple result for the width of final-state broadening holds in several modern theories for FSE. However, there is much debate over the details of the lineshape [30]. For example, the Lorentzian broadening model cannot be rigorously correct since it violates the second-moment ω^2 sum rule obeyed by $S(Q,\omega)$. However, the simple result (11) indicates the property which is the most important in determining the FSE in real systems: the strength of the interparticle interactions.

An important consequence of the final-state broadening given by (11) is that the scattering will approach the IA very slowly with increasing Q, because $\sigma_{\text{tot}}(Q)$ decreases logarithmically with Q for He–He scattering [31]. Therefore, helium is very similar to a hard-sphere system [29] where approximate Y-scaling behavior is obtained without the IA being valid. The final-state broadening interferes with observation of the features in $n(p)$ which are the most interesting, i.e. the Bose-condensate peak in ^{4}He.

Any attempt to determine $n(p)$ must somehow take FSE into account. At present, we depend on theoretical calculations to provide the appropriate corrections. Unfortunately, this adds another layer of complication and uncertainty to the interpretation of the experimental results. Fortu-

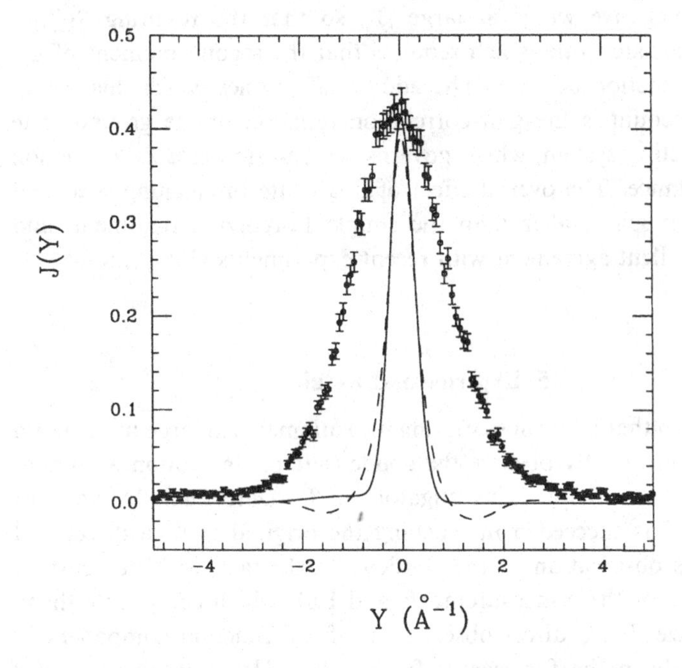

Fig. 6. The observed scattering converted to $J(Y)$, at a Q of 23 Å^{-1} and a temperature of 4.2 K. The calculated instrumental resolution (solid line) and final-state broadening (dashed line) are shown. The instrumental and final-state broadening have similar widths and are both much less than the intrinsic width of the observed scattering.

nately, recent theoretical work has provided an accurate description of the FSE broadening that can be used in extracting information about $n(p)$.

The recent non-Lorentzian broadening theory by Silver [32] for ^{4}He, based on earlier work by Gersch and Rodrigue [33], yields the final-state broadening function shown in Fig. 6 which is to be convoluted with the impulse approximation prediction to obtain the predicted scattering. Similar results have been obtained by Carraro and Koonin [34] based on a variational approach. The instrumental broadening, calculated using a Monte Carlo simulation [35], is also shown in Fig. 6 for reference. As can be seen, the instrumental resolution and final-state broadening are of comparable width, and both are much narrower than the intrinsic width of the helium scattering.

The FWHM of the function in Fig. 6 is comparable to that from the simple Lorentzian broadening theory. However, the modern theories [32,

34] predict negative wings at large $|Y|$ so that the resulting $S(Q, \omega)$ satisfies the ω^2 sum rule which requires that the second moment of the broadening function be zero. The additional physics which this theory takes into account is the pair-correlation function of the ground state of the interacting system, which governs the collision rate as a function of recoil distance. The overall effect of final-state broadening predicted by Silver is much smaller than the simple Lorentzian treatment, and produces excellent agreement with recent experiments [36].

5 Experimental Results

The suggestion that neutron scattering measurements at large momentum transfers could directly observe the condensate set in motion a number of studies by a variety of investigators [37]. Unfortunately, none of these studies has succeeded in reaching the original goal: a direct and unambiguous observation of the condensate. In fact, with our current understanding of these measurements and FSE which complicate them, we now realize that a direct observation of a δ-function component in $n(p)$ is unlikely in the foreseeable future [38]. However, these studies of liquid helium have provided a wealth of detailed information on the momentum distribution in this strongly interacting quantum liquid. With the use of some well-founded theoretical results, information on the magnitude of the condensate fraction can be extracted, including its temperature and density dependence.

The earliest attempts to measure $n(p)$ and n_0 were carried out at reactor based sources [39]. These measurements provided general information on $n(p)$ and pointed the way toward techniques to extract the condensate. However, they were limited by the large FSE at the relatively low Q ($\approx$ 5–15 Å^{-1}) attainable at reactor based sources. Some measurements were carried out at larger Q, but only at the expense of resolution making it difficult to obtain accurate information on $n(p)$.

More recently, measurements [40, 41, 42, 43, 44] have been carried out using spallation neutron sources, such as the Intense Pulsed Neutron Source (IPNS) at Argonne National Laboratory. These sources have a high flux of epithermal (high energy) neutrons allowing high resolution measurements to be carried out at much larger Q than are obtainable with reactors. The higher Q have the advantage that the scattering is consistent with the predictions of the IA, in terms of the location and symmetry of the observed scattering. In addition, FSE are more amenable

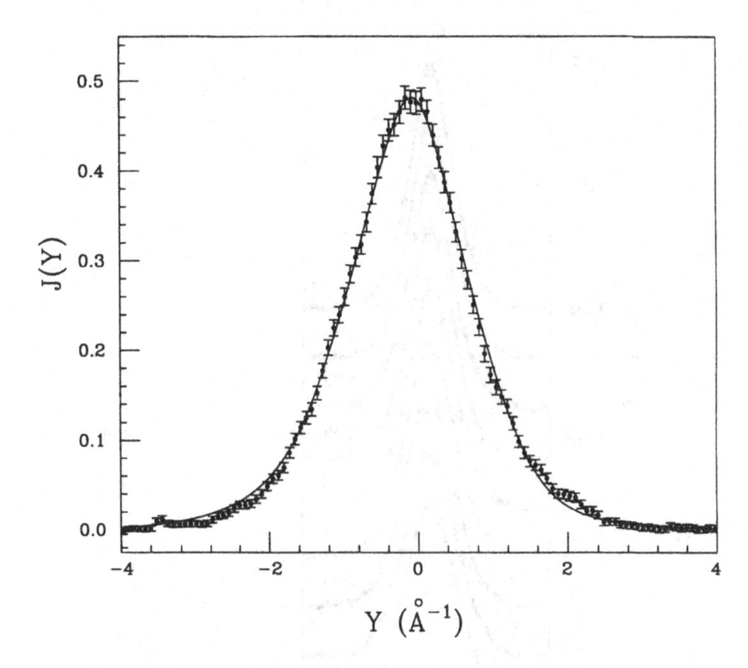

Fig. 7. Observed scattering at $T = 0.35$ K. The line is a fit of the model scattering function discussed in the text broadened by the instrumental resolution function and the final-state broadening function of Silver [32].

to theoretical treatment [30] at these higher Q. Therefore, information on the underlying $n(p)$ can be more readily obtained.

To illustrate the results obtainable at spallation sources, we will discuss recent measurements [40, 41, 42] using the high resolution PHOENIX spectrometer at the IPNS at Argonne. Fig. 6 shows the measured scattering at a Q of 23 Å^{-1}, converted to $J(Y)$, at 4.2 K in the normal liquid. The scattering shown in Fig. 6 is in qualitative agreement with the predictions of the IA. It is centered at, and symmetric about, $Y = 0$. In addition, measurements [28] at a variety of Q have shown that the shape of $J(Y)$ is independent of Q above approximately 15 Å^{-1}. Thus, at least at this qualitative level, the scattering is well described by the IA.

The observed scattering in the normal liquid is broad and featureless. $J(Y)$ is nearly Gaussian, as in classical liquids [52]. Upon cooling into the superfluid phase the scattering, shown in Fig. 7, becomes visibly more peaked near $Y = 0$. However, no distinct condensate peak is observed. The increase in scattering at small Y is consistent with the appearance of a condensate peak broadened by the finite instrumental resolution and

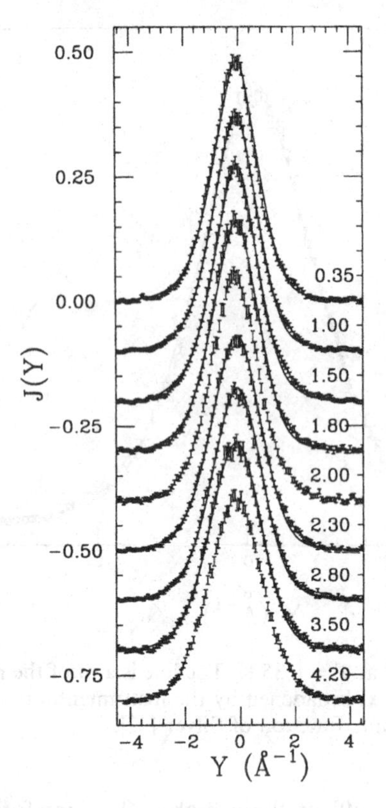

Fig. 8. Observed $J(Y)$ at a variety of temperatures in the normal and superfluid phases at a density of 0.147 g cm^{-3}.

FSE. However, as we discuss below, it is also consistent with many other forms for $n(p)$ as well [41], some of which do not have a finite n_0.

The temperature dependence of the neutron scattering is displayed in Fig. 8. The scattering is nearly Gaussian and temperature independent above the superfluid transition. Below the transition, the scattering becomes non-Gaussian with an increase in intensity around $Y = 0$, consistent with the appearance of a condensate. There is a rapid change in the scattering from just below the transition to 1.5 K, after which the scattering changes little. This is consistent with Path Integral Monte Carlo predictions [25] that the condensate fraction is largely temperature independent below 1.5 K but decreases rapidly above that temperature.

The ultimate goal of these studies is to obtain information on $n(p)$ in general and the condensate in particular. Towards that end, we would like

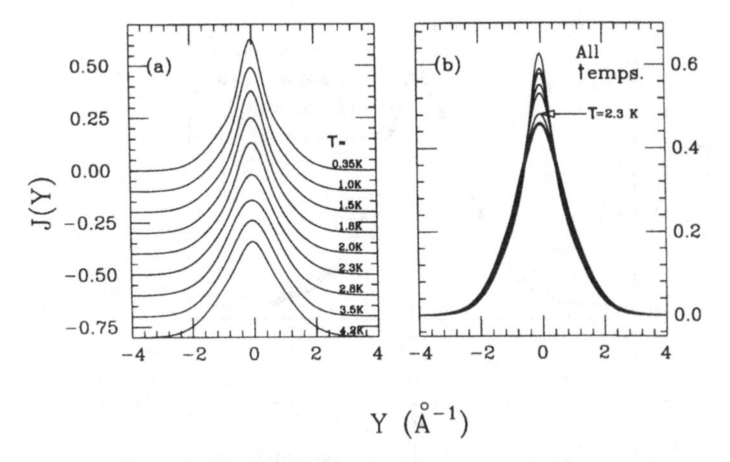

Fig. 9. $J_{\text{Model}}(Y)$ with instrumental resolution and FSE deconvoluted obtained from the observed scattering shown in Fig. 8.

to remove the effects of instrumental resolution and FSE and to invert the transformation between $n(p)$ and $J(Y)$. However, as discussed previously, these are ill-defined procedures [45] that are strongly affected by statistical noise present in the experimental data. Therefore, rather than attempting to deconvolute the instrumental resolution, it is more appropriate to fit an expression for $J(Y)$, broadened by instrumental resolution and FSE, to the observed scattering data. Theoretical predictions for $J(Y)$ are discussed in the following section. Here we discuss a simple expression for $J(Y)$ which has sufficient flexibility to reflect the behavior of the true scattering accurately, yet which is also sufficiently constrained so that unphysical behavior is not introduced due to the finite statistical accuracy of the data.

The scattering function that we have found most convenient [41] for describing the observed scattering is a sum of two Gaussians

$$J_{\text{Model}}(Y) = \sum_{i=1}^{2} \frac{a_i}{(2\pi\sigma_i^2)^{\frac{1}{2}}} e^{-(Y-Y_c)^2/2\sigma_i^2}, \tag{12}$$

whose amplitudes, widths and common center may be varied. This form is not unique and many other forms could be used to fit the data. Nevertheless it has sufficient flexibility that it can represent the scattering in both the normal liquid, where the scattering is nearly Gaussian, and the superfluid.

Fig. 9 shows the resultant model $J(Y)$, which represents the scattering

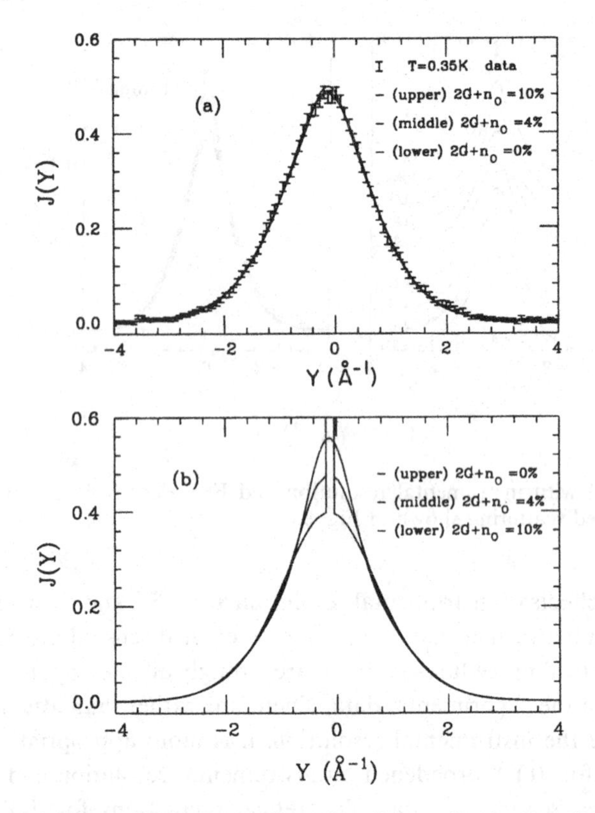

Fig. 10. (a) Three different two-Gaussian fits to the data at 0.35 K. (b) The corresponding model momentum distributions $J_{Model}(Y)$. One is a two-Gaussian fit with a condensate fraction of 10% , another is a two-Gaussian fit with a condensate fraction of 4% , and the third is a two-Gaussian fit with no condensate fraction.

after the removal of instrumental resolution and FSE, over a broad range of temperatures. As can be seen, the corrected scattering is nearly Gaussian above T_λ. Below T_λ the scattering becomes non-Gaussian, with a significant increase in the strength at small Y. However, in no cases do the fits indicate that we need a δ-function component to describe the scattering adequately.

The above results indicate that a δ-function component is not necessary to explain the scattering in the superfluid. This does not, however, mean that the scattering is not consistent with a momentum distribution containing a sharp peak at $p = 0$. As an illustration of the range of model scattering functions which can fit the data equally well, consider the three different two-Gaussian fits to the scattering at $T = 0.35$ K shown in Fig. 10. The dashed and dotted lines show the underlying distributions which best fit the data if they are constrained to possess a narrow component by fixing the width of one of the Gaussians to be very narrow ($\sigma = 0.03$ in this case). The dashed line has a narrow component with 10% of the total area, and the dotted line has a narrow component with 4% of the total area. Finally, the solid line is a two-Gaussian fit in which both the amplitudes and widths are allowed to vary. The χ^2 values of these fits are very close to each other. If we think of the narrow component of the fits as representing the contribution from the condensate and the broad component as representing the momentum distribution of the uncondensed atoms, then this example shows that, with an appropriate choice for the uncondensed component, the scattering data is consistent with a condensate fraction ranging from zero (no condensate) to values above 10 %.

6 Comparison to Theory

A direct model-independent determination of n_0 is beyond current experimental capabilities for the reasons discussed in Section 5. However, we may still obtain information on the condensate by comparing theoretical calculations of $n(p)$ with the experimental data. Such comparisons provide a direct test of the theoretical predictions and, indirectly, give information about the magnitude of the condensate.

The dashed line in Fig. 11(a) shows the theoretical prediction for $J_{IA}(Y)$ in the normal liquid using the PIMC calculations [25] of $n(p)$. The theoretical $n(p)$ has been converted to $J(Y)$ using the IA and broadened by the instrumental resolution. The agreement between the theoretical predication and the experiment is excellent. In this case the IA, calculated using the theoretical $n(p)$, provides an excellent description of the scattering in the normal liquid.

FSE have little effect on the observed scattering in the normal liquid at these Q values. However, they may be included by convoluting the theoretical predictions with the broadening function shown in Fig. 6. The solid lines in Fig. 11(a) and (b) are obtained when FSE are included. Within the statistical accuracy of the measurements, there is no observable change when FSE are included. Since, based on the ω^2 sum rule, FSE do not change the second moment of the scattering, they have little effect on the broad, nearly Gaussian $J(Y)$ of the normal liquid.

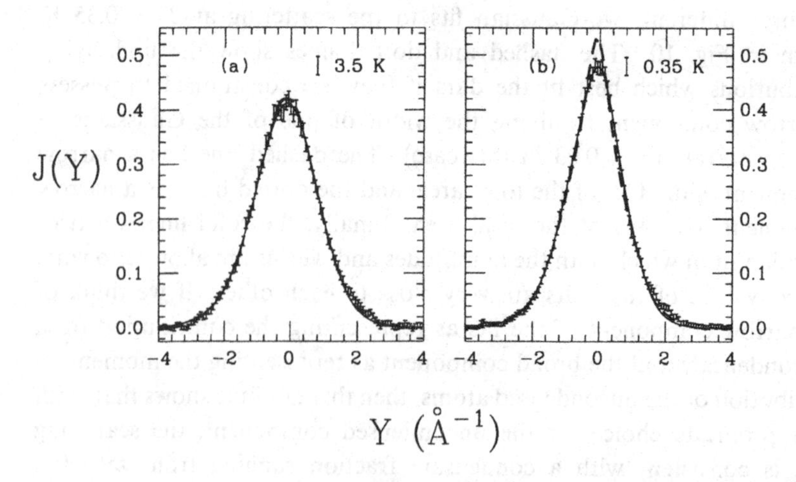

Fig. 11. Comparison of the observed scattering in (a) the normal liquid at 3.5 K and (b) the superfluid at 0.35 K, with the theoretical predictions from GFMC and PIMC calculations. The dashed lines are the theoretical predictions with only instrumental broadening included. The solid lines are the theoretical predictions with instrumental resolution and final-state effects included.

The dashed line in Fig. 11(b) shows a similar comparison of the GFMC calculations [21] to the scattering in the superfluid. The agreement between the theoretical and experimental results is quite poor, particularly in the region of the peak center, where the condensate would have the largest contribution. Based on this comparison alone, which has neglected FSE, we would conclude that n_0, if not identically zero, would have a much smaller value than the theoretical predictions.

The inclusion of FSE in the comparison between the experimental and theoretical results is essential in the superfluid phase. This is due to the appearance of a sharp feature in $n(p)$ for the superfluid: the condensate. While FSE have little effect on the broad component of the scattering, as observed in the normal liquid, they significantly broaden the contribution from the condensate. Taking FSE into account, the agreement between theory and experiment is now excellent! For the first time, *ab initio* numerical calculations of $n(p)$ in the superfluid are in good agreement with experiment [40].

An important point regarding the final-state corrections is in order here. The final-state broadening prediction of Silver, as shown in Fig. 6, has a narrow central peak and *negative* tails at high Y. The negative tails are essential if the broadening function is to satisfy the second-moment sum rule. Thus, final-state effects not only broaden the condensate peak, they also shift intensity throughout the entire spectrum. For the particular form of the FSE used here, the negative tails will cause a *depletion* of the scattering at intermediate Y when a condensate is present.

In view of the discussion in Section 5 regarding the relationship between $J(Y)$ and $n(p)$, it is appropriate to examine how sensitive the observed scattering is to the theoretical $n(p)$. Inevitably there is a finite statistical accuracy attached to the experimental results, and a whole range of different $n(p)$s may give equally good agreement with the data. If the statistical accuracy of the results is high then only a limited range of $n(p)$s, all with very similar shapes, will be consistent with the data. Alternatively, if the statistical accuracy is poor then the experimental results will only place very weak constraints on the underlying shape of $n(p)$.

Unfortunately, it is not possible to vary the condensate fraction in numerical calculations to see what effect it has on the comparison with the experimental data. The numerical calculations use as input only the known interatomic potential. Based on this potential, the full many body state of the liquid is calculated. There are no parameters that one can adjust to make n_0 larger or smaller. The result of the simulation, just as in the real system, is determined solely by the interaction between the atoms. Short of adjusting the interaction, which would undoubtedly change the entire $n(p)$, we cannot adjust n_0.

We can, however, attempt to make a limited test of the sensitivity of the experimental results to the value of the condensate fraction. The momentum distribution can be decomposed into a condensate contribution and a non-condensed contribution. We may adjust the relative weights of the condensate component and the non-condensed component such that they still satisfy the normalization of $n(p)$. However, once we choose a value for the condensate different from the theoretical value the resultant momentum distribution no longer corresponds to any *ab initio* calculation for helium. Thus, while we are testing the sensitivity of the scattering to the value of n_0, it is only in the limited range where $n(p)$ has a very special shape for the non-condensed component.

With the above caveats in mind, we may replace the condensate δ-function with a Gaussian of variable width and amplitude, keeping the shape of the non-condensate distribution the same as the theoretical prediction. The best agreement is obtained when the width of the Gaussian is less than 0.05 Å^{-1} and n_0 is 10%, in agreement with the best numerical predictions for the condensate. Significant deviations are observed when the width is greater than 0.2 $Å^{-1}$ and n_0 is less than 8% or greater than 12%. For this particular model for the non-condensate part of $n(p)$ provided by the GFMC calculations, we find that there is indeed a condensate with $n_0 = 10 \pm 2\%$. However, we note that changing the shape of the uncondensed $n(p)$ would also have an effect on the value for the condensate fraction.

In a similar fashion, the sensitivity to the expected singular behavior in $n(p)$ at small p can be examined. The GFMC results, which give excellent agreement with the observed scattering, do not contain the expected singular contribution. (Because of finite-size effects, these results are only valid for $p \gtrsim \pi/L$, where L is the system size.) The variational $n(p)$, discussed previously, explicitly includes this behavior. Both results are in excellant agreement with experiment. This is not very surprising, since the weak singular behavior at small p is suppressed when $n(p)$ is transformed to $J(Y)$, as discussed earlier. Thus, the predicted small p singular behavior makes little contribution to the observed scattering and, with the experimental techniques now available, will be difficult, if not impossible, to observe.

Thus, the experimental results in the superfluid provide a clear indication of a narrow component in $n(p)$ containing approximately 9–10% of the intensity, which is precisely that expected for the condensate. Unfortunately, due to the finite statistical error inherent in any experiment, they cannot definitely prove the existence of a condensate in the form of a δ-function. Some other singular behavior not associated with a condensate could be responsible for the increase in the scattering at small p observed in the superfluid. As seen in the comparison with the variational $n(p)$, however, this would have to be very singular behavior, much more than the $1/p$ singularity, to agree with the experimental results. Thus, while the experimental results cannot rule out a ground state $n(p)$ which does not contain a δ-function condensate, they do provide strong evidence for a very narrow feature containing 10$\pm$2% of the total area. The excellent agreement with the numerical results suggests that this very narrow feature is indeed due to the Bose condensate [3].

Similar comparisons have been carried out at a variety of temperatures and several different densities in both the normal and superfluid phases. GFMC and variational calculations are used for comparison with low temperature measurements and PIMC results are used for comparison to measurements above 1 K. The agreement is excellent over the entire temperature range! Theory and experiment appear to have converged for $n(p)$ in liquid ^{4}He at low densities (SVP).

7 Empirical Estimates of the Condensate Fraction

Due to its importance in understanding the superfluid phase, considerable emphasis has been placed on determining n_0. Since the width of FSE broadening follows the very slow decrease of the He–He cross section at higher Q, it is very unlikely that a condensate peak can be resolved in deep inelastic neutron scattering experiments [38].

The excellent agreement between theoretical calculations of $n(p)$ and the scattering data is strong (though not conclusive) evidence for the presence of a condensate. It is important to realize, however, that there are a number of physical systems, such as ^{3}He–^{4}He mixtures and ^{4}He in confined geometries, for which accurate theoretical calculations of $n(p)$ are difficult to carry out. For these systems, it may not be possible to carry out direct comparisons to theory.

Therefore, it is important to develop alternative techniques to extract information on the magnitude of n_0 from experimental data. All of these, in one form or another, involve modeling $n(p)$ in the superfluid to extract a value for n_0. Early models used for extracting the condensate usually consisted of simple functions that could be fit easily to the experimental data, such as a sum of Gaussians. Results for n_0 ranging from 0 (no condensate) to 20% were obtained, leading to considerable confusion about the condensate [46].

A major improvement in analysis of experimental results was introduced by the work of Sears and coworkers [47], which attempted to incorporate known behavior of the momentum distribution as a constraint in extracting the value of n_0. The momentum distribution in the superfluid phase was modeled as a δ-function due to the condensate, a singular term induced by the condensate which had been predicted theoretically, and an uncondensed component which was assumed to have the same shape as $n(p)$ in the normal liquid. Using this technique, values for the condensate fraction of 10–13% (when extrapolated to $T = 0$) were obtained [5]. These results are in reasonable agreement with current theoretical predictions. However, when an improved form for the singular term is used [48], the inferred values of n_0 are 4–5%, considerably below the theoretical estimates.

The Sears method suffers from some important limitations. For example, it assumes that the shape of the non-singular part of $n(p)$ in the superfluid is the same shape as the full $n(p)$ in the normal liquid just above the superfluid transition. At best this is probably only a reasonable first approximation. In addition, the necessity of measuring $n(p)$ in the normal liquid phase (above T_λ) introduces other uncertainties which will affect the determination of the condensate from the measurements below T_λ.

An alternative method [50] for determining the magnitude of n_0 was recently introduced by Snow, Wang and Sokol. The expression used for $n(p)$ explicitly incorporates the known analytic behavior of $n(p)$ at small p and places minimal restrictions on the shape of $n(p)$ in other regions. This expression for $n(p)$, converted to $J(Y)$ and broadened by instrumental resolution and FSE, is compared directly to the scattering and the model parameters adjusted to obtain the best fit.

The expression used by Snow *et al.* separated $n(p)$ into two components at an arbitrary cutoff momentum p_c. Above p_c a single Gaussian was used to model $n(p)$. Below this cutoff a δ-function, the known singular behavior and, to allow matching at p_c, a Gaussian were used to model $n(p)$. When the normalization of $n(p)$ and continuity above and below p_c are enforced, this model has four adjustable parameters: the widths of the two Gaussians σ_1 and σ_2, the cutoff momentum p_c and the condensate fraction n_0.

The fitting parameters are all related to various unknown aspects of the momentum distribution. The cutoff momentum p_c is the momentum at which the condensate-induced singularity becomes negligible in comparison to the non-singular part of the momentum distribution. However, the results are not sensitive to the choice of p_c as long as it is in the range $0.1 \leq p \leq 0.7$ Å^{-1}. This is just the phonon region where the theoretical expression for the condensate-induced singularity is expected to be valid [18, 19]. The value of σ_2 is determined primarily by the width of the momentum distribution in the intermediate-to-large Y region. The value of σ_1, on the other hand, is not tightly constrained by the scattering data. It represents the unknown effect near $Y = 0$ of the condensate on the non-singular part of the momentum distribution. Finally, the value for n_0 is determined primarily by the scattering at small p. Thus, in practice, this fit has just two adjustable parameters, n_0 and σ_2, which are determined by different regions of the observed scattering.

Fig. 12 shows n_0 as a function of temperature at low density as determined by the above procedure, along with the results of numerical simulations. The extracted value of the $T = 0$ condensate fraction of

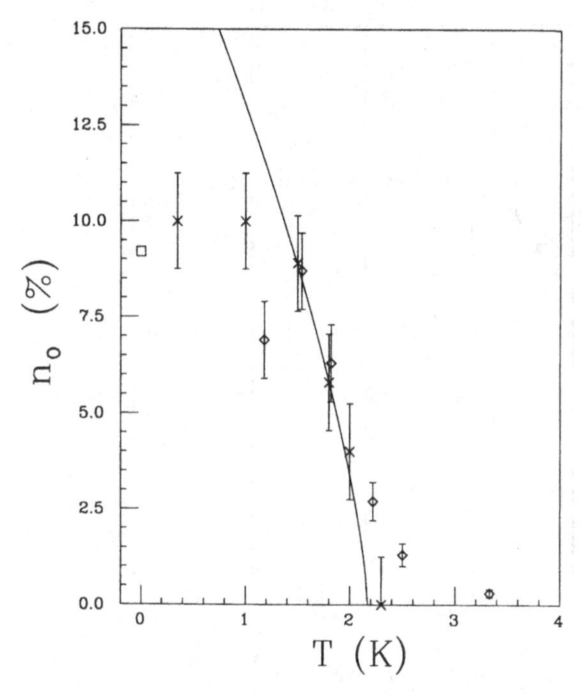

Fig. 12. The value of n_0 as a function of temperature at a constant density of 0.147 g cm^{-3}, obtained from fits using the method of Snow *et al.*[50] (crosses). Also shown are the theoretical calculations of n_0 at $T = 0$ from GFMC (square) and at finite temperatures from PIMC calculations (diamonds). The solid line is a plot of the renormalization group theory prediction.

10% is in excellent agreement with the GFMC result of 9.2%. The weak temperature dependence of n_0 seen in these fits at low temperatures is also consistent with the PIMC results [25] and the finite-temperature extensions of variational calculations.

The condensate wave function must vanish at the second-order superfluid–normal fluid phase transition, since it is the microscopic order parameter for the superfluid phase. According to the renormalization group theory for second-order phase transitions, the temperature dependence of n_0 near T_λ is expected to take the form $n_0 = At^{2\beta}$, where $t = (1 - T/T_\lambda)$. Unfortunately, the errors in the current experimental estimates are too large to place any useful constraints on the critical exponent which controls the behavior of the condensate close to the superfluid transition. However, we may compare this prediction with our experimental results using the accepted value of β=0.35 obtained from

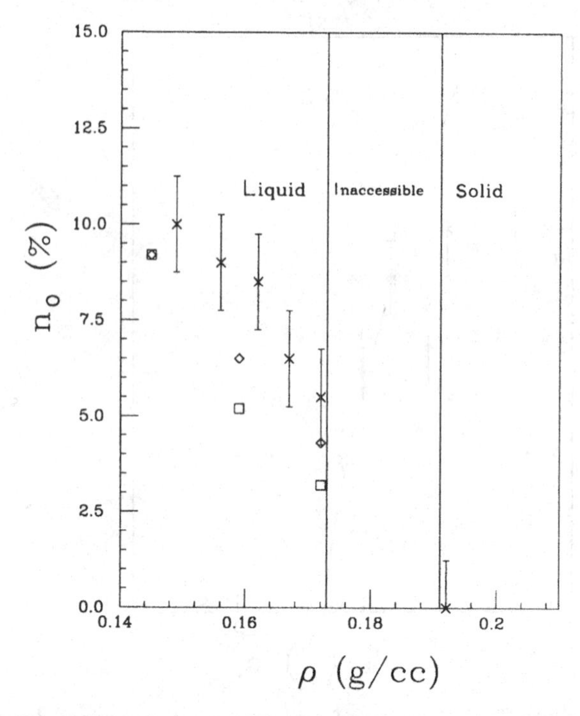

Fig. 13. The value of n_0 as a function of density at 0.75 K obtained from fits using the method of Snow *et al.* [50] (crosses). The GFMC results for n_0 (squares) and the HNC/S results (diamonds) are also shown.

superfluid fraction [49] and heat capacity [51] measurements. Such a comparison is shown in Fig. 12 and, as can be seen, the agreement is excellent over a surprisingly large temperature range.

Fig. 13 shows the density-dependence results at $T = 0.75$ K. The estimates of n_0 are systematically somewhat higher than the theoretical results using the GFMC method, but show the same density dependence within the experimental error. The condensate fraction decreases with increasing density. Notice that the estimate for the condensate remains finite close to the superfluid–solid phase transition. Since the superfluid–solid phase transition is first order, there is no reason for the condensate to vanish continuously near the critical density. We see no evidence for the presence of a condensate in the solid phase, a further indication of the adequacy of our method for extracting n_0.

As a final note, it is important to keep in mind that all these different methods have one feature in common – they implicitly assume the existence of a condensate. As discussed previously, the actual scattering data is also consistent with an increase in scattering at small p without the appearance of a condensate, but then one would have to come up with a plausible theory of the origin of this anomalous scattering.

8 Conclusions

We have now reached a stage where there is excellent agreement between experiment and theory for $n(p)$ in both the normal and superfluid phases of liquid ^{4}He. Direct comparison of the experimental results to theory and to various empirical expressions provides convincing evidence for a Bose condensate containing 10% of the atoms at $T = 0$. In addition, the temperature and density dependence of n_0 have been measured and are also in agreement with theoretical predictions obtained by numerical simulations.

Unfortunately, the original goal [4] of a direct observation of the condensate fraction has not come to pass. In view of our current understanding of FSE in helium, it is unlikely that this goal will ever be reached in deep inelastic neutron scattering experiments. Thus, while the current experiments provide convincing evidence for a condensate, they do not provide the sought-after direct observation of the condensate δ-function. The values inferred for n_0 are dependent on the current theories and models that are used to interpret experimental results.

The difficulties of observing Bose condensation directly in liquid helium are somewhat surprising considering the ease with which the effects of the condensate can be observed. Many easily measured quantities, such as the superfluid fraction, second sound and thermal conductivity exhibit large changes when Bose condensation occurs. However, due to the lack of a detailed microscopic theory, none of these changes can be used to determine the magnitude of the condensate directly. We note, however, that the superfluid fraction and the specific heat have been calculated using the PIMC method in conjunction with the condensate fraction [25]. Effectively, these results give us $\rho_s(T)$ as a function of $n_0(T)$.

The situation is similar for many microscopic quantities which change when Bose–Einstein condensation occurs. For example, the static structure factor $S(Q)$ changes slightly when the liquid enters the superfluid phase. Phenomenological theories attempting to relate these changes to n_0 provide reasonable results [53]. However, the predictions of these theories have been shown to be incorrect [25, 54]. Similarly, attempts to relate changes in the average kinetic energy per atom to the condensate fraction [5] have also been shown to be incorrect.

Lacking the ability to observe the condensate directly and unambiguously, attention has focused on semi-empirical expressions for the momentum distribution. As we have seen, fits using these expressions are capable of providing information on the magnitude of the condensate, and they have been employed to study the temperature and density dependence of the condensate.

It is essential to remember, however, that if the underlying assumptions of the expressions used in the fits are incorrect, then the results inferred will not have much meaning. It is interesting to note that the observed scattering in liquid helium can be fit extremely well simply by using the momentum distribution of the ideal Bose gas [55]. This holds true in both the normal liquid phase and in the superfluid phase. The best description of the scattering in the superfluid is obtained by using an ideal Bose gas $n(p)$ with *no* condensate. Based on this naïve model we might conclude that while Bose statistics alter $n(p)$ from the classical result, as indicated by the large increase in intensity at small p, a condensate is not necessary to explain the observed scattering. This illustrates that it is essential that only expressions with a sound physical basis must be employed if meaningful information on the condensate is to be obtained. Moreover, it warns that even for weakly interacting Bose gases, e.g. excitons, fits to the *ideal* Bose gas momentum distribution should not be overinterpreted, since interactions may significantly alter and obscure the appearance of a condensate peak in the spectral data.

Acknowledgements. I would like to acknowledge the contributions of T. R. Sosnick, W. M. Snow, Y. Wang, K. Herwig, and R. Blasdell in the experimental measurements described here. I would also like to acknowledge many useful conversations with R. N. Silver, A. Griffin, H. R. Glyde, and W. G. Stirling.

This work was supported by NSF grant DMR-9123469 and OBES/DMS support of the Intense Pulsed Neutron Source at Argonne National Laboratory under DOE grant W-31-109-Eng-38.

References

[1] The cessation of bubbling as ^{4}He passed through the superfluid transition was first noted by J. C. McLennan, H. D. Smith, and J. O. Wilhelm, Phil. Mag. **14**, 161 (1932).

[2] The first experiments showing superfluidity explicitly were reported by J.F. Allen and A.D. Misener, Nature **141**, 75 (1938), and P. Kapitza, Nature **141**, 74 (1938).

[3] F. London, Nature **141**, 643 (1938).

[4] This method was first proposed by P.C. Hohenberg and P.M. Platzman, Phys. Rev. **152**, 198 (1966).

[5] H.R. Glyde and E.C. Svensson, in *Methods of Experimental Physics*, K. Sköld and D. L. Price, eds. (Academic, N.Y., 1987), Vol.23, Part B, p. 303.

[6] P.E. Sokol, T. R. Sosnick, and W.M. Snow, in *Momentum Distributions*, R. N. Silver and P. E. Sokol, eds. (Plenum, N.Y., 1989), p.139.

[7] R.K. Pathria, *Statistical Mechanics* (Pergamon, Oxford, 1972).

[8] R.A. Aziz, V.P.S. Nain, J.S. Cerley, W.L. Taylor, and G.T. McConville, J. Chem. Phys. **70**, 4330 (1979).

[9] W.E. Keller, *Helium-3 and Helium-4* (Plenum, N.Y., 1969).

[10] P. E. Sokol, R.N. Silver and J. W. Clark, in *Momentum Distributions*, R.N. Silver and P. E. Sokol, eds. (Plenum, N.Y., 1989).

[11] el Hadi *et al.*, Physica **41**, 289–304 (1969); Phys. Chem. Ref. Data **2**, 923 (1973); **6**, 51 (1977).

[12] O. Penrose and L. Onsager, Phys. Rev. **104**, 576 (1956).

[13] C.N. Yang, Rev. Mod. Phys. B **34**, 694 (1962).

[14] P.W. Anderson, Rev. Mod. Phys. B **38**, 298 (1966).

[15] G. Baym, private communication.

[16] J. W. Clark and M.L. Ristig, in *Momentum Distributions*, R. N. Silver and P. E. Sokol, eds. (Plenum, N.Y., 1989), p. 39.

[17] A. Griffin, Can. J. Phys. **65**, 1368 (1987).

[18] P.C. Hohenberg and P.C. Martin, Ann. Phys. (NY) **34**, 291 (1965).

[19] J. Gavoret and P. Nozières, Ann. Phys. (NY) **28**, 349 (1964).

[20] E. Manousakis, V. R. Pandharipande, and Q. N. Usmani, Phys. Rev. B **33**, 7022 (1985); E. Manousakis and V. R. Pandharipande, Phys. Rev. B **30**, 5064 (1984); E. Manousakis, in *Momentum Distributions*, R. N. Silver and P. E. Sokol, eds. (Plenum, N.Y., 1989), p. 81.

[21] P.A. Whitlock and R. Panoff, Can. J. Phys. **65**, 1409 (1987); R.M. Panoff and P.A. Whitlock, in *Momentum Distributions*, R. N. Silver and P. E. Sokol, eds. (Plenum, N.Y., 1989), p. 59.

[22] S.A. Vitiello, K.J. Runge, G.V. Chester and M.H. Kalos, Phys. Rev. B **42**, 228 (1990).

[23] E. Feenberg, *Theory of Quantum Fluids* (Academic, N.Y., 1969); C.E. Campbell, *Progress in Liquid Physics*, C.A. Croxton, ed. (Wiley, N.Y., 1968), ch. 6.

[24] C.E. Campbell, in *Excitations in Two-Dimensional and Three-Dimensional Quantum Fluids*, A.G.F. Wyatt and H.J. Lauter, eds. (Plenum, N.Y., 1991), p. 159.

[25] D.M.Ceperly and E.L.Pollock, Phys. Rev. Lett. **56**, 351 (1986); D.M. Ceperly, and E.L. Pollock, Can. J. Phys. **65**, 1416 (1987); E.L. Pollock and D.M. Ceperley, Phys. Rev. B **36**, 8343 (1987); D.M. Ceperley, in *Momentum Distributions*, R.N. Silver and P.E. Sokol, eds. (Plenum, N.Y., 1989), p. 71.

[26] A.D.B. Woods and V.F. Sears, Phys. Rev. Lett. **39**, 415 (1977); H.A. Mook, Phys. Rev. Lett. **32**, 1167 (1974).

[27] D.L. Price and K. Skold in *Methods in Experimental Physics*, Vol. 23A, D.L. Price and K. Sköld, eds. (Academic, N.Y., 1987), p. 1.

[28] P.E. Sokol, Can. J. Phys. **65**, 1393 (1987).

[29] J.J. Weinstein and J.W. Negele, Phys. Rev. Lett. **49**, 1016 (1982).

[30] R.N. Silver, in *Momentum Distributions*, R.N. Silver and P.E. Sokol, eds. (Plenum, N.Y., 1989), p. 111.

[31] R. Feltgen, H. Kirst, K.A. Koehler, and F. Torello, J. Chem. Phys. **26**, 2630 (1982).

[32] R.N. Silver, Phys. Rev. B **37**, 3794 (1988); Phys. Rev. B **39**, 4022 (1989).

[33] H. A. Gersch and L. J. Rodriguez, Phys. Rev. A **8**, 905 (1973); L. J. Rodriguez, H. A. Gersch and H. A. Mook, Phys. Rev. A **9**, 2085 (1974).

[34] C. Carraro and S.E. Koonin, Phys. Rev. Lett. **65**, 2792 (1990); Phys. Rev. B **41**, 6741 (1990).

[35] P.E. Sokol, to be published.

[36] P.E. Sokol, T.R. Sosnick, W.M. Snow, and R.N. Silver, Phys. Rev. B **43**, 216 (1991); K.W. Herwig, P.E. Sokol, W.M. Snow, and R.C. Blasdell, Phys. Rev. B **44**, 308 (1991).

[37] For a brief review, see E.C. Svensson, in *75th Jubilee Conference on Helium-4*, J.G.M. Armitage, ed. (World Scientific, Singapore, 1983), p. 10.

[38] P.E. Sokol and W.M. Snow, in *Excitations in Two-Dimensional and Three-Dimensional Quantum Fluids*, A.G.F. Wyatt and H.J. Lauter, eds. (Plenum, N.Y., 1991) p. 49.

[39] R.A. Cowley and A.D.B. Woods, Phys. Rev. Lett. **21**, 787 (1968).

[40] T.R. Sosnick, W.M. Snow, P.E. Sokol, and R.N. Silver, Europhys. Lett. **9**, 707 (1989).

[41] T.R. Sosnick, W.M. Snow, and P.E. Sokol, Phys. Rev. **B41**, 11185 (1990).

[42] K.W. Herwig, P.E. Sokol, T.R. Sosnick, W.M. Snow, and R. C. Blasdell, Phys. Rev. B **41**, 103 (1990).

[43] K.H. Anderson, W.G. Stirling, A.D. Taylor, S.M. Bennington, Z.A. Bowden, I. Bailey, and H.R. Glyde, Physica B **180**, 865 (1993).

[44] J. Mayers and A.C. Evans, Rutherford Appleton Laboratory Report RAL-91-048.

[45] D.S. Sivia and R.N. Silver, in *Momentum Distributions*, R.N. Silver and P.E. Sokol, eds. (Plenum, N.Y., 1989), p. 377.

[46] H. W. Jackson, Phys. Rev. A **10**, 278 (1974).

[47] V.F. Sears, E.C. Svensson, P. Martel and A.D.B. Woods, Phys. Rev. Lett. **49**, 279 (1982).

[48] A. Griffin, Phys. Rev. B **32**, 3289 (1985).

[49] J.A. Lipa and T.C.P. Chui, Phys. Rev. Lett. **51**, 2291 (1983).

[50] W. M. Snow, Y. Wang and P. E. Sokol, Europhys. Lett. **19**, 403 (1992).

[51] D.S. Greywall and G. Ahlers, Phys. Rev. A **7**, 2145 (1973).

[52] V.F. Sears, Can. J. Phys. **63**, 68 (1985).

[53] G. J. Hyland, G. Rowlands, and F. W. Cummings, Phys. Lett **31A**, 465 (1970); F. W. Cummings, G. J. Hyland, and G. Rowlands, Phys. Lett **86A**, 370 (1981).

[54] G.L. Masserini, L. Reatto, and S.A. Vitiello, J. Low Temp. Phys. **89**, 335 (1992).

[55] H.R. Glyde and W.G. Stirling, in *Momentum Distributions*, R.N. Silver and P.E. Sokol, eds. (Plenum, N.Y., 1989), p. 123.

5

Sum Rules and Bose–Einstein Condensation

S Stringari
Dipartimento di Fisica
Universitá di Trento
I-38050 Povo
Italy

Abstract

Various sum rules, accounting for the coupling between density and particle excitations and emphasizing in an explicit way the role of Bose–Einstein condensation, are discussed. Important consequences on the fluctuations of the particle operator as well as on the structure of elementary excitations are reviewed. These include a recent generalization of the Hohenberg–Mermin–Wagner theorem holding at zero temperature.

1 Introduction

The sum rule approach has been employed extensively in the literature in order to explore various dynamic features of quantum many body systems from a microscopic point of view (see [1] and references therein). An important merit of the method is its explicit emphasis on the role of conservation laws and of the symmetries of the problem. Furthermore, the explicit determination of sum rules is relatively easy and often requires only a limited knowledge of the system. Usually the sum rule approach is, however, employed without giving special emphasis to the possible occurrence of (spontaneously) broken symmetries. For example, the most famous f-sum rule [2] holding for a large class of systems is not affected by the existence of an order parameter in the system.

The purpose of this paper is to discuss a different class of sum rules which are directly affected by the presence of a broken symmetry. These sum rules can be used to predict significant properties of the system which are the consequence of the existence of an order parameter. In this work we will make explicit reference to Bose systems and to the consequences of Bose–Einstein condensation (BEC). Most of these results can however be generalized to discuss other systems exhibiting spontaneously broken symmetries.

In Section 2 we use a sum rule due to Bogoliubov in order to derive important constraints on the fluctuations of the particle operator as well as to obtain Goldstone-type bounds for the energy of elementary excitations. In Section 3 we explore the consequences of BEC on the long range behavior of the half diagonal two-body density matrix and discuss the coupling between the *density* and *particle* pictures of elementary excitations in Bose superfluids.

2 The Bogoliubov Sum Rule

The Bogoliubov sum rule [3, 4]

$$\int_{-\infty}^{+\infty} \mathscr{A}_{a-a^\dagger,\rho}(\omega) d\omega = \langle [a_{\mathbf{q}} - a^\dagger_{-\mathbf{q}}, \rho_{\mathbf{q}}] \rangle = \langle a_0 + a^\dagger_0 \rangle = 2\sqrt{Nn_0} \quad (1)$$

is the most remarkable among the sum rules depending explicitly on the order parameter $n_0 = \frac{1}{N} \mid \langle a_0 \rangle \mid^2$. In (1) $\mathscr{A}_{A^\dagger,B}(\omega)$ is the spectral function

$$\mathscr{A}_{A^\dagger,B}(\omega) = \frac{1}{Z} \sum_{m,n} (e^{-\beta E_m} - e^{-\beta E_n}) \langle m \mid A^\dagger \mid n \rangle \langle n \mid B \mid m \rangle \times \delta(\omega - E_n - E_m) \quad (2)$$

relative to the operators A, B and we have made the choice $A^\dagger = a_{\mathbf{q}} - a^\dagger_{-\mathbf{q}}$ and $B = \rho_{\mathbf{q}}$ ($\rho_{\mathbf{q}}$, $a_{\mathbf{q}}$ and $a^\dagger_{-\mathbf{q}}$ are the usual density, particle annihilation and creation operators, respectively, and we have taken $\langle a_0 \rangle = \langle a^\dagger_0 \rangle = \sqrt{Nn_0}$). In deriving the result (1), we have used the completeness relation and the Bose commutation relations for a and $a^\dagger$. Despite its simplicity, (1) is at the basis of very important results characterizing the macroscopic behavior of Bose superfluids. Its richness is mainly due to the fact that it exploits in a microscopic way the effects of the "phase" operator, proportional to the difference $a_{\mathbf{q}} - a^\dagger_{-\mathbf{q}}$. Some important applications of (1) are discussed below.

2.1 The Hohenberg–Mermin–Wagner (HMW) Theorem

This theorem, of fundamental importance in condensed matter physics, states that no long range order can exist ($n_0 = 0$ in the Bose case) in a significant class of one- and two-dimensional systems at finite temperature [5, 6, 7]. The theorem is based on the Bogoliubov inequality [3, 4]

$$\langle \{A^\dagger, A\} \rangle \langle [B^\dagger, [H, B]] \rangle \geq k_B T \mid \langle [A^\dagger, B] \rangle \mid^2 \quad (3)$$

for the fluctuations of the operator A (we assume $\langle A\rangle = \langle B\rangle = 0$). Inequality (3) can be derived starting from the inequality [3, 4]

$$\chi_{A^\dagger,A}\langle[B^\dagger,[H,B]]\rangle \geq |\langle[A^\dagger,B]\rangle|^2 \tag{4}$$

for the static response relative to the operator A

$$\chi_{A^\dagger,A} = \int_{-\infty}^{+\infty} \mathscr{A}_{A^\dagger,B}(\omega)\frac{1}{\omega}d\omega \tag{5}$$

and the result $\langle\{A^\dagger,A\}\rangle \geq 2k_BT\chi_{A^\dagger,A}$ following from the fluctuation dissipation theorem.

Choosing in the Bose case $A^\dagger = a_{\mathbf{q}}$ and $B = \rho_{\mathbf{q}}$ we obtain the inequality [5]

$$n(q) \equiv \langle a^\dagger_{\mathbf{q}}a_{\mathbf{q}}\rangle \geq n_0 m\frac{k_BT}{q^2} - \frac{1}{2} \tag{6}$$

for the momentum distribution $n(q)$ of the system. In order to derive result (6) we have used the f-sum rule [2]

$$\int \omega S(q,\omega)d\omega = \frac{1}{2}\langle[\rho^\dagger_{\mathbf{q}},[H,\rho_{\mathbf{q}}]]\rangle = N\frac{q^2}{2m} \tag{7}$$

holding for Galilean invariant potentials. In (7), $S(q,\omega) = \mathscr{A}_{\rho^\dagger,\rho}(\omega)/(1 - \exp(-\beta\omega))$ is the usual dynamic structure function. Inequality (6) points out the occurrence of an important infrared $1/q^2$ divergence in the momentum distribution which originates from the thermal fluctuations of the phase of the condensate. This behavior permits proof of the absence of BEC in one- and two-dimensional systems at finite temperature. In fact, due to such a divergence, the normalization condition for the momentum distribution $\sum_{\mathbf{q}} n(q) = N$ cannot be satisfied in one- and two-dimensions, unless $n_0 = 0$ [5].

2.2 Extension of the HMW Theorem at T=0

As clearly revealed by Eqs. (3) and (6), the HMW theorem does not apply at zero temperature. Actually there are important examples of two-dimensional systems obeying the HMW theorem at $T \neq 0$ and exhibiting long range order in the ground state. The physical reason is that quantum fluctuations have in general a weaker effect with respect to thermal fluctuations in destroying the order parameter.

An extension of the Hohenberg–Mermin–Wagner theorem holding

at zero temperature has been recently formulated by Pitaevskii and Stringari [8]. To this purpose one uses the uncertainty principle inequality

$$\langle\{A^\dagger, A\}\rangle\langle\{B^\dagger, B\}\rangle \geq |\langle[A^\dagger, B]\rangle|^2 \tag{8}$$

rather than the Bogoliubov inequality (3). Both inequalities provide rigorous constraints on the fluctuations of the operator A. However, while the Bogoliubov inequality is sensitive to thermal fluctuations and becomes less and less useful as $T \to 0$, the uncertainty principle is particularly powerful in the low temperature regime dominated by quantum fluctuations. By making the choice $A^\dagger = a_{\mathbf{q}}$, $B = \rho_{\mathbf{q}}$ one obtains the following non-trivial inequality for the momentum distribution [8]:

$$n(q) \geq \frac{n_0}{4S(q)} - \frac{1}{2} \tag{9}$$

of a Bose system where $S(q) = \frac{1}{N}\langle\rho_{\mathbf{q}}^\dagger \rho_{\mathbf{q}}\rangle$ is the static structure function. The low q behavior of $S(q)$ is fixed, at zero temperature, by the further bound

$$S(q) \leq \sqrt{\int \omega S(q,\omega) d\omega \int \frac{1}{\omega} S(q,\omega) d\omega} \tag{10}$$

that can be easily calculated at small q. In fact, in this limit, the inverse energy weighted sum rule (compressibility sum rule) $\int \frac{1}{\omega} S(q,\omega) = \frac{1}{2}\chi(q)$ approaches the compressibility parameter of the system ($lim_{q\to 0}\chi(q) = 1/mc^2$), while the energy weighted sum rule (f-sum rule) is given by (7). As a consequence $S(q) \leq q/2mc$ at small q and the momentum distribution $n(q)$ diverges at least as

$$n(q) = n_0 \frac{mc}{2q}. \tag{11}$$

It is worth pointing out that the $1/q$ law for the momentum distribution, already known in the literature [9, 10], has been obtained here without any assumption on the nature of the elementary excitations of the system (phonons in a neutral Bose superfluid) and follows uniquely from the existence of BEC, the validity of the f-sum rule and the finiteness of the compressibility. By imposing the proper normalization on the momentum distribution one can then rule out the existence of BEC in one-dimensional systems at zero temperature. Note that the ideal Bose Gas does not violate the theorem since in this case the compressibility sum rule for $\chi(q)$ diverges as $1/q^2$.

Starting from the same inequality (8), it is also possible to rule out the existence of long range order in other one-dimensional systems at $T = 0$

(isotropic antiferromagnets, crystals) [8]. The theorem does not apply to one-dimensional ferromagnets since in this case the inverse energy weighted sum rule (magnetic susceptibility) diverges as $1/q^2$.

Inequality (9) has been used recently [11] to rule out the existence of BEC in the bosonic representation of the electronic wave function

$$\Psi_B(\mathbf{r}_1, ..., \mathbf{r}_N) = \exp(\frac{i}{\nu} \sum_{i,j} \alpha_{i,j}) \Psi_F(\mathbf{r}_1, ..., \mathbf{r}_N) \tag{12}$$

proposed by Girvin and MacDonald [12] for the fractional quantum Hall effect. In Eq. (12), Ψ_F is the fermionic wave function of electrons, $\nu = 1/(2k+1)$, where k is an integer, ν is the usual filling factor and $\alpha_{i,j}$ is the angle between the vector connecting particles i and j and an arbitrary fixed axis. Using Laughlin's expression [13] for the ground state wave function Ψ_F, the authors of Ref. [12] were able to show that there is no Bose–Einstein condensation in the bosonic wave function Ψ_B, but only algebraic long range order. The same result was obtained in Ref. [14] starting directly from the Chern–Simons–Landau–Ginzburg theory. In Ref. [11] the absence of long range order was proven starting directly from the uncertainty principle inequality (9). The result follows from the fact that a charged liquid in an external magnetic field is characterized by a suppression of density fluctuations resulting in a quadratic law

$$S(q) = \frac{q^2}{2m\omega_c} \tag{13}$$

for the static structure function (ω_c is the cyclotronic frquency). This behaviour reflects the incompressibility of the system and is consistent with Kohn's theorem [15]. Inequality (9) implies a $1/q^2$ divergence for the momentum distribution of the bosonic wave function of this two-dimensional problem. This is incompatible with the normalization of $n(q)$, fixed by the total number of electrons, unless the Bose condensate relative to Ψ_B identically vanishes.

2.3 BEC and Excitation Energies

The sum rule technique can be used to obtain useful constraints on the energy of elementary excitations. In particular, use of the Bogoliubov sum rule (1) makes it possible to emphasize the role of the order parameter.

Given a pair of operators A and B one can derive the following rigorous inequality, holding at zero temperature, for the energy ω_0 of the lowest state excited by the operators A and B:

$$\omega_0^2 \leq \frac{\langle [A^\dagger, [H, A]] \rangle \langle [B^\dagger, [H, B]] \rangle}{| \langle [A^\dagger, B] \rangle |^2}. \tag{14}$$

Bounds of the form (14) were first considered by Wagner [4]. Result (14) can be obtained using the inequality (holding at zero temperature)

$$\omega_0^2 \leq \frac{\langle [A^\dagger, [H, A]] \rangle}{\chi_{A^\dagger, A}}, \tag{15}$$

where the rhs coincides with the ratio between the energy weighted and the inverse energy weighted sum rules relative to the operator A. Use of the Bogoliubov inequality (4) then yields (14).

Result (14) has the important merit of providing a rigorous upper bound for ω_0 in terms of quantities involving commutators. This is an advantage, for example, with respect to the so-called Feynman bound:

$$\omega_0 \leq \frac{\langle [A^\dagger, [H, A]] \rangle}{\langle \{A^\dagger, A\} \rangle} \tag{16}$$

involving the anticommutator $\langle \{A^\dagger, A\} \rangle$ in the denominator and hence requiring direct knowledge of the fluctuations of the operator A. The occurrence of the anticommutator makes it difficult to exploit the low momentum regime of elementary excitations. Vice versa, (14) can be directly employed for this purpose.

Result (14) is particularly interesting when the expectation value of the commutator $[A^\dagger, B]$ is proportional to the order parameter of the problem. This happens in a Bose superfluid with the choice $A^\dagger = a_{\mathbf{q}} - a^\dagger_{-\mathbf{q}}$ and $B = \rho_{\mathbf{q}}$, already considered in this work. One then finds

$$\omega_0^2(q) \leq \frac{1}{n_0} \frac{q^2}{2m} [\frac{q^2}{2m} - \mu + NV(0) + \sum_{\mathbf{p}} V(p)(n(\mathbf{p}+\mathbf{q}) + \overline{n}(\mathbf{p}+\mathbf{q}))], \tag{17}$$

where we have carried out the commutators using the grand canonical hamiltonian $H' = H - \mu N$ (μ is the chemical potential) and taken a central potential with Fourier transform $V(p)$. The quantitities $n(p)$ and $\overline{n}(p)$ are defined by $n(p) = \langle a^\dagger_{\mathbf{p}} a_{\mathbf{p}} \rangle$ and $\overline{n}(p) = \frac{1}{2} \langle a^\dagger_{\mathbf{p}} a^\dagger_{-\mathbf{p}} + a_{\mathbf{p}} a_{-\mathbf{p}} \rangle$.

Result (17) is a rigorous inequality holding for any Bose system interacting with central potentials. It has the form of a Goldstone theorem. In fact, since its rhs behaves as q^2 when $q \to 0$, it proves the existence of gapless excitations, provided the order parameter n_0 is different from zero.

It is worth noting that in the limit of a dilute Bose gas ($\mu = NV(0)$,

$n(p) = \bar{n}(p) = N\delta_{\mathbf{p},0}, n_0 = 1$), this upper bound coincides with the Bogoliubov dispersion law

$$\omega_B^2(q) = \frac{q^2}{2m}\left[\frac{q^2}{2m} + 2NV(q)\right] \tag{18}$$

for any value of $\mathbf{q}$. This result follows from the fact that the sum rules entering inequality (14) are exhausted by a single excitation, multi-particle states playing a negligeable role in a dilute gas. Of course in a strongly interacting system, such as liquid ^{4}He, multiparticle excitations are much more important and the bound (17) turns out to be significantly higher than the lowest excitation energy $\omega_0(q)$.

It is useful to compare the Goldstone-type inequality (17) with another, also rigorous, upper bound, which is derivable from (14) by choosing

$$A = \mathbf{q}\cdot\mathbf{j}_\mathbf{q} = [H, \rho_\mathbf{q}] = \frac{1}{2m}\sum_\mathbf{k}(q^2 + 2\mathbf{q}\cdot\mathbf{k})a^\dagger_{\mathbf{k}+\mathbf{q}}a_\mathbf{k} \tag{19}$$

and $B = \rho_\mathbf{q}$. In (19), $\mathbf{j}_\mathbf{q}$ is the usual current operator. The resulting bound then coincides with the ratio between the cubic energy weighted and the energy weighted sum rules relative to the density operator $\rho_\mathbf{q}$ [16,17]

$$\omega_0^2(q) \leq \frac{\langle[[\rho_{-\mathbf{q}}, H], [H, [H, \rho_\mathbf{q}]]]\rangle}{\langle[\rho_{-\mathbf{q}}, [H, \rho_\mathbf{q}]]\rangle}. \tag{20}$$

The explicit calculation of the triple commutator yields (we choose $\mathbf{q}$ along the z-axis) [16, 17]

$$\langle[[\rho_{-\mathbf{q}}, H], [H, [H, \rho_\mathbf{q}]]]\rangle \tag{21}$$

$$= N\frac{q^2}{m}\left[(\frac{q^2}{2m})^2 + \frac{2q^2}{m}\langle E_K\rangle + \frac{\rho}{m}\int d\mathbf{s}(1 - \cos qz)g(s)\nabla_z^2 V(s)\right], \tag{22}$$

where $\langle E_K\rangle$ is the ground state kinetic energy and $g(s)$ is the pair correlation function. The bound (20) then becomes:

$$\omega_0^2(q) \leq (\frac{q^2}{2m})^2 + \frac{2q^2}{m}\langle E_K\rangle + \frac{\rho}{m}\int d\mathbf{s}(1 - \cos qz)g(s)\nabla_z^2 V(s) \tag{23}$$

and exhibits a quadratic behaviour in q as $q \to 0$. It is worth noting that in the limit of a dilute Bose gas ($\langle E_K\rangle = 0, g(s) = 1$) both bounds (23) and (17) coincide with the Bogoliubov dispersion relation (18).

The fact that it is possible to prove, via (23), the existence of gapless excitations without assuming the existence of a broken symmetry is directly connected (see Eq. (20)) with the conservation of the total current $\mathbf{j}_{\mathbf{q}=0}$, holding in translationally invariant systems. This discussion also suggests that Goldstone-type inequalities of form (17) are particularly useful in sytems where the current is *not* conserved and where consequently only the existence of a spontaneously broken symmetry can ensure in a simple way the occurrence of gapless excitations. This is the case, for example, of spin excitations in magnetic systems (spin current is not conserved) or Bose systems in random external potentials.

Let us discuss, for example, the effects of an external potential of the form (we take $U_{-\mathbf{k}} = U_{\mathbf{k}}^*$)

$$V_{\text{ext}} = \sum_{\mathbf{k}\neq 0} U_{\mathbf{k}} \rho_{\mathbf{k}} \tag{24}$$

on the sum rules discussed above. A first important result is that the Goldstone-type upper bound (17) is not directly affected by the external field because of the exact commutation property

$$[a_{\mathbf{q}} - a^{\dagger}_{-\mathbf{q}}, [V_{\text{ext}}, a^{\dagger}_{\mathbf{q}} - a_{-\mathbf{q}}]] = 0. \tag{25}$$

Vice versa, the cubic energy weighted sum rule for the density operator $\rho_{\mathbf{q}}$ obtains an extra contribution from the external force given by

$$\langle [[\rho_{-\mathbf{q}}, H], [V_{\text{ext}}, [H, \rho_{\mathbf{q}}]]] \rangle = \frac{q^2}{m^2} \langle -\sum_{\mathbf{k}} k_z^2 U_{\mathbf{k}} \rho_{\mathbf{k}} \rangle. \tag{26}$$

At small q the new term provides the leading contribution to the triple commutator (21). Since the quantity $\langle [\rho_{-\mathbf{q}}, [H, \rho_{\mathbf{q}}]] \rangle$ is not changed by the external force, being still given by the f-sum rule (7), the upper bound (20) no longer vanishes with q. This different behaviour is due to the fact that the current $\mathbf{j}_{\mathbf{q}=0}$ is not conserved in the presence of the external field (24). This result also reveals that the relationship $\mathbf{j}_{\mathbf{q}} = \mathbf{q}\sqrt{Nn_0}/2m(a^{\dagger}_{\mathbf{q}} - a_{-\mathbf{q}})$ between the current and the gradient of the phase operator, holding in a translationally invariant dilute Bose gas, is no longer valid in the presence of an external potential and gives rise to a normal (non-superfluid) component of the density of the system at zero temperature. This is discussed, for example, in Huang's article in this volume.

3 The Half Diagonal Two-body Density Matrix

In this section we discuss another sum rule, also sensitive to the presence of Bose–Einstein condensation, given by [18]

$$\int_{-\infty}^{+\infty} \frac{1}{1-e^{-\beta\omega}} \mathscr{A}_{a+a^\dagger,\rho}(\omega) d\omega = \langle (a_{\mathbf{q}} + a^\dagger_{-\mathbf{q}})\rho_{\mathbf{q}} \rangle, \tag{27}$$

where $\mathscr{A}$ is the spectral function already introduced in Section 1 with the choice $A^\dagger = a_{\mathbf{q}} + a^\dagger_{-\mathbf{q}}$ and $B = \rho_{\mathbf{q}}$.

In contrast to (1), which describes the commutation relation between the density and the phase operators, this sum rule cannot be expressed in terms of a commutator. Physically it accounts for the coupling between the density of the system and the modulus of the condensate, proportional to $a_{\mathbf{q}} + a^\dagger_{-\mathbf{q}}$. An interesting feature of this sum rule is that it fixes the long range behaviour of the half diagonal two-body density matrix

$$\rho^{(2)}(\mathbf{r}_1, \mathbf{r}_2; \mathbf{r}'_1, \mathbf{r}_2) = \langle \Psi^\dagger(\mathbf{r}_1)\Psi^\dagger(\mathbf{r}_2)\Psi(\mathbf{r}'_1)\Psi(\mathbf{r}_2) \rangle. \tag{28}$$

The occurrence of BEC is in fact not only relevant for the long-range behaviour of the one-body density matrix

$$\rho^{(1)}(\mathbf{r}, \mathbf{r}') = \langle \Psi^\dagger(\mathbf{r})\Psi(\mathbf{r}') \rangle \tag{29}$$

fixed by the condition

$$\lim_{\mathbf{r}' \to \infty} \rho^{(1)}(\mathbf{r}, \mathbf{r}') = \rho n_0 , \tag{30}$$

but also for the one of the two-body matrix (28). The long-range order (LRO) in the two-body density matrix (28) is naturally defined by [19]

$$\lim_{\mathbf{r}'_1 \to \infty} \rho^{(2)}(\mathbf{r}_1, \mathbf{r}_2; \mathbf{r}'_1, \mathbf{r}_2) = n_0 \rho^2 (1 + F_1(\mid \mathbf{r}_1 - \mathbf{r}_2 \mid)) \tag{31}$$

and is characterized by the condensate fraction n_0 as well as by the function $F_1(r)$. The properties of this function have been recently investigated in Refs. [18, 19]. The link between the sum rule (27) and the long-range behaviour of $\rho^{(2)}$ is fixed by the relation

$$\langle \rho_{\mathbf{q}}(a_{\mathbf{q}} + a^\dagger_{-\mathbf{q}}) \rangle = \sqrt{N n_0}(1 + 2F_1(q)), \tag{32}$$

where $F_1(q)$ is the Fourier transform of $F_1(r)$. At low q the behaviour of this function is fixed by the density dependence of the condensate according to [18]

$$\lim_{q \to 0} \frac{1 + 2F_1(q)}{q} = \frac{1}{2n_0 mc} \frac{\partial (n_0 \rho)}{\partial \rho} . \tag{33}$$

The occurrence of LRO in the two-body density matrix implies the existence of a non-trivial relation for the chemical potential that can be obtained starting from the following expression holding in systems exhibiting Bose–Einstein condensation:

$$\mu = E(N) - E(N-1) = \langle H \rangle - \frac{\langle a_0^\dagger H a_0 \rangle}{\langle a_0^\dagger H a_0 \rangle}. \tag{34}$$

Result (34) for the chemical potential follows from the property that the $\mathbf{p} = 0$ state in a Bose superfluid plays the role of a *reservoir* and that consequently, adding (or destroying) a particle in this state yields the equilibrium state relative to the $N+1$ $(N-1)$ system. Starting from (34) one easily derives the equation:

$$\mu = -\frac{\langle a_0^\dagger [H, a_0] \rangle}{\langle a_0^\dagger a_0 \rangle} = \frac{1}{\sqrt{N n_0}} \sum_{\mathbf{p}} \langle a_{-\mathbf{p}}^\dagger \rho_{\mathbf{p}} \rangle V(p). \tag{35}$$

Using (1) and (32), one then finds the exact non-trivial relationship [18]

$$\mu = \rho \int d\mathbf{r} V(r)(1 + F_1(r)) \tag{36}$$

relating the LRO function $F_1(r)$ to the chemical potential of the system. The result in (36) holds for any system exhibiting Bose condensation and interacting with central potentials.

The LRO function $F_1(q)$ plays a crucial role in the coupling between the *density*

$$| F \rangle = \frac{1}{\sqrt{NS(q)}} \rho_{\mathbf{q}} \mid 0 \rangle \tag{37}$$

and *particle*

$$| P \rangle = \frac{1}{\sqrt{n(q)}} a_{-\mathbf{q}} \mid 0 \rangle \tag{38}$$

states that provide natural approximations to the elementary excitations of a Bose system. The expression in (37) is the famous Bijl–Feynman approximation [20]. Both the *density* and *particle* pictures coincide with the exact eigenstates in the limit of a dilute Bose gas. For a strongly correlated liquid, they provide only an approximate description. The coupling between the two states is different from zero because of the occurrence of the Bose–Einstein condensation and is given by [1]:

$$\langle F \mid P \rangle = \sqrt{\frac{n_0}{S(q) n(q)}} F_1(q). \tag{39}$$

The coupling turns out to be complete when $q \to 0$ (in fact in this limit one has $S(q) = q/2mc, n(q) = n_0 mc/2q$ and $F_1(q) = -\frac{1}{2}$), showing in an explicit way that in the hydrodynamic limit the *density* and *particle* picture of elementary excitations of a Bose condensed system coincide. The coupling between the *density* and *particle* pictures is at the basis of fundamental properties exhibited by Bose superfluids (for a recent exhaustive discussion see Ref. [21]).

It is finally interesting to compare the average excitations energies of the states (37) and (38). Both energies provide a rigorous upper bound to the energy of the lowest excited state of the system. The energy of the Feynman state (37) is given by the famous result [20]

$$\epsilon_F(q) = \frac{\langle F \mid H \mid F\rangle}{\langle F \mid F\rangle} = \frac{q^2}{2mS(q)} \tag{40}$$

and is expressed in terms of static structure function. It is well known that in liquid ^{4}He, (40) provides the exact dispersion relation in the low q phonon regime, while it gives only a poor description at higher momenta.

The energy of the particle state (38) takes instead the form [1]

$$\epsilon_P(q) = \mu - \frac{q^2}{2m} - \frac{W(q)}{n(q)}, \tag{41}$$

where μ is the chemical potential and the quantity $W(q)$ is given by

$$W(q) = \int d\mathbf{r}_1 d\mathbf{r}'_1 \rho^{(2)}(\mathbf{r}_1, 0; \mathbf{r}'_1, 0) e^{-i\mathbf{q}(\mathbf{r}'_1 - \mathbf{r}_1)} V(r_1), \tag{42}$$

with $\rho^{(2)}(\mathbf{r}_1, 0; \mathbf{r}'_1, 0)$ defined in (28).

It is also useful to note that, due to relations (35)–(36) for the chemical potential, the particle energy (41) has no gap at $q = 0$ and vanishes linearly with q [1].

It is possible to obtain an explicit expression for the average value of the particle energy (41) in momentum space:

$$\overline{\epsilon}_P = \frac{\sum_{\mathbf{q}} n(q)\epsilon_P(q)}{\sum_{\mathbf{q}} n(q)}. \tag{43}$$

This average is sensitive to the values of $\epsilon_P(q)$ in the interval of momenta where the quantity $q^2 n(q)$ has a significant weight. In superfluid ^{4}He, this corresponds to the range $q = 1 - 3\text{Å}^{-1}$ including the maxon and roton region. Using the operator identity $-\sum_{\mathbf{p}} a^\dagger_{\mathbf{p}}[H, a_{\mathbf{p}}] = E_K + 2V$, the average in (43) takes the form

$$\overline{\epsilon}_P = \mu - \frac{1}{N}(\langle E_K\rangle + 2\langle V\rangle), \tag{44}$$

where $\langle E_K\rangle$ and $\langle V\rangle$ are the kinetic energy and the potential energy relative to the ground state. At zero pressure, where $\mu = \frac{1}{N}(\langle E_K \rangle + \langle V \rangle)$, (44) yields $\bar{\epsilon}_P = -\langle V \rangle / N = 21 \sim 22K$ in superfluid ^{4}He. It is instructive to compare the above value with the corresponding average of the Feynman energy:

$$\bar{\epsilon}_F = \frac{\sum_{\mathbf{q}} n(q)\epsilon_F(q)}{\sum_{\mathbf{q}} n(q)}. \tag{45}$$

Using microscopic estimates for $S(q)$ and $n(q)$, we find $\bar{\epsilon}_F = 24 \sim 25K$, a value rather close to $\bar{\epsilon}_P$.

The fact that the energies of the *density* and *particle* states (37)–(38) turn out to be comparable in the most relevant region, $q = 1 - 3\text{Å}^{-1}$ reveals the importance of a careful microscopic investigation of the coupling (induced by BEC) between the two pictures in order to obtain a better understanding of excitations in Bose superfluids.

Acknowledgement. It is a great pleasure to thank L. Pitaevskii for a stimulating collaboration and discussion.

References

[1] S. Stringari, Phys. Rev. B **46**, 2974 (1992).
[2] D. Pines and Ph. Nozières, *The Theory of Quantum Liquids* (Benjamin, New York, 1966), Vol.I; Ph. Nozières and D. Pines, *The Theory of Quantum Liquids* (Addison-Wesley, 1990), Vol.II.
[3] N.N. Bogoliubov, Phys. Abh. SU **6**, 1 (1962).
[4] H. Wagner, Z. Physik **195**, 273 (1966).
[5] P.C. Hohenberg, Phys. Rev. **158**, 383 (1967).
[6] N.D. Mermin and H. Wagner, Phys. Rev. Lett. **17**, 1133 (1966).
[7] N.D. Mermin, Phys. Rev. **176**, 250 (1968).
[8] L. Pitaevskii and S. Stringari, J. Low Temp. Phys. **85**, 377 (1991).
[9] T. Gavoret and Ph. Nozières, Ann. Phys. (NY) **28**, 349 (1964).
[10] L. Reatto and G.V. Chester, Phys. Rev. **155**, 88 (1967).
[11] L. Pitaevskii and S. Stringari, Phys. Rev. B **47**, 10915, (1993).
[12] S. Girvin and A. MacDonald, Phys. Rev. Lett. **58**, 1252 (1987).
[13] R.B. Laughlin, Phys.Rev.Lett. **50**, 1395 (1983).
[14] S.C. Zhang, Int. J. Mod. Phys. **6**, 25 (1992).
[15] W. Kohn, Phys.Rev. **123**, 1242 (1961).
[16] R.D. Puff, Phys. Rev. A **137**, 406 (1965).
[17] D. Pines and C.-W. Woo, Phys. Rev. Lett. **24**, 1044 (1970).
[18] S. Stringari, J. Low Temp. Phys. **84**, 279 (1991).

[19] M.L. Ristig and J. Clark, Phys. Rev. B **40**, 4355 (1989).
[20] A. Bijl, Physica **8**, 655 (1940); R.P. Feynman, Phys. Rev. **94**, 262 (1954).
[21] A. Griffin, *Excitations in a Bose-Condensed Liquid* (Cambridge University Press, 1993).

6

Dilute-Degenerate Gases

F. Laloë

Laboratoire de Physique de l'ENS†
24, rue Lhomond
F75231 Paris, Cedex 05
France

Abstract

We study the properties of quantum gases which are dilute in terms of the interactions but degenerate; this may be the case if the thermal wavelength of the particles is much larger than the interatomic potential range. We assume that the potential contains a short range repulsive part, which creates strong correlations between particles that cannot be treated by ordinary perturbation theory, even if the region of configuration space where they are dominant is relatively small. Our method combines two basic ingredients: Ursell operators, which generalize the Ursell functions that are usually used to generate quantum cluster expansions, but are not in themselves symmetrized (the operators can be related to an auxiliary system of distinguishable particles); in a distinct step, an exact treatment of particle indistinguishability is given through the inclusion of exchange cycles of arbitrary length. For non-condensed boson systems, as well as for fermion systems, the method provides a compact expression of the partition function that generalizes the well-known Beth–Uhlenbeck formula to arbitrary degeneracy. For the study of the Bose–Einstein condensation, our treatment retains more information on the short range correlations than, for instance, the Hartree–Fock method. It also introduces a natural and explicit distinction between particles and quasiparticles; the latter may condense into a single one-particle state, which obeys a non-linear equation somewhat similar to the Gross–Pitaevskii equation. Condensation of pairs (bosons or fermions) appears naturally in this formalism as a competing transition process, as well as condensation of more complex objects (triplets, etc.),

† Laboratoire de Spectroscopie Hertzienne de l'ENS, UA associée au CNRS 18, associé à l'Université Pierre et Marie Curie.

without requiring any new ingredient, such as some variational wave function adequate for each channel of condensation.

1 Introduction

Recent progress in experiments looking for Bose–Einstein condensation (BEC) in dilute systems, in spin polarized hydrogen [1, 2, 3], laser-cooled clouds of atoms [4] and excitons[5, 6, 7], provides a natural stimulus towards a more detailed understanding of the properties of degenerate quantum gases. In these gases at very low temperatures, unusually large values of the thermal wavelength λ_T may be obtained, and one can reach situations where $n\lambda_T^3$ is of order one while the condition $na^3 \ll 1$ is still satisfied (n is the number density of particles and a is the range of the interatomic potential); these systems are degenerate, but they remain dilute in terms of the interactions. A motivation for this work is the hope that BEC of these "dilute-degenerate gases" will be observed experimentally and studied in detail in the near future.

A second motivation arises from the fact that dilute systems, such as those studied in atomic physics, often provide useful model systems, allowing calculations that give general views on physical mechanisms and processes can be extended to denser systems. But the spirit of atomic physics also implies a detailed treatment of two-body correlations (as opposed to mean-field approximations); it is well-known that binary collisions can be treated exactly, and there is no special reason to believe that the occurrence of degeneracy will destroy this possibility. Moreover, in the study of the Bose–Einstein and superfluid transitions, or processes occurring in the build up of degeneracy, a microscopic approach may be particularly useful because, after all, several basic questions in this domain are still not answered very clearly. For instance, even at the level of the one-body density operator, one can ask precisely how the momentum distribution function is affected when one adds repulsive hard cores to an ideal Bose condensed gas. By a continuity argument, it is clear that the distribution will resemble that of the ideal gas, which contains a delta function of momentum centred at zero velocity. But two different scenarios are then conceivable: either the delta function will survive the effects of the interaction, just changing its weight, or it will acquire a finite width and become a "non-monochromatic peak". The first scenario is commonly accepted, and indeed at zero temperature there is a general argument by Penrose and Onsager [8] that proves its validity; the argument basically uses only the fact that the ground state wave function of a system of bosons has no node and, moreover, gives a good order of magnitude estimate of the "condensate fraction" in liquid ^{4}He. Nevertheless, at non-zero temperature, the argument does not apply anymore since the N-body wave functions of the excited states oscillate around zero, and the existence of a momentum delta function becomes less clear. Generally speaking, in most treatments of the Bose–Einstein condensation in dilute systems, the zero width of the peak is not proved, but, rather, it is assumed. This is in particular true of all first order perturbation calculations, or in the various forms of mean-field theories where, by construction, the wave functions remain unaffected under the effect of the interactions, and certainly do not vanish on any of the contact hypersurfaces mentioned in footnote.† The zero width of the peak at the origin is more an assumption than a result.‡

If one now goes to the level of two-particle density operators (correlations), more questions arise, maybe more important conceptually because they are related to the exact relation between the Bose–Einstein condensation and superfluidity. We know that the former occurs in ideal gases while the latter does not, but in interacting systems are they fundamentally the same phenomenon, always taking place at exactly the same temperature? Or is it conceivable that, for some class of systems, two distinct transitions will occur? The very notion of superfluidity implies that there is a mobile condensate that is robust against the effects of dissipation, and one would like to understand in detail what kind of correlations, occurring presumably at the same time in the space of momenta and coordinates of the particles, physically create this robustness.

Of course these questions have been discussed in the literature, and there is already an impressive body of work on the properties of the Bose–Einstein and superfluid transitions in dilute systems. This includes various contributions made after the important ideas introduced by Bogoliubov [10], the method of the pseudopotentials developed by Huang, Lee, Yang and collaborators [11, 12, 13], the "binary collision approximation" of Lee and Yang [14, 15], just to refer to some of the best-known

† Actually, reasoning in the configuration space of the system, where hard cores make all N-body wave functions of the system vanish on a large number of contact hypersurfaces [11], gives the impression that the peak centred at zero velocity might indeed have a finite width: one could expect some kind of Heisenberg relation between the distance over which the wave function can propagate in a $3N$ dimension space without vanishing (of the order of the classical mean-free-path for hard cores in the system) and the width of the momentum peak at the origin. If that were the case, the peak would not be really infinitely narrow, except in the limit of vanishing densities.

‡ For a critical discussion of the experimental evidence of a condensate fraction in superfluid helium 4, see [9].

contributions. In most cases the present work will, of course, not do more than recover the same physical results from another point of view, where two-body correlations are emphasized in more detail. In particular, in our effort to treat short range correlations as precisely as possible, we will avoid any perturbation series in the interatomic potential V, even if resummations are possible, for instance by selecting ladder diagrams in order to reconstruct the T collision matrix from V. Rather, we shall use cluster expansions which, as is well-known (see, for instance, [16]), are a tool of choice for obtaining density expansions instead of V expansions; they contain exponential expressions of the hamiltonians which do not create any divergence problem for infinite potentials (hard cores). For this purpose, we will make use of generalized quantum Ursell functions, actually Ursell operators, which intrinsically do not contain statistics (as opposed to the usual Ursell functions, which are already symmetrized [16, 17]): their action is defined, not only in the space of states of the real system of indistinguishable particles, but also in the larger space of an auxiliary system of distinguishable particles. If the system is dilute in terms of the interactions but not of statistics, it is possible to limit the calculation to low order Ursell operators† while, with the usual (fully symmetrized) Ursell functions, the order would tend to diverge with increasing degeneracy.

This method means that we have to give up the formalism of second quantization, despite its well-known usefulness, but this is the price to pay for the introduction of interactions and statistics in completely independent steps. It also implies that, at some point, an explicit symmetrization of the wave functions becomes indispensable and, moreover, that no approximation whatsoever can be made at this step: otherwise the possibility of treating degenerate systems would be lost. Fortunately it turns out that the symmetrization operation can indeed be performed exactly without great difficulties; this is done by introducing simple products of operators, corresponding to exchange cycles, or, more generally, simple functions of operators that correspond to summations over the size of cycles up to infinity. In this way, interactions and degeneracy are treated at completely different orders – actually all orders for the latter – and relatively simple and concise expressions of the partition function are obtained. When applied to dilute systems (with respect to the interactions), the method can be seen as complementary to the method of

† For instance, when a Bose–Einstein condensation takes place, the Ursell operators in themselves have no singular variation.

Belyaev for bosons [18] and that of Galitskii for fermions [19], extended to non-zero temperatures.

The study of the Bose–Einstein condensation process itself in its various forms (condensation of particles, of pairs, etc.) is delicate, because the phase transition introduces a sensitivity to terms that would otherwise remain small corrections. We propose to include them within a self consistent equation for the "dressed Ursell operator". This leads to results that remain mathematically simple and physically satisfying; for instance, the repulsive part of the interactions tends to increase the value of the chemical potential at which the condensation takes place; more generally one determines qualitative predictions on the effects of the repulsive or attractive part of the potential on the critical density. The modification of the Ursell operator is somewhat similar to a Hartree–Fock theory, but it is free of divergences for hard core potentials since it incorporates short range correlations; it also includes some of the physics contained in the non-linear Schrödinger equation [20], for instance the modification of the condensate wave function under the effects of the interactions. Finally, we will also see how condensation of pairs, triplets, etc., appears naturally in this point of view. The present text does not give the details of the calculations, which will be published elsewhere [21]; it just attempts to summarize the physical ideas and the main results.

2 Exchange Effects and Products of Operators

The purpose of this section is to show how the operation of symmetrization introduced by quantum statistics may be performed easily by the introduction of products of operators or of simple functions of operators. This mechanism will be basic for all calculations in the following sections, and will allow us to obtain partition functions as well as reduced density matrices. As an illustration, at the end of this section we treat the ideal gas by this method and recover known results.

2.1 Two-Particle System

In the study of the effects of exchange in atomic collisions, it is often found that particle indistinguishability effects introduce, in the density operator of an atom after collision, the product of the density operators initially associated with the internal states of the atoms; examples are spin exchange collisions [22], metastability exchange collisions [23] or identical spin–rotation effects [24] which are the origin of nuclear spin waves occurring in spin polarized gases. Here we will see how products of one body operators also occur in the evaluation of the partition function for identical atoms and, after summation over the size of exchange cycles, simple functions of these density operators. Our first example deals with the simplest possible case: a two-atom system which we will define by a one-particle density operator ρ_F, describing external as well as internal (spin) variables; the notation "F" is for "free", for reasons that will be explained shortly.

Assume for a moment that the two atoms are distinguishable and that the system is described by:

$$\rho(1,2) = \rho_F(1) \otimes \rho_F(2). \tag{1}$$

If one is interested in the reduced density operator ρ_I describing one atom only, one takes the partial trace:

$$\rho_I(1) = \operatorname*{Tr}_2 \{\rho(1,2)\} \tag{2}$$

and one trivially obtains:

$$\rho_I = \rho_F. \tag{3}$$

Now, if the two particles are identical, equation (1) must be replaced by:

$$\rho(1,2) \propto [1 + \eta P_{ex}]\ \rho_F(1) \otimes \rho_F(2)\ [1 + \eta P_{ex}], \tag{4}$$

where $\eta = \pm 1$ (bosons or fermions) and P_{ex} is the exchange operator (we are not interested here in the normalization factor of ρ and do not write it explicitly). A simple calculation shows that the one-body operator, still defined by (2), now has a slightly more complicated expression than (3):

$$\rho_I \propto \rho_F + \eta \rho_F^2. \tag{5}$$

The reason is that the two terms that are linear in the exchange operator turn out to take the simple form of squares of ρ_F. If, for simplicity, we assume that the particles do not have internal states, ρ_F can be defined through its Wigner transform $\rho_F^W(\mathbf{r}, \mathbf{p})$, and it easy to show that the Wigner transform of relation (4) is

$$\rho_W(\mathbf{r}_1, \mathbf{p}_1, \mathbf{r}_2, \mathbf{p}_2) = \textit{direct term} + \eta \times \textit{exchange term}, \tag{6}$$

where the *direct term* is simply the product:

$$\rho_F^W(\mathbf{r}_1, \mathbf{p}_1) \times \rho_F^W(\mathbf{r}_2, \mathbf{p}_2), \tag{7}$$

while the *exchange term* has the more complicated form, which is the integral over $\mathbf{r}'$ and $\mathbf{k}'$ of the expression:

$$(2\pi)^{-3} e^{i(\mathbf{p}_1-\mathbf{p}_2)\cdot\mathbf{r}'/\hbar} e^{2i\mathbf{k}'\cdot(\mathbf{r}_1-\mathbf{r}_2)} \times \rho_F^W\left(\frac{\mathbf{r}_1+\mathbf{r}_2}{2} - \frac{\mathbf{r}'}{4}, \frac{\mathbf{p}_1+\mathbf{p}_2}{2} + \hbar\mathbf{k}'\right)$$
$$\times \rho_F^W\left(\frac{\mathbf{r}_1+\mathbf{r}_2}{2} + \frac{\mathbf{r}'}{4}, \frac{\mathbf{p}_1+\mathbf{p}_2}{2} - \hbar\mathbf{k}'\right) \tag{8}$$

The former has no limitation concerning the distance at which the two particles can be found, while the integral over $\mathbf{k}'$ in the latter puts a limit on this distance (of the order of the Planck constant divided by the spread over relative momenta of the two particles); in other words the exchange term has a range for the relative position which is of the order of the De Broglie wavelength. This shows that the interpretation of ρ_F is simple: while ρ_I gives the "full" one particle density operator, defined independently of the position of the second particle, the operator ρ_F characterizes a "free" particle when it is far from the other (too far away for exchange effects to play a role).

The calculations which lead to equation (5) also apply when the particles have any kind of internal state. For instance one can assume that the particles have $(2I+1)$ spin states, and that they are unpolarized (equal probability to be in any of these states). It is then easy to see that the exchange term is reduced by a factor $(2I+1)$, which corresponds to the reduction of effects of exchange when internal states of the particles can be used to "tag" them. As expected, the factor disappears if all particles are in the same spin state: all particles are then fully indistinguishable.

2.2 Many Particles

The same physical ideas are true for a many particle system. One may consider the expression:

$$\rho_I(1) = \left[Z_N^F\right]^{-1} \operatorname*{Tr}_{2,3,\ldots,N} \left\{ {}^S_A\, \rho_F(1) \otimes \rho_F(2) \otimes \ldots \otimes \rho_F(N)\, {}^S_A \right\} \tag{9}$$

where the symmetrization operator S applies for bosons, the antisymmetrization operator A for fermions and where Z_N^F is defined by:

$$Z_N^F = \operatorname*{Tr}_{1,2,3,\ldots,N} \left\{ {}^S_A\, \rho_F(1) \otimes \rho_F(2) \otimes \ldots \rho_F(N)\, {}^S_A \right\} \tag{10}$$

The operator ρ_I is obtained from ρ_F after two successive, opposite, operations:† first one goes from one particle to N by a tensor product, secondly one comes back to one particle by a partial trace; but in between symmetrization (or antisymmetrization) has been applied so that ρ_I is not simply equal to ρ_F. The operators S and A can be expressed as a sum of all permutation operators of the numbered particles:

$$S = \frac{1}{N!}\sum_{\alpha} P_{\alpha}, \qquad A = \frac{1}{N!}\sum_{\alpha} \epsilon_{\alpha} P_{\alpha}, \tag{11}$$

where N is the number of particles and ϵ_α is the parity ± 1 of the permutations P_α. A calculation of expression (10) can be made by expressing every permutation as a product of cycles C_k of k particles:

$$P_{\alpha} = \underbrace{C_1(i)C_1(j)C_1(n)}_{m_1 \text{ factors}} \times \underbrace{C_2(i',j')C_2(n',q')C_2(s',t')}_{m_2 \text{ factors}} \times C_3(.,.,.)..., \tag{12}$$

with m_1 cycles of length one (identity operators acting on the particles that do not change place), m_2 cycles of length two (transpositions of two particles), m_3 cycles of three particles, etc. Here $C_k(i, j, k, ..)$ denotes the cycle where particle i replaces particle j, particle j replaces particle k, etc. One obviously has:

$$\sum_k k m_k = N. \tag{13}$$

The trace over the N particles in (9) is taken in a space that is simply the tensor product of N single particle spaces of state, so that every cycle of length k in (12) corresponds to a trace taken in a different subspace which introduces a separate factor Γ_k. Because the numbering of the particles does not affect the value of Γ_k (it just changes the names of dummy variables), we can for convenience renumber the relevant particles from 1 to k. The effect of C_k is then to move particle 1 into the place initially occupied by particle 2, particle 2 into the place occupied by particle 3, and so on, until one comes back to the place of particle 1. Introducing a complete set of states $\{| 1 : \varphi_n >\}$ in the space of states‡ of particle 1,

† In (10), one can suppress one of the operators S (or A) inside the braces by using a circular permutation of the operators under the trace, combined with the relation $S^2 = S$. In (9) the trace is only a partial trace so that the argument does not apply anymore; but, because all one-particle operators inside the brace are the same, their product commutes with S (or A), so that the same simplification can be made.

‡ If the particles have internal states, the index n symbolizes at the same time the orbital quantum numbers as well as those characterizing the internal state. For instance, if the particles have spin I, a summation such as $\sum_n$ contains in fact two summations, one over orbital quantum numbers, and a second over $(2I + 1)$ spin states.

$\{| 2 : \varphi_n >\}$ for particle 2, etc., one can then write:

$$\Gamma_k = \sum_{n_1, n_2, \ldots n_k} < 1 : \varphi_{n_1} \mid \rho_F(1) \mid 1 : \varphi_{n_2} >< 2 : \varphi_{n_2} \mid \rho_F(2) \mid 2 : \varphi_{n_3} >$$
$$\times \ldots \times < k : \varphi_{n_k} \mid \rho_F(k) \mid k : \varphi_{n_1} > = \mathrm{Tr}\left\{[\rho_F]^k\right\} \tag{14}$$

The contribution of the cycle is therefore merely the trace of the kth power of the operator ρ_F. Consequently, a given permutation P_α which corresponds to a given set of values of the m_k introduces into the summation the following term:

$$\prod_k \left[\mathrm{Tr}\left\{[\rho_F]^k\right\}\right]^{m_k}. \tag{15}$$

At this point, if we call $\sum_{\{m_k\}}$ a summation over all possible ways to break the number N into m_k according to (13), and $c\{m_k\}$ the number of permutations P_α which correspond to any particular decomposition of this type, we obtain:

$$Z_N^F = \frac{1}{N!} \sum_{\{m_k\}} c\{m_k\} \prod_k \left[\eta^{1+k}\right]^{m_k} \left[\mathrm{Tr}\left\{(\rho_F)^k\right\}\right]^{m_k}. \tag{16}$$

This formula uses the fact that cycles of an even number of particles have parity -1 so that the parity of a cycle of length k is $(-1)^{1+k}$. For instance, if $m_1 = N$ (all the other m_k being equal to zero), there is only one permutation (the identity) that fits into this decomposition, and c is equal to one. The other extreme happens when $m_N = 1$ (all the other m_k being equal to zero): one is then dealing with all cycles of maximum length N, and it is easy to see that their number is $(N-1)$. More generally, if we choose any combination of m_k obeying (13), we can distribute in $N!$ different ways the N particles into the N places contained inside cycles, but all distributions do not generate different permutations P_α for two reasons: first every group of cycles of length k can be found in $m_k!$ different orders; second, inside each cycle, any of the k particles can be put arbitrarily in the first position, generating k different ways to recover the same permutation. Therefore:

$$c\{m_k\} = N! \prod_k \frac{1}{k \times m_k!}. \tag{17}$$

The summation over the m_k is difficult when equation (13) correlates the variations of these indices, but becomes easy if this condition is removed. This can be done by going to the grand canonical ensemble, and defining:

$$Z_{gc}^F = \sum_N z^N Z_N^F \tag{18}$$

(where z is the fugacity, which would be equal to $e^{\beta\mu}$ at thermal equilibrium); all operators of interest can be obtained from this generating function. The introduction of z^N amounts to multiplying in (16) ρ_F by z, according to (13). Then, summing over m_k introduces the product for all values of k of the exponential of the quantities $\eta k^{-1}\,\mathrm{Tr}\left\{(\eta z \rho_F)^k\right\}$, that is, the exponential of a series over k that can be summed by the function $-\eta\,\mathrm{Log}\,(1-\eta z\rho_F)$. Finally one is led to the simple result for the partition function:

$$Z_{gc}^F = \exp\left[-\eta\,\mathrm{Tr}\left\{\mathrm{Log}\left[1-\eta z\rho_F\right]\right\}\right]. \tag{19}$$

Now, if we wish to calculate operators, in the canonical ensemble we can introduce infinitesimal variations of the operator ρ_F defined by:

$$d\rho_F = d\epsilon \mid \varphi\rangle\langle\varphi \mid \rho_F, \tag{20}$$

where $d\epsilon$ is an arbitrarily small number. One then obtains for the variation of Z_N:

$$dZ_N^F = d\epsilon \underset{1,2,3,\ldots,N}{\mathrm{Tr}} \left\{ \underset{A}{\overset{S}{}} \rho_F(1)\otimes\rho_F(2)\otimes\ldots\rho_F(N)\sum_{i=1}^{N} \mid i:\varphi\rangle\langle i:\varphi\mid \right\} \tag{21}$$

or:†

$$\frac{d}{d\epsilon}\,\mathrm{Log}\,Z_N^F = N\,\mathrm{Tr}\left\{\mid\varphi\rangle\langle\varphi\mid\rho_I\right\} = N\langle\varphi\mid\rho_I\mid\varphi\rangle \tag{22}$$

(all derivatives with respect to ϵ are taken at the value $\epsilon = 0$). If the grand canonical ensemble is used, the N particle density operator is defined by:

$$\rho_{gc} = \left[Z_{gc}^F\right]^{-1}\sum_N z^N \left[\underset{A}{\overset{S}{}}\rho_F(1)\otimes\rho_F(2)\otimes\ldots\rho_F(N)\underset{A}{\overset{S}{}}\right]. \tag{23}$$

A variation of ρ_F according to (20) creates a variation of Z_{gc}^F given by:

$$dZ_{gc}^F = \sum_N z^N dZ_N^F. \tag{24}$$

If we insert (21) into this equation and compare with (23), we see that this sum over N is nothing but the product of Z_{gc}^F by the sum of the contributions of every subspace $N = 1,2,3,\ldots.$ of the Fock space to the

† Here we take the convention where the operator ρ_I is normalized to the number N of particles (instead of one).

average value of a "one-particle operator", as usually defined in second quantization. But this average value can also be expressed as a function of the one particle density operator in the grand canonical ensemble ρ_I^{gc} (normalized to the average number of particles), and we obtain:

$$\frac{d}{d\epsilon} \operatorname{Log} Z_{gc}^F = \operatorname{Tr}\{\rho_I^{gc} \mid \varphi\rangle\langle\varphi \mid\} = \langle\varphi \mid \rho_I^{gc} \mid \varphi\rangle. \tag{25}$$

This formula shows that ρ_I^{gc} is the "operator derivative" of Z_{gc}^F (in (20) we have only considered diagonal variations of ρ_F, but a generalization to off-diagonal variations is easy). By choosing in succession for $\mid \varphi\rangle$ all eigenvectors of ρ_F, and by taking the derivative of (19), one obtains:

$$\rho_I^{gc} = \frac{z\rho_F}{1 - \eta z \rho_F}. \tag{26}$$

The two particle density operator may be obtained by a similar method from a variation of ρ_F of the form:

$$d\rho_F = \left[d\epsilon_1 \mid \varphi\rangle\langle\varphi \mid + d\epsilon_2 \mid \chi\rangle\langle\chi \mid\right] \rho_F \tag{27}$$

but, for brevity, we do not give the calculation here.

We can now assume that ρ_F takes the simple form (thermal equilibrium for an ideal gas):†

$$\rho_F = e^{-\beta H_o}, \tag{28}$$

where H_0 is some one particle hamiltonian, for instance $P^2/2m$ for a particle in a box; by replacing in (19) the trace by a sum over d^3k, and in (26) the operator ρ_F by its diagonal element $e^{-\beta\hbar^2 k^2/2m}$, one immediately recovers the usual formulas that are found in textbooks on statistical mechanics. The method that we have used is more indirect than the traditional method, but it gives a physical interpretation to the term in $[\rho_F]^k$ that is obtained by expanding the fraction in (26): it corresponds to the contribution of all cyclic exchange of k particles in the system. The validity of (26) is also more general: it is not limited to any particular choice for ρ_F or to thermal equilibrium;‡ moreover, exactly as in Section 2.1, it is easy to include the case where particles have internal states, for instance spin states. The formalism gives a good idea of the general properties of a system of identical particles when the correlations have the minimum value that remains compatible with statistics. A way to

† The operator ρ_F does not have to have the same trace as ρ_I, and we do not need to insert into (28) any additional constant to reconstruct the partition function through (10).

‡ Actually the complete flexibility in the choice of ρ_F is important for taking the operator derivative that has provided us with the reduced density operator ρ_I^{gc}.

describe $z\rho_F$ is as the operator generalization of the chemical potential μ: while μ is a convenient (but indirect) way to characterize the number of particles N, the operator $z\rho_F$ may be used to indirectly characterize all properties of the full one particle operator ρ_I^{gc}. It is therefore not surprising that, at thermal equilibrium, there should be a simple relation between the chemical potential and the trace of $z\rho_F$ [25].

3 Equation of State of a Dilute-Degenerate Gas

3.1 General Considerations

We now assume that the gas is at thermal equilibrium. In the canonical ensemble, the partition function Z_N of the system is given by:

$$Z_N = \mathrm{Tr}\left\{K_N \begin{matrix} S \\ A \end{matrix}\right\} \tag{29}$$

where S applies for bosons while A applies for fermions, and where the operators K_N are defined by:

$$\begin{aligned} K_1 &= \exp -\beta H_0(1) \\ K_2 &= \exp -\beta\left[H_0(1) + H_0(2) + V_{12}\right] \\ \dots & \quad \dots \dots \\ K_N &= \exp -\beta\left[\sum_{i=1}^{N} H_0(i) + \sum_{i>j} V_{ij}\right] \end{aligned} \tag{30}$$

in an obvious notation. We use cluster techniques to expand the operators K_i into Ursell operators U_l according to:

$$\begin{aligned} K_1 &= U_1 \\ K_2(1,2) &= U_1(1)U_1(2) + U_2(1,2) \\ K_3(1,2,3) &= U_1(1)U_2(2)U_1(3) + U_2(1,2)U_1(3) + U_2(2,3)U_1(1) \\ &\quad + U_2(3,1)U_1(2) + U_3(1,2,3). \end{aligned} \tag{31}$$

In terms of the Ursell operators, the N particle operator K_N can be written in the form:

$$K_N = \sum_{\{m_l'\}} \sum_{\{D'\}} \underbrace{U_1(.)U_1(.)...U_1(.)}_{m_1' \text{ factors}} \times \underbrace{U_2(.,.)U_2(.,.)...U_2(.,.)}_{m_2' \text{ factors}} \times U_3(.,.,.)... \tag{32}$$

where the first summation is made on all possible ways to decompose the number of particles as:

$$N = \sum_l l m_l' \tag{33}$$

The second summation corresponds to all non-equivalent ways to distribute the N particles into the variables of the Ursell operators, symbolized by dots in (32). The only difference between our definitions and those of Section 4.2 of [16] is that we use operators instead of symmetrized functions; these operators are defined, not only within the state space that is appropriate for bosons or fermions, but also in the larger space obtained by the tensor product that occurs for distinguishable particles. Hence the need for an explicit inclusion of S or A in (29).

As in the preceding section we decompose these operators into a sum of permutations P_α that, in turn, we decompose into independent cycles of particles C:

$$\begin{matrix} S \\ A \end{matrix} = \frac{1}{N!} \sum_{\{m_k\}} \sum_{\{D\}} \underbrace{C_1(.)C_1(.)C_1(.)}_{m_1 \text{ factors}} \times \eta^{m_2} \underbrace{C_2(.,.)C_2(.,.)C_2(.,.)}_{m_2 \text{ factors}} \times C_3(.,.,.)... \tag{34}$$

where the first summation is similar to that of (32), while the second corresponds to all non-equivalent ways to distribute numbers ranging from one to N into the variables of the various Cs.

We can now insert (32) and (34) into (29) and obtain, inside a four index summation, numbers that are traces calculated in the space of distinguishable particles, i.e. in the ordinary tensor product of N single particle state spaces. Inside most of these traces, factorization into simpler terms occur.† For instance, if the term in question contains particle number i contained at the same time in a U_1 operator as well as in a C_1, that particle completely separates by introducing the simple number $\mathrm{Tr}\{U_1\}$. Or, if n particles are all in separate U_1s but contained inside the same cycle C_n (or conversely inside the same U_n but all in separate C_1s), this group of particles contributes by a factor $\mathrm{Tr}\{[U_1]^n\}$ (or $\mathrm{Tr}_{1...n}\{U_n\}$). More generally, in each term of the four index summation, particles group into clusters (U–C clusters), which associate together all particles that belong either to the same cycle C_k (with $k > 1$), or the same U_l (with $l > 1$). The general term is therefore the product of the contributions of all the clusters that it contains:

$$Z_N = \frac{1}{N!} \sum_{\{m_i\}} \sum_{\{D\}} \sum_{\{m_l'\}} \sum_{\{D'\}} \prod_{\text{clusters}} \Gamma_{\text{cluster}} \tag{35}$$

(the number of clusters into which each term is factorized depends, in

† Here we give only the outline of the reasoning; a more detailed article will be published [21].

general, on this term). It is also clear that clusters differing only by the labelling of particles which they contain give the same contribution. It is therefore useful to work in terms of diagrams (U–C diagrams) which emphasize the way particles are connected through exchange cycles and Ursell operators, rather than their labelling. For instance, the first diagram will correspond to any particle in a U_1 and in a C_1, and contribute the value $\mathrm{Tr}\{U_1\}$ as mentioned above; the second to $\mathrm{Tr}\{[U_1]^2\}$, the third to $\mathrm{Tr}_{1,2}\{U_2\}$, etc. Now, inside each term of the multiple summation, a given diagram Γ_{diag} may occur several times, and we call m_{diag} the number of times it is repeated. If, moreover, we simplify the notation $\sum_{\{m_i\}}\sum_{\{D\}}$ into a single sum $\sum_{\{P_\alpha\}}$ over the permutations, and similarly $\sum_{\{m_i'\}}\sum_{\{D'\}}$ into $\sum_{\{U\}}$, we obtain:

$$Z_N = \frac{1}{N!}\sum_{\{P_\alpha\}}\sum_{\{U\}}\prod_{\text{diag}}\left[\Gamma_{\text{diag}}\right]^{m_{\text{diag}}}. \tag{36}$$

If we call n_{diag} the number of particles that a given diagram contains, we obviously have:

$$N = \sum_{\text{diag}} m_{\text{diag}} \times n_{\text{diag}}. \tag{37}$$

The next step is to remark that identical diagrams appear not only in the same term of the double summation but also in many different terms. Therefore, if $\sum_{\{m_{\text{diag}}\}}$ symbolizes a summation over all possible ways to decompose N according to (37), we can also write:

$$Z_N = \frac{1}{N!}\sum_{\{m_{\text{diag}}\}} c\{m_{\text{diag}}\} \times \prod_{\text{diag}}\left[\Gamma_{\text{diag}}\right]^{m_{\text{diag}}}, \tag{38}$$

where $c\{m_{\text{diag}}\}$ is the numbers of terms in the double summation of (36) that correspond to this particular decomposition of N. This number can be obtained by distributing the N particles inside the sites of the diagrams, which can be done in $N!$ different ways, and counting how many times the same term of the double summation is obtained. For instance, since there are $m_{\text{diag}}!$ ways to interchange the order of all U–C clusters arising from the same diagram, there is a redundancy factor equal to $\prod_{\text{diag}}\left(m_{diag}!\right)$ that comes in. Moreover, there are also equivalent ways to put n_{diag} particles into one given diagram which lead to the same U–C cluster. The detail of this counting depends on the precise way in which the diagrams are defined, which we do not specify in detail here (see Ref. [21]). This introduces an extra factor f_{diag} that depends only on the topology and the size of the "branches" of the given diagram. The net result is:

$$c\left\{m_{\text{diag}}\right\} = N! \prod_{\text{diag}} \frac{1}{m_{\text{diag}}!} \frac{1}{f_{\text{diag}}}. \tag{39}$$

At this point, it is convenient to remove condition (37), which limits the choice on the values of the possible numbers m_{diag}; this can be done by using the grand canonical ensemble and its partition function:

$$Z_{g.c.} = \sum_{N} e^{\beta\mu N} Z_N. \tag{40}$$

Then a great simplification occurs because the sums over m_{diag}, which are now independent, contain factorials according to (39) which introduce simple exponentials; moreover the factors $e^{\beta\mu N}$ can be included by multiplying every number Γ_{diag} by $e^{\beta\mu n_{\text{diag}}}$, so that:

$$Z_{g.c.} = \prod_{\text{diag}} \exp\left[\exp\left(\beta\mu n_{\text{diag}}\right) \times \frac{\Gamma_{\text{diag}}}{f_{\text{diag}}}\right]. \tag{41}$$

Finally, we obtain the grand potential in the form:†

$$-\frac{1}{\beta} \operatorname{Log} Z_{g.c.} = -\frac{1}{\beta} \sum_{\text{diag}} e^{\beta\mu n_{\text{diag}}} \times \frac{\Gamma_{\text{diag}}}{f_{\text{diag}}}. \tag{42}$$

This is an exact formula, which gives the value of the pressure of the system multiplied by its volume (and divided by the temperature), which is an extensive quantity (as opposed to $Z_{g.c.}$). It is therefore well adapted to approximations, for instance truncations of the summation, and this is the subject of the rest of this article.

3.2 Non-condensed Dilute Gas

We first check our result by assuming that the gas is ideal, so that all operators U_k are zero except U_1. In this case, only one class of diagram exists, corresponding to cycles C_k of arbitrary length k. We then just need to calculate the numerical contribution Γ_k of these cycles, as well as the counting factors f_k. But this calculation has already been done in Section 2, with the replacement of ρ_F by U_1; we have seen that $\Gamma_k = Tr\left\{[U_1]^k\right\}$, while f_k is equal to $1/k$. We therefore obtain for the grand potential of

† The logarithm of Z contains only connected U–C diagrams, a general well-known property of diagram expansions.

the ideal gas, by a simple change in notation in (19):

$$-\frac{1}{\beta}\,\mathrm{Log}\,Z^0_{g.c.} = \frac{\eta}{\beta}\,\mathrm{Tr}\left\{\mathrm{Log}\left[1-\eta e^{\beta\mu}\,U_1\right]\right\}. \tag{43}$$

This formula is has already been discussed in Section 2.2.

What happens now if we add interactions? In this section, we limit ourselves to the terms that correspond to second virial corrections for the interactions, but we treat statistical effects exactly. In (42) there are two different classes of diagrams that contain only one U_2: those where the two particles in the U_2 operator belong to two different exchange cycles, and those for which they belong to the same cycle.

We begin with the first class of diagrams. In the simplest of them, only two particles are involved, and both of them are contained in the C_1 functions (they remain unaffected by exchange); this simply introduces the contribution:

$$\Gamma_{11} = \mathrm{Tr}_{1,2}\left\{U_2(1,2)\right\}. \tag{44}$$

The next diagram in this series corresponds to three clustered particles, two contained in the same U_2 and two in one permutation operator C_2. The numerical value of this second diagram is:

$$\begin{aligned}\Gamma_{2,1} &= \eta\,\mathrm{Tr}_{1,2,3}\left\{U_2(1,2)U_1(3)C_2(1,3)C_1(2)\right\} \\ &= \eta\,\mathrm{Tr}_{1,2,3}\left\{U_2(1,2)U_1(3)P_{ex}(1,3)\right\}\end{aligned} \tag{45}$$

or:

$$\Gamma_{2,1} = \eta \sum_{n_1,n_2,n_3} <1:\varphi_{n_1}|<2:\varphi_{n_2}\mid U_2(1,2)\mid 1:\varphi_{n_3}>|\,2:\varphi_{n_2}> \\ \times <\varphi_{n_3}\mid U_1\mid \varphi_{n_1}>, \tag{46}$$

that is:

$$\Gamma_{2,1} = \eta\,\mathrm{Tr}_{1,2}\left\{U_2(1,2)U_1(1)\right\}. \tag{47}$$

Similarly, one would calculate a contribution $\Gamma_{1,2}$ arising from the exchange of particles 2 and 3, and obtained by replacing $U_1(1)$ by $U_1(2)$ in (47). More generally, when U_2 clusters together k_1 particles, belonging to the same permutation cycle of length k_1, with k_2 particles belonging to another cycle of length k_2, the calculation of the effect of each of these cycles remains very similar to that of Section 2.2: now we have two particles that exchange separately with others, but the algebra of operators remains the same for each of them. We therefore obtain the contribution:

$$\Gamma_{k_1,k_2} = \eta^{k_1-1}\eta^{k_2-1} \operatorname*{Tr}_{1,2} \left\{ U_2(1,2) \left[U_1(1)\right]^{k_1-1} \left[U_1(2)\right]^{k_2-1} \right\}. \tag{48}$$

For this class of diagrams, the value of the counting factor $1/f$ is simply $1/2$; this is because, while the numbering of the particles inside U_2 can be used to distinguish the various clusters corresponding to the same diagram, $U_2(m,n)$ is not distinct from $U_2(n,m)$. Now the final step, according to (42), is to make a summation over all possible values of k_1 and k_2 after inserting an exponential of β times the chemical potential multiplied by the number of particles contained in the diagram. We denote the sum by:

$$\Delta Z_{\text{direct}} = \frac{1}{2} \sum_{k_1,k_2} e^{\beta\mu(k_1+k_2)} \Gamma_{k_1,k_2}. \tag{49}$$

The summation can be done by introducing a simple function of the U_1's operators (a fraction), which results in a total contribution that is equal to:

$$\Delta Z_{\text{direct}} = \frac{1}{2} \operatorname*{Tr}_{1,2} \left\{ U_2(1,2) \frac{e^{\beta\mu}}{1 - \eta e^{\beta\mu} U_1(1)} \frac{e^{\beta\mu}}{1 - \eta e^{\beta\mu} U_1(2)} \right\}. \tag{50}$$

For the second class of diagrams, which we will call exchange diagrams, the two particles contained in U_2 are intermixed inside the same circular permutation, which slightly complicates the situation. The first exchange diagram corresponds to the two particles in question contained in the same transposition:

$$\Gamma^{ex}_{1,1} = \operatorname*{Tr}_{1,2} \left\{ U_2(1,2) C_2(1,2) \right\} = \operatorname*{Tr}_{1,2} \left\{ U_2(1,2) P_{ex} \right\}. \tag{51}$$

The second will contain three particles:

$$\Gamma^{ex}_{1,2} = \operatorname*{Tr}_{1,2,3} \left\{ U_2(1,2) U_1(3) C_3(1,2,3) \right\}, \tag{52}$$

which is equal to:

$$\Gamma^{ex}_{1,2} = \sum_{n_1,n_2,n_3} < 1 : \varphi_{n_1} | < 2 : \varphi_{n_2} \mid U_2(1,2) \mid 1 : \varphi_{n_2} > | \, 2 : \varphi_{n_3} > \\ \times < \varphi_{n_3} \mid U_1 \mid 1 : \varphi_{n_1} > . \tag{53}$$

Now, we can use the equality:

$$< 1 : \varphi_{n_1} | < 2 : \varphi_{n_2} \mid U_2(1,2) \mid 1 : \varphi_{n_2} > | \, 2 : \varphi_{n_3} > \\ = < 1 : \varphi_{n_1} | < 2 : \varphi_{n_2} \mid U_2(1,2) P_{ex} \mid 1 : \varphi_{n_3} > | \, 2 : \varphi_{n_2} >, \tag{54}$$

which allows us to make the same summation over indices as in equation (46) to obtain:

$$\Gamma_{1,2}^{ex} = \operatorname*{Tr}_{1,2}\{U_2(1,2)P_{ex}U_1(1)\}. \tag{55}$$

A similar term occurs if the circular permutation $C_3(1,2,3)$ of (52) is replaced by $C_3(1,3,2)$; the calculation can easily be repeated to give:

$$\Gamma_{2,1}^{ex} = \operatorname*{Tr}_{1,2}\{U_2(1,2)P_{ex}U_1(2)\}. \tag{56}$$

We will not give further details here, but from the preceding equations it is not difficult to see that the generic term of this second class of diagrams will be obtained from (48) by a simple replacement of U_2 by the product U_2P_{ex}. Finally, the value of the grand potential (multiplied by $-\beta$), to first order in U_2, will be

$$\operatorname{Log} Z_{g.c.} = \operatorname{Log} Z_{g.c.}^0 + \Delta Z, \tag{57}$$

with

$$\Delta Z = \operatorname*{Tr}_{1,2}\left\{e^{2\beta\mu}U_2(1,2)\frac{[1+\eta Pex]}{2}\frac{1}{1-\eta e^{\beta\mu}U_1(1)}\frac{1}{1-\eta e^{\beta\mu}U_1(2)}\right\}. \tag{58}$$

This result has been obtained within an approximation which is basically a second virial treatment of the interactions, while it contains all statistical corrections (we have summed the exchange cycles to all orders). The formula therefore remains valid if the degree of degeneracy of the gas is significant. Nevertheless, as pointed out for instance in Section 2.1 of Ref. [26] and in Ref. [27], virial series (even summed to infinity) are no longer appropriate beyond values where the density exceeds that of a phase transition. This is because singularities in the thermodynamic quantities occur at a transition, in the limit of infinite systems. Therefore, for bosons, the validity of (58) is limited to non-condensed systems; the discussion of what happens when a Bose–Einstein condensation takes place is postponed until Section 4.

Equation (58) can be seen as a generalization of the Beth–Uhlenbeck formula: it reduces to it if the two denominators containing U_1 functions are replaced by one, an operation which is valid in the limit of low densities where the chemical potential is very negative and $e^{\beta\mu}$ is small. The details of this calculation are given in [21].

3.3 A Fictitious Imperfect Bose Gas

In the same spirit as Huang, Yang and Luttinger, who discuss the properties of a "fictitious imperfect Bose gas" in their classic article on the effect of hard sphere interactions on the properties of a gas of bosons [28], we now examine the properties of a physical system, assuming that its grand potential is exactly given by (57) and (58). In fact, we will see in Section 4 that there is no reason to believe that any real system will obey this equation exactly but, for future reference, it is nevertheless useful to discuss the implications of (58) in more detail.

For a given value of the chemical potential μ, the number of particles N of the system is given by:

$$N = \beta^{-1} \frac{d}{d\mu} \operatorname{Log} Z_{g.c.} \tag{59}$$

What happens in an ideal gas of bosons is well-known; when μ increases, starting from large negative values where the gas is classical, the density of particles also increases and degeneracy builds up progressively; when the critical value $\mu = 0$ (in the thermodynamic limit) is reached, the density of excited particles saturates and all additional particles have to accumulate into a single quantum ground state. More precisely, what happens is the following: as long as μ remains much smaller than $-e_1$, the opposite of the energy associated with the first excited state†, the summation over the discrete energy states which occurs in the trace giving the logarithm of $Z^0_{g.c.}$ can be replaced by an integral over momenta with a good approximation. The density of particles is then proportional to the integral over d^3k of the derivative (with respect to μ) of the function $\operatorname{Log}\left[1 - \exp\beta\left(\mu - \hbar^2 k^2/2m\right)\right]$. When $k \to 0$, this function only has a logarithmic divergence, its derivative diverges only as k^{-2} at the origin, and thus the integral over d^3k remains finite (in three dimensions); this limits the density of particles that can be obtained within this integral approximation. But if μ increases more and becomes comparable to $-e_1$, or even crosses this value and approaches zero, the discrete sum can no longer be approximated by an integral; actually the population of the ground state increases without limit, while that of the first excited state does not; in other words, the ground state absorbs a significant proportion of the particles. The transition between the two regimes occurs when $|\mu|$ is of the order of e_1, an infinitesimal energy in the limit where

† We choose the origin of the energies so that the energy of the ground state is equal to zero.

the size L of the box containing the gas becomes very large.† When L increases, the crossover becomes sharper and sharper and, in the thermodynamic limit, a singularity occurs at $\mu = 0$, which is the origin of the Bose–Einstein transition. Clearly, its very existence is related to the slow divergence of a logarithmic function at the origin.

Before Bose–Einstein condensation takes place, the second term in the right hand side of (57) remains an arbitrarily small correction if the range of the interaction potential is reduced. Nevertheless, however small the new term is, it completely changes the nature of the transition. From the beginning, it is obvious that it does not contain logarithms, but fractions which diverge more rapidly at the origin. It remains finite when $\mu \to 0$, because the behaviour at the origin of the integral involved in the trace is of the type:

$$\int d^3k_1 \int d^3k_2 < \mathbf{k}_1 = 0, \mathbf{k}_2 = 0 \mid U_2 \mid \mathbf{k}_1 = 0, \mathbf{k}_2 = 0 > [k_1]^{-2} [k_2]^{-2}, \quad (60)$$

which is finite. However, this integral becomes divergent when a derivative with respect to μ in the denominators of the fractions in (58) is taken, since variations at the origin in $[k_1]^{-4,2} [k_2]^{-2,4}$ are introduced. In other words, the second term will always end up dominant over the first term for some value of the chemical potential $\mu_{\text{crossover}}$. For very dilute systems, or for systems where the interaction potential is either sufficiently weak or with a sufficiently small range, $\mid \mu_{\text{crossover}} \mid$ is also very small but, because its value does not depend on the size of the container but only on microscopic quantities, in the thermodynamic limit it will still be much larger than e_1. The divergence in density introduced by the interactions will always occur before that of the population of the ground state, and therefore quench it. No Bose–Einstein transition then remains possible: it is replaced by a continuous change of regime that simulates the transition to some extent since, at some point, the system starts to accumulate particles into a narrow peak of momentum centred at the origin. The smaller the potential range, the narrower this peak, but its width will never completely vanish: the situation is exactly that of a "non-monochromatic peak" discussed in the introduction.

4 Bose–Einstein Condensation in a Dilute Gas

Equation (58) is an approximation which does not contain corrections including higher powers of U_2, as well as higher order operators U_3, U_4,

† The single particle energy e_1 is proportional to L^{-2}.

etc. We begin by a study of the effects of the terms containing binary interactions only, i.e. U_2 functions only. We will see that quadratic, cubic, etc., terms in U_2 can be incorporated by a "dressing" of the operators by the interactions; this actually allows a summation of an infinite series and restores, but for quasiparticles instead of particles, the phenomenon of a "monochromatic Bose–Einstein condensation" that had disappeared in the lowest order calculation of Section 3.3.

4.1 Condensation in the Presence of Interactions

In this section we study bosons only. How can we add U_2 functions into the first order U–C diagrams that we have studied and summed in Section 3.2? If, for instance, we start from the direct diagram, where a first U_2 incorporates two particles belonging to two different cycles, one can choose one particle in one of the cycles, remove its U_1 and replace it by a second U_2, grouping this particle together with another particle. There are three ways in which this can be done:

(i) The second particle in the added U_2 may belong to a new, third exchange cycle of length k_3; this will correspond to a "direct–direct diagram". To calculate its contribution, one can use reasoning that is similar to that of Section 3.2, basically because the two U_2 functions play a symmetrical role in the diagram. One therefore expects contributions where, in (48), U_1 will be replaced as follows:†

$$U_1(n) \Longrightarrow \frac{1}{2} e^{\beta\mu k_3} \operatorname*{Tr}_q \left\{ U_2(n,q) \left[U_1(q)\right]^{k_3-1} \right\} \tag{61}$$

(the exponential in front corresponds to the number of additional particles; there is no η factor since we are dealing with bosons).

(ii) The new U_2 may connect two particles belonging to the same cycle, which can be obtained from the original cycle by adding k_3 extra particles between those in the new U_2 (the new cycle then has a total length $k_2 + k_3$); this is a mixed "direct-exchange diagram" and one can convince oneself that its contribution will be obtained by the substitution:

$$U_1(n) \Longrightarrow \frac{1}{2} e^{\beta\mu k_3} \operatorname*{Tr}_q \left\{ U_2(n,q) P_{ex}(n,q) \left[U_1(q)\right]^{k_3-1} \right\}. \tag{62}$$

† All the present discussion is more intuitive than rigorous; in particular the presence of additional factors 1/2 in (61) and (62), arising from the counting factors f of the new diagrams, is not obvious and requires a detailed study [21].

(iii) The new U_2 may also reconnect the two initial cycles, creating a "U_2 direct loop" and generating a term that can be expressed as a trace over two particles only (and not three as in the two preceding cases). For the moment we leave aside this class of diagrams.

Now, if we start from the exchange diagrams, the situation is similar, and there are also three possibilities:

(i) Adding a new cycle again gives a "direct-exchange" diagram.
(ii) One can also lengthen the initial diagram and, as in (ii) above, introduce a new U_2 that contains two particles; if these particles are not entangled with the two particles in the other U_2 (running along the permutation cycle, one obtains the two particles of one of the U_2 functions and then the two of the other), we obtain what we will call an "exchange–exchange diagram".
(iii) Finally, one can do as in (ii) but entangle the two couples of particles inside the U_2 functions (along the cycle one gets alternatively particles belonging to each U_2; this provides a diagram with a "U_2 exchange loop" which we will also leave aside for the moment.

We therefore see that substitutions (61) and (62) again apply. If now we group these results and sum over all values of k_3, we obtain:

$$U_1(n) \Longrightarrow \frac{1}{2}\operatorname*{Tr}_{q}\left\{U_2(n,q)\,[1+P_{ex}]\,\frac{e^{\beta\mu}}{1-e^{\beta\mu}U_1(2)}\right\}. \tag{63}$$

It is clear that the above procedure is iterative, and that it allows us to build higher and higher order terms in the series of U–C diagrams. In particular, when applied successively to all U_1 functions that belong to the pure exchange cycles that occur for the ideal gas, it generates from the zero order terms the first order U_2 terms that are contained in (58). The ideal gas cycles have a counting factor $1/f = 1/k$ but, because each of the U_1 operators is modified in turn according to the above equations, giving the same numerical result, an extra factor k arises that cancels the first; this is why no logarithm is obtained in the first order correction. On the whole, when (63) is iterated from zero up to infinity, one is led to the substitution (we now use the variable 1 instead of n):

$$U_1(1) \Longrightarrow \widetilde{U}_1(1) = U_1(1)+\frac{1}{2}\operatorname*{Tr}_{2}\left\{U_2(1,2)\,[1+P_{ex}]\,\frac{e^{\beta\mu}}{1-e^{\beta\mu}\widetilde{U}_1(2)}\right\}. \tag{64}$$

It is useful to notice that, in this equation, the "dressed operator" $\widetilde{U}_1$ has an implicit definition, since it also appears in the denominator of (64); in other words, $\widetilde{U}_1$ is defined by a self-consistent equation.

Now, when this substitution is done, the effects of the interactions are contained in $\widetilde{U}_1$ itself and the value of the grand potential can be obtained by the same calculation as for an ideal gas, but just replacing U_1 by $\widetilde{U}_1$. This gives the result†, similar to (43):

$$\operatorname{Log} Z_{g.c.} \simeq -\operatorname{Tr}\left\{\operatorname{Log}\left[1 - e^{\beta\mu}\widetilde{U}_1(1)\right]\right\}. \tag{65}$$

This equation contains a summation of diagrams to infinite order in U_2so that it is clearly more general than (57) and (58). Nevertheless, since the diagrams containing loops are not included, it is not exact either. The approximation behind (65) bears some resemblance with that of a mean-field theory, but not in terms of the potential: here, Ursell operators are used instead as the basic objects, so that the two body correlations are included more precisely than in usual Hartree–Fock-type theories. For instance, as in the preceding section, infinitely repulsive hard cores do not introduce any divergence, the physical reason being that our method incorporates the fact that the particles are prevented from coming close to each other.

How does the Bose–Einstein condensation takes place in this scheme? The general scenario will be similar to that of the ideal gas, but here condensation will take place when the function inside the logarithm of (65) diverges, that is, when the product of $e^{\beta\mu}$ by the largest eigenvalue of the operator $\widetilde{U}_1$ becomes equal to one. If the range of the interaction potential is small, the two operators U_1 and $\widetilde{U}_1$ are not very different, so that this largest eigenvalue will be one plus some small correction. What matters is the sign of this correction, which according to (64) depends in turn on the sign of the matrix elements of U_2. Because this operator contains the difference between two exponentials, one involving the negative of the sum of the kinetic energy of two particles added to their interaction energy, the other involving the negative of the kinetic energy only, we see that repulsive interactions will tend to give negative values to its matrix elements, while attractive interactions will favour positive values. We conclude that, for repulsive interactions, the Bose–Einstein condensation phenomenon will take place when $e^{\beta\mu} > 1$, that is, for positive values of the chemical potential; for attractive interactions, for negative values of the chemical potential. These results are physically satisfactory and correspond to those found in the method used by Huang and coworkers [12, 26], but here we do not approximate the potential

† It can easily be checked, for instance, that, if a lowest order approximation is made by replacing $\widetilde{U}_1(1)$ by $U_1(1)$ in the right hand side of (64), relation (58) is recovered.

by a scattering length; in particular, no special instability is predicted if the potential becomes attractive, while keeping a small range (or a small scattering length) .

Another feature of our theory is that condensation takes place when a divergence occurs in a function which depends on the operator $\widetilde{U}_1(1)$, which is not directly related to the one-body density operator as for the ideal gas, but contains density corrections. The eigenstates of $\widetilde{U}_1(1)$ are the energy states of the quasiparticles, which include the effects of the interactions averaged through U_2 operators, and not those of the particles. We therefore find a scheme that incorporates the notion of quasiparticles, and relates real particles to quasiparticles in a quantitative way, since through (64) it is possible to express the one-particle density operator as a function of $\widetilde{U}_1(1)$.

4.2 Condensation of Pairs

We now come back to the general case where particles may be fermions or bosons; we study the effect of another class of diagrams which we left aside in the preceding section, where the emphasis was put on the diagrams where the U_2 functions do not connect the various cycles in too intricate a way. Here, on the other hand, we study what happens if we introduce the maximum connectivity possible between two different cycles, or inside a given cycle. If, for instance, we start from the first "direct U_2 loop" diagram that we ignored in our study of the Bose–Einstein condensation of interacting particles, we have to consider two U_2 functions connecting two pairs of particles that are both placed inside cycles of size two, that is, transpositions. It is easy to see that this term, which now contains a summation over two independent indices, has a contribution:

$$e^{4\beta\mu} \operatorname*{Tr}_{1,2} \left\{ [U_2(1,2)]^2 \right\} . \tag{66}$$

If three U_2 functions connect two triplet of particles that belong to two three-particle cycles, one obtains in the same way the term:

$$e^{6\beta\mu} \operatorname*{Tr}_{1,2} \left\{ [U_2(1,2)]^3 \right\} \tag{67}$$

or, more generally, with two cycles of equal length k where all particles are linked two by two by U_2 functions:

$$e^{2k\beta\mu} \operatorname*{Tr}_{1,2} \left\{ [U_2(1,2)]^k \right\} . \tag{68}$$

None of the above terms contains an η factor, even for fermions, because two cycles of the same parity are always involved.

Now, starting from the diagram that we called the "U_2 exchange loop" in the preceding section, we can proceed in a similar way as before, but this time we do not consider two cycles of equal length, but only one cycle of even length. The latter is saturated at all points by U_2 functions, each taking two particles from similar places in the two different halves of the cycle. The first term which was ignored in the preceding section, is obtained from a four particle cycle and two U_2 functions, and simple reasoning shows that its value is:

$$\eta e^{4\beta\mu} \operatorname*{Tr}_{1,2} \left\{ [U_2(1,2)P_{ex}]^2 \right\} \tag{69}$$

(the factor η arises because a cycle of an even number of particles is odd). More generally, one obtains from a cycle acting on $2k$ particles the term:

$$\eta e^{2k\beta\mu} \operatorname*{Tr}_{1,2} \left\{ [U_2(1,2)P_{ex}]^k \right\}. \tag{70}$$

Obviously both kinds of terms (68) and (70) can easily be grouped by introducing the operator $U_2\,[1 + \eta P_{ex}]$, but we also note that another modification of U_2 is necessary as well. The reason is that, if we want to avoid a limitation of the validity of the results to lowest order in the degree of degeneracy of the system, when calculating the contribution of any class of diagrams we must include all those that contain additional arbitrarily long exchange cycles. In other words, as in Sections 3.2 and 4.1, we must incorporate an arbitrarily large number of intermediate U_1 functions at all places where they can be inserted into this new class of diagrams, without changing their topology. Moreover, after inserting these chains of U_1 functions at any place between the U_2 functions, we must take the sum over the lengths of the chains. But we already know from Section 3.2 what the effect of this process is: the summation merely introduces operators $\left[1 - \eta e^{\beta\mu} U_1\right]^{-1}$. More precisely, since in the present case this operation has to be carried out twice, once for each particle contained in the U_2 functions, two fractions are introduced. Therefore, to incorporate all diagrams that we have discussed, we have to replace U_2 in (68) by the new operator $\widehat{U}_2$ defined as:

$$\widehat{U}_2(1,2) = U_2(1,2)\frac{[1 + \eta P_{ex}]}{2}\left[\frac{1}{1 - \eta e^{\beta\mu} U_1(1)} \times \frac{1}{1 - \eta e^{\beta\mu} U_1(2)}\right]. \tag{71}$$

Finally we have to sum over all values of k. The situation, then, becomes reminiscent of that for the ideal gas: because all pairs of particles appearing in the U_2 functions play the same role, it turns out that the counting factor of the diagrams is simply given by $1/f = 2k$. This implies that one again obtains a logarithmic function,† so that the term to be added to the logarithm of z is:

$$-\operatorname*{Tr}_{1,2}\{\text{Log}[1 - e^{2\beta\mu}\widehat{U}_2(1,2)]\}. \tag{72}$$

The physical interpretation of this new contribution to the grand potential is simple: we are now dealing with the condensation of pairs of particles, which is indeed possible in principle for both fermions and bosons [29]. The mechanism is similar to that studied in Section 4.1, and what determines the possible existence of a phase transition is the largest eigenvalue of the operator $\widehat{U}_2$. If it is positive, which requires some attraction between the particles, a Bose–Einstein condensation channel becomes available for a critical value of the chemical potential which is proportional to the logarithm of this eigenvalue; if it is negative, equation (72) becomes similar to that describng the grand potential of an ideal gas of fermions, and no new transition occurs. In what follows, we assume that the eigenvalue is indeed positive in order to discuss the influence of the new channel of condensation. If this is the case, there is still a difference with the usual condensation of single particles, since $\widehat{U}_2$ acts at the same time on two kinds of variables, those associated with the centre of mass of the two interacting particles, and those associated with the relative motion. The effect on the centre of mass is almost the same as the effect of U_1 on one particle; the only difference is that the mass is twice as large, decreasing the de Broglie wavelength. On the other hand, the effect on the variables of relative motion is significantly distinct, since the difference which appears in $\widehat{U}_2$ gives a finite (microscopic) range to the action of the operator; the interaction potential may or may not create bound states.

The two fractions containing the operators $U_1(1,2)$ in (71) play a role which is similar those occurring in (58): if the gas is non-degenerate, they introduce small corrections, but if this is not the case, they can completely change the properties of U_2. They also have a simple physical interpretation: using (26), which gives the one particle density operator ρ_I to zero order in the interaction (with the notation U_1 in place of ρ_F), one easily sees that they are equal to $[1 + \eta\rho_I]$. The latter is nothing

† In fact, we have already incorporated the terms $k = 0$ in the summation leading to (65) so that, to be precise, one should not merely add (71) to that value of the grand potential, but also remove afterwards the term that has been double counted, which is in fact simply given by (58).

but the usual statistical factor that appears in the Uehling–Uhlenbeck transport equation for instance, or in the Landau theory of Fermi liquids (more precisely the factors are an operator generalization, because U_1 and U_2 do not commute in general). For highly degenerate fermions, for instance, these statistical factors remove all the contributions to $\widehat{U}_2$ which correspond to matrix elements of U_2 between one particle states below the Fermi level, and only those close to the surface of the Fermi sphere contribute in practice.

Will the pair Bose–Einstein condensation really take place, or will it be quenched by other processes? We begin the discussion with bosons. If the potential is only weakly attractive, or if it has a very small range, the largest eigenvalue of $\widehat{U}_2$ will be smaller than that of $\widetilde{U}_1$ (which is close to one). Then the Bose–Einstein condensation of particles (or, more precisely, of quasiparticles) will take place before the condensation of pairs, that is, at a smaller value of μ. In this case, the critical value of μ at which pair condensation occurs will never be attained (with repulsive interactions, the chemical potential continues to increase after Bose–Einstein condensation has occurred, but here we assume attractive interaction †); the presence of pairs will then just bring a small correction to the calculations of the preceding section. But, if the interaction potential becomes more and more attractive, able to sustain one or more bound states with negative energies, large eigenvalues of $\widehat{U}_2$ appear, and then it becomes perfectly possible that the condensation of pairs will correspond to a smaller value of the chemical potential than that of single particles. In this case, the condensation of pairs occurs first‡ and the condensation of single particles never takes place. Our model gives a physical picture based on a competition between these two channels of condensation, among which the channel which dominates is that for which the critical density (of particles, or of molecules) is reached first.

For fermions, no condensation of particles can take place, so that no process will stop the chemical potential from rising to positive values; the limit is the Fermi energy at low temperatures. Now, what we have to compare is the largest eigenvalue of $\widehat{U}_2$ with the exponential $e^{-2\beta\mu}$. Since at low temperatures the chemical potential remains roughly

† Also, for attractive interactions, the critical values of μ are negative, so that we do not have to worry about singularities in (71).

‡ Of course, it is perfectly possible that another condensation process involving more than two particles will, in turn, take over and quench the condensation of pairs; this possibity includes ordinary condensation into a liquid or even a solid phase. For a brief discussion of other condensation processes see next section.

constant and positive, when β increases the exponential varies rapidly and always becomes smaller, at some point, than the eigenvalue. A singularity will then necessarily occur: even weakly attractive potentials will allow a Bose–Einstein condensation of pairs. We therefore recover a well-known property of fermions: at sufficiently low temperature, arbitrarily small attractions destabilize the Fermi sphere, an essential ingredient of the BCS theory [30]. If the temperature is not very low, and if there is no well-formed Fermi sphere with a sharp limit (the usual assumption in the BCS theory), formulas (71) and (72) still apply without any particular modification; the fractions in (71) automatically take care of the cross-over between the degenerate and classical regime, as noted above. We note in passing that, instead of the potential or of the scattering length which often appears in the theory of pairing in Fermi systems [31], here the relevant matrix element is that of an operator which depends both directly on the temperature through the definition of U_2 and indirectly through the factors $[1-\rho_I]$ in the definition of $\widehat{U}_2$; moreover it contains all partial scattering waves and not only the *s*-wave. In addition, one may use (71) and (72) to discuss the cross-over between the two regimes where condensation takes place, either for strongly bound binary molecules, or weakly bound BCS pairs [32, 33, 34].†

A final remark is that what we have discussed, for simplicity, is the condensation of bare, non-interacting, pairs. But by introducing the contribution of U_3 operators (and by replacing U_1 by $\widetilde{U}_1$), it would be possible to extend the theory in the same spirit as that of Section 4.1 and define a dressing of the operator $\widehat{U}_2$ under the effect of the interactions of pairs with a third partner which would lead to the condensation of "quasipairs of particles".

4.3 Other Phase Transitions

A generalization of the preceding considerations and of (71) leads to condensation of triplets; the contribution to the partition function is similar to (72), except that U_3 replaces U_2 in a trace over three particles, and more complicated exchange operators are present. More generally, channels corresponding to the condensation of groups of k particles can be written formally in terms of operators acting in the space of k particles.

† For more specific discussions including films, see Refs. [35] and [36].

In nuclei, four fermion condensate has been discussed in the context of the condensation of α particles [37].

But even all these condensation channels do not necessarily exhaust all the diagrams contained in the partition function. If all sorts of diagrams can indeed be grouped, as above, in condensation channels where they generate logarithms, then all the physics is contained in competing Bose–Einstein condensation processes of composite particles. If some other diagrams cannot be regrouped in this way, then there will be a channel for formation of a "non-monochromatic momentum peak", as discussed in Section 3.3, in competition with the ordinary ("monochromatic") condensation channels. Of course there is no proof at this stage that such processes can exist, for a given class of interaction potentials, so that this remark remains speculative.

Finally, we note that the self-consistent equation (64) which defines the $\widetilde{U}$ operator of the quasiparticles is non-linear, and may contain instabilities if the coupling is strong enough and the temperature sufficiently low. It may happen that a spontaneous symmetry breaking process, analogous to ferromagnetism, will allow one to obtain lower values of the grand potential than the values obtained without this process; this would correspond to a superfluid instability.

5 Conclusion

In this article, we have used another version of perturbation theory, not in terms of the interaction potential, but in terms of Ursell operators which give rise to various contributions in the form of U–C diagrams. This leads to the exact expression (42) for the grand potential of the system, expressing it as a sum of various terms arising from these diagrams. Each of these terms is obtained as an integral (a trace) over a finite number of variables. The expression is valid for dilute or dense systems such as liquids or solids, but for gases it becomes simpler because it can be truncated conveniently. A feature of the formalism is that the size of exchange cycles appears explicitly, which provides a natural connection with the numerical work of Ceperley and Pollock [38], emphasizing the role of cyclic exchange of all orders in Bose–Einstein condensation and superfluidity. In our case, the summations over the lengths of cycles are straightforward, introducing simple functions (fractions) of the operators. Several mechanisms can lead to a divergence of the size of the exchange cycles in the relevant diagrams; each of them introduces a separate channel for Bose–Einstein condensation. In other words, the formalism provides information on various physical processes which occur in parallel, such as the condensation of particles (or quasiparticles), that of pairs, or the contribution of larger clusters, without requiring any *ad hoc* additional assumption. Refining the results on a given condensation channel by including interactions through a renormalization of the operators is also possible, at least formally, by the introduction of integrals of finite dimension. For instance, if necessary, one could improve the theory of condensation of pairs by replacing, in the statistical factors appearing in the definition of $\widehat{U}_2$, the bare operators U_1 by dressed operators $\widetilde{U}$. The method gives simple formulas that remain valid at both low and high temperatures (degenerate or classical systems); in particular, it includes automatically all s,p,d,... collision channels. It could be adapted relatively easily to the inclusion of external potentials (inhomogeneous BEC), since this additional interaction is entirely contained in U_1. We plan to use our method for calculating condensate fractions in this kind of problem. A more systematic comparison with the results of the literature is also desirable: energy of the ground state for bosons [39, 18] and fermions [40, 19], as well as pseudopotential calculations of the magnetic susceptibility for fermions. Finally, the study of superfluid gases, especially in the case considered by Siggia and Ruckenstein where there are two internal states which may condense simultaneously [41], is another natural domain of application of our formalism.

Acknowledgment. The author has benefited from very helpful discussions with Edouard Brézin, Roger Balian and Philippe Nozières. More recently he has also had stimulating discussions with A. Stringari, K. Huang and W. Mullin and is very grateful to Peter Grüter and to W. Mullin for a careful reading of the manuscript and suggesting many improvements.

References

[1] T.J. Greytak, this volume.
[2] I. Silvera, this volume.
[3] T.J. Greytak and D. Kleppner, in *New Trends in Atomic Physics* Vol. II, G. Grynberg and R. Stora, eds. (North-Holland, N.Y., 1984), p.1127.
[4] Y. Castin, J. Dalibard, and C. Cohen-Tannoudji, this volume.

[5] J.P. Wolfe, Jia Ling Lin, and D.W. Snoke, this volume.

[6] A. Mysyrowicz, this volume.

[7] L.V. Keldysh, this volume.

[8] O. Penrose and L. Onsager, Phys. Rev. **104,** 576 (1956).

[9] P.E. Sokol, this volume.

[10] N. Bogoliubov, Journal of Physics USSR **11**, 23 (1947).

[11] Kerson Huang, *Statistical Mechanics*, (Wiley, N.Y., 1963) Section 13.3; sec. ed. (1987), Section 10.5.

[12] K. Huang and C.N. Yang, Phys. Rev. **105**, 767 (1957); K. Huang, C.N. Yang and J.M. Luttinger, Phys. Rev. **105**, 776 (1956); T.D. Lee, K. Huang and C.N. Yang, Phys. Rev. **106**, 1135 (1957).

[13] T.D. Lee and C.N. Yang, Phys. Rev. **112**, 1419 (1958); **113**, 1406 (1959).

[14] T.D. Lee and C.N. Yang, Phys. Rev. **113**, 1165 (1959); **116**, 25 (1959); **117**, 12, 22 and 897 (1960).

[15] R.K. Pathria, *Statistical Mechanics*, (Pergamon, Oxford, 1972), Section 9.7.

[16] K. Huang, loc. cit., Chapter 14; second edition (1987), Chapter 10.

[17] R.K. Pathria, Ref.[15], Section 9.6.

[18] S.T. Belyaev, Soviet Phys. JETP **34**, 289 (1958).

[19] V.M. Galitskii, Soviet Phys. JETP **34**, 104 (1958).

[20] E.P. Gross, Journ. Math. Phys. **4**, 195 (1963).

[21] F. Laloë, J. Physique, in preparation

[22] M. Pinard and F. Laloë, J. Physique **41**, 769 (1980).

[23] M. Pinard and F. Laloë, J. Physique **41**, 799 (1980).

[24] C. Lhuillier et F. Laloë, J. Physique **43,** 197 and 225 (1982).

[25] P.J. Nacher, G. Tastevin and F. Laloë, J. Physique I **1**, 181 (1991).

[26] K. Huang, in *Studies in Statistical Mechanics,* Vol. II, J. De Boer and G.E. Uhlenbeck, eds. (North-Holland, Amsterdam, 1964).

[27] C.N. Yang and T.D. Lee, Phys. Rev. **87**, 404 (1952).

[28] K. Huang, C.N. Yang and J.M. Luttinger, Phys. Rev. **105**, 776 (1957), Section 6.

[29] C.N. Yang, Rev. Mod. Phys. **34**, 694 (1962).

[30] J. Bardeen, L.N. Cooper and J.R. Schrieffer, Phys. Rev. **108**, 1175 (1957).

[31] E.M. Lifshitz and L.P. Pitaevskii, *Statistical Physics,* Vol. II (Pergamon, Oxford, 1980), Sections 39–40.

[32] A.J. Leggett, in *Modern Trends in Condensed Matter Physics*, Proceedings 16th Karpacz Winter School, A. Pekalski and J. Przystawa, eds. (Springer, Berlin, 1980); A.J. Leggett, J. Physique coll. **C7**, 19 (1980).

[33] See review by M. Randeria, this volume.

[34] P. Nozières and S. Schmitt-Rink, J. Low Temp. Phys. **59**, 195 (1985).

[35] D.M. Eagles, Phys. Rev. **186**, 456 (1969).

[36] K. Miyake, Progr. Theor. Phys. **69**, 1794 (1983).

[37] M. Apostol, Phys. Lett. **110A**, 141 (1985).

[38] D.M. Ceperley and E.L. Pollock, Phys. Rev. Lett. **56**, 351 (1986); Phys. Rev. B**36**, 8343 (1987); E.L. Pollock and D.M. Ceperley, Phys. Rev. B **39**, 2084 (1989).

[39] Lifshitz and Pitaevskii, Ref.[31], Section 25.

[40] Lifshitz and Pitaevskii, Ref.[31], Section 6.

[41] E.D. Siggia and A.E. Ruckenstein, Phys.Rev.Lett. **44**, 1423 (1980); J. Physique **40**, C7 15 (1980); Phys. Rev. B **23**, 3580 (1981).

7

Prospects for Bose–Einstein Condensation in Magnetically Trapped Atomic Hydrogen

Thomas J. Greytak
Physics Department
Massachusetts Institute of Technology
Cambridge, MA 02139
USA

Abstract

Atomic hydrogen would be an ideal substance in which to observe Bose–Einstein condensation. Recombination of the atoms into molecules can be slowed dramatically by polarizing the electronic spins in a high magnetic field at low temperature, but achieving the necessary density in samples confined by material walls has not been possible. Evaporative cooling of magnetically trapped spin-polarized hydrogen is a more promising approach. It has already produced temperatures as low as 100 μK (well below the laser cooling limit) at a density of 8×10^{13} cm^{-3}. *At this density, the transition temperature is 30 μK. Reaching the transition by evaporative cooling appears to be possible. Laser techniques are being developed to detect the transition and to study the properties of the resulting gas.*

1 Introduction

The first discussion of atomic hydrogen as a candidate for Bose–Einstein condensation was published by Hecht [1] in 1959. He used the quantum theory of corresponding states to conclude that atomic hydrogen, prevented from recombining into molecules by a strong magnetic field, would remain a gas down to the absolute zero of temperature. He gave the expression for the transition temperature in such a gas in the absence of interactions,

$$
\begin{aligned}
T_c &= \frac{h^2}{2\pi k_B M}\left(\frac{n}{2.612}\right)^{2/3} \\
&= \left(\frac{n}{4.965 \times 10^{20}}\right)^{2/3}, \qquad (1)
\end{aligned}
$$

and pointed out that the gas would be a mixture of two of the hyperfine states of hydrogen. The idea was raised again by Stwalley and Nosanow [2] in 1976. By this time more complete ground state energy calculations had been done confirming the absence of a liquid phase for atomic hydrogen. They pointed out that the presence of the interaction between the atoms could lead to the Bose condensed state being a superfluid as well. The weakness of the interaction and the ability to change the density would allow a detailed comparison between theory and experiment. Although Stwalley and Nosanow saw BEC as the ultimate long term goal of spin-polarized atomic hydrogen research, they also expected that "these systems should yield exciting and fundamental new information about the properties of quantum systems".

The ultimate goal of BEC in hydrogen has yet to be realized, though we seem to be quite close at the current time. On the other hand, since spin-polarized atomic hydrogen was first created in the laboratory by Silvera and Walraven [3], the prediction of exciting results from the "new" quantum system has certainly been fulfilled. Once the atomic hydrogen has been electron spin polarized, it spontaneously develops nuclear polarization [4, 5]. The doubly polarized gas was found to support sharply defined nuclear spin waves [6–8], even though the mean separation between the atoms is much larger than their DeBroglie wave length. The doubly polarized gas, when compressed to higher densities [9, 10], became a model system for the study of three-body recombination processes [11]. Cryogenic atomic hydrogen masers [12–14] offer the the possibility of significant increases in frequency stability over similar room temperature devices [15]. Most recently, the cold hydrogen atoms have been used to study atom–surface interactions, in particular the sticking probability of very cold atoms on well-understood surfaces [16–18]. These aspects of spin-polarized atomic hydrogen have been the subject of two extensive reviews [19, 20]. In this article I will concentrate on the possibilities for attaining BEC. I will review some of the early prospects and explain why they did not come to fruition. I will discuss magnetic trapping and evaporative cooling of the hydrogen, which seems to be the most promising avenue toward BEC at present. And finally, I will examine what the condensed state might look like in the trapped gas, and how it could be studied.

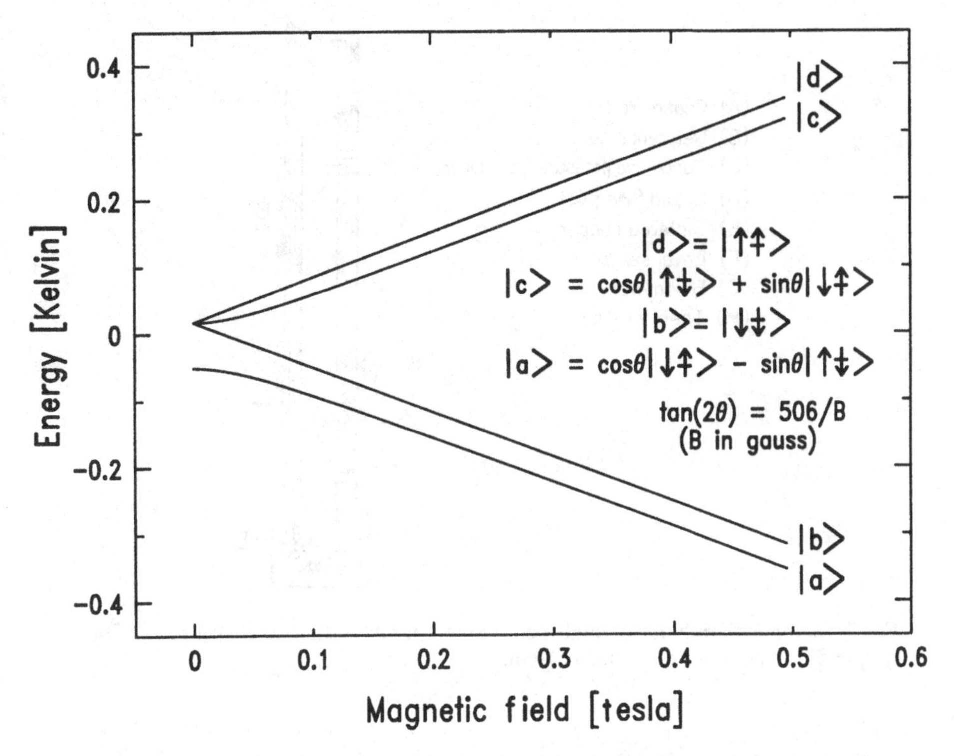

Fig. 1. Hyperfine states of atomic hydrogen. ↑ and ⇞ indicate, respectively, the spin of the electron and the proton. Simple fills with high field seekers are done at fields of the order of 8 T. Magnetic trapping of low field seekers involves fields from 1 T down to 1 mT.

2 Early Approaches

Figure 1 shows the splitting of the ground state of atomic hydrogen due to the hyperfine interaction. The two lowest energy states, labeled **a** and **b**, are called 'high field seekers' since they are drawn into a region of high magnetic field. Similarly, the two highest energy states, **c** and **d**, are 'low field seekers'. All four states are created in roughly equal amounts when molecular hydrogen is dissociated in an RF discharge. In the simplest experiments the atomic hydrogen develops electron polarization when the atoms impinge on a region of high magnetic field (typically the center of a superconducting magnet at about 8 Tesla) and low temperature (typically 300 mK obtained by a dilution refrigerator). Atoms in the **c** and **d** states are excluded from the high field region; those in the **a** and **b**

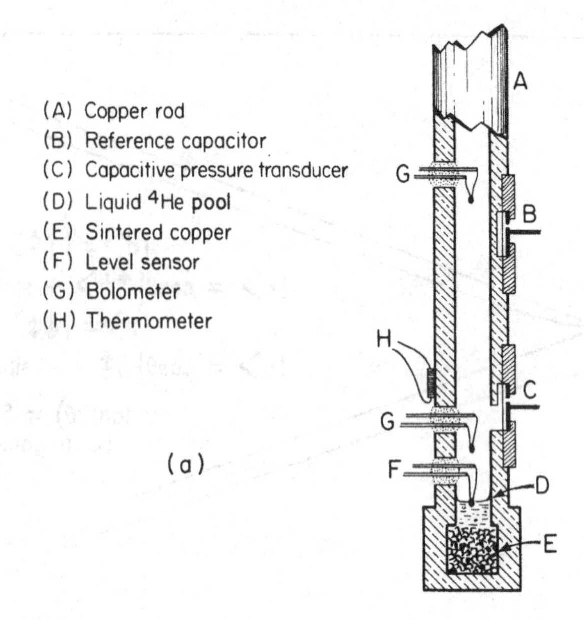

Fig. 2. Sample chambers for high field seekers used by the MIT group. (a) For simple fills. The inner diameter is 1 cm.

states enter, lose their excess energy through wall collisions, and become trapped.

It is unfortunate that Maxwell's equations do not allow a maximum in the magnitude of the magnetic field in a source free region. Otherwise, the high field seekers could be held in a pure magnetic trap. In practice the atoms are usually confined radially and from below by physical walls, as is shown in Fig. 2(a). One could say that the magnetic field acts as a cork in an otherwise open bottle.

The electron-polarized gas of **a** and **b** atoms is not stable against recombination. The finite admixture of the 'wrong' direction of electron spin in the **a** state due to hyperfine mixing allows the reactions $\mathbf{a} + \mathbf{b} \rightarrow$ ortho H_2 and $\mathbf{a} + \mathbf{a} \rightarrow$ para H_2. The recombination of two hydrogen atoms releases 4.476 eV of energy. Initially most of this energy goes into rotation and vibration of the molecule, leaving less than 1% for translational motion [11]. Since a two-body collision cannot conserve momentum while simultaneously releasing kinetic energy, these reactions require the presence of a third body. The third body could be another hydrogen atom, but at the densities achieved in simple

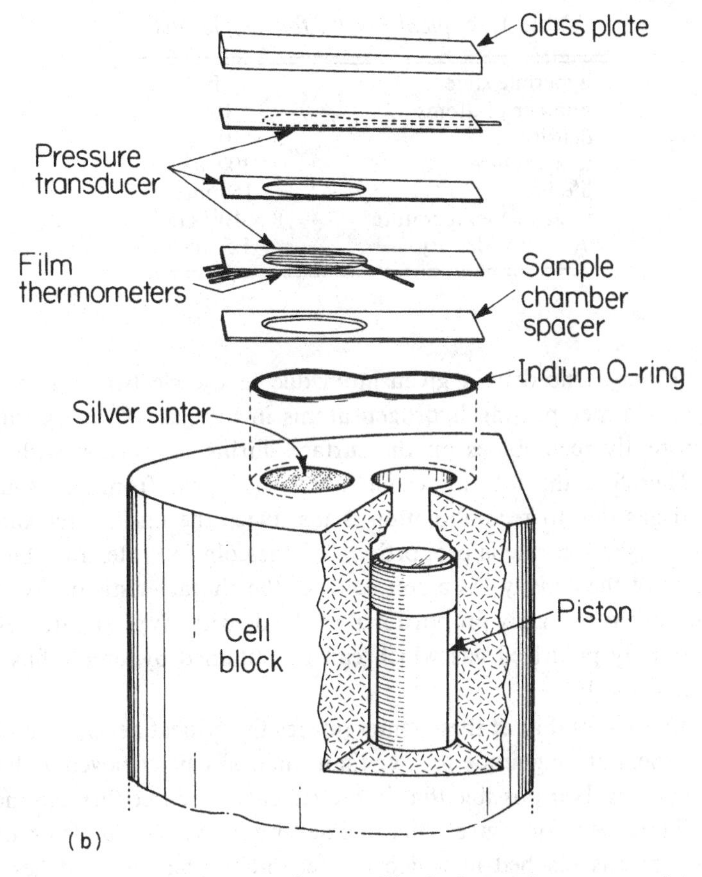

Fig. 2. (b) For compression experiments. The hydrogen is compressed into the region above the sintered silver, 90 μm high and 0.8 cm in diameter.

fill experiments the reaction takes place while the atoms are adsorbed on the walls of the cell with the wall acting as the third body. Recombination is slowed, but not eliminated, by coating the walls of the cell with superfluid helium to minimize the van der Waals attraction between the atoms and the surface. The result of these reactions is to remove all of the **a** atoms while leaving a finite fraction of the **b** atoms. The resulting gas of **b** atoms is nuclear as well as electron-spin-polarized.

Although the gas of pure **b** atoms is stable against recombination, nuclear relaxation can occur during collisions [4], causing a transition to the **a** state. The nuclear relaxation is driven by the fluctuating mag-

Table 1. *Typical results for simple fills*

hyperfine state	**b**
number of atoms	10^{17}
density	3×10^{16} cm^{-3}
temperature	300 *m*K
lifetime	150 min
n_c at this temperature	8×10^{19} cm^{-3}
T_c at this density	1.5 mK

netic field experienced by a given atom due to the electronic magnetic moments of other passing hydrogen atoms in the gas. The resulting **a** atom promptly recombines on the surface during a collision with a **b** atom. Therefore the rate at which atoms disappear from the doubly polarized gas due to recombination is just twice the nuclear relaxation rate. The relaxation rate is proportional to the collision rate, and thus to the square of the density. As a consequence the characteristic decay time of the sample is inversely proportional to the density. Typical parameters for the doubly polarized gas which can be obtained by simple fills are shown in Table 1.

The table shows that at these temperatures the critical density for BEC is three orders of magnitude greater than can easily be achieved with the current sources. Is it possible that improvements in source flux can make up the difference? For a given flux, the density in the cell builds up until a steady state is reached in which the recombination rate matches the flux. Since the recombination rate is proportional to the square of the density, the steady state density varies as the square root of the flux. To increase the density by three orders of magnitude, the flux would have to be increased by a factor of a million. The heat load on the refrigerator would increase by a similar factor. This is not a feasible approach to attaining the phase transition.

Several groups tried to reach the transition by compressing the doubly polarized gas. The MIT group used a piston [9] (see Fig. 2(b)). Other groups compressed the hydrogen inside a bubble [10, 21]. At the critical density of 8×10^{19} cm^{-3} the decay time for the sample due to nuclear relaxation in the bulk gas would be about three seconds. Careful studies, however, showed that the compressed gas recombined more rapidly with increasing density than was expected from relaxation. At high enough densities, the samples simply ex-

Table 2. *Results obtainable by compression*

hyperfine state	**b**
number of atoms	2×10^{16}
density	4.5×10^{18} cm^{-3}
temperature	550 mK
lifetime	5 sec
n_c at this temperature	2×10^{20} cm^{-3}
T_c at this density	44 mK

ploded. It was determined that the recombination was caused by a three-body process in the gas, predicted by Kagan, Vartanyantz and Shlyapnikov [11], which takes place even when the three atoms are all in the **b** state. Typical results that can be obtained by compression are given in Table 2.

The Bose–Einstein transition occurs when the thermal DeBroglie wavelength of the atoms becomes comparable to their mean separation. The compression experiments were an attempt to approach this condition by decreasing the separation at fixed temperature. Another approach would be to increase the DeBroglie wavelength by cooling the atoms at fixed density. This approach fails in the presence of surfaces due to three-body recombination. As the temperature is lowered, the density of atoms adsorbed on the surface increases to the point where the three-body rate *on the surface* becomes prohibitive.

Two other possibilities have been suggested for approaching BEC with high field seeking states in the presence of surfaces. Theories suggest that very high magnetic fields will decrease the three-body recombination rate [11, 22]. However, experiments done at fields as high as 20 T showed that the three-body rate had not yet fallen below its value at 8 T [23]. Kagan and Shlyapnikov [24] have suggested a novel experiment in which a strongly inhomogeneous field is used to compress the spin-polarized hydrogen against a small region of the surface in an otherwise open geometry. The hope is that the heat of recombination can be dissipated in a region of the cell well removed from the small, high density cloud, while the lost atoms are replenished from the low density gas filling the greater part of the cell. This experiment has been tried [25], but the problem of sample heating has not yet been resolved.

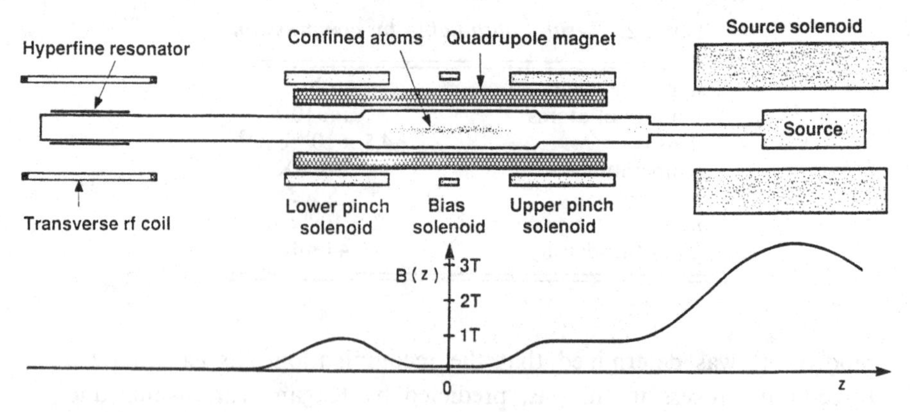

Fig. 3. Magnetic trapping and cooling geometry proposed by Hess.

3 Magnetic Trapping and Evaporative Cooling

The obstacles to BEC caused by three-body recombination inspired Harald Hess to propose an alternative approach [26]. He recognized that magnetic field configurations are possible which possess a local minimum in the magnitude of the field and could be used to trap the low field seeking states of hydrogen, **c** and **d**. In addition, he pointed out that it would be possible to cool the trapped atoms evaporatively to temperatures well below that of the surrounding cell walls. The one disadvantage of working with the low field seekers is that they are not as stable against relaxation due to collisions as are the high field seekers. The number of low field seekers per unit time converted to high field seeking states by electronic relaxation is proportional to the square of the gas density. The relaxation time at a density of 10^{14} cm^{-3} is about 10 seconds. Thus experiments with magnetically trapped low field seekers will typically be limited to modest densities and will require cooling below 50 microkelvin in order to observe BEC.

The configuration originally proposed by Hess is shown in Fig. 3. Radial magnetic confinement of the atoms is accomplished with a cylindrical quadrupolar field. The magnitude of this field is zero on the axis and increases linearly with increasing radius. Axial confinement is provided by a solenoid at either end of the trap. If the field on the axis were allowed to remain at zero, atoms passing through the zero-field region could undergo non-adiabatic transitions to the lower hyperfine states. To avoid this, a small uniform bias field is applied by a solenoid about

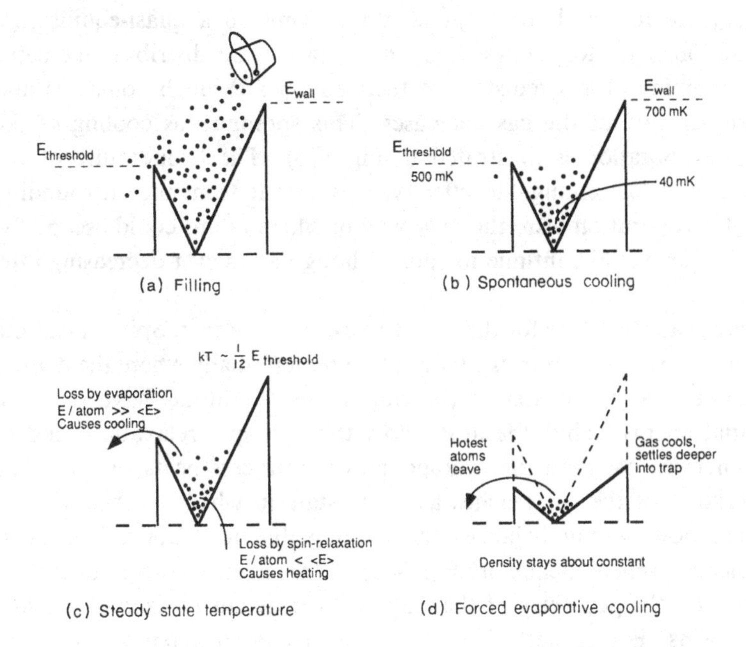

Fig. 4. Stages in the trapping and cooling of atomic hydrogen.

the trap center. The atomic hydrogen source is located in a high field solenoid.

Figure 4 illustrates the trapping and cooling process. In Fig. 4(a) two important energies are introduced. The potential energy difference between the center of the trap and the wall of the cell is represented by E_{wall}. The radial trapping fields are proportional to E_{wall}. In practice the lowest barrier to escape from the trap, $E_{threshold}$, occurs on axis at the lower pinch solenoid. $E_{threshold}$ governs the evaporative cooling process. Atoms which manage to reach a helium covered surface, either by going radially out to the wall or (much more likely) by escaping over the threshold and entering the bottom part of the cell, can be adsorbed and recombine into molecules.

Collisions between the atoms are essential to the loading of the trap. Otherwise, atoms from the source would accelerate into the trapping region, move across unimpeded, ride up the far side of the potential and leave the trap over $E_{threshold}$. Radially, the atoms would bounce back and forth between the walls. It is only through atom–atom collisions that a fraction of the atoms lose enough energy to become trapped near the center of the cell. Soon all atoms with energies greater than

$E_{\text{threshold}}$ are lost and the trapped atoms come to a quasi-equilibrium distribution. The atoms in the high energy tail of this distribution escape. The remaining atoms redistribute their energies through collisions and the temperature of the gas decreases. This spontaneous cooling of the gas by evaporation is illustrated in Fig. 4(b). The temperature of the gas becomes low enough to effectively isolate it from the surrounding walls. If evaporation were the only way in which atoms could escape the trap, the gas would continue to cool, although at an ever decreasing rate, forever!

There is another way for the atoms to escape, however. Spin relaxation, since it is a two-body process, takes place preferentially where the density is highest, near the center of the trap where the atoms have the least potential energy. Thus the atoms lost through spin relaxation tend to have energies less than the average, and the process heats the gas. The temperature of the gas reaches a steady state in which the heating due to relaxation exactly balances the cooling due to evaporation. This is indicated schematically in Fig. 4(c). The steady state temperature depends on the geometry of the trap and the details of each of the loss mechanisms. For an early MIT trap, the steady state temperature was about one-twelfth of $E_{\text{threshold}}$ over a wide range of operating conditions.

Since the steady state temperature is proportional to $E_{\text{threshold}}$, the gas can be cooled further by slowly decreasing $E_{\text{threshold}}$. The rate at which $E_{\text{threshold}}$ is changed must be slow compared to the equilibration rate of the atoms within the trap. Although the total number of atoms decreases in this process, the density need not. In fact, if $E_{\text{threshold}}$ (the lower pinch field) were decreased without changing E_{wall} (the quadrupole field), the cooled atoms would fall deeper into the trap and the density would actually increase! Since a higher density means increased electronic relaxation, there is an advantage in keeping the density roughly constant during the forced evaporative cooling. This can be done by decreasing E_{wall} simultaneously with $E_{\text{threshold}}$. This is the situation depicted in Fig. 4(d). By controlling the two energies separately, one can change the temperature and the density independently during the cooling process.

The trap shown in Fig. 3 was built at MIT and cooled atoms to less than 3 mK [27, 28]. An apparatus similar in design was built in Amsterdam [29]. Experiments done there confined the atoms while keeping them in thermal contact with the wall; they achieved gas temperatures down to 80 mK. A second generation magnetic trap was designed at MIT by John Doyle. The aims were to increase the trapping volume, obtain lower temperatures and provide for optical access to the low temperature

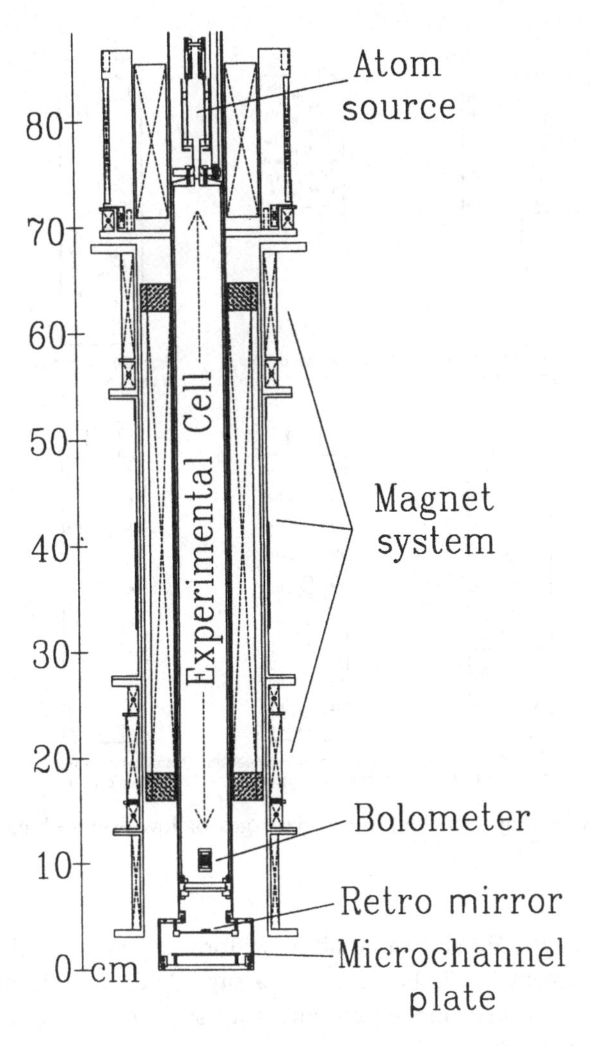

Fig. 5. The magnetic trap in current use at MIT.

gas. A scale drawing of the new trap is shown in Fig. 5. I will discuss its operation and capabilities in some detail since this trap, or one like it, will be used to try to achieve BEC in atomic hydrogen.

The loading of the trap and the subsequent development of electronic and nuclear polarization in the gas can be understood with the help of the simplified sketches in Fig. 6. Molecular hydrogen is condensed as a solid on the walls of the source region during the initial cool down of

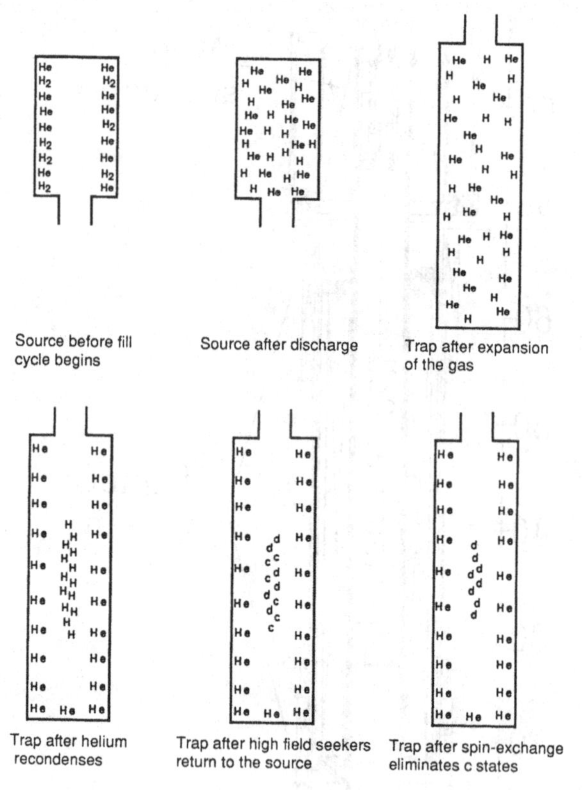

Fig. 6. Steps leading to a doubly polarized gas of low field seeking hydrogen atoms in the MIT trap.

the cell. Enough ^{4}He is then admitted to form a puddle on the bottom of the experimental cell. Because it is a superfluid, the ^{4}He provides a self healing saturated film which covers all surfaces in the source and trapping regions of the cell.

Loading the trap begins with an RF discharge in the source region which evaporates the helium and vaporizes and dissociates some of the hydrogen. The resulting gas cloud expands into the larger trapping region below. The helium aids in the transport of the hydrogen and in its subsequent equilibration to the wall temperature. The helium acts as a buffer which slows the diffusion of the hydrogen to the walls. The wall temperature at this point is kept high enough (about 250 mK) that hydrogen atoms which happen to stick to the surface equilibrate and desorb before they can recombine. Since the helium is unaffected by the trapping potential, helium–hydrogen collisions bring the hydrogen gas *in the trap* to the wall temperature, which is below the 800 mK trap depth. The helium then quickly recondenses on the walls. The pressure in the gas puff is sufficient to drive even the high field seekers out of the high magnetic field present in the source region. After the helium recondenses, electronic polarization develops as the **a** and **b** state atoms are drawn back into the source. Less than a second has elapsed since the discharge was turned off.

Nuclear polarization develops in the trapped gas due to spin-exchange collisions: $\mathbf{c} + \mathbf{c} \rightarrow \mathbf{b} + \mathbf{d}$. All the **c** atoms are eliminated by this process; the resulting **b** atoms are ejected from the trap. After about three seconds the only atoms left in the trap are in the **d** state. The wall temperature is then lowered to less than 80 mK. Now any atom that escapes from the trap is permanently adsorbed on the surface and cannot return to heat the gas. Thus the **d** atoms are thermally isolated. Atoms promoted by collisions to energies high enough to reach the walls will escape over $E_{\text{threshold}}$. The vapor pressure of the helium is too low for helium-hydrogen collisions to be important. Similarly, thermal radiation from the walls has a negligible effect. The **d** atoms are left to decay slowly due to electronic spin relaxation, the dominant process being $\mathbf{d} + \mathbf{d} \rightarrow \mathbf{a} + \mathbf{a}$.

In the above method of trap loading using a single short RF discharge, the various physical processes occur sequentially. Other methods using different patterns of discharges have been tried in order to maximize the final **d** state density. In particular if the discharge is pulsed continuously, all the above processes occur simultaneously and the ultimate density of **d** atoms is determined by a balance between the flux of **d** atoms from the source and spin relaxation taking place either in the trap or on the surface. Table 3 shows typical conditions in the trapped gas soon after the trap has been filled but before forced evaporative cooling has commenced.

Determining the properties of the trapped gas poses a particular problem since the introduction of a solid probe, such as a thermometer or a pressure transducer, would cause the atoms to be adsorbed and recombine. Optical probes that avoid this problem are being developed and will be discussed below. However, almost all measurements to date have required that the atoms be released from the trap. The hyperfine states in the gas are determined by zero-field hyperfine resonance in the gas after it flows into a field free region [27] (see Fig. 3). The total number of atoms can be determined from the strength of

Table 3. *Magnetic trap after filling*

hyperfine state	**d**
number of atoms	2×10^{14}
density on axis	2×10^{13} cm^{-3}
temperature	50 mK
lifetime	200 sec
n_c at this temperature	6×10^{18} cm^{-3}
T_c at this density	12 μK

the magnetic resonance signal [27, 28] or, more easily, from the energy liberated when the released atoms recombine [30]. After nuclear polarization has been established, the time evolution of the total number of atoms in the trap, $N(t)$, is governed by spin relaxation and depends on the trap shape, the electronic spin relaxation rate constant g, and the temperature of the gas. Studies of $N(t)$ for isolated atoms [27] (where T was estimated) and for atoms in weak thermal contact with the walls [29] (where T was known) confirmed the values of g calculated theoretically [31, 32]. Once g was established, the temperature of the gas and its density could be determined indirectly from an analysis of $N(t)$ [28].

A more direct method of measuring the gas temperature, developed by Doyle [30], is illustrated in Fig. 7. A highly sensitive bolometer is used to measure the flux of atoms from the trap as $E_{\text{threshold}}$ is ramped down to zero at a uniform rate. If the time necessary to drop the field is short compared to the thermal equilibration time in the gas, the flux versus time gives the number density versus energy, $n(E)$. The energy distribution $n(E)$ is important in itself; for example it can be used to study the approach to equilibrium during and after evaporative cooling. The temperature of the gas is determined by comparing $n(E)$ to the calculated distribution expected for the known magnetic field profile. Figure 8 shows $n(E)$ measured after different degrees of evaporative cooling in the MIT trap [17]. The lowest temperature achieved in that experiment, 100 μK, is well below the recoil limit for Lyman-α laser cooling of hydrogen, $T = (h\nu)^2/2mc^2k_B \approx 650$ μK. Some of the features of this gas are given in Table 4.

Other properties of this coldest hydrogen gas may be of interest. The mean thermal speed, $\bar{v} = \sqrt{8k_BT/\pi M}$, is 145 cm/s. The gravitational height, k_BT/Mg, is 8.4 cm and is beginning to play a role in the vertical

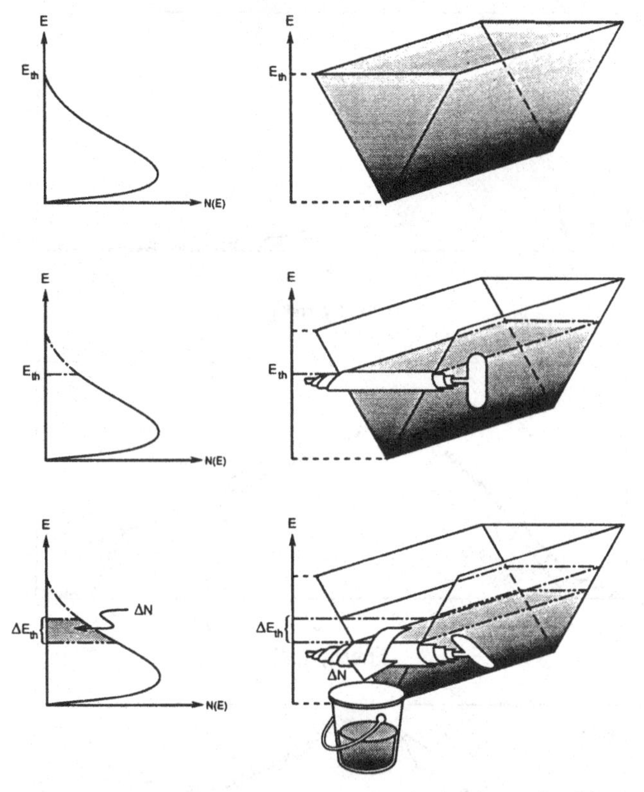

Fig. 7. The energy distribution in the gas, $n(E)$, can be determined by measuring the flux of atoms out of the trap as the threshold for escape is lowered quickly to zero.

distribution of atoms in the trap. The thermal DeBroglie wavelength,

$$\begin{aligned} \Lambda(T) &= \left(\frac{h^2}{2\pi M k_B T}\right)^{1/2} \\ &= 1.739 \times 10^{-7} \frac{1}{\sqrt{T}}, \end{aligned} \tag{2}$$

is 174 nm; it is getting close to the mean separation between atoms, $n^{-1/3}$, which is equal to 232 nm. These properties depend only on the temperature. The mean free path, $\lambda = \sqrt{\pi/8}/n\sigma$, depends on the density as well. The total scattering cross section for this cold gas can be expressed in terms of the s-wave scattering length, a: $\sigma = 4\pi a^2$. The s-wave scattering length for spin-polarized hydrogen is 0.72 Å [33]. Thus

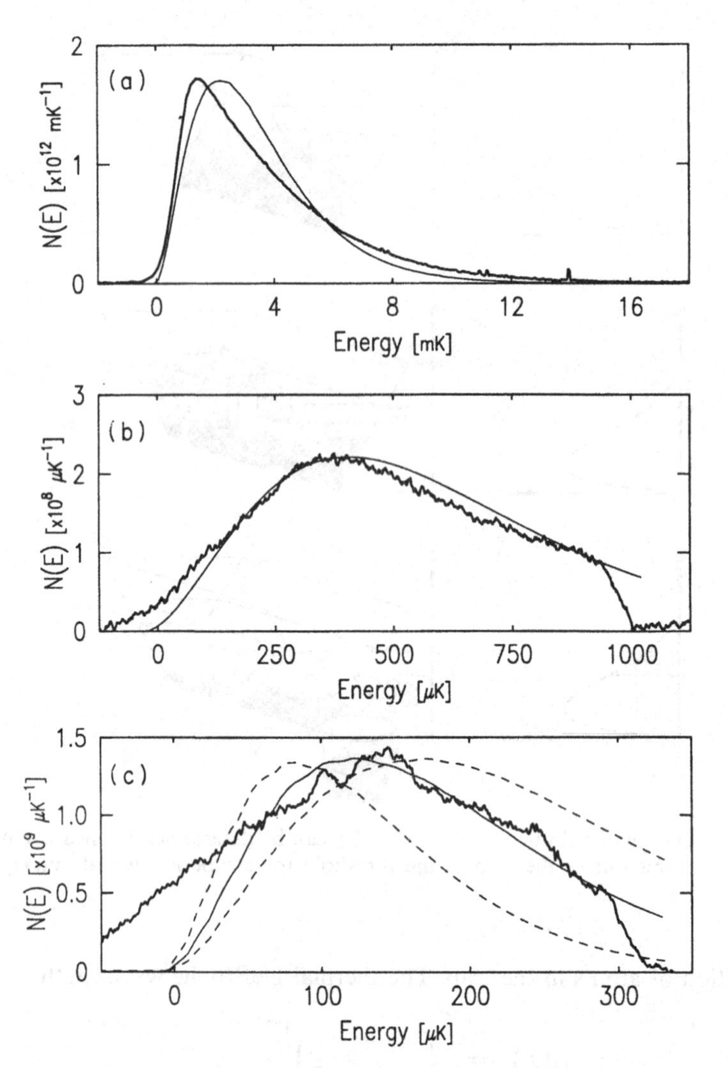

Fig. 8. Energy distributions for evaporatively cooled atomic hydrogen. The solid curves are the calculated distributions for atoms at (a) 1.1 mK, (b) 190 μK, and (c) 100 μK. The dashed curves in (c) are for temperatures 30% higher and lower than the best fit value. The finite atom signals from $E < 0$ are due to a small axial field inhomogeneity (< 1 G) that slightly deforms the bottom of the trap.

Table 4. *Coldest gas attained by evaporation*

hyperfine state	**d**
number of atoms	3×10^{11}
density on axis	8×10^{13} cm^{-3}
temperature	100 μK
lifetime	50 sec
n_c at this temperature	5×10^{14} cm^{-3}
T_c at this density	30 μK

for a uniform gas at a density of 8×10^{13} cm^{-3} the mean free path would be 12 cm and the mean free time, $\lambda/\bar{v}$, would be 83 ms. It is clear, then, that the trapped atoms traverse the confining potential hundreds of times between collisions.

As Table 4 indicates, one would have to cool this gas by only a little more than a factor of three in order to attain BEC. Is this feasible? For evaporative cooling to be effective the thermalization rate for the gas (proportional to the collision rate) must be greater than the spin relaxation rate. Both of these rates are proportional to the density, but the collision rate is also proportional to the square root of the temperature whereas the relaxation rate is independent of temperature for low temperatures and fields. Therefore the efficiency of evaporative cooling decreases as the temperature is lowered [26]. Nevertheless, model calculations [34] indicate that evaporative cooling using the current MIT trap can cross the phase boundary for BEC at densities between about 5×10^{13} and 5×10^{14} cm^{-3}. In order to observe the phase transition, it would be very useful to have a non-destructive means of studying the gas. The following section discusses the expected properties of the hydrogen gas as it undergoes BEC in the trap and a possible unambiguous way of observing the phase transition.

4 Bose Condensation in a Magnetic Trap

Bose–Einstein condensation of hydrogen in a magnetic trap will differ in two fundamental ways from the simple case treated in textbooks. The gas will not be homogeneous since it is confined by a spatially varying potential. Also, the interactions between the atoms, although weak, have a marked effect on the properties of the gas.

BEC in non-uniform potentials has been treated by Bagnato, Pritchard and Kleppner [35]. They find that the density at the onset of BEC is given by the same expression that applies to the homogeneous case,

$$\begin{aligned} n_c(T) &= 2.612\Lambda(T)^{-3} \\ &= 4.965 \times 10^{20} T^{3/2}, \end{aligned} \qquad (3)$$

irrespective of the nature of the trapping potential. For a given trap the condensate will begin to form at the location of the minimum in the potential energy. As more atoms are added at fixed temperature, or as the temperature is lowered at fixed N, the phase boundary expands outward, coinciding with surfaces of constant potential energy. The density is equal to the critical density $n_c(T)$ on the boundary and is greater than the critical density inside the boundary.

The properties of a weakly interacting Bose gas have been covered in detail by Huang [36]. The interaction between two particles is normally described by an interaction potential $V(|\mathbf{r}_1 - \mathbf{r}_2|)$, but it can alternatively be described by the wavevector dependent partial-wave phase shifts, $\delta_l(k)$. These phase shifts determine the form of the scattering cross section. At sufficiently low energies, such as those associated with hydrogen in magnetic traps, only s-wave scattering need be considered. This means that only a single number, the s-wave scattering length a, is necessary to parameterize the interactions between the hydrogen atoms. If, in addition, the density of the gas is low enough that three-body collisions can be neglected ($an^{1/3} \ll 1$, which is certainly true in the trap), then the entire many-body behavior of the system depends solely on a single parameter, a. The value $a = 0.72$ Å calculated for spin-polarized hydrogen by Friend and Etters [33] was based on the precise potential between hydrogen atoms determined by Kolos and Wolniewicz [37]. It is interesting to note, though, that exactly the same value would result from hypothetical hard-sphere atoms of diameter a.

Although the fundamental quantity which determines the properties of the weakly interacting Bose gas is the s-wave scattering length, there are several derived quantities which are useful in developing a physical understanding of the behavior of the gas. The healing length, η, is the characteristic length associated with the many-body wave function and indicates how rapidly the density can change in space [38]

$$\begin{aligned} \eta &= (8\pi an)^{-1/2} \\ &= 2.35 \times 10^3 \frac{1}{\sqrt{n}}. \end{aligned} \qquad (4)$$

The interaction strength V_o has the units of energy times volume

$$\begin{aligned} V_o &= \frac{h^2}{\pi M} a \\ V_o/k_B &= 4.36 \times 10^{-22}\ \text{K cm}^3. \end{aligned} \tag{5}$$

It sets the energy scale for various spatial configurations of the density. In particular, the ground state energy per particle in a uniform gas is given by $E(0)/N = nV_o/2$.

The sound speed in a uniform weakly interacting Bose gas is given by

$$u = \frac{\hbar}{M}(4\pi a n)^{1/2}. \tag{6}$$

This quantity by itself is of little use in understanding the trapped hydrogen. However, it allows one to motivate the characteristic time for changes in the wave-function, τ_c, as being related to the time necessary for a sound wave to propagate over a distance equal to the healing length:

$$\begin{aligned} \tau_c &= \frac{\hbar}{nV_o} = \sqrt{2}\,\frac{\eta}{u} \\ &= \frac{1.75 \times 10^{10}}{n}. \end{aligned} \tag{7}$$

There has long been concern about the time necessary to nucleate the Bose–Einstein phase transition in a low density gas once the critical density or temperature has been reached. Recent theoretical calculations by Stoof [39] and Kagan, Svistunov and Shlyapnikov [40] indicate that the time is short compared to the expected lifetime of the trapped gas. The details of the two theories differ (see the articles by Kagan and Stoof in this book.) However, there seems to be agreement that an observable peaking of the energy density $n(E)$ in a group of states close to $E = 0$ will occur on a timescale of the order of τ_c. For a particle density of $n = 8 \times 10^{13}$ cm^{-3}, the characteristic time is $\tau_c = 0.2$ ms.

Let us consider a specific example of a configuration in which BEC might occur in the trapped hydrogen. By the time the magnetic fields have been reduced to relatively low values during the evaporative cooling process, the potential near the center can be approximated by a parabolic minimum. There is a great deal of freedom in choosing the exact shape of the potential in which the coldest atoms will reside. One possible potential which is both attainable and convenient is

$$U(r,z)/k_B = 1.4 \times 10^{-2} r^2 + 3.0 \times 10^{-5} z^2, \tag{8}$$

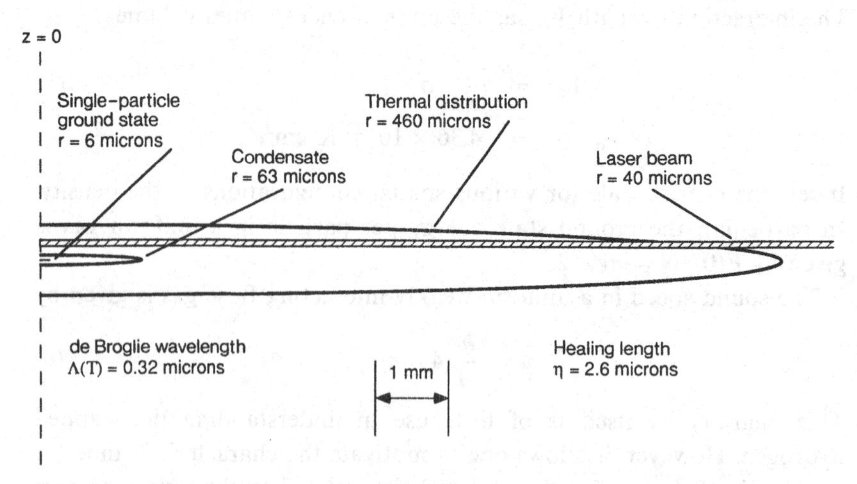

Fig. 9. Spatial profiles in a realistic trap potential. The temperature is 30 μK, and the corresponding critical density is 8×10^{13} cm^{-3}. The phase boundary for BEC has been drawn for a condensate fraction of 2.5%.

where the radial (r) and axial (z) coordinates are expressed in centimeters and U/k_B is expressed in units of degrees Kelvin. In an actual experiment the trap shape, temperature and number of atoms will all be changing simultaneously. I will consider here a simpler hypothetical situation where the trap shape and the temperature are held fixed and the total number of atoms in the trap is slowly increased. I will take the temperature to be 30 μK. At this temperature the critical density is $n_c(T) = 8 \times 10^{13}$ cm^{-3}.

When N is so small that the density at the center of the trap is much less than $n_c(T)$, the density profile is proportional to e^{-U/k_BT}, as in the case of a classical gas. The surface on which the density has fallen to e^{-1} of its maximum value is ellipsoid of revolution extending to $r = 460$ μm at $z = 0$ and $z = \pm 10\,000$ μm on axis (22 times longer than it is wide). This is shown in Fig. 9. As N increases the density distribution develops more weight at the low values of U than would be expected from the classical form. Critical density is reached at the center of the trap when $N = 4.32 \times 10^{11}$. Consider the various distance scales. The thermal DeBroglie wavelength $\Lambda(T) = 0.32$ μm at this temperature. The healing length $\eta = 2.6$ μm at this density. If there were no interactions between the particles, the atoms would begin to condense into the single particle ground state associated with this harmonic potential. That wave function would have a radial extent of about 6 μm, too narrow to be resolved in Fig. 9. Because of the repulsive interactions, the actual condensate is spread over a wider region. For example, when N reaches 4.55×10^{11} the fraction of atoms in the condensate is 2.5% and the phase boundary extends out to $r = 63\ \mu$m.

Once a condensed state has formed in the gas, the density at the center of the trap increases rapidly above $n_c(T)$ as more atoms are added. Since the atom loss rate due to electronic relaxation is proportional to the square of the density, the gas should begin to decay more rapidly once BEC has been established. Hijmans *et al.* [41] have suggested that the density at the trap center builds up so quickly with increasing condensate fraction that the resulting rapid relaxation and subsequent recombination can be likened to an 'explosion' (see the contribution by Hijmans et al. in this book.) In order to study this possibility for the specific trap discussed above it is necessary to compute the complete density distribution, $n(\mathbf{r})$.

A simple thermodynamic approach to calculating $n(\mathbf{r})$ has been introduced by Oliva [42]. In this approach the chemical potential μ is taken to be constant throughout the system. At each point μ is assumed to be the sum of the 'internal chemical potential' of the gas, $\mu_o(n(\mathbf{r}))$, and the potential energy at that location:

$$\mu = \mu_o(n(\mathbf{r})) + U(\mathbf{r}). \tag{9}$$

The internal chemical potential is simply the chemical potential of a homogeneous gas of the same density in the absence of an applied potential. For the weakly interacting Bose gas†

$$\begin{aligned} \mu_o(n(\mathbf{r})) &= V_o(n + n_c(T)), \quad n \geq n_c(T), && (10) \\ &= 2nV_o + k_B T \ln z, \quad n < n_c(T). && (11) \end{aligned}$$

In the normal state the expression for $\mu_o(n(\mathbf{r}))$ involves the fugacity, $z \equiv e^{\mu_o/k_B T}$, which must be determined from the relation

$$g_{3/2}(z) = 2.612 \frac{n}{n_c(T)}, \tag{12}$$

where $g_{3/2}(z)$ is the familiar function from the theory of the ideal Bose gas. To find $n(\mathbf{r})$ one first chooses a value for μ. Then at each point in the gas μ_o is found from Eq. (9) and either Eq. (11) or Eq. (12) is inverted to find the density at that point. Finally, one integrates $n(\mathbf{r})$ over the trap to find the total number of atoms N corresponding to that value of μ. Since N is a monotonically increasing function of μ, one can watch $n(\mathbf{r})$ evolve

† Expressions for most of the thermodynamic variables in a weakly interacting Bose gas are given in the review by Greytak and Kleppner [19]. The chemical potential is most easily found from the fact that it is the Gibbs function per particle: $\mu_o = G/N = (E - TS + PV)/N$.

as the trap is filled by systematically increasing the values of μ used in the procedure. This 'local density approximation' approach to finding $n(\mathbf{r})$ should be accurate except in those regions where $n(\mathbf{r})$ is changing on a scale comparable to the healing length.

The expression for the internal chemical potential in the condensed state, Eq. (11), can easily be inverted and used with Eq. (9) to give an analytic expression for the density:

$$n(\mathbf{r}) = n(0) - U(\mathbf{r})/V_o \qquad n \geq n_c(T). \tag{13}$$

This expression assumes $U(\mathbf{r})$ has its minimum value of 0 at $\mathbf{r} = 0$. The density at the origin, $n(0)$, is related to the overall chemical potential of the gas: $\mu = V_o(n(0) + n_c(T))$. This is an attractively simple result. The density inside the phase boundary is directly proportional (with a negative sign) to $U(\mathbf{r})$; the density profile will be an inverted replica of the trapping potential.

Figure 10 demonstrates how rapidly the density increases near the center of the trap as additional atoms are added, once the critical density has been exceeded. The isothermal compressibility increases discontinuously by a factor of two upon entering the condensed state, causing a similar increase in the gradient of the density at the phase boundary [19]. This factor of two change in the slope has been observed in earlier density profile calculations [43–45] and is reproduced in the local density approximation approach [42]. However, at the low gas densities encountered in traps, the slope in the normal state flattens so rapidly just above the transition that the *apparent* slope change upon entering the transition region is much larger than a factor of two, as can be seen in the Fig. 10.

Figure 11 shows how the condensate fraction, density at the origin and decay time vary as more atoms are added to the trap. For a homogeneous Bose gas the fraction of atoms in the condensate is $1 - n_c(T)/n$. To determine the number of 'condensate atoms' in the trap this fraction is multiplied by the density and integrated over the condensed phase. The condensate fraction rises rapidly with increasing N. The density at the origin is a particularly steep function of the condensate fraction. Fortunately very little of the volume of the trap is associated with the region near the origin so the effect on the decay time is mitigated. The decay rate of the gas was calculated simply by integrating gn^2 over the trap, neglecting any possible changes in the dipolar rate constant g in the condensed state, or possible coherence effects due to long range correlations. The resulting decrease in the lifetime of the gas

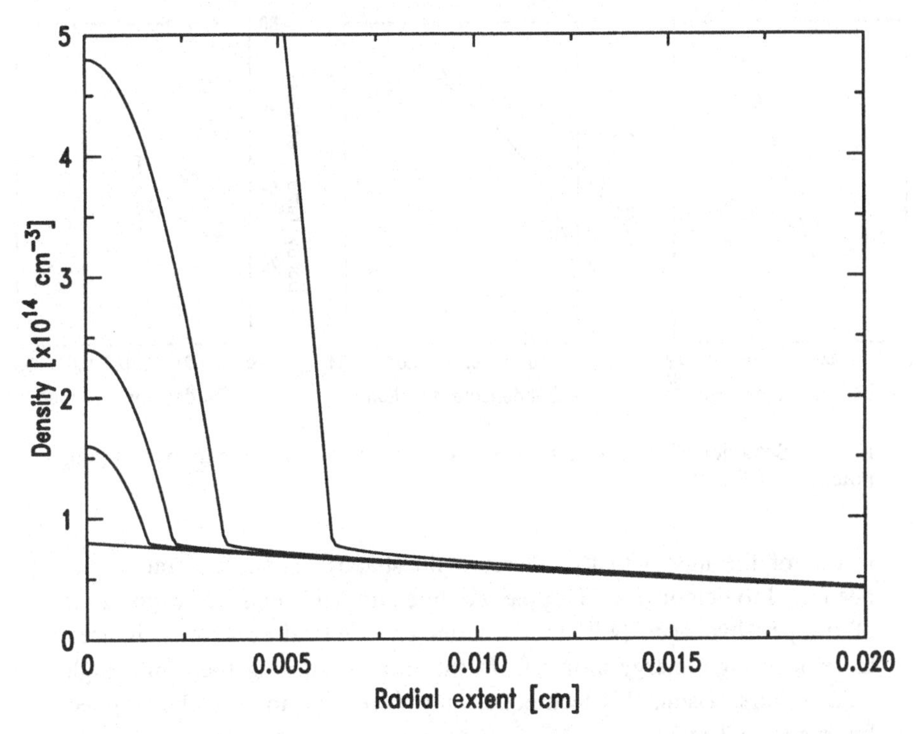

Fig. 10. Density profiles showing the formation of the condensed phase in the trap of Fig. 9. The lowest curve is the density just before condensation begins. The other curves correspond to condensate fractions of 2.6×10^{-5}, 1.5×10^{-4}, 1.5×10^{-3} and 2.5×10^{-2}.

would certainly be observable, but may not prove to be an experimental obstacle, at least up to condensate fractions of a few percent.

The detection of BEC in a magnetic trap will most likely require the development of new experimental probes. Optical techniques offer the most promise. The Amsterdam group has developed a pulsed Lyman-α source which can be used to measure and manipulate the trapped hydrogen *in situ* through single-photon absorption. They have used 1S–2P Lyman-α transmission spectroscopy to obtain information on the temperature and density of the gas [46]. The absorption lines are broadened by the Zeeman effect which depends on the magnetic field at the atom. Thus the lineshape depends on $n(r)$ through the field profile $B(r)$. A model was made of the lineshape that includes T and the density on axis, n_o. These two parameters were then determined by fitting the

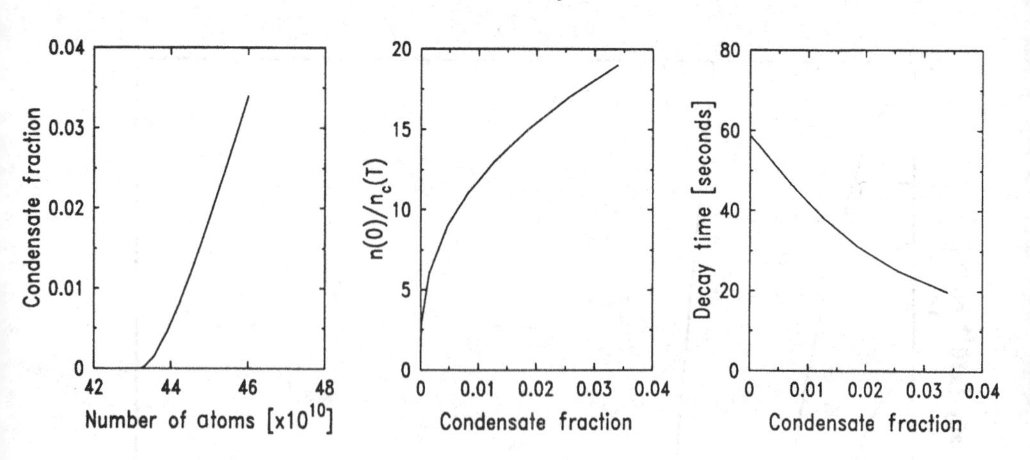

Fig. 11. Behavior of critical parameters during Bose condensation in the trapping potential of Fig. 9.

results of the model to the observed lineshapes. The same source has been used to demonstrate Doppler cooling and light-induced evaporation of trapped hydrogen [47]. In the latter case optical absorption is used to remove high energy atoms from the trap by turning them into high field seekers. Using this technique, they were able to cool the trapped hydrogen to 3 mK.

Jon Sandberg at MIT developed a different optical probe, based on two-photon absorption. The process is illustrated in Fig. 12. A continuous source of radiation at one half the 1S–2S frequency is focused to a thin line in the trapped gas (see also Fig. 9). The light is reflected back along the same path by a mirror located at low temperature. Some atoms undergo a two-photon transition from the 1S to the 2S state. The 2S state is metastable, so a small electric field is applied to mix the 2S and the 2P states. The atoms then decay rapidly from the 2P state by emitting a Lyman-α photon. The number of Lyman-α photons detected by a micro-channel plate at low temperature is directly proportional to the number of hydrogen atoms along the light beam path. The wavelength dependence of the detector sensitivity strongly discriminates against light from the source itself. To further decrease the sensitivity to the illuminating radiation, the source can be chopped and the Stark mixing of the 2S and 2P states delayed until the 'off' interval.

One way of using the two-photon absorption to study the gas is to map out the radial density profile by translating the beam across the gas parallel to the axis (in practice, it may be simpler to move the gas across

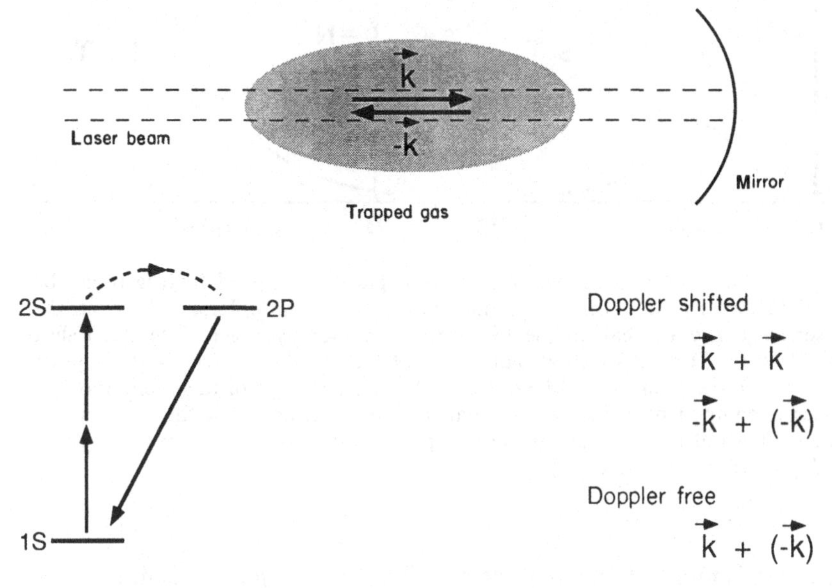

Fig. 12. Two-photon absorption in a cold hydrogen gas.

the beam). If the trap potential is known, the radial extent of $n(r)$ gives a measure of the temperature. BEC could be identified by the sudden appearance of a sharp peak in the density at the center of the trap, as in Fig. 10.

A more direct signature of the phase transition can be obtained by measuring the spectrum of the absorption as the frequency of the source is scanned about the atomic frequency. Figure 12 shows that there are two types of processes which contribute to the two-photon absorption. If one photon is taken from each of the two counter propagating waves, there will be no first order Doppler shift associated with the atom's motion. This process gives rise to an extremely narrow absorption line (of the order of 3 KHz wide due primarily to the time of flight of an atom across the beam [48]), centered at half the atomic frequency $v_o/2$, with a strength proportional to the total number of atoms illuminated. On the other hand, if the two absorbed photons were traveling in the same direction, there would be a Doppler shift of the absorption frequency

$$v - v_o/2 = \frac{\hbar k^2}{2\pi M} + \frac{kp\cos\theta}{2\pi M}, \tag{14}$$

where k is the wavevector of the photon and $p\cos\theta$ is the component of

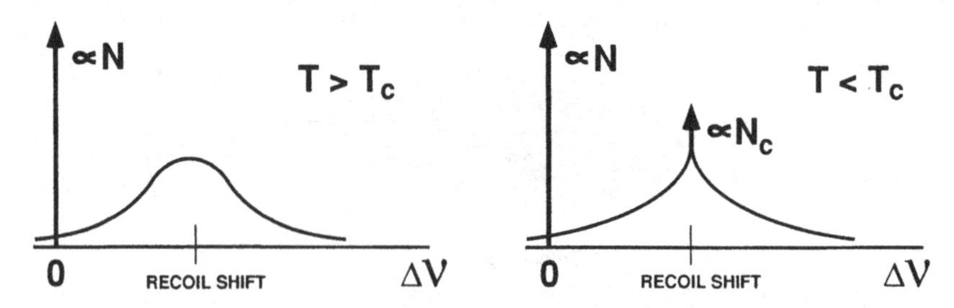

Fig. 13. The spectrum of two-photon absorption in trapped hydrogen can be used to identify Bose–Einstein condensation. $\Delta\nu$ is the difference between the laser frequency and half of the 1S–2S frequency for hydrogen. The recoil shift is 6.7 MHz. The full width at half height of the low density (classical) Doppler part of the spectrum is 5.7 MHz at $T = 30\ \mu$K. The ratio of the strength of the discrete component in the Doppler part of the spectrum below the transition to the strength of the Doppler free part is proportional to the condensate fraction. Adapted from Sandberg [48].

the atom's momentum along the axis. The familiar p dependent Doppler term (which mirrors the probability distribution for a given component of the atom's momentum) is displaced to higher frequencies by a recoil term. In most gases the recoil shift is negligible compared to the width of the Doppler spectrum; but in trapped hydrogen, because of the light mass and low temperature, the width and the shift are comparable. This is shown schematically in Fig. 13.

By sweeping the frequency of the source one can measure the momentum distribution and, of course, deduce the temperature. The signal will be quite weak compared to the Doppler free case, however, since this part of the absorption spectrum is much broader than the line width of the source. The situation changes below the transition when a finite fraction of the atoms come almost to rest. In this case a sharp feature will appear in the absorption centered at the recoil shift with a strength proportional to the condensate fraction (see Fig. 13). Kagan, Svistunov and Shlyapnikov [40] estimate that the states which develop high occupation below the transition are within a range $\Delta E \sim nV_0$ of $E = 0$. This would give rise to a frequency spread of the order of one KHz in the recoil shifted condensate line, which is less than the time of flight broadening expected for the Doppler free component. These considerations have led Sandberg [48] to suggest that measuring the two-photon absorption spectrum may be the simplest and least ambiguous way of detecting BEC in trapped hydrogen.

The continuous wave laser source for the two-photon absorption has been completed. It produces 20 mW of tunable light at $\lambda = 243$ nm with a spectral width of about 2 KHz. Work is now underway to see the absorption signal in the cold hydrogen.

Acknowledgments. The research on spin-polarized hydrogen at MIT has been, since its inception, a collaborative effort between the author (a condensed matter physicist) and Daniel Kleppner (an atomic physicist). Many talented graduate students and post doctoral fellows have contributed to the success of the work, as can be seen by the multitude of names on our papers. Two individuals who deserve special recognition for the most recent advances are John Doyle, who was responsible for the second generation magnetic trap and Jon Sandberg, who built up the entire UV laser system starting with an empty room. Our current support is provided by the National Science Foundation under Grant No. DMR-91-19426 and the Air Force Office of Scientific Research under Grant No. F49620-92-J-0356.

References

[1] C. E. Hecht, Physica **25**, 1159 (1959).

[2] W. C. Stwalley and L. H. Nosanow, Phys. Rev. Lett. **36**, 910 (1976).

[3] I. F. Silvera and J. T. M. Walraven, Phys. Rev. Lett. **44**, 164 (1980).

[4] B. W. Statt and A. J. Berlinsky, Phys. Rev. Lett. **45**, 2105, (1980).

[5] R. W. Cline, T. J. Greytak and D. Kleppner, Phys. Rev. Lett. **47**, 1195 (1981).

[6] E. P. Bashkin, Pis'ma Zh. Eksp. Teor. Fiz. **33** , 11 (1981) [Soviet Physics JETP Lett. **33**, 8 (1981)].

[7] C. Lhuillier and F. Laloë, J. Phys. (Paris) **43**, 197, 225 (1982).

[8] B. R. Johnson, J. S. Denker, N. Bigelow, L. P. Levy, J. H. Freed and D. M. Lee, Phys. Rev. Lett. **52**, 1508 (1984).

[9] D. A. Bell, H. F. Hess, G. P. Kochanski, S. Buchman, L. Pollack, Y. M. Xiao, D. Kleppner and T. J. Greytak, Phys. Rev. B **34**, 7670 (1986).

[10] R. Sprik, J. T. M. Walraven and I. F. Silvera, Phys. Rev. B **32**, 5668 (1985).

[11] Yu. Kagan, I. A. Vartanyantz and G. V. Shlyapnikov, Sov. Phys. JETP **54**, 590 (1981).

[12] H. F. Hess, G. P. Kochanski, J. M. Doyle, T. J. Greytak and D. Kleppner, Phys. Rev. A **34** ,1602 (1986).

[13] M. D. Hürlimann, W. N. Hardy, A. J. Berlinsky and R. W. Cline, Phys. Rev. A **34**, 1605 (1986).

[14] R. L. Walsworth Jr, I. F. Silvera, H. P. Godfried, C. C. Agosta, R. F. C. Vessot and E. M. Mattison, Phys. Rev. A **34**, 2550, (1986).

[15] A. J. Berlinsky and W. N. Hardy, 13th Annual Precision Time and Time Interval Applications and Planning Meeting, 1981, NASA Conference Publication 220, p. 547.

[16] J. J. Berkhout, E. J. Wolters, R. van Roijen and J. T. M. Walraven, Phys. Rev. Lett. **57**, 2387 (1986).

[17] J. M. Doyle, J. C. Sandberg, I. A. Yu, C. L. Cesar, D. Kleppner and T. J. Greytak, Phys. Rev. Lett. **67**, 603 (1991).

[18] I. A. Yu, J. M. Doyle, J. C. Sandberg, C. L. Cesar, D. Kleppner and T. J. Greytak, Phys. Rev. Lett. **71**, 1589 (1993).

[19] T. J. Greytak and D. Kleppner, in *New Trends in Atomic Physics*, G. Grynberg and R. Stora, eds. (North-Holland, Amsterdam, 1984).

[20] I. F. Silvera and J. T. M. Walraven, Spin polarized atomic hydrogen, in *Progress in Low Temperature Physics*, D.F. Brewer, ed. (North-Holland, Amsterdam, 1986), Vol. X, p. 139.

[21] T. Tommila, E. Tjukanov, M. Krusius and S. Jaakkola, Phys. Rev. B **36**, 6837 (1987).

[22] H. T. C. Stoof, L. P. H. de Goey, B. J. Verhaar and W. Glockle, Phys. Rev. B **38**, 11221 (1988).

[23] J. D. Gillaspy, I. F. Silvera and J. S. Brooks, Phys. Rev. B **40**, 210 (1989).

[24] Yu. Kagan and G. V. Shlyapnikov, Phys. Lett. A **130**, 483, (1988).

[25] E. Tjukanov, A. Ya. Katunin, A. I. Safonov, P. Arvela, M. Karhunen, B. G. Lazarev, G. V. Shlyapnikov, I. I. Lukashevich and S. Jaakkola, Physica B **178**, 129 (1992).

[26] H. F. Hess, Phys. Rev. B **34**, 3476 (1986).

[27] H. F. Hess, G. P. Kochanski, J. M. Doyle, N. Masuhara, D. Kleppner and T. J. Greytak, Phys. Rev. Lett. **59**, 672 (1987).

[28] N. Masuhara, J. M. Doyle, J. Sandberg, D. Kleppner, T. J. Greytak, H. F. Hess and G. P. Kochanski, Phys. Rev. Lett. **61**, 935 (1988).

[29] R. van Roijen, J. J. Berkhout, S. Jaakkola and J. T. M. Walraven, Phys. Rev. Lett. **61**, 931 (1988).

[30] J. M. Doyle, J. C. Sandberg, N. Masuhara, I. A. Yu, D. Kleppner and T. J. Greytak, J. Opt. Soc. Am. B **6**, 2244 (1989).

[31] A. Lagendijk, I. F. Silvera and B. J. Verhaar, Phys. Rev. B **33**, 626 (1986).

[32] H. T. C. Stoof, J. M. V. A. Koelman and B. J. Verhaar, Phys. Rev. B **38**, 4688 (1988).

[33] D. G. Friend and R. D. Etters, J. Low Temp. Phys. **39**, 409 (1980).

[34] J. M. Doyle, Ph.D. thesis, MIT (1991).

[35] V. Bagnato, D. E. Pritchard and D. Kleppner, Phys. Rev. A **35**, 4354 (1987).

[36] K. Huang, *Studies in Statistical Mechanics II*, J. DeBoer and G. E. Uhlenbeck, eds. (North-Holland, Amsterdam, 1964).

[37] W. Kolos and L. Wolniewicz, J. Chem. Phys. **43**, 2429 (1965), and Chem. Phys. Lett. **24**, 457 (1974).

[38] E. P. Gross, J. Math. Phys. **4**, 195 (1963).

[39] H. T. C. Stoof , Phys. Rev. Lett. **66**, 3148 (1991).

[40] Yu. M. Kagan, B. V. Svistunov and G. V. Shlyapnikov, Sov. Phys. JETP **75**, 387 (1992).

[41] T. W. Hijmans, Yu. Kagan, G. V. Shlyapnikov and J. T. M. Walraven, contribution to this volume.

[42] J. Oliva, Phys. Rev. B **39**, 4197 (1989).

[43] V. V. Goldman, I. F. Silvera and A. J. Leggett, Phys. Rev. B **24**, 2870 (1981).

[44] C. A. Condat and R. A. Guyer, Phys. Rev. B **24**, 2874 (1981).

[45] D. A. Huse and E. D. Siggia, J. Low Temp. Phys. **46**, 137 (1982).

[46] O. J. Luiten, H. G. C. Werij, I. D. Setija, M. W. Reynolds, T. W. Hijmans and J. T. M. Walraven, Phys. Rev. Lett. **70**, 544 (1993).

[47] I. D. Setija, H. G. C. Werij, O. J. Luiten, M. W. Reynolds, T. W. Hijmans and J. T. M. Walraven, Phys. Rev. Lett. **70**, 2257 (1993).

[48] J. C. Sandberg, Ph.D. thesis, MIT (1993).

8

Spin-Polarized Hydrogen: Prospects for Bose–Einstein Condensation and Two-Dimensional Superfluidity

Isaac F. Silvera
Lyman Laboratory of Physics
Harvard University
Cambridge MA 02138
USA

Abstract

Spin-polarized atomic hydrogen continues to be one of the most promising candidates for Bose condensation of an atomic system. In contrast to liquid helium, hydrogen is gaseous and therefore its density can be changed to vary its behavior from the weakly to the strongly interacting Bose gas. Until now, efforts to Bose condense hydrogen have been thwarted by recombination and relaxation phenomena. After a long introduction to the subject, two promising approaches to observe quantum degenerate behavior are discussed: the microwave trap and a two-dimensional gas of hydrogen.

1 Introduction

Since the stabilization of atomic hydrogen as a spin-polarized gas (H↓), reported in 1980 [1] , there has been a continuous effort to observe Bose–Einstein condensation (BEC) or other effects of quantum degeneracy in this Bose gas. This challenge has not yet been realized due to the instability of H↓ towards recombination to H_2 or relaxation among hyperfine states as the conditions for BEC are approached. Although this difficulty is formidable, hydrogen presents a unique opportunity to study BEC and the related superfluidity, as it remains a gas to $T = 0$ [2]. By comparison, ^{4}He, the only experimentally observed boson fluid, is a strongly interacting superfluid Bose liquid, with little flexibility for varying its density. As a result of the gaseous nature of hydrogen, its density, n, can be varied over several orders of magnitude with the possibility of studying BEC and its relationship to superfluidity, ranging from the weakly to the strongly interacting boson gas.

The critical temperature for BEC is given by

$$T_c = 3.31(\hbar^2/mk_B)n^{2/3} \quad (1)$$

When evaluated at $T_c = 100$ mK, n has a value of $1.57 \times 10^{19}/\text{cm}^3$. Hydrogen has been cooled to such temperatures but not at this density. The difficulty in achieving the conditions for BEC is the rapid intrinsic recombination rate to molecular hydrogen. This has two effects: (i) it limits the density, as the recombination rate is proportional to n^3 in the gas phase, and (ii) the 4.6 eV per recombining pair is usually dissipated in the gas and can result in severe overheating of the sample, preventing the attainment of T_c.

Hydrogen was initially stabilized in a cell with a thin film of superfluid ^{4}He covering the copper walls of the container. Hydrogen atoms reside in states in the gas phase or states on the surface, and the ratio depends exponentially on the adsorption energy. Helium presents a surface to the hydrogen atoms with a very small adsorption energy, $\varepsilon_a/k_B \approx 1$K. In the high temperature limit the surface coverage σ of hydrogen is given by

$$\sigma = n\lambda \exp(\varepsilon_a/k_BT) , \quad (2)$$

where λ is the thermal de Broglie wavelength. From this we see that as n increases or T decreases, the coverage builds up. The build up of surface density can be very detrimental, as on the surface recombination can take place when two hydrogen atoms collide in the presence of a helium atom. The rate for this process is proportional to σ^2. At very high density the surface recombination is dominated by three-body hydrogen collisions, proportional to σ^3; ultimately two- and three-body recombination can prevent the achievement of low temperatures in a cell or the build up of gas phase density at low temperatures. Due to this problem some effort has been expended to use magnetic traps to isolate hydrogen from the container walls and completely eliminate wall recombination.

A more precise description of the recombination problem shows that the rates depend on the hyperfine states of the colliding atoms. Due to the spin 1/2 of the electron and spin 1/2 of the proton, there are four hyperfine states, delineated a, b, c and d in order of increasing energy, as shown in Fig. 1. In a high magnetic field the a and b states are predominantly electron spin-down states, while the c and d states are spin-up states. The recombination rate constants depend sensitively on the hyperfine states of the atoms. For example, if two of the atoms in a recombination collision are in one of the electron spin-up states and one of the spin-down states, we have the situation for unpolarized hydrogen

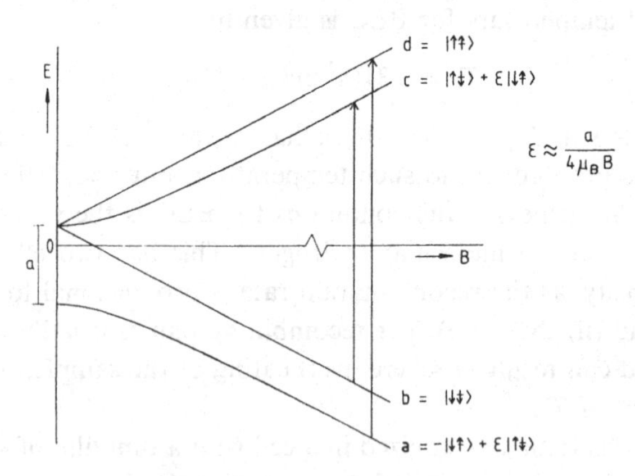

Fig. 1. The hyperfine energy levels of hydrogen as a function of magnetic field. High field states are shown in terms of the projections of the electron and nuclear spins; a is the hyperfine coupling constant.

and the recombination rate constant is very large. The rate constant is reduced by several orders of magnitude when all atoms are in either the up-spin states or down-spin states, i.e. spin-polarized. The a and c states have admixtures of spin-up and spin-down. This admixture becomes very small in a large magnetic field, which enhances the stability. However, the b and d states are pure electron spin states and gases made of either one of these states have still higher stability, as a recombination channel is closed. Thus, the stability of the gas can be greatly increased by placing all of the atoms in hyperfine states b or d.

Two-body collisions can also result in relaxation amongst the hyperfine states. This can have beneficial or detrimental effects. For example, in the static magnetic trap, the gas very rapidly relaxes to a single hyperfine state, d, due to spin-exchange processes. This state is very stable. Ultimately, however, as the conditions for BEC are approached, magnetic dipolar relaxation dominates the behavior [3]. The d-state atoms relax to the spin-down states which are ejected from the trap. These processes have prevented the attainment of BEC in the static trap, as will be discussed elsewhere in this volume [4].

Spin-polarized hydrogen also exists as an almost two-dimensional gas when condensed on the surface of helium. The atoms are bound to the ^{4}He surface by an adsorption energy $\varepsilon_a/k_B \approx 1$K and are in two-dimensional free particle momentum states; in thermodynamic equilibrium the coverage σ is given by Eq. (2). At high densities this expression is modified according to the Bose distribution; due to effective repulsive interactions between hydrogen atoms on the surface, the coverage saturates at $\sigma_{sat} \approx 1 \times 10^{14}/\text{cm}^2$. The surface gas of hydrogen is expected to demonstrate the behavior of a degenerate Bose gas. Although it can be shown rigorously that Bose- Einstein condensation does not exist in two dimensions, it is believed that a quasi-condensate with short-range correlations will exist and the correlation length rapidly becomes the size of a laboratory sample as temperature is reduced below the critical value. The critical temperature for the Kosterlitz–Thouless transition to two-dimensional superfluidity is

$$T_c^{2D} = \pi \hbar^2 \sigma_s / 2 k_B m \,, \tag{3}$$

where $\sigma_s \simeq \sigma$ is the superfluid fraction [5]. For $\sigma_s = \sigma_{sat}$, the saturation coverage, the calculated critical temperature is 756 mK. It is believed that coverages of hydrogen in this range can be achieved. Thus, it is quite realistic to seek the KT transition in hydrogen. Until recently most efforts have concentrated on achieving BEC in three dimensions; however, it now seems that it may be easier to deal with the recombination problems in two dimensions, so new efforts are underway.

In the preceding we have given a brief introduction to spin-polarized hydrogen and Bose condensation. In the remainder of this paper we shall concentrate the discussion on two promising approaches: the microwave trap and two-dimensional quasi-BEC. There are a number of other interesting ideas for BEC which have been reviewed recently by Silvera [6] and Silvera and Reynolds [7]. An earlier review by Silvera and Walraven gives a detailed description of the general development of the field up to about 1985, with descriptions of the experimental techniques and an exhaustive treatment of recombination and relaxation processes. The review by Greytak and Kleppner [8] discusses many of the properties of the weakly interacting Bose gas in the region of BEC.

2 The Microwave Trap

In the introduction we discussed the static magnetic trap which has advanced to within a factor of two or three of the critical density for BEC. Temperatures as low as a few hundred microkelvin have been achieved in this trap [9] by means of evaporative cooling of atoms trapped in the *d*-state (lowering of the potential walls of the trap to allow hot atoms to escape, with thermalization of the remaining atoms at a lower temperature). At these low temperatures, as BEC is approached, the density peaks up in the center of the trap and the relaxation rate, quadratic in density, increases [3]. As a result the further increase of density ceases and the path to BEC is blocked. Nevertheless, the MIT group feels that they can map out a trajectory in n–T space which will allow them to Bose condense a sample. If this occurs, a second problem recently studied by Hijmans *et al.* [10] arises. This is a so-called relaxation explosion, not actually an explosion, but a large enhancement of the relaxation rate. As soon as the BEC phase line is crossed, the peak density rapidly increases with a concomitant increase in the relaxation rate and loss of sample. As a result the sample is confined to the BEC line and the order parameter cannot develop in magnitude.

The microwave trap (MWT) does not suffer from these problems as it is the ground a or b states which are trapped. The MWT was proposed by Agosta and Silvera [11], with further considerations by Agosta *et al.* [12]. At high frequencies it is possible to produce a magnetic field which has either a maximum or a minimum. A very convenient geometry is a spherical microwave cavity in which one of the fundamental modes has a field maximum. This can be tuned to the region of the a–d transition or the b–c transition of hydrogen. An analysis in terms of dressed-atom states shows that in the region of resonance a potential well is developed in the microwave field gradient. The depth of this well for trapping of atoms is of the order of several millikelvin for reasonable microwave powers and cavity quality factor. Thus, to load this trap very cold atoms must be used.

The eigenstates of the Hamiltonian (dressed states) near resonance between the a and d states are

$$\begin{aligned} |1N\rangle &= \cos\theta|d\rangle\otimes|N-1\rangle + \sin\theta|a\rangle \otimes |N\rangle \text{(a)}, \\ |2N\rangle &= -\sin\theta|d\rangle\otimes|N-1\rangle + \cos\theta|a\rangle \otimes |N\rangle \text{(b)}, \end{aligned} \tag{4}$$

where $N = 0, 1, 2, \ldots$, and $|N\rangle$ denotes the state of the magnetic mode with N photons before coupling to the atom. The angle θ is given by $\tan 2\theta = \omega_r/\delta$, where ω_r is the Rabi frequency ($\omega_r = \gamma_e B_{rf}$ for a circularly polarized microwave field of amplitude B_{rf}) and $\delta = \omega - \omega_0$ is the detuning of the microwave frequency from the resonance frequency, $\omega_0 = B_0/\gamma_e$ (B_0 is the applied static field). The energies are

$$\begin{aligned} E_{1N} &= N\hbar\omega + h\Omega/2 \text{(a)}, \\ E_{2N} &= N\hbar\omega - h\Omega/2 \text{(b)}, \end{aligned} \tag{5}$$

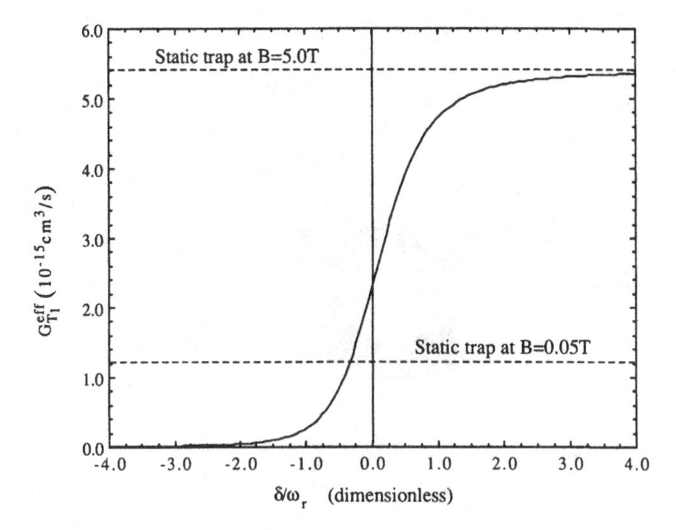

Fig. 2. The spin relaxation loss rate in a microwave trap for hydrogen. The dashed lines show the loss rate in static traps for two different magnetic field amplitudes. By using a negative detuning the loss rate can be made negligible in the microwave trap; for large positive detuning the microwave rate approaches the static rate.

where $\Omega = (\omega_r^2 + \delta^2)^{1/2}$. The energy levels vary with position through the variation of Ω on position. From Eq. (5) we see that state $|2N\rangle$ is a trapped state and $|1N\rangle$ is an antitrapped state. By varying δ, the admixture of states can be varied, and, with a reasonable negative detuning, the character of the state $|2N\rangle$ becomes almost completely $|a\rangle$, the ground state. Agosta *et al.* [12] have calculated the loss rate due to relaxation and find that it becomes negligibly small, as shown in Fig. 2. Thus, with the MWT, the principle loss mechanism of the static trap, spin relaxation, can be suppressed to a very low level. Moreover, the relaxation explosion is not a characteristic of this trap. After loading the trap, by further detuning, the potential walls are lowered and evaporative cooling can be used to achieve very low temperatures; alternatively atoms can be selectively pumped from the trapped to the antitrapped state for forced evaporative cooling, which does not change the shape of the trap. If BEC can be achieved the principle loss mechanism will be due to three-body recombination in the center of the trap which will become important in the density region of 10^{17} to $10^{18}/\text{cm}^3$. Since the initial densities for BEC in this trap will be of order $10^{14}/\text{cm}^3$, there should be no problem with a recombination "explosion" when BEC is achieved.

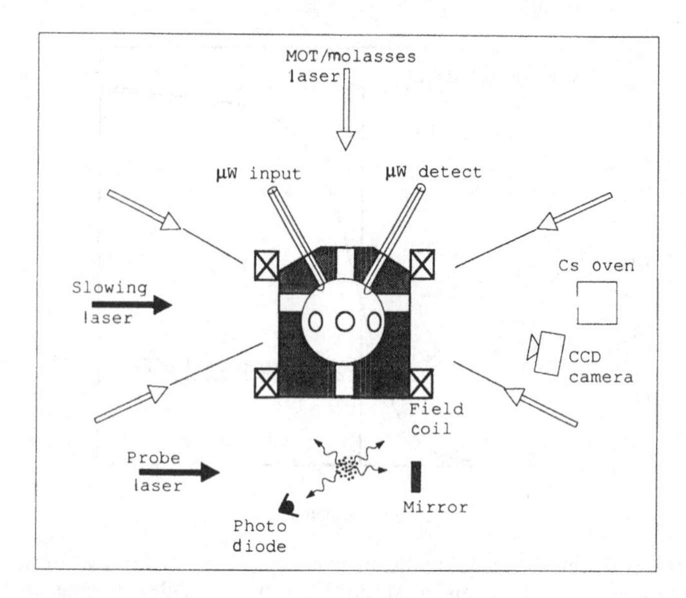

Fig. 3. A schematic diagram of the cesium trap, showing the microwave cavity and the several laser beams used for cooling the atoms to be loaded into the trap. Atoms could be detected by releasing them from the trap and observing their fluorescence as they passed through a probe beam. An alternative means of observation is to laser illuminate trapped atoms and observe them in fluorescence with a ccd camera.

Silvera *et al.* [13] have briefly discussed the problem of atomic losses due to microwave noise; this appears to be a technically resolvable problem.

The principle difficulty with the MWT is that it can only be loaded with very cold atoms. Silvera and Reynolds [14] proposed a hybrid trap in which H atoms would first be cooled in a static trap (in, say, the *d*-state), evaporatively cooled to a few hundred microkelvin, and then transferred to the MWT. Since this is a very demanding experiment, the MWT was recently demonstrated experimentally [6, 15] with cesium atoms in an experimental setup shown in Fig. 3. The atoms are first laser cooled to a few microkelvin using a magneto-optical trap and polarization laser cooling; the laser is then turned off and the microwave power, up to 80 Watts in an x-band microwave cavity, is turned on. A complication in this experiment was that the microwave potential was insufficiently deep to hold atoms against the gravitational field. This problem was solved by using a static gradient magnetic field

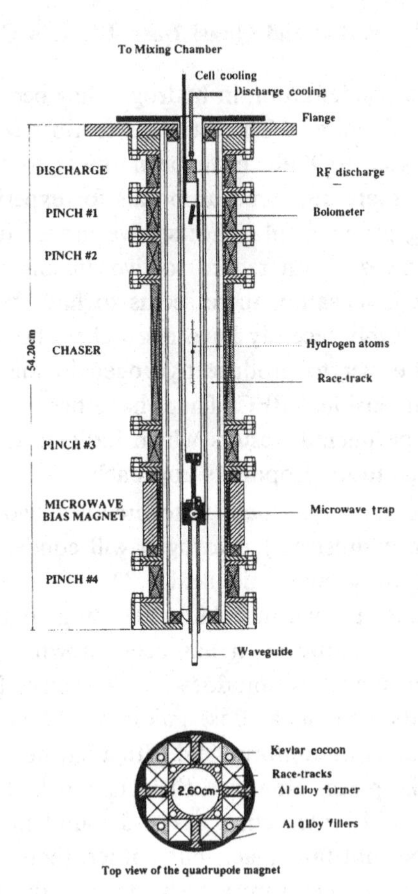

Fig. 4. The hybrid static microwave trap under construction at Harvard. The static trap consists of 11 separately controlled superconducting magnets, used for trapping, cooling and spatially manipulating the atoms. After being cooled, atoms will be transferred to the microwave trap by varying the magnetic field gradients.

to levitate the Cs atoms, with the microwave field gradient confining them. The trapped atoms could be observed visually by inducing optical fluorescence with a pulsed resonant laser beam. Zhao *et al.* [16] are constructing a MWT for hydrogen, shown in Fig. 4. This trap should pave the way to BEC in a three-dimensional gas of hydrogen.

3 The KT Transition and Quasi Bose–Einstein Condensation

The quest for Bose condensation in hydrogen has been focused strongly on the three-dimensional gas. The attempts to Bose condense hydrogen as a *high density* gas, to date, have been thwarted by recombination heating. Although there are some proposals for experiments which can handle the heating problem [6] , efforts have moved in the direction of very low density traps. With regard to two-dimensional KT superfluidity, quasi-Bose condensation there seems to have been few efforts in this direction [17]. Only recently have researchers become aware that it is probably much easier to produce hydrogen in this state experimentally than three-dimensional BEC. There have been a number of recent theoretical and experimental results which indicate that this avenue of research may be the more propitious approach.

The principle of the experiment is to create a two-dimensional surface (of helium on a substrate); hydrogen will condense on the surface and be stabilized by a magnetic field. The hydrogen behaves as a two-dimensional gas, populating states acording to the Bose–Einstein distribution function. However, it has been shown that in two dimensions the Bose–Einstein transition does not take place [18]. Bagnato and Kleppner [19] studied the ideal Bose gas in two dimensions and found that, for example, in a non-uniform field with a harmonic potential a BEC transition does take place. However, Shevchenko [20] studied the same problem for the weakly interacting gas and found that in this case the BEC transition does not take place, but, rather, there is a KT transition and that the critical temperature is very close to the BEC temperature in the absence of interactions. He then studied the quasi-condensate and its development in a finite system [21]. The quasi-condensate had been studied earlier by Kagan *et al.* [22]. Stoof and Bijlsma [23] have also studied the thermodynamic properties of the weakly interacting Bose gas and arrive at similar conclusions. Thus, on theoretical grounds we conclude that in two dimensions in an inhomogeneous magnetic field there is a KT transition and a quasicondensate which becomes macroscopic at a temperature close to T_c^{2D}.

Earlier, it was believed that it would be very difficult to study two-dimensional BEC because of the very rapid recombination at the surface coverages which are required and the resultant heating of the sample. However, recent experiments have shown that when molecules recombine on a surface there is a high probability that they will form in highly excited states of H_2 and fly off the surface. The fraction F of the recombination energy deposited on the surface is found to have an upper bound of 0.04 for two-body recombination, and there is probably a similar bound for three-body surface recombination [17, 24]. Thus, only a very small fraction of the recombination energy goes into the surface to heat the sample and the heating problem of three-dimensional gases can be strongly suppressed.

On the theoretical side there are a few additional advantages for the experimentalist in studying properties of a quantum degenerate system in two dimensions. Kagan *et al.* [22] had shown that the recombination rate constant, K, is reduced by a factor of 6 for three-body recombination between atoms in the condensate; for a fixed surface coverage the recombination rate should decrease by this amount as T/T_c^{2D} decreases and almost all of the atoms are in the quasi-condensate or superfluid fraction. The origin for this decrease is in the very simple form of the wavefunction for the zero-momentum or ground state particles, which leads to a $\sqrt{3!}$ reduction in the matrix element for the process. For the same reason, the two-particle interaction potential is reduced by 2! As a result, if there is thermodynamic equilibrium between surface state atoms and a thermal reservoir of gas atoms in three-dimensions, then at the transition there is an increase in surface density by about a factor of two [25]. Since the recombination rate is proportional to $K\sigma^3$, the rate is expected to have a slight increase at T_c^{2D} rather than a decrease. The most recent development is a calculation by Stoof and Bijlsma [26], who go beyond the approach of Kagan *et al.* [22] and, in addition to wavefunction symmetry effects, they consider the many-body states of the gas in greater detail. They find an enormous reduction of the rate constant, more than 400, and an increase in the density step to about 3 at the critical temperature. Together, this predicts that the rate will decrease by about $3^3/400 \approx 15$ at the transition. This is likely an overestimate since the calculation is probably not valid in the critical region. Nevertheless, outside the critical region there is still a large predicted reduction in the recombination rate in the superfluid, according to this result. From an experimental point of view these predictions suggest a very nice property of the sample which can be used to identify the transition, the recombination rate or the recombination heating of the sample. It will certainly be interesting if such enormous changes (400) can arise due to Bose effects.

Experimental attempts to study the KT transition are so meager that it is not meaningful to assess how close experiments have approached this transition. An experiment was started at the University of Amsterdam to observe the KT transition by optical methods using a Lyman-alpha laser source. The experimental plan was to introduce a gas of hydrogen to a small, cold helium surface on the surface of a large reservoir of atoms stored at a higher temperature, with higher temperature walls. In this way the main recombination losses would be on the cold surface and could be maintained at a manageable level. The density of atoms on the surface would be determined using the Lyman-alpha radiation to optically excite the system such that the surface coverage could be measured. However, it was realized that the hydrogen interacts with the surface via the surface ripplons of the helium and at the low temperature of the experiment these are weakly coupled to the helium phonon bath [27]. As a result the ripplons heat up and the hydrogen would not be at a sufficiently low temperature required for the KT transition; the experiment was aborted. An experiment at the University of Kyoto [17] using a cold surface has been carried out at a measured cell temperature of 87 mK; however the coverages were much too low for the KT transition. This experiment also determined that the F factor described above is about 0.015. An experiment is currently being developed at Harvard University [28] using a two-dimensional magnetic field gradient with a maximum field at the center of a disc-shaped surface. In this way very high surface coverages can be achieved in a very small area so that the total sample loss is very small. The atoms that cover the surface are fed from a large buffer volume at a higher temperature in such a way that the total number of atoms on the surface is maintained constant; this inhibits the density step increase at the critical temperature and should amplify sensitivity to changes in recombination energy. The means of detection of the KT transition will be a measurement of the recombination heating on the cell surface. Similar experiments are planned at the University of Amsterdam [29] and the University of Turku in collaboration with the Kurchatov Institute [30], using optical detection or thermal detection respectively.

4 Conclusions

For more than 13 years, since the initial stabilization of atomic hydrogen, researchers have been devising schemes to observe Bose–Einstein condensation in the weakly interacting Bose gas. In the first several years a number of clever experiments were attempted, only to be thwarted by the rapid loss of sample, and heating due to recombination and relaxation among hyperfine levels. In this period, there has been an enormous growth in the basic theoretical and experimental understanding of this seemingly simple system. Fortified with this knowledge, new experiments and approaches have been developed, bringing us within reach of the goal of BEC, which at times has seemed unattainable. The coming years promise exciting new results in this simple Boson gas.

Acknowledgment. I thank the Department of Energy, Grant Number DE-FGO2-85ER45190 for support.

References

[1] I.F. Silvera and J.T.M. Walraven, Phys. Rev. Lett. **44**, 164 (1980).

[2] I.F. Silvera and J.T.M. Walraven, *Progress in Low Temp. Physics*, Vol. X, D.F. Brewer, ed. (Elsevier, Amsterdam, 1986), p. 139.

[3] A. Lagendijk, I.F. Silvera, and B.J. Verhaar, Phys. Rev. B **33**, 626 (1986).

[4] See T.J. Greytak, this volume.

[5] D.O. Edwards, Physica B **109** and **110**, 1531 (1982).

[6] I.F. Silvera, J. Low Temp. Phys. **89**, 287 (1992).

[7] I.F. Silvera and M. Reynolds, J. Low Temp. Phys. **87**, 343 (1992).

[8] T.J. Greytak and D. Kleppner, *New Trends in Atomic Physics*, Vol. II, G. Grynberg and R. Stora, eds. (North Holland, Amsterdam, 1984), p. 1127.

[9] J.M. Doyle, J.C. Sandberg, I.A. Yu, C.L. Cesar, D. Kleppner, and T.J. Greytak, Phys. Rev. Lett. **67**, 603 (1991).

[10] T.W. Hijmans, Y. Kagan, G.V. Shlyapnikov, and J.T.M. Walraven, preprint (1993).

[11] C.C. Agosta and I.F. Silvera, *Spin Polarized Quantum Systems*, S. Stringari, ed. (World Scientific, Singapore, 1989), p. 254.

[12] C.C. Agosta, I.F. Silvera, H.T.C. Stoof, and B.J. Verhaar, Phys. Rev. Lett. **62**, 2361 (1989).

[13] I.F. Silvera, C. Gerz, L.S. Goldner, W.D. Phillips, M.W. Reynolds, S.L. Rolston, R.J.C. Spreeuw, and C.I. Westbrook, LT20, Physica B **194–196**, in press (1994).

[14] I.F. Silvera and M. Reynolds, Physica B **169**, 449 (1991).

[15] L. Goldner, R.J.C. Spreeuw, C. Gerz, W.D. Phillips, M.W. Reynolds, S.L. Rolston, I.F. Silvera, and C.I. Westbrook, LT20, Physica B **194–196**, in press (1994).

[16] Z. Zhao, I.F. Silvera, and M. Reynolds, J. Low Temp. Phys. **89**, 703 (1992).

[17] A. Matsubara, T. Arai, S. Hotta, J.S. Korhonen, T. Mizusaki, and A. Hirai, this volume.
[18] P.C. Hohenberg, Phys. Rev. **158**, 383 (1967).
[19] V. Bagnato and D. Kleppner, Phys. Rev. A **44** , 7439 (1991).
[20] S.I. Shevchenko, Sov. Phys. JETP **73**, 1009 (1991).
[21] S.I. Shevchenko, Sov. J. Low Temp. Phys. **18**, 223 (1992).
[22] Y. Kagan, B.V. Svistunov, and G.V. Shlyapnikov, Sov. Phys. JETP **66**, 314 (1987).
[23] H.T.C. Stoof and M. Bijlsma, Phys. Rev. B **47** (2), 939 (1993).
[24] Z. Zhao, E.S. Meyer, B. Freedman, J. Kim, J.C. Mester, and I.F. Silvera, LT20, Physica B **194–196**, in press (1994).
[25] B.V. Svistunov, T.W. Hijmans, B.V. Shlyapnikov, and J.T.M. Walraven, this volume.
[26] H. Stoof and M. Bijlsmaa, preprint (1993).
[27] M.W. Reynolds, I.D. Setija, and G.V. Shlyapnikov, Phys. Rev. B **46**, 575 (1992).
[28] I.F. Silvera, unpublished.
[29] M. Reynolds, private communication.
[30] S. Jaakkola, private communication.

9

Laser Cooling and Trapping of Neutral Atoms

Y. Castin, J. Dalibard and C. Cohen-Tannoudji
Laboratoire de Spectroscopie Hertzienne de l'E.N.S.†
et Collège de France
24, rue Lhomond
F-75231 Paris Cedex 5
France

Abstract

We present a simple review of the basic physical processes allowing one to control, with laser beams, the velocity and the position of neutral atoms. The control of the velocity corresponds to a cooling of atoms, that is, a reduction of the atomic velocity spread around a given value. The control of the position means a trapping of atoms in real space. The best present performances will be given, in terms of the lowest temperatures and the highest densities. The corresponding highest quantum degeneracy will also be estimated. It is imposed by fundamental limits, which will be described briefly . We also give the general trends in this field of research and outline the new directions which look promising for observing quantum statistical effects in laser cooled atomic samples, but which are for the moment restricted by unsolved problems.

1 Introducing the Simple Schemes

The radiative forces acting on atoms in a light field can be split into two parts, a reactive one and a dissipative one. The dissipative force (radiation pressure), which basically involves scattering processes, is velocity dependent. We will see that this dependence leads to the Doppler cooling scheme and to the concept of optical molasses, and we will give the corresponding minimal achievable temperature. The dissipative force can be made position dependent, through a gradient of the magnetic field, so that the atoms are also trapped in the so-called magneto-optical trap.

† Unité de recherche de l'Ecole Normale Supérieure et de l'Université Paris 6, associée au CNRS

1.1 Effect of Light on the Atomic Internal State

Since the atoms in laser cooling are illuminated by quasi-resonant light, they can be considered as two-level atoms, with a metastable state g, called "ground state" in what follows, and an excited state e. The transition between g and e has a resonant frequency ω_A and an electric dipole d. The excited state can decay by spontaneous emission with rate Γ. In this section, we consider the simple case of atoms with no sublevels in the ground state g. The influence of the presence of Zeeman sublevels in g on laser cooling is the subject of Section 2.

The laser electric field is a superposition of travelling waves of frequency ω_L close to ω_A and of complex amplitude $\mathscr{E}_0$. An important parameter is $\delta = \omega_L - \omega_A$, the detuning of the laser frequency ω_L from the atomic frequency ω_A. The dipolar coupling between atom and light leads to a time dependent amplitude of transition from g to e, denoted by $(\Omega/2)e^{-i\omega_L t}$, and from e to g, $(\Omega^*/2)\, e^{i\omega_L t}$. The parameter $\Omega = -d\mathscr{E}_0/\hbar$ is called the Rabi frequency. The amplitude of transition can be made time independent by the following time dependent unitary transformation $S(t)$ changing the excited state:

$$S(t) = e^{i\omega_L t|e\rangle\langle e|} \tag{1}$$

which gives the $|g\rangle \longrightarrow |e\rangle$ transition the effective Bohr frequency $\omega_A - \omega_L = -\delta$ instead of ω_A. This leads to the time independent effective hamiltonian H_{eff} for the evolution of the internal atomic state:

$$H_{\text{eff}} = \hbar \begin{pmatrix} -\delta - i\Gamma/2 & \Omega/2 \\ \Omega^*/2 & 0 \end{pmatrix}. \tag{2}$$

This matrix is given in the $\{|e\rangle, |g\rangle\}$ basis and the $-i\Gamma/2$ term accounts for the instability of the excited state.

The effect of the coupling will be considered in the perturbative regime:

$$s = \frac{|\Omega|^2/2}{\delta^2 + \Gamma^2/4} \ll 1. \tag{3}$$

The dimensionless quantity s is called the saturation parameter and gives the relative amount of time spent by the atoms in the excited state. In this regime, one of the eigenvectors of H_{eff} remains close to g, with the eigenvalue $\hbar(\delta - i\Gamma/2)s/2$. The corresponding real part $\hbar\delta' = \hbar\delta s/2$ is the signature of a reactive effect: it describes the shift of the energy of the ground state by the incoming light (light shift). The imaginary part is the signature of a dissipative effect: the ground state g becomes unstable by contamination by the excited state through the Rabi coupling; the corresponding rate $\Gamma' = \Gamma s/2$ is the excitation rate of the ground state by the laser light. Note that the following discussion can be generalized to arbitrary laser intensities (see e.g. the so-called "dressed atom" picture in [1]).

1.2 Effect of Light on the Atomic External State

In contrast to the internal atomic variables, which correspond to the electronic motion relative to the nucleus, the external variables are related to the motion of the atomic center of mass; the important ones for laser cooling are the position $\mathbf{r}$ and the momentum $\mathbf{p}$ of the atom. These external variables are also changed by the interaction with the laser light. For example, the momentum $\mathbf{p}$ shifts to $\mathbf{p} \pm \hbar\mathbf{k}$ after the absorption or the emission of one photon of momentum $\hbar\mathbf{k}$. The corresponding atomic recoil velocity $\hbar k/M$, where M is the mass of the atom, is an important parameter in laser cooling, as we shall see; it is of the order of 3 mm/s for cesium. The effect of light on atomic motion can be described in terms of radiative forces. These forces can be dissipative or reactive, depending on the contribution or non-contribution of the momentum of spontaneously emitted photons.

For the simple case of an atom in a plane wave, the dissipative force corresponds to the usual radiation pressure; it is due to the absorption of one laser photon of momentum $\hbar\mathbf{k}_L$ and the spontaneous emission of one fluorescence photon of momentum $\hbar\mathbf{k}_S$. The corresponding average change in atomic momentum is $\langle\delta\mathbf{p}\rangle = \langle\hbar\mathbf{k}_L - \hbar\mathbf{k}_S\rangle = \hbar\mathbf{k}_L$; the mean contribution of $\hbar\mathbf{k}_S$ is zero because spontaneous emission occurs with the same probability in two opposite directions. The mean radiative force is then simply $\hbar\mathbf{k}_L$ times the rate of absorption-spontaneous emission cycles. This rate, of the order of $\Gamma s/2$ at low saturation, cannot exceed $\Gamma/2$ at high saturation. The dissipative force thus saturates at high laser intensity to $\hbar k_L \Gamma/2$, a limit almost 10^4 higher than the gravitational force Mg for cesium.

The reactive forces derive from an effective potential, which is the position dependent light shift of the atomic level g; it is due to successive absorption-stimulated emission cycles. Consider, for example, the light shift U of the atomic ground state in the vicinity of a gaussian laser beam. When the laser light is detuned to the red ($\delta < 0$), $U(\mathbf{r})$ is negative and maximal in absolute value in the laser beam, and close to zero outside the laser beam. The effect of $U(\mathbf{r})$ leads to a trapping of the atoms around the maxima of intensity of the laser light. This has been used to make a trap, at a detuning large enough for the dissipative effects (heating due to spontaneous emission) to be negligible [2]. An improved version of the optical trap, using stabilization by radiative pressure forces, is presented in [3].

1.3 Doppler Cooling and the Doppler Limit

The Doppler cooling scheme has been proposed independently by Hänsch and Schawlow as well as Wineland and Dehmelt in 1975 [4, 5]. It relies on the atomic velocity dependence of the radiation pressure due to the Doppler effect. In this scheme, the cooling of the atomic velocity along a given direction z is obtained by the superposition of two counterpropagating running waves along z. The two waves have the same weak amplitude $\mathscr{E}_0$ and the same frequency ω_L, detuned to the red ($\delta = \omega_L - \omega_A < 0$). Consider an atom of velocity $\mathbf{v}$, and call $\mathbf{k}$ the wave vector of the laser wave copropagating with the atom ($\mathbf{k} \cdot \mathbf{v} > 0$). In the atomic rest frame, this wave has an apparent frequency $\omega_L - \mathbf{k} \cdot \mathbf{v} < \omega_L < \omega_A$, so it is moved farther from resonance by the Doppler effect $-\mathbf{k} \cdot \mathbf{v}$. The counterpropagating wave, with wave vector $-\mathbf{k}$, is on the contrary, is brought closer to resonance by the Doppler effect ($\mathbf{k} \cdot \mathbf{v} > 0$) (see Fig. 1). The atom will then absorb photons preferentially in the counterpropagating wave, so that it feels a mean radiative force opposed to its velocity along z. This force vanishes for an atom at rest, and can be shown to behave as a linear friction force:

$$F_z = -\alpha v_z, \tag{4}$$

where α is a friction coefficient, for sufficiently slow atoms. Cooling is provided along the other directions x and y by use of two additional standing waves. The corresponding light field acts as a viscous medium on atomic motion, the so-called "optical molasses" [6].

Spontaneous emission plays an essential role in the cooling mechanism: it allows the dissipation of energy by emission of fluorescence photons with a frequency $\omega_S > \omega_L$. However, the random momentum recoil of the atoms after spontaneous emission is responsible for a heating, i.e. an increase of the mean atomic kinetic energy, which can be described, as in standard brownian motion theory, by a momentum diffusion coefficient D:

$$\frac{d}{dt} \langle p_i^2 \rangle \Big|_{\text{heating}} = 2D \, , \qquad i = x, y, z \tag{5}$$

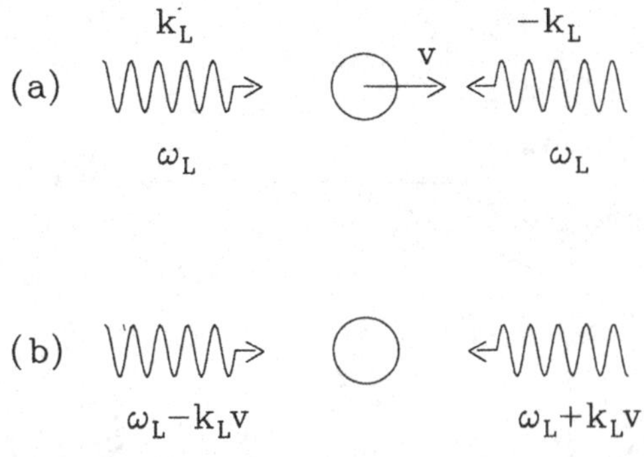

k_L $-k_L$

(a) v

ω_L ω_L

(b)

$\omega_L - k_L v$ $\omega_L + k_L v$

Fig. 1. The laser waves of the Doppler cooling configuration, in the laboratory (a) and in the atomic rest frame (b).

This heating can counterbalance the effect of the damping force:

$$\frac{d}{dt}\langle p_i^2\rangle\Big|_{\text{cooling}} = -\frac{2\alpha}{M}\langle p_i^2\rangle \ , \qquad i = x, y, z \tag{6}$$

so that equilibrium is reached. The corresponding stationary velocity distribution is gaussian, with an effective temperature T given by Eintein's law:

$$k_B T = \frac{D}{\alpha}, \tag{7}$$

where k_B is the Boltzmann constant. After an explicit derivation of α and D for the two-level atom model, one finds that the minimal temperature T_D, called the Doppler limit, is obtained for a detuning $\delta = -\Gamma/2$ and is given by [7, 8]:

$$k_B T_D = \frac{\hbar\Gamma}{2}. \tag{8}$$

It corresponds to $T_D = 120\ \mu$K for cesium atoms.

1.4 The Magneto-Optical Trap (MOT)

The effect of Doppler cooling is a compression in momentum space, but it has no confining effect in real space, the atoms performing simply a spatial diffusive random walk in the optical molasses. In order to trap the atoms in real space, the radiation pressure is also made position dependent via a gradient of magnetic field.

The principle of the MOT can be explained using the following model (see Fig. 2). Consider an atomic transition occurring between a ground

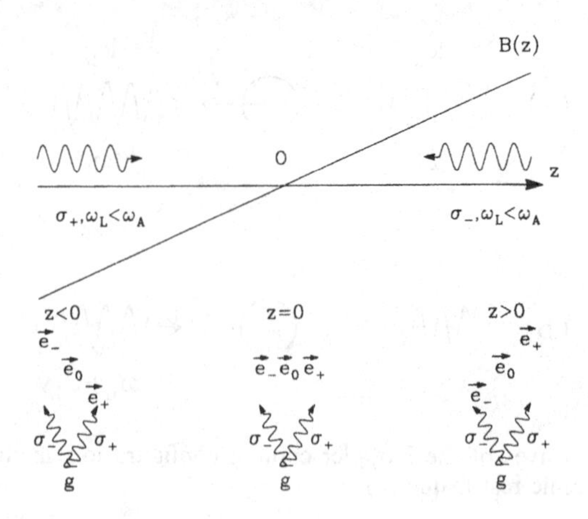

Fig. 2. The principle of the magneto-optical trap on a $j_g = 0 \rightarrow j_e = 1$ transition.

state of angular momentum $j_g = 0$ and an excited state of angular momentum $j_e = 1$. Atoms are excited by two counterpropagating laser waves which are, respectively , σ_+ and σ_- polarized along z. The fact that the excited state has several Zeeman sublevels $|e, m = -1\rangle_z$, $|e, m = 0\rangle_z$ and $|e, m = 1\rangle_z$ of angular momentum $m\hbar$ along z is essential. First, these sublevels undergo different Zeeman shifts due to the external magnetic field, say $-\mu B$, 0 and μB, respectively, for a magnetic field B along the quantization axis z, $\mu > 0$ being proportional to the magnetic dipole moment in the excited state. Second, the ground state is coupled to $|e, 1\rangle$ only by absorption of photons in the σ_+ laser wave, and to $|e, -1\rangle$ only by absorption of photons in the σ_- laser wave. Suppose now that B varies linearly along z and vanishes in $z = 0$. An atom at rest in $z = 0$ sees no mean radiative force, the two laser beams being exactly equally detuned far from resonance. In $z > 0$, however, because of the Zeeman shifts, the resonance frequency for the $|g, 0\rangle \longleftrightarrow |e, 1\rangle$ transition is larger than ω_A, and is smaller than ω_A for the $|g, 0\rangle \longleftrightarrow |e, -1\rangle$ transition. The σ_- laser wave is then closer to resonance than the σ_+ laser wave, and the atom at rest feels a net radiative force pushing it towards $z = 0$. The conclusion is reversed for negative values of z. In this configuration, there is thus, in addition to the damping force, a trapping force, linear with z around $z = 0$, so that the net mean radiative force along z reads:

$$F_z = -\alpha v_z - \kappa z. \tag{9}$$

The atomic motion is damped in an effective external harmonic potential, and the atomic cloud develops a finite spatial extension given in steady state by:

$$\kappa \langle z^2 \rangle = k_B T = \frac{D}{\alpha}. \tag{10}$$

The first experimental demonstration of the MOT has been reported in [9].

2 Beating the Doppler Limit

Since 1988, precise experimental measurements of the temperatures of optical molasses have been performed [10–16]. The lowest temperatures are found to be well below the Doppler limit (8). For a heavy atom such as cesium, the lowest measured temperatures are of the order of 3 μK, a factor of 40 smaller than the Doppler limit, and are observed at large atom-laser detuning: $|\delta| \gg \Gamma$, instead of $\delta = -\Gamma/2$ expected from the Doppler cooling theory. All these results are experimental evidence of very efficient new cooling mechanisms.

These mechanisms have been identified theoretically on simple one-dimensional (1D) models [17, 18]. They rely strongly on the existence of several degenerate sublevels in the atomic state g, a feature left out in the two-level atom model used for Doppler cooling but present in experimental optical molasses, because of the hyperfine and Zeeman structures of the atomic ground state. A second important feature is the spatial variation of the laser field polarization on the optical wavelength scale, a condition which is automatically fulfilled in three-dimensional (3D) laser configurations, and which is obtained in the simple one-dimensional models by giving different polarizations to the two counterpropagating waves.

In this section, we will analyze in some detail only one of the presently known polarization gradient cooling mechanisms, relying on the so-called "Sisyphus effect" [17]. The corresponding minimal kinetic energies no longer scale as $\hbar\Gamma$, as for Doppler cooling, but as the recoil energy $E_R = \hbar^2 k^2/2M$, i.e. the mean increase of the atomic kinetic energy after the spontaneous emission of a single photon. Note that E_R is well below the Doppler limit $\hbar\Gamma/2$ for the typical atomic transitions used in optical molasses. In the case of cesium atoms, for example, $\hbar\Gamma/2$ is one thousand times larger than the recoil energy. In this very cold regime, it has been predicted theoretically and checked experimentally that quantization of atomic motion plays an important role in Sisyphus cooling. This has

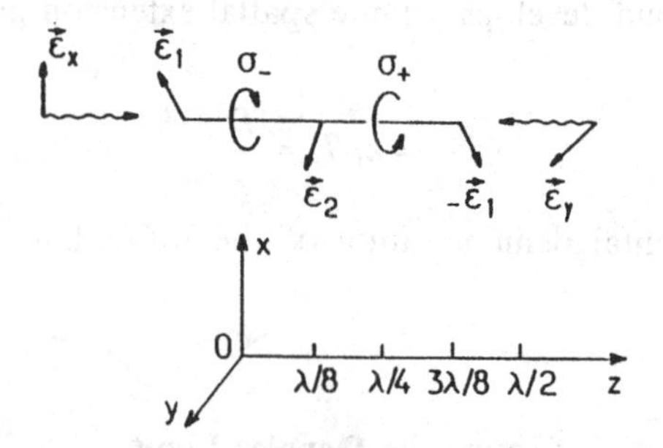

Fig. 3. Resulting electric field in the x–y laser configuration.

led to a new picture for the laser cooled atomic samples, the so-called "optical lattices".

2.1 *Optical Pumping and Light Shifts*

As seen in Section 1.1 for the two-level atom case in the low saturation regime the effect of light on the internal atomic state is to shift the energy levels (reactive effect) and to give a finite lifetime to the ground state (dissipative effect). When the atom has several Zeeman sublevels in the ground state, the reactive effect can lead to different light shifts of the sublevels, and the dissipation effect can lead to real transitions between the various sublevels, through optical pumping cycles [19].

The last two features play an important role in laser cooling. They are present in the simple one-dimensional scheme depicted in Fig. 3. The light field is obtained by superposition of two coherent counterpropagating running waves along z, with the same amplitude $\mathcal{E}_0$ and frequency ω_L, but with orthogonal linear polarizations. For an appropriate phase choice, the total electric field is given by:

$$\mathcal{E}(z,t) = \mathcal{E}^+(z)e^{-i\omega_L t} + \text{c.c} \tag{11}$$

$$\mathcal{E}^+(z) = \mathcal{E}_0\left[\mathbf{e}_x e^{ikz} - i\mathbf{e}_y e^{-ikz}\right], \tag{12}$$

where $\mathbf{e}_x$, $\mathbf{e}_y$ are unit vectors along x and y. One can check that the polarization of $\mathcal{E}(z,t)$ depends on z, with a periodicity $\lambda_{\text{opt}}/2$, where $\lambda_{\text{opt}} = 2\pi/k$ is the optical wavelength. For example, at $z = 0, \lambda_{\text{opt}}/2, ..., \mathcal{E}(z,t)$ is σ_- circularly polarized along z; it is σ_+ polarized at $z = \lambda_{\text{opt}}/4, 3\lambda_{\text{opt}}/4, ...$; it is linearly polarized along $(\mathbf{e}_x - \mathbf{e}_y)/\sqrt{2}$ at $z = \lambda_{\text{opt}}/8, 5\lambda_{\text{opt}}/8, ...$ and along $(\mathbf{e}_x + \mathbf{e}_y)/\sqrt{2}$ at $z = 3\lambda_{\text{opt}}/8, 7\lambda_{\text{opt}}/8$.

Consider an atom with an angular momentum $j_g = 1/2$ in the ground

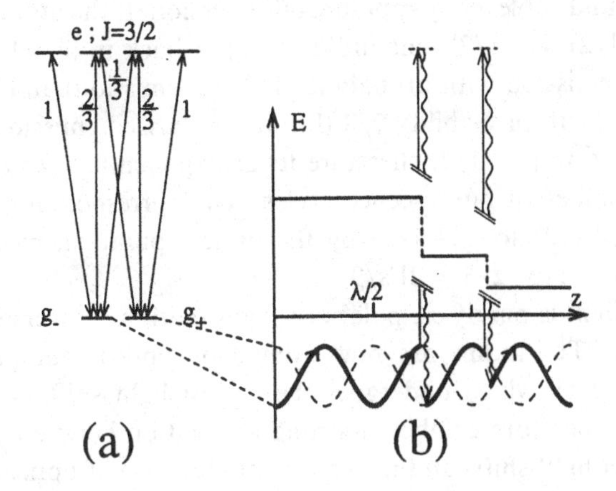

Fig. 4. (a) Atomic-level scheme and intensity factors (square of the Clebsch-Gordan coefficients) for a $j_g = 1/2 \to j_e = 3/2$ atomic transition. (b) In the Sisyphus cooling configuration, the position-dependent light shifts of the ground state sublevels and a typical atomic trajectory in the position-energy space.

state and an angular momentum $j_e = 3/2$ in the excited state. This atom is illuminated by the previous field configuration, with a laser frequency ω_L below the resonance frequency ω_A:

$$\delta = \omega_L - \omega_A < 0 \tag{13}$$

The transition amplitudes between the atomic energy levels of a given angular momentum along z, $\{|g,m\rangle,\ m = \pm 1/2\}$ and $\{|e,m'\rangle, m' = \pm 3/2, \pm 1/2\}$, by absorption or stimulated emission of laser photons are now weighted by the appropriate Clebsch–Gordan coefficients (see Fig. 4(a)) and thus depend on the polarization of the electric field.

Where the light is purely σ_- polarized, the $|g,-1/2\rangle$ sublevel is coupled to the electric field with an amplitude $\sqrt{3}$ larger than the $|g,+1/2\rangle$ sublevel. It therefore experiences a light shift three times larger, $\hbar\delta s$ instead of $\hbar\delta s/3$ for $|g,+1/2\rangle$, with a saturation parameter given by (3).† In the presence of pure σ_- light, the sublevel $|g,-1/2\rangle$ not only has a well defined light shift, but is also a trap for the internal atomic state. It is coupled by laser light only to the $|e,-3/2\rangle$ excited state sublevel, which by spontaneous emission can only decay to $|g,-1/2\rangle$. If the atom is initially in the $|e,+3/2\rangle$ excited sublevel, it is eventually decays to the

† At first sight, there seems to be a discrepancy of a factor of 2 between the light shift (and also the excitation rate) derived here and the one given in Section 1.1. In fact, at $z = 0$, the electric field amplitude is $\sqrt{2}\mathscr{E}_0$ instead of $\mathscr{E}_0$ in Section 1.1.

$|g, +1/2\rangle$ ground sublevel by spontaneous emission. If the atom is initially in the $\{|g, +1/2\rangle, |e, -1/2\rangle\}$ manifold, it is put back in $|g, +1/2\rangle$ after a spontaneous emission with probability 1/3 (σ_- emission) and is trapped into $|g, -1/2\rangle$ with probability 2/3 (linearly polarized emission along z). The population in $|g, +1/2\rangle$ therefore tends exponentially to zero for an increasing number of fluorescence cycles. The corresponding transition rate from $|g, +1/2\rangle$ to $|g, -1/2\rangle$ by the optical pumping mechanism is found to be $\Gamma s \cdot 1/3 \cdot 2/3 = 2\Gamma s/9$.

When the light is purely σ_+ polarized, the situation is reversed between the sublevels. The atoms are now excited by optical pumping to the $|g, +1/2\rangle$ sublevel, which undergoes the largest light shift. The present configuration therefore exhibits a strong correlation between the spatial modulation of light shifts and the spatial modulation of optical pumping rates. For a negative atom-laser detuning δ, the optical pumping rates are the largest, from the highest energy ground state sublevel to the lowest one.

2.2 Sisyphus Cooling Mechanism

We now explain how cooling occurs in the previously described atom-laser configuration. The key point is that the position dependent light shifts of the sublevels $|g, \pm 1/2\rangle$ appear as effective potentials $U_\pm(z)$ for a moving atom. It will thus be convenient to introduce the mechanical energies $E_\pm = Mv^2/2 + U_\pm(z)$, where v is the atomic velocity along z.

A typical trajectory in $(z, E_\pm)$ space is shown in Fig. 4(b). It starts in the sublevel $|g, -1/2\rangle$, in a minimum of $U_-(z)$, with a kinetic energy larger than the modulation depth:

$$U_0 = -\frac{2}{3}\hbar\delta s \tag{14}$$

of the potential. If the optical pumping time τ_p from $|g, -1/2\rangle$ to $|g, +1/2\rangle$ is long enough, the atom can climb the potential hill and reach the top of $U_-(z)$ before changing its internal state. During this part of the motion, the mechanical energy E_- remains constant and there is a conversion of kinetic energy into potential energy. In the vicinity of the maxima of $U_-(z)$, the laser light is σ_+ polarized and the atom has the maximal probability to be optically pumped into the $|g, +1/2\rangle$ sublevel by absorption of a laser photon followed by a spontaneous emission. If such a process takes place, the atom is goes back into a valley, but for the $U_+(z)$ potential this time. The resulting decrease of potential energy leads to a corresponding excess of energy for the spontaneously emitted photon with respect to $\hbar\omega_L$. Dissipation of energy thus originates in this cooling scheme from the spontaneous emission of anti-Stokes Raman photons.

Thus the atoms do not stop climbing potential hills, as Sisyphus did in Greek mythology, until their kinetic energy becomes of the order of the modulation depth U_0. This intuitive reasoning is confirmed by a theoretical analysis, similar to the one used already for Doppler cooling, based on introducing a friction coefficient α and a momentum diffusion coefficient D, as in brownian motion theory. The derivation of the coefficients D and α leads to the equilibrium temperature given in [17]:

$$k_B T = \frac{\langle p^2 \rangle}{M} \propto U_0 \sim -\hbar\delta s. \tag{15}$$

2.3 Limit of Sisyphus Cooling

The equilibrium temperature given in the previous section is proportional to the laser intensity, and scales as $1/|\delta|$ for large detuning ($|\delta| \gg \Gamma$). This has been confirmed experimentally in 3D on a wide range of detunings and laser intensities, and for several atomic species [15, 16]. The precise value of the slope for the 3D experiments cannot of course be deduced from the simple 1D model that we described. Actually, such a scaling law for Sisyphus cooling cannot be realistic for any values of the parameters. For example, it leads to a vanishing equilibrium temperature in the limit of vanishing laser intensity. In this regime of very small light shifts, the experimental results do indeed deviate from (15).

In order to have a more complete understanding of Sisyphus cooling, one has to take into careful account the heating of the atoms due to random recoil after spontaneous emission. The Sisyphus effect, corresponding to a decrease of the potential energy of the order of U_0 after an optical pumping cycle, is then counterbalanced by an increase of the kinetic energy of the order of the recoil energy $E_R = \hbar^2 k^2/2M$. Cooling works only when U_0 is above a threshold of a few E_R, and the minimal achievable temperature is expected from (15) to scale as the recoil energy E_R.

The precise value of the Sisyphus cooling limit has been derived from a quantum treatment of both the internal atomic state and the atomic motion in laser light, for the simple 1D model presented in [20]. For a given atom-laser detuning δ/Γ, the dependence of the steady state kinetic energy on the modulation depth U_0 of the potential is shown in Fig. 5.

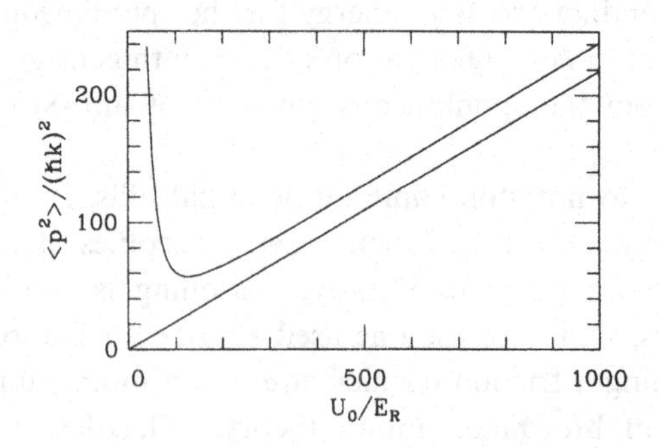

Fig. 5. In a full quantum treatment of Sisyphus cooling, mean kinetic energy $\langle p^2 \rangle / 2M$ in units of recoil energy $E_R = \hbar^2 k^2 / 2M$ as a function of the modulation depth U_0 of the optical potential wells in units of E_R. The atom-laser detuning is $\delta = -5\Gamma$.

The scaling law $k_B T \sim U_0$ is recovered in the limit of large U_0/E_R. The existence of a threshold on U_0 corresponds to a rapid increase of the average kinetic energy, when U_0/E_R becomes too small. The smallest possible value of the root mean square atomic momentum is of the order of 5.5 $\hbar k$, reached for $U_0 \sim 100 E_R$ in the limit of large detunings $|\delta| \gg \Gamma$.

2.4 *Quantization of Motion in the Optical Potential Wells*

It has been found to be fruitful to try to obtain more insight into the coldest regime of Sisyphus cooling. One can check that in this regime the following inequality holds:

$$\Omega_{\mathrm{osc}} \tau_p \gg 1, \tag{16}$$

where $\Omega_{\mathrm{osc}} \sim k\sqrt{U_0/M}$ is the oscillation frequency of the atoms in the bottom of the potential wells and $\tau_p \sim 1/\Gamma s$ is the optical pumping time. This leads to a level spacing $\sim \hbar\Omega_{\mathrm{osc}}$ for the quantized atomic motion in $U_\pm(z)$, much larger than the radiative width $\hbar/\tau_p$ of these levels due to optical pumping.

Therefore, a good basis for the analysis of the cooling is provided by the stationary solutions of the Schrödinger equation for the atomic motion in $U_\pm(z)$ [21]. Since the potentials $U_\pm(z)$ are periodic, of period $\lambda_{\mathrm{opt}}/2$, the corresponding energy spectrum has a band structure. The lowest energy bands, corresponding to quasi-harmonic motion in the bottom of the potential wells (see Fig. 6 taken from [22]), have a negligible tunnelling width. On the contrary, the energy gaps become very small for the quasi-

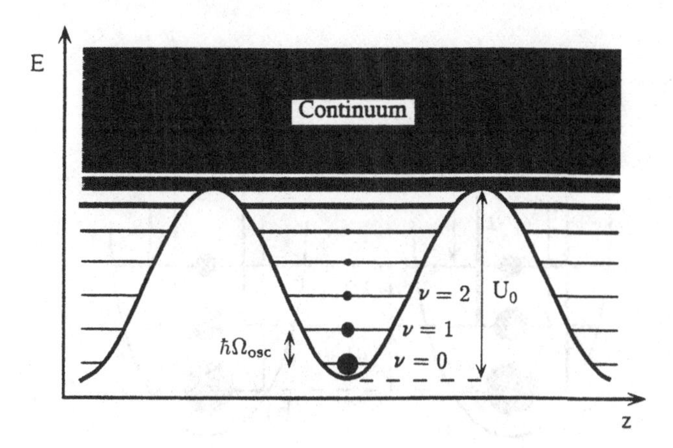

Fig. 6. Band structure of the energy spectrum for the quantum motion in $U_+(z)$ optical potential wells. The vibrational levels are labelled $v = 0, 1, ...$ The dark circular areas are proportional to the steady state populations of the vibrational levels.

free motion well above the potential hills. The effect of optical pumping in this basis can be described simply by transition rates between the various energy levels, when the inequality (16) holds. The corresponding rate equations allow one to calculate the steady state populations of the energy bands. The maximal population in the most populated band $v = 0$ is found to be 0.34 in [21], for $U_0 \simeq 60E_R$.

The existence of well resolved external quantum levels in 1D Sisyphus cooling has been demonstrated experimentally. A probe laser beam, with a very small intensity, is sent through the optical molasses along the direction of the laser cooling beams. Two side-bands are clearly observed on the probe absorption spectrum, for a probe frequency $\omega_p = \omega_L \pm \Omega_{osc}$. They correspond to stimulated Raman transitions between two successive vibrational levels in the potential wells. Since the lowest level is the most populated one, Raman cycles starting with atoms in the state $v = 0$ are expected to be dominant. When ω_p is below ω_L ($\omega_p \simeq \omega_L - \Omega_{osc}$), the dominant process is therefore the absorption of one photon in the cooling beams followed by stimulated emission of one photon in the probe beam, and the probe beam is amplified. When ω_p is above ω_L ($\omega_p \simeq \omega_L + \Omega_{osc}$), the Raman processes are mainly the absorption of one photon in the probe beam followed by the stimulated emission of one photon in the cooling beams, and the probe beam is absorbed (see Fig. 7 taken from [22]). The so-called "overtones" (Raman transitions

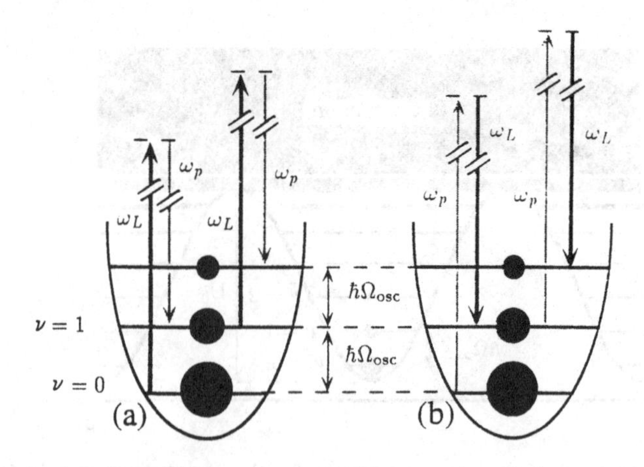

Fig. 7. Examples of Raman processes induced between the atomic vibrational levels by a weak probe beam of frequency ω_p passed through the one-dimensional optical molasses. The populations of the energy levels are represented by dark areas. In (a), the Raman processes lead to an amplification of the probe beam ($\omega_p \simeq \omega_L - \Omega_{osc}$). In (b), the Raman processes lead to an absorption of the probe beam ($\omega_p \simeq \omega_L + \Omega_{osc}$).

with $\delta v = 2$, $\delta v = 3, ...$) can also be identified. A stimulated Rayleigh line is also observed at $\omega_p = \omega_L$; it can be shown to give evidence for a spatial antiferromagnetic order of the atoms [23, 24]. These experiments are the first observations of quantized external levels of a neutral atom in an optical field. An important point is that about half of the atoms are expected to be in minimum uncertainty states [22]. These observations have been confirmed by studies of the fluorescence spectra of 1D optical molasses [25].

2.5 Optical Lattices in 2D and 3D

In the preceding sections, we have seen that 1D Sisyphus cooling in a laser field can lead to a strong accumulation of the atoms around the points where the field is circularly polarized. Such a property can be generalized to two-dimensions (2D) and three-dimensions (3D).

Theoretical results are available in 2D [26, 27]. The quantum approach is similar to the one used in 1D. In this paper, the main results concern the population of the energy levels in steady state. They are found to be strongly reduced: the maximum population of the deepest band shifts from 0.34 in 1D to 0.09 in 2D. This does not mean that cooling is less efficient in 2D, but it is simply a signature of degeneracy effects among the vibrational levels. Let us indeed consider the atomic motion in the bottom of the 2D potential wells. It is found to be almost harmonic, and the energy bands are labeled by the two quantum numbers n_x and n_y of the corresponding 2D isotropic harmonic oscillator. If the motions along x and y are considered as decoupled, the probability of finding the atoms in the ground state $|n_x = 0, n_y = 0\rangle$ of their motion is the product of the probabilities of finding them in the ground states $|n_x = 0\rangle$ and $|n_y = 0\rangle$ of their motion along x and y respectively. This simple reasoning leads to the estimate:

$$\Pi^{2D}_{n_x=n_y=0} \simeq \left(\Pi^{1D}_{n=0}\right)^2, \tag{17}$$

not far from the exact results.

To obtain experimental evidence for vibrational levels in $\sigma_\pm$ optical potential wells, one has to eliminate phase fluctuations between the orthogonal running waves in order to avoid distorsion of the electric field polarization away from $\sigma_\pm$ in the minima of the wells. This has been realized in Münich, with two and three orthogonal standing waves with well controlled phases [28, 29]. In Paris, an alternative to this approach has been developed. The idea is to use in nD the minimal number of running waves leading to cooling, which is $n+1$. In this case, as is evident for the particular case $n = 1$, the periodic pattern of potential wells is only shifted by phase fluctuations. A hexagonal lattice in 2D, and a cubic body centered lattice in 3D, have been obtained and have been analyzed by a probe beam [30].

All these results lead to a completely different physical picture of the laser cooled atomic samples. The concept of optical molasses is replaced by the one of optical lattices, which consist of periodic arrays of micro-traps, with vibrational levels for the atomic motion. This description is well suited to the discussion of statistical effects, the quantum degeneracy (i.e. the population of the most populated state of the one-atom density matrix) being derived simply from the population of the lowest energy band. However, these optical lattices are for the moment quite empty, only a small percentage of the sites being occupied. In order to increase the quantum degeneracy, one has to increase the atomic density, using an extra trapping mechanism.

2.6 Experimental Results from the Magneto-Optical Trap

The geometry of the magneto-optical trap has been described via a simple 1D model in Section 1.4. It can be extended to a 3D configuration in the following way. The magnetic field gradient is provided by two coaxial coils travelled by opposite electric currents. Close to $\mathbf{r} = \mathbf{0}$, where the resulting magnetic field is vanishing, one has the approximate spatial dependence for $\mathbf{B}$:

$$\mathbf{B}(\mathbf{r}) \simeq \frac{b_0}{a}(-x/2, -y/2, z). \tag{18}$$

Three pairs of counterpropagating σ_+ and σ_- circularly polarized laser waves, along x, y and z, provide 3D trapping and cooling in the presence of the magnetic gradient. Note that the laser beams propagating along z and perpendicularly to z have opposite helicities because magnetic field gradients are of opposite signs along z and x, y.

Experimental studies of the magneto-optical trap for atoms in a cell have provided evidence for two different regimes (see, for example, [31] for cesium atoms).

First, a low density regime, with a density n less than 10^{10} atoms/cm^3, is obtained when a small number N of atoms (up to 10^6) are present in the trap. The important point is that the temperatures in the trap are the same (down to 3 μK) as the ones measured in the $\sigma_+ - \sigma_-$ optical molasses in the absence of magnetic field. The sub-Doppler cooling mechanisms, which rely strongly on the light shifts of the various Zeeman ground sublevels, are thus not destroyed by the magnetic field in the trap [31–34]. This is due to the fact that the trapping mechanism accumulates the atoms in a small region around $\mathbf{B} = 0$. The corresponding minimal radii of the atomic cloud in the trap are found experimentally to be of the order of 40 μm for a magnetic field gradient of 10 gauss/cm.

When a large number N of atoms are present in the trap (up to $3 \cdot 10^8$), a high density regime is observed, with maximal densities n around $3 \cdot 10^{12}$ atoms/cm^3. However, the temperatures in the trap are then found to be higher than the ones in the $\sigma_+ - \sigma_-$ optical molasses, the corresponding excess of temperature ΔT scaling simply as $N^{1/3}$ for a given laser detuning and intensity. The highest quantum degeneracies in the trap are actually not obtained at the highest densities.

In order to estimate this quantum degeneracy, we suppose that the mean number of atoms in the most populated quantum levels inside the trap is given by:

$$\Pi_{\max} = n\Lambda_{\mathrm{DB}}^3 \qquad \text{with} \qquad \Lambda_{\mathrm{DB}} = \frac{h}{\sqrt{2\pi M k_{\mathrm{B}} T}}. \tag{19}$$

This formula, valid only for a gas of non-interacting particles in thermal equilibrium in the low degeneracy regime, is found to be in reasonably good agreement, when transposed to the 2D case, with the theoretical results mentioned in Section 2.5. At moderate density $n = 8 \cdot 10^{11}$ atoms/cm^3 and quite low temperature $T = 10\mu$K, a situation which is a compromise between the search for low temperatures and high densities in the trap, equation (19) leads to a degeneracy $\Pi_{\max}$ of the order of 10^{-4}.

We now give a qualitative account of the collective effects responsible for the limits on the maximal densities n and the corresponding temperatures T achievable in the magneto-optical trap.

At long range, the exchange of fluorescence photons between atoms leads to a repulsive force, proportional to the flux of photons emitted from one of the atoms through a fictitious surface located around the other atom and of area σ_{A}, the resonant absorption cross section. It thus scales as $1/r_{12}^2$, where r_{12} is the distance between the two atoms. This force has a mean antitrapping effect, which limits the compression in real space of the atomic cloud by the trap. It also fluctuates, which introduces extra heating and increases the temperature [35, 36]. Fortunately, the effect of this long range interaction should be reduced at high laser detuning.

At short range (i.e. for r_{12} smaller than the optical wavelength λ_{opt}), the dipolar interaction between an atom in excited state and an atom in ground state has non-negligible effects. It shifts the energy levels by an amount of the order of the natural width $\hbar\Gamma$ of the excited state, for $r_{12} \sim \lambda_{\mathrm{opt}}/2\pi$. Such a shift is larger than the light shifts leading to the low laser intensity Sisyphus effect and can destroy the cooling mechanism. At high atomic densities, it can lead to the formation of molecules, with a loss term of atoms from the trap scaling as n^2 [37, 38].

The existence of such strong atomic interactions in the presence of laser light thus leads to the following intuitive conclusion, that the density in the magneto-optical trap cannot exceed a value such that the mean distance between atoms becomes smaller than λ_{opt}, which leads to a maximal density n given by

$$n\lambda_{\mathrm{opt}}^3 \sim 1. \tag{20}$$

This corresponds to $n \sim 2 \cdot 10^{12}$ atoms/cm^3 on cesium, of the order of the experimental results.

3 Beating the Recoil Limit

One of the conclusions of the previous section was that it seemed difficult, in the presence of quasi-resonant laser light of wavelength $\lambda_{\rm opt}$, to achieve atomic densities n so high that $n\lambda_{\rm opt}^3 \gg 1$, while keeping the very low temperatures realized at low densities. In the search for quantum degeneracy, the regime where the atomic DeBroglie wavelength $\lambda_{\rm DB}$ is much larger than the optical wavelength $\lambda_{\rm opt}$ is thus particularly interesting. In this case, the parameter $n\lambda_{\rm DB}^3$ can be on the order of unity, with $n\lambda_{\rm opt}^3$ still being very small, so that collisional effects between atoms should remain small.

This interesting regime is called "subrecoil", because the condition $\lambda_{\rm DB} = h/\Delta p \gg \lambda_{\rm opt}$ implies an equilibrium temperature $T = \Delta p^2/Mk_{\rm B}$ much smaller than the so-called recoil temperature $T_{\rm R} = \hbar^2 k^2/Mk_{\rm B}$, where $k = 2\pi/\lambda_{\rm opt}$. The problem is that subrecoil temperatures cannot be reached by the "usual" cooling scheme described in the previous sections. In these schemes, indeed, spontaneous emission of fluorescence photons, which is essential for carrying away the atomic mechanical energy, never stops. The atoms thus continuously experience random recoils of size $\hbar k$ in momentum space, so that the equilibrium temperature cannot be much smaller than the recoil temperature.

In Section 3.1, we present a 1D subrecoil cooling scheme called "velocity-selective coherent population trapping" [39, 40]. It is not the only one in the laser cooling field. Another scheme has been shown experimentally to beat the recoil limit [41], and there are also several theoretical proposals [42, 43]. The general idea of all these schemes is, however, the same: it consists of accumulating the atoms into states that are very weakly coupled to laser light and with a very narrow width in momentum space. The fundamental limits of subrecoil cooling addressed in Section 3.3 can thus be considered as being quite general.

3.1 1D Velocity-Selective Coherent Population Trapping

Consider the so-called "Λ type" atomic system illuminated by two mutually coherent laser waves, of opposite wave vectors $k\mathbf{e}_z$ and $-k\mathbf{e}_z$ along z and of same frequency $\omega_{\rm L}$, but with orthogonal circular polarizations, respectively σ_+ and σ_- (see Fig. 8). We describe in a quantum manner

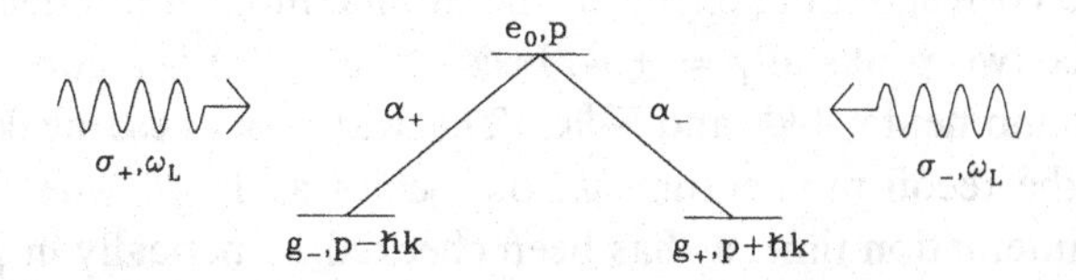

Fig. 8. The Λ atomic system in the σ_+–σ_- laser configuration. In the absence of spontaneous emission, note the existence of closed families of atomic states labelled by p, the atomic momentum along z in the excited state.

the atomic motion along z by introducing the external states $|p\rangle$ corresponding to the plane waves of momentum $p\mathbf{e}_z$. The atomic dynamics can then be analyzed as follows [40].

In the absence of spontaneous emission, an atom initially in the state $|e_0, p\rangle$, i.e. in the internal (electronic) state $|e_0\rangle$ and with a center of mass motion of momentum p along z, is coupled only to $|g_-, p - \hbar k\rangle$ (by stimulated emission of a photon in the σ_+ wave) with an amplitude α_+, and to $|g_+, p + \hbar k\rangle$ (by stimulated emission of a photon in the σ_- wave) with an amplitude α_-. The space of atomic states can then be split into a direct sum of an infinite number of closed families $\mathscr{F}(p) = \{|e_0, p\rangle, |g_-, p - \hbar k\rangle, |g_+, p + \hbar k\rangle\}$ labelled by p. Since there is only one internal excited state $|e_0\rangle$ for two ground levels $|g_\pm\rangle$, we can form, inside each family $\mathscr{F}(p)$, a particular linear combination $|\psi_{\text{NC}}(p)\rangle$ uncoupled to the laser:

$$|\psi_{\text{NC}}(p)\rangle = \frac{1}{\sqrt{|\alpha_+|^2 + |\alpha_-|^2}} [\alpha_- |g_-, p - \hbar k\rangle - \alpha_+ |g_+, p + \hbar k\rangle]. \qquad (21)$$

Note that this combination is not a stationary state, because it has no well defined kinetic energy:

$$(p - \hbar k)^2 \neq (p + \hbar k)^2, \qquad (22)$$

except for the particular case $p = 0$, for which $|\psi_{\text{NC}}\rangle$ is a perfectly trapping state, called the "dark state".

Let us now take into account spontaneous emission. Starting from $|e_0, p\rangle$, the atom now eventually decays into $|g_\pm, p - p_z^{\text{S}}\rangle$, where p_z^{S} is the momentum of the fluorescence photon along z and is randomly distributed between $-\hbar k$ and $+\hbar k$. Spontaneous emission thus couples family $\mathscr{F}(p)$ to the "neighbouring" families $\mathscr{F}(p')$, with $p' = p \pm \hbar k - p_z^{\text{S}}$ in the range $[p - 2\hbar k, p + 2\hbar k]$. This random change of p after a fluorescence cycle may transfer the atoms into the dark state $|\psi_{\text{NC}}(0)\rangle$ or into a quasi-dark state for p very close to 0, where they remain trapped and where they pile up. The corresponding effect on the atomic momentum distribution is to produce two peaks at $p = \pm\hbar k$, because $|\psi_{NC}(0)\rangle$ is a superposition of states of momenta $+\hbar k$ and $-\hbar k$. The width of these peaks is not limited by the recoil momentum $\hbar k$, but scales as $1/\sqrt{t}$, where t is the atom-laser interaction time, as has been checked numerically in [40]. For an idea of the proof and an analysis of the height of the peaks as a function of time, we refer to [44].

3.2 Experimental Results Below the Recoil Limit

The 1D cooling scheme introduced in Section 3.1 has been demonstrated experimentally on metastable helium [39], on a transition between a ground state of angular momentum $j_g = 1$ and an excited state of angular momentum $j_e = 1$. To see why the Λ structure of Section 3.1 is an acceptable model in this geometry, let us introduce the internal atomic states of angular momentum $m\hbar$ around z, $\{|g,m\rangle_z, |e,m\rangle_z, m = 0, \pm 1\}$. Since the laser field has no component along z, the V system formed by $\{|e,-1\rangle_z, |g,0\rangle_z, |e,1\rangle_z\}$, and the Λ system $\{|g,-1\rangle_z, |e,0\rangle_z, |g,1\rangle_z\}$, are not coherently coupled. The key point then is that the excited sublevel of the Λ system $|e,0\rangle_z$ cannot decay by spontaneous emission into $|g,0\rangle_z$, because the corresponding $\Delta m = 0$ Clebsch–Gordan coefficient is vanishing. Therefore, after a few spontaneous emissions from the excited sublevels of the V system, $|e,\pm 1\rangle_z$, the atoms are pumped into the Λ system, from which they cannot escape.

Experimental results on the atomic momentum distribution after an interaction time $t = 350\Gamma^{-1}$, where $\hbar\Gamma$ is the natural width of the excited state, are shown in Fig. 9. The two peaks in $\pm\hbar k$ are visible above the initial momentum distribution. Their half width at half maximum δp allows one to define an effective temperature† T_{eff} of the order of 0.5 recoil temperature T_{R}. Note that the velocity selective coherent population trapping can be generalized on a $j_g = 1 \rightarrow j_e = 1$ transition to multidimensional cooling [40, 45, 46]. A 2D experiment on helium is in progress in Paris.

As mentioned in the introductory remarks of this section, another subrecoil cooling scheme has been demonstrated experimentally [41]. Stimulated Raman transitions, between two hyperfine ground states $|1\rangle$ and $|2\rangle$, are produced by two counterpropagating synchronized laser

† More precisely, δp is defined as the half width of the peaks at the relative height $e^{-1/2}$, so that the equation $T_{\text{eff}} = \delta p^2/M$ defining T_{eff} as a function of δp is exact for a Maxwell–Boltzmann distribution.

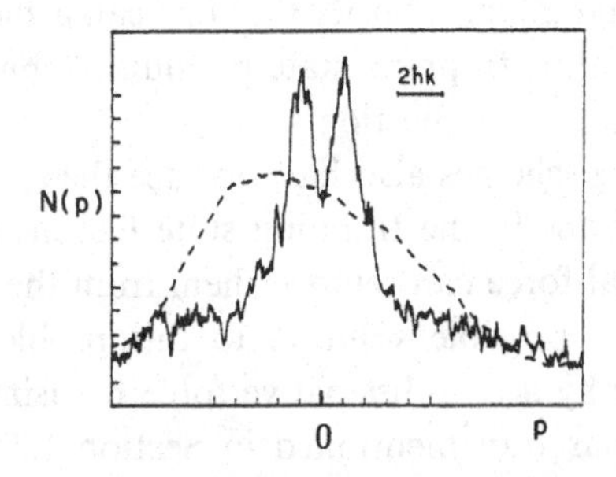

Fig. 9. Experimental results on the momentum distribution for 1D subrecoil cooling. The two peaks of the final distribution in $\pm\hbar k$ (solid line) are above the initial distribution (dashed line).

pulses and provide a high selectivity in atomic velocity. The atoms whose velocity v meets the Raman resonance condition $v = v_0$ are transferred into $|2\rangle$ by the pulses and are then "pushed" towards zero velocity back into $|1\rangle$ through a pumping resonant traveling wave. Successive cooling cycles of this type are then performed, with a scanning of v_0 from high velocities towards low velocities, in order to collect most of the atoms. Since v_0 is always different from 0, there is a small region around $v = 0$ where the atoms never meet the Raman resonance condition, and where they pile up. Very sharp momentum distributions around $p = 0$ have thus been obtained and observed in 1D, with effective temperatures down to $T_R/10$.

3.3 The Limits of Subrecoil Cooling

In order to obtain observable effects depending on the quantum statistics of the atoms, it is necessary to provide three-dimensional (3D) cooling. For the moment, there is no experimental evidence of an efficient velocity-selective coherent population trapping in 3D or even in 2D. In the low density regime, new theoretical methods have been developed to try to predict the evolution of the effective temperature as a function of the interaction time [44]. They rely on Monte-Carlo wave function simulations, describing both the internal and external atomic state quantum mechanically. This procedure has, first, the advantage of using much smaller objects than the full atomic density matrix ρ in 2D or 3D (a discretized wave function has $\mathcal{N}$ components, whereas ρ has $\mathcal{N}^2$ components). Since it deals with stochastic objects, it has also allowed a connection with efficient tools of statistical physics, the so-called Lévy flights [47]. The maximal atomic density that one can achieve without destroying the darkness of the trapping state by multiple photon scattering between atoms is still an open question.

The subrecoil cooling schemes also have to face the severe problem of gravity. Indeed, the atoms in the trapping state feel no radiative force, so that the gravitational force can remove them from the dark state and let them fall down! A possible solution to this problem is to try to counterbalance gravity by a non-dissipative force by using, for example, the non-resonant dipolar trap mentioned in Section 1.2. One can also perform cooling in a free falling frame. This has been done already, for ordinary optical molasses, in an airplane in parabolic flight [48].

A different and promising type of solution is adressed in the next section. The idea is to use gravity as an ally for making a trap, instead of trying to eliminate it.

4 What Next?

As explained in Section 3, the subrecoil cooling schemes known at present cannot be applied efficiently in 3D because of gravity. In this section, we explain how to make an atomic cavity by using gravity and a parabolic atomic mirror (Section 4.1). A quantum picture for the atomic motion inside this cavity is briefly given in Section 4.2: as light waves do in usual laser cavities, matter waves have well defined modes in the gravitational cavity, and one may hope to obtain quantum statistical effects if several atoms occupy the same quantum modes.

4.1 A Gravitational Atomic Cavity

The realization of an atomic cavity requires non-dissipative bounding of the atomic motion in all directions.

Bounding of vertical motion can be obtained through a horizontal plane atomic mirror. The principle of such a device has been given in [49]. By total internal reflection of a travelling laser wave inside a piece of glass, an evanescent wave is produced in the vacuum above the glass, with an intensity $I(z)$ that is exponentially decreasing with increasing z. When the laser frequency ω_L is larger than the atomic resonance frequency ω_A, the atomic ground state experiences a positive light shift in vacuum, which is a decreasing function of z and thus gives rise to a repulsive potential for the atoms. The laser is detuned far from resonance so that dissipative effects during reflection are small: fluorescence cycles

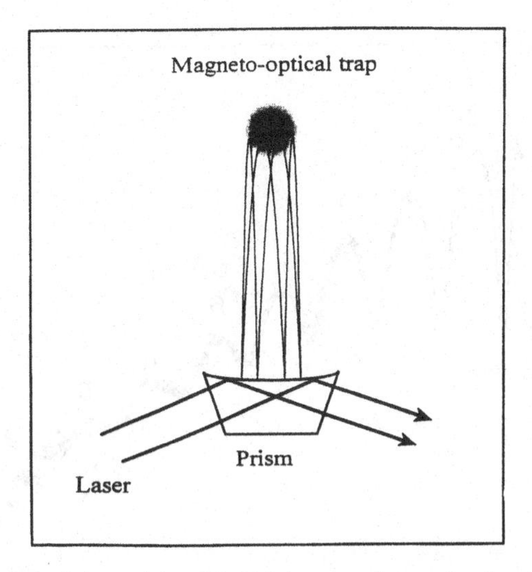

Fig. 10. Principle and loading of the atomic gravitational cavity.

indeed have a rate scaling as $I/(\omega_L - \omega_A)^2$. However, ω_L has to be not too far from resonance so that the laser has a sufficiently high intensity to provide reactive effects (i.e. light shifts, proportional to $I/(\omega_L - \omega_A)$) able to reflect the atoms. This last condition is very restrictive for thermal atomic velocities, and in this case only reflection for grazing atomic incidence is observed [50]. On the contrary, it is easily fulfilled using laser cooling techniques, and reflection under normal atomic incidence can then be obtained [51].

In order to bound the transverse atomic motion also, the plane mirror of the previous scheme is replaced by a parabolic mirror, with an evanescent wave still present at its surface. The atoms are released with negligible initial velocities from a switched-off magneto-optical trap and bounce off the evanescent wave (see Fig. 10). The corresponding classical orbits are paraxial and are shown to be stable [52]. Experimental realizations are in progress in Paris (at the ENS) and in Gaithersburg (at NIST). Recent observation of about ten bounces has been reported in Paris for cesium atoms with the following experimental evidence. At time t after the drop of the atoms, a probe beam is sent through a small volume δV around the initial position of the atomic cloud, and the fluorescence light is collected. This procedure gives a signal proportional to the number $N(t)$ of atoms in the volume δV. Such a measurement is destructive because atoms are pushed away and heated by the probe

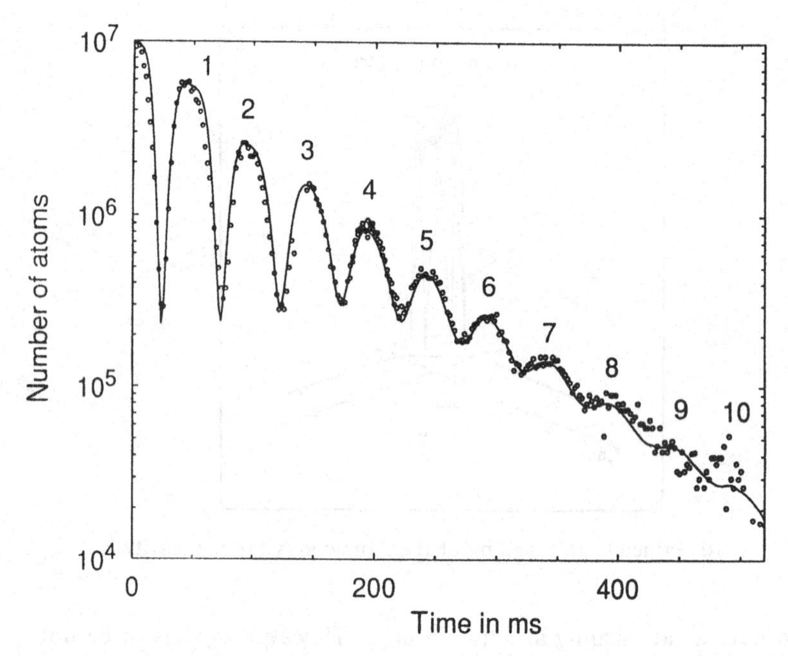

Fig. 11. The number of atoms in the gravitational cavity in Paris is measured as a function of time using the fluorescence induced by a probe beam (see text). The points on the diagram give the number of atoms in the probe beam for different times after their release. The curve is a fit calculated by a Monte-Carlo simulation. The successive bounces are labelled 1,2... .

beam so that the whole process has to be restarted from the beginning if one wishes to obtain $N(t')$ at a different time t'. The result of numerous such cycles is shown in Fig. 11 (see [53]; less recent results, with only four bounces, are given in [54]).

4.2 Quantum Degeneracy Effects in the Cavity

When the atomic motion is treated quantum mechanically, the dynamics in the cavity is analogous to that of light in an optical cavity. The corresponding quantum modes for the matter waves have been investigated theoretically [52]. They correspond mathematically to the stationary solutions of the Schrödinger equation for the atomic wave function $\psi(\mathbf{r},t)$:

$$i\hbar\partial_t\psi = -\frac{\hbar^2}{2M}\Delta\psi + Mgz\psi \tag{23}$$

$$\psi(\mathbf{r},t) = \phi(\mathbf{r})e^{-iEt/\hbar} \tag{24}$$

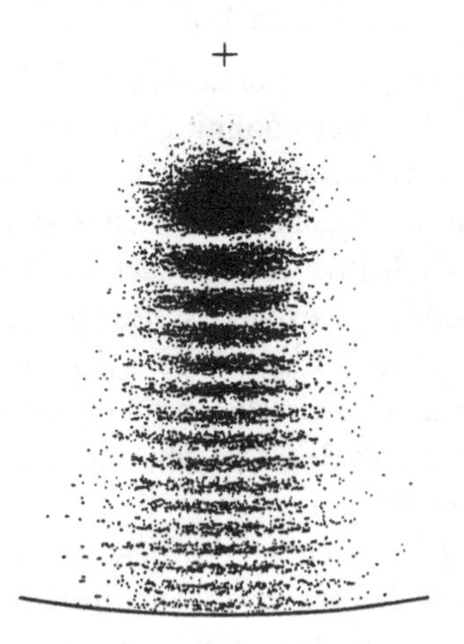

Fig. 12. The calculated probability distribution in real space of a mode in the gravitational atomic cavity. The cross corresponds to the focus of the parabolic surface of the mirror.

with appropriate boundary conditions, responsible for the discrete nature of the allowed energies E. In an idealized model, the atomic mirror is considered a perfect one and ψ is required to vanish on the surface of the mirror. The problem is then readily separated in parabolic coordinates, so that one can solve separately for "transverse" and "vertical" motions. As an illustrative example, a vertical cut of the probability distribution in real space $|\phi(\mathbf{r})|^2$ is given in Fig. 12. The numerous horizontal node lines indicate a highly excited vertical motion. On the other hand, the transverse motion is in the ground state. Estimates of the density of modes in the cavity for realistic experimental conditions are discussed in [52].

The basic idea to obtain statistical effects in the gravitational cavity is simply to put several (bosonic!) atoms in the same quantum mode. It appears to be more promising to try to populate a highly excited mode rather than the ground state. Such a mode has a much larger spatial volume, so that the atomic density in real space is lower and collisional effects are reduced for a given quantum degeneracy. The method of filling one particular mode of the cavity is still a difficult open question. One has to find efficient selective processes, and the subsequent atomic dynamics, far from thermal equilibrium, should be close to that of photons in a laser cavity, rather than to that of Bose–Einstein condensation.

5 Conclusion

In this article, we have given a brief survey of what are currently quite standard techniques in the laser cooling field. Ultracold atomic samples are obtained in optical molasses, optical lattices and in the magneto-optical traps, but quantum degeneracy remains small, for the following fundamental reasons. Collective effects, such as the dipolar interaction between atoms in ground and excited states, prevent us from reaching densities n much larger than $1/\lambda_{\mathrm{opt}}^3$, where λ_{opt} is the optical wavelength. On the other hand, the spontaneously emitted photons limit the maximal atomic de Broglie wavelength λ_{DB} to a few $\lambda_{\mathrm{opt}}/10$, so that $n\lambda_{\mathrm{DB}}^3$ should not be much larger than 10^{-3}.

We have also described some of the new potential offered by laser manipulation of atoms in the search for quantum degeneracies.

First, subrecoil cooling could beat the previous simple reasoning. It leads to de Broglie wavelengths larger than the optical wavelength, so that condition $n\lambda_{\mathrm{DB}}^3 \sim 1$ could be satisfied even though $n\lambda_{\mathrm{opt}}^3 \ll 1$. However, one has to find a way to counterbalance gravity, and more detailed investigations of the effect of multidimensionality are required.

Second, the search for degeneracy effects among the highly excited states of a gravitational atomic cavity is a challenging alternative to Bose–Einstein condensation. It should have the advantage of reducing the collisional constraints. But one first has to find an efficient way of selecting such excited modes!

References

[1] J. Dalibard and C. Cohen-Tannoudji, J. Opt. Soc. Am. B **2**, 1707 (1985).

[2] S. Chu, J. Bjorkholm, A. Ashkin and A. Cable, Phys. Rev. Lett. **57**, 314 (1986).

[3] P. Gould, P. Lett, P. Julienne, W.D. Phillips, W. Thorsheim and J. Weiner, Phys. Rev. Lett. **60**, 788 (1988).

[4] T. Hänsch and A. Schawlow, Opt. Commun. **13**, 68 (1975).

[5] D. Wineland and H. Dehmelt, Bull. Am. Phys. Soc. **20**, 637 (1975).

[6] S. Chu, L. Hollberg, J. Bjorkholm, A. Cable and A. Ashkin, Phys. Rev. Lett. **55**, 48 (1985).

[7] D.J. Wineland and W.M. Itano, Phys. Rev. A **20**, 1521 (1979).

[8] J.P. Gordon and A. Ashkin, Phys. Rev. A **21**, 1606 (1980).

[9] E.L. Raab, M. Prentiss, A. Cable, S. Chu and D.E. Pritchard, Phys. Rev. Lett. **59**, 2631 (1987).

[10] P. Lett, R. Watts, C. Westbrook, W.D. Phillips, P. Gould and H. Metcalf, Phys. Rev. Lett. **61**, 169 (1988).

[11] P. Lett, W.D. Phillips, S. Rolston, C. Tanner, R. Watts and C. Westbrook, J. Opt. Soc. Am. B **6**, 2084 (1989).

[12] J. Dalibard, C. Salomon, A. Aspect, E. Arimondo, R. Kaiser, N. Vansteenkiste and C. Cohen-Tannoudji, *Atomic Physics 11*, S Haroche, J.C Gay and G. Grynberg, eds. (World Scientific, Singapore, 1989), p.199.

[13] Y. Shevy, D.S. Weiss, P.J. Ungar and S. Chu, Phys. Rev. Lett. **62**, 1118 (1989).

[14] D.S. Weiss, E. Riis, Y. Shevy, P.J. Ungar and S. Chu, J. Opt. Soc. Am. B **6**, 2072 (1989).

[15] C. Salomon, J. Dalibard, W.D. Phillips, A. Clairon and S. Guellati, Europhys. Lett. **12**, 683 (1990).

[16] C. Gerz, T.W. Hodapp, P. Jessen, K.M. Jones, W.D. Phillips, C.J. Westbrook and K. Mølmer, Europhys. Lett. **21**, 661 (1993).

[17] J. Dalibard and C. Cohen-Tannoudji, J. Opt. Soc. Am. B **6**, 2023 (1989).

[18] P.J. Ungar, D.S. Weiss, E. Riis and S. Chu, J. Opt. Soc. Am. B **6**, 2058 (1989).

[19] C. Cohen-Tannoudji, in *Fundamental systems in Quantum optics*, Proceedings of Session LIII of Les Houches Summer School, J. Dalibard, J.-M. Raimond and J. Zinn-Justin, eds. (North-Holland, Amsterdam, 1992).

[20] Y. Castin, J. Dalibard and C. Cohen-Tannoudji, *Light Induced Kinetic Effects*, L. Moi, S. Gozzini, C. Gabbanini, E. Arimondo and F. Strumia, eds. (ETS Editrice, Pisa, 1991).

[21] Y. Castin and J. Dalibard, Europhys. Lett. **14**, 761 (1991).

[22] J.Y. Courtois, thèse de doctorat de l'École Polytechnique, Paris, France, 1993.

[23] P. Verkerk, B. Lounis, C. Salomon, C. Cohen-Tannoudji, J.Y. Courtois and G. Grynberg, Phys. Rev. Lett. **68**, 3861 (1992).

[24] J.Y. Courtois and G. Grynberg, Phys. Rev. A **46**, 7060 (1992).

[25] P.S. Jessen, C. Gerz, P.D. Lett, W.D. Phillips, S.L. Rolston, R.J.C. Spreeuw and C.I. Westbrook, Phys. Rev. Lett. **69**, 49 (1992).

[26] K. Berg-Sørensen, Y. Castin, K. Mølmer and J. Dalibard, Europhys. Lett. **22**, 663 (1993).

[27] Y. Castin, K. Berg-Sørensen, K. Mølmer and J. Dalibard, to appear in *Fundamentals of Quantum Optics III*, Ehlotzky, ed. (Springer, Berlin, 1993).
[28] A. Hemmerich, Zimmerman and T.W. Hänsch, Europhys. Lett. **22**, 89 (1993).
[29] A. Hemmerich and T.W. Hänsch, Phys. Rev. Lett. **70**, 410 (1993).
[30] G. Grynberg, B. Lounis, P. Verkerk, J.Y. Courtois and C. Salomon, Phys. Rev. Lett. **70**, 2249 (1993).
[31] A. Clairon, P. Laurent, A. Nadir, M. Drewsen, D. Grison, B. Lounis and C. Salomon, in the *Proceedings of the 6th European Frequency and Time Forum*, held at ESTEC, Noordwijk (ESA SP-340, June 1992).
[32] A.M. Steane and C.J. Foot, Europhys. Lett. **14**, 231 (1991).
[33] A.M. Steane, M. Chowdhury and C.J. Foot, J. Opt. Soc. Am. B **9**, 2142 (1992).
[34] C.D. Wallace, T.P. Dinneen, K.Y.N. Tan, A. Kumarakrishnan, P.L. Gould and J. Javanainen, to be published (1993).
[35] T. Walker, D. Sesko and C. Wieman, Phys. Rev. Lett. **64**, 408 (1990).
[36] A.M. Smith and K. Burnett, J. Opt. Soc. Am. B **9**, 1240 (1992).
[37] A. Gallagher and D.E. Pritchard, Phys. Rev. Lett. **63**, 957 (1989).
[38] P.S. Julienne and J. Vigué, Phys. Rev. A **44**, 4464 (1991).
[39] A. Aspect, E. Arimondo, R. Kaiser, N. Vansteenkiste and C. Cohen-Tannoudji, Phys. Rev. Lett. **61**, 826 (1988).
[40] A. Aspect, E. Arimondo, R. Kaiser, N. Vansteenkiste and C. Cohen-Tannoudji, J. Opt. Soc. Am. B **6**, 2112 (1989).
[41] M. Kasevich and S. Chu, Phys. Rev. Lett. **69**, 1741 (1992).
[42] H. Wallis and W. Ertmer, J. Opt. Soc. Am. B **6**, 2211 (1989).
[43] K. Mølmer, Phys. Rev. Lett. **66**, 2301 (1991).
[44] F. Bardou, J.P. Bouchaud, O. Emile, A. Aspect and C. Cohen-Tannoudji, Phys. Rev. Lett. **72**, 203 (1994).
[45] F. Mauri, F. Papoff and E. Arimondo, in *Light Induced Kinetic Effects*, L. Moi, S. Gozzini, C. Gabbanini, E. Arimondo and F. Strumia, eds. (ETS Editrice, Pisa, 1991).
[46] M.A. Ol'shanii and V.G. Minogin, in *Light Induced Kinetic Effects*, L. Moi, S. Gozzini, C. Gabbanini, E. Arimondo and F. Strumia, eds. (ETS Editrice, Pisa, 1991).
[47] J.P. Bouchaud and A. Georges, Phys. Rep. **195**, 125 (1990).
[48] B. Lounis, J. Reichel and C. Salomon, C. R. Acad. Sci. Paris, Série II, **316**, 739 (1993).
[49] R. Cook and R. Hill, Opt. Comm. **43**, 258 (1982).

[50] V. Balykin, V. Letokhov, Y. Ovchinikov and A. Sidorov, Phys. Rev. Lett. **60**, 2137 (1988).

[51] M. Kasevich, D. Weiss and S. Chu, Opt. Lett. **15**, 607 (1990).

[52] H. Wallis, J. Dalibard and C. Cohen-Tannoudji, Appl. Phys. B **54**, 407 (1992).

[53] C. Aminoff, A. Steane, P. Bouyer, P. Desbiolles, J. Dalibard and C. Cohen-Tannoudji, to be published.

[54] C. Aminoff, P. Bouyer and P. Desbiolles, C. R. Acad. Sci. Paris Série II, **316**, 1535 (1993).

10

Kinetics of Bose–Einstein Condensate Formation in an Interacting Bose Gas

Yu. Kagan

Department of Superconductivity and Solid State Physics
Kurchatov Institute
123182 Moscow
Russia

Abstract

The kinetics of the formation of coherent correlation properties associated with Bose condensation is studied in detail. The evolution of a nonequilibrium state with no "condensate-seed" is related to a hierarchy of relaxation times. At the first stage, a particle flux in energy space toward low energies sets in. The evolution in this case is described by a nonlinear Boltzmann equation, with a characteristic time given by interparticle collisions. When the particles which will later form the condensate have a kinetic energy which is less than the potential energy, a quasicondensate starts to form. In this stage, fluctuations of the density (but not of the phase) are suppressed and short-range coherent correlation properties are governed by the equation of motion for a quasiclassical complex field. The next stage is connected with the formation of the long-range order. The time for forming topological order and therefore genuine superfluidity proves to be dependent on the system size. The off-diagonal long-range order, arising after the attenuation of long-wave phase fluctuations, has a size-dependent relaxation time as well.

1 Introduction

The problem of Bose–Einstein Condensation (BEC) kinetics, being interesting in itself, has acquired a special significance in connection with experimental efforts to observe this condensation in a number of systems with particles with a finite lifetime. Such systems include spin-polarized atomic hydrogen [1], excitons [2] and biexcitons [3] in semiconductors and, more recently, laser-cooled atomic systems [4]. In all cases, the density is supposed to be low so that we always deal with a weakly interacting Bose gas. This makes the investigation of these systems especially attractive, because it allows the possibility of revealing the main features which are characteristic for a Bose system in general, and, at the same time, provides a chance for direct comparison with theory.

In all conceivable metastable systems whose evolution should lead to BEC, the initial state contains no seed of condensate. If the final equilibrium state is one with a sufficiently high condensate density, then we are faced with the time evolution of an essentially nonequilibrium system. Under these conditions the question of the time for attaining BEC, or, more precisely, of the formation time of the appropriate correlation properties, becomes very nontrivial. The answer to this question is important not only for comparison with the lifetime of a system, but it can also be crucial for the possibility of detecting the condensate.

In Ref. [5] it has been shown that the rate for inelastic processes decreases substantially when a Bose condensate appears in the system. This effect opens up an interesting opportunity for the observation of BEC, particularly in an atomic hydrogen gas. Although there is no true Bose condensate for $T \neq 0$ in the two-dimensional case, it was shown in Ref. [6] that the change in the probability for the inelastic processes persists below the Kosterlitz–Thouless transition, due to specific properties of the two-dimensional quasicondensate (i.e. a condensate with a fluctuating phase). However, in all cases this effect manifests itself only after the development of specific quantum correlations in the process of evolution of the system.

In a nonuniform space the opposite effect is realized: the appearance of a condensate causes a sharp increase of the rate of any density-dependent process (see, e.g. Ref. [7]). It allows BEC to be revealed even when the condensate is localized in a very small volume. The latter is characteristic, in particular, of a trap-type experiment (see, e.g. Ref. [8]). Again, we need a certain time for the necessary quantum correlations to form.

One promising way of studying BEC in an excitonic gas is based on the measurement of the transport motion of an expanding gas with the aim of revealing the excitonic superfluidity [2,9]. The use of this method assumes indirectly that the time of the formation of topological order is short enough.

Several theoretical papers have addressed the question of the time evolution of a nonequilibrium interacting Bose gas. Levich and Yakhot [10], considering the evolution of an ideal Bose gas coupled to a heat bath, came to the conclusion that the formation of a condensate peak requires an infinite time. Nozières [11] also obtained the same result. Recently, studying the kinetics of BEC in an ideal Bose gas, Tikhodeev [12] has obtained the same answer, provided the size of the system goes to infinity.

Later, Levich and Yakhot [13] took into consideration the interparticle interactions which should naturally play a dominant role in the temporal evolution. Within the framework of the nonlinear Boltzmann kinetic equation, they found a finite time for BEC formation. However, these authors made a drastic modification of the collision integral in order to simplify the calculation, which led to such a strong change of the character of the evolution of the particle energy distribution that the authors themselves doubted the validity of the results. Moreover, they realized correctly that the kinetic equation is not applicable in the low energy region where the kinetic energy becomes less than the potential energy. The attempt to improve the kinetic equation was not adequate because it still made use of the random-phase approximation, which fails precisely in this region.

Snoke and Wolfe [14] undertook a direct numerical evaluation of the Boltzmann kinetic equation for the weakly interacting Bose gas. The calculations demonstrated the formation of the Bose distribution with chemical potential μ close to zero on the time scale of the classical scattering time of the particles. However, a condensate did not appear even after a much longer period of time. Simultaneously, the conservation of the particle number was violated. As we shall see below, both results are not accidental and have a common origin.

Another extreme result was recently reported by Stoof [15], who asserted that the time required for the formation of a Bose condensate was $\sim \hbar/T_c$, where T_c is the temperature of BEC. That result seems surprising since it does not depend on the interparticle interaction, and moreover this time is shorter than any characteristic interaction time in a dilute Bose gas.

Actually, the question of the formation time of a Bose condensate does not have an unambiguous answer. The real answer depends strongly on the particular problem under consideration (see Ref. [16]). The ambiguity arises at the stage of evolution when a substantial fraction of the particles which should form the condensate in equilibrium is concentrated in the energy interval $\epsilon \leq n_0\tilde{U}$, where $\tilde{U}$ is the effective particle interaction vertex, and n_0 is the equilibrium density of the condensate. At this stage the kinetic energy of the particles becomes smaller than their average potential energy and a pronounced mixing of states with different momenta $\mathbf{k}$ arises. The state of the system, taking into account that in this case the occupation numbers obey the condition $n_{\mathbf{k}} \gg 1$, can be characterized by a quasiclassical complex-valued field Ψ (see e.g. [17]) whose dynamical evolution is described by the nonlinear Schrödinger-type equation ($\hbar = 1$)

$$i\frac{\partial \Psi}{\partial t} = -\frac{\nabla^2}{2m}\Psi + \tilde{U} \mid \Psi \mid^2 \Psi, \tag{1.1}$$

where m is the mass of the particle. At this stage the random phase approximation is not valid and (1.1) cannot be reduced to an equation for occupation numbers. (There is a well-known analogy: the classical electric field and the Maxwell equations as a limit of quantum electrodynamics when the photon occupation numbers are much larger than unity.)

In the limit $t \to \infty$, the solution of (1.1) approaches the value of the complex order parameter with a macroscopically defined modulus and phase which characterize the equilibrium state of an interacting Bose gas at $T < T_c$. In the process of evolution, the modulus and the phase of the function Ψ fluctuate strongly. Only with the relaxation of these fluctuations do the proper coherent correlations start to set in. The attenuation of the modulus or the density fluctuations begins first. This leads to the formation of a short-range order with correlation properties that are close to those of an equilibrium system with a true condensate (see below). The characteristic time scale for this stage of the evolution is given by

$$\tau_c = \frac{1}{n_0 \tilde{U}}, \tag{1.2}$$

and, correspondingly, the length scale of the region where this short-range order comes into play has the value of the correlation length

$$r_c = \frac{1}{(2mn_0\tilde{U})^{1/2}}. \tag{1.3}$$

Thus, even at this stage of evolution, the system should demonstrate a number of physical properties identical to those of a Bose gas with a real condensate. If we are interested in the processes governed by the local correlation properties at distances $r < r_c$, we can conditionally regard τ_c in (1.2) as a time scale for BEC (for the particles which are already in the energy interval $\epsilon \sim n_0\tilde{U}$). Since, however, the relaxation of the phase fluctuations requires a much longer time, a genuine condensate with a δ-function momentum distribution of condensate particles does not appear even at $t \gg \tau_c$. In a sense, we obtain a quasicondensate or a condensate with a fluctuating phase. The organization of long-range order, which is necessary for the δ-function peak to build up, requires a long time, depending on the size of a system.

The formation of the quasicondensate is inevitably accompanied by the appearance of a huge number of topological defects. This is easy to understand if one takes into account the absence of phase correlations at a distance $r \gg r_c$ [18]. Hence, the quasicondensate does not imply macroscopic superfluidity. Due to the short-range order, one can describe the nonequilibrium state in terms of vortex lines and phonons, both being characterized by essentially nonequilibrium distributions. The evolution in this stage is related to the removal of the topological defects from the system and the relaxation of anomalous phase fluctuations preventing the formation of long-range order. The former occurs first, due to a dissipative interaction of vortices with elementary excitations (see, e.g. [19,20]). The decay of the vortex state with some characteristic time τ_v should result in the formation of topological long-range order (TLRO) and, thus, of macroscopic superfluidity. The appearance of TLRO does not imply in general the appearance of the off-diagonal long-range order (OLDRO) and, therefore, a δ-function peak in the particle distribution, since phase fluctuations can survive independently of topological defects. Then, one can expect that after the relaxation of topological defects, the state of system will be similar to the state of a two-dimensional superfluid at finite temperature (the existence of TLRO even though the condensate is absent). Since the phase fluctuations are coupled to the fluctuations of the phonon subsystem, the relaxation in this case and the time τ_ϕ for approaching OLDRO can be studied via the attenuation of nonequilibrium phonon fluctuations [18]. The times τ_v and τ_ϕ are both sensitive to the temperature and on the size of the system.

Returning to the formation of the quasicondensate, we note that the time τ_c in (1.2) is not the only time scale and generally not the longest one. Usually, the future condensate particles are initially located in an energy interval far from the coherent region $\epsilon < n_0\tilde{U}$. As a result, the initial evolution of the nonequilibrium subsystem can be described by the Boltzmann kinetic equation. In the most interesting energy region, where the nonlinearity of the collision integral plays a decisive role, the occupation numbers obey the condition $n_{\mathbf{k}} \gg 1$. This means that (1.1) is correct in this region as well. But now all modes can be regarded as independent and the use of the random-phase approximation is justified. This leads to the relation $< a^*_{\mathbf{k}} a_{\mathbf{k}'} >= n_{\mathbf{k}} \delta_{\mathbf{k},\mathbf{k}'}$, where $a_{\mathbf{k}}$ is a Fourier component of the field Ψ. (This relation is not correct in the coherent region.) Naturally, under these conditions the coherent correlations are absent. It is important to emphasize that the region where (1.1) is correct proves to be much wider than the coherent one, not to mention a narrow region where (1.1) is used for the description of the superfluid state (the Ginsburg–Gross–Pitaevskii equation – see, e.g. [21]). In the framework of the random-phase approximation, (1.1) can be transformed into a kinetic equation. In the homogeneous case, the latter assumes the form (see, e.g. [22])

$$\dot{n}_{\mathbf{k}}(t) = -J_{\text{coll}}(\mathbf{k}, [n]), \tag{1.4}$$

$$\begin{aligned} J_{\text{coll}}(\mathbf{k}_1, [n]) = \tfrac{\tilde{U}^2}{16\pi^5} \int d^3\mathbf{k}_2 d^3\mathbf{k}_3 d^3\mathbf{k}_4 [n_1 n_2 (n_3 + n_4) - (n_1 + n_2) n_3 n_4] \\ \times \delta(\mathbf{k}_1 + \mathbf{k}_2 - \mathbf{k}_3 - \mathbf{k}_4) \delta(\epsilon_1 + \epsilon_2 - \epsilon_3 - \epsilon_4), \end{aligned} \tag{1.5}$$

where $n_i \equiv n_{\mathbf{k}_i}, \epsilon_i = k_i^2/2m$. Equation (1.4) differs from the general Boltzmann equation for an interacting Bose gas only by the absence of a spontaneous contribution to the scattering ($n_{\mathbf{k}} \gg 1$!).

If one tries to use (1.4) and (1.5) in the entire energy interval including the coherent region, one faces a serious difficulty. This becomes clear if one compares (1.1) and (1.4). The kinetic equation (1.4) is independent of the sign of U, whereas (1.1) is very sensitive to this sign. When U is negative the evolution of Ψ in the coherent region leads to collapse (see Ref. [22]) with the absorption of particles at $\epsilon = 0$, but when U is positive, a condensate should form. The kinetic equation itself cannot distinguish between these two cases. In numerical studies of (1.4) where the additional special condition at $\epsilon = 0$ is not introduced, it becomes impossible to prevent the solution from collapsing in the above sense (see the discussion in [23]). Evidently, Ref. [14] has this problem.

The evolution in the energy region where the employment of (1.4) is justified is characterized by a time scale (see Ref. [16] and below)

$$\tau_{\text{kin}} = \frac{1}{n\sigma v_T}, \tag{1.6}$$

which is the usual kinetic time of interparticle collision. Here σ is the scattering cross section, v_T is the thermal velocity and n is the total density of the gas. Comparing (1.6) and (1.2), we see that the condition

$$\tau_{\text{kin}} \gg \tau_c \tag{1.7}$$

essentially always holds.

Thus, the kinetics of BEC are characterized by a hierarchy of time scales and the overall scenario is as follows. Let us assume that after fast cooling, the particles which will subsequently from the condensate are initially concentrated at high energies, of the order of the equilibrium value T_c. Several collisions per particle in this case result in the formation of a particle flux in energy space toward lower energies. The self-evolution in the interval between T_c and $n_0\tilde{U}$ (kinetic region) along the energy axis is described by the Boltzmann kinetic equation. A certain energy ϵ_* divides this interval in two parts. For $\epsilon < \epsilon_*$, the occupation numbers $n_{\mathbf{k}}$ become greater than the equilibrium ones (as a result, $n_{\mathbf{k}} \gg 1$) when the particle flux approaches this region, and one can take advantage of (1.4). At $\epsilon > \epsilon_*$, the general Boltzmann equation must be used. As the analysis below shows, the time scale for the passage through both kinetic regions is determined by (1.6).

According to (1.7), the initial stage of transformation in the coherent region takes place much faster. Thus, the evolution in the kinetic region turns out to be the limiting step for the formation time of a quasicondensate. Further steps in the BEC scenario are related to the relaxation of the topological defects and the nonequilibrium phase fluctuation, with times τ_v and τ_ϕ extremely long in comparison to the kinetic time τ_{kin} given by (1.6).

2 Coherent Regime. Formation of the Quasicondensate

We now study the evolution in the coherent region $\epsilon < n_0 U$, assuming that the bulk of the particles which are about to form the condensate are in this energy interval. As we have discussed, we have to use (1.1) for the complex field Ψ in this case.

In the limit $t \to \infty$, the function Ψ should reach its equilibrium value

$$\Psi_0 = n_0^{1/2} \exp(i\Phi_0 - i\mu_0 t), \qquad \mu_0 = n_0\tilde{U}, \tag{2.1}$$

where Φ_0 is the equilibrium value of the condensate phase. We are ignoring the particles outside the condensate in this energy region, since their total number per unit volume is small in comparison with n_0. Writing the function Ψ in the form

$$\Psi_0 = n_0^{1/2} \exp(-i\mu_0 t) f(\mathbf{r}, t), \tag{2.2}$$

we can rewrite (1.1) as

$$i\frac{\partial f}{\partial t} = -\frac{\nabla^2}{2m} f + n_0\tilde{U}(|\, f\,|^2 f - f).$$

In terms of dimensionless variables

$$\bar{t} = t/\tau_c, \qquad \bar{r} = r/r_c \tag{2.3}$$

(see (1.2) and (1.3)), this equation becomes

$$i\frac{\partial f}{\partial \bar{t}} = -\Delta_{\bar{r}} f + | f |^2 f - f. \qquad (2.4)$$

As $\bar{t} \to \infty$, we have $f \to \exp(i\Phi_0)$.

It follows from (2.4) that the evolution on length scales $\sim r_c$ occurs over times $\sim \tau_c$. With increasing $\bar{t}$, the gradients of the phase and the density are smoothed out. This smoothing is equivalent to a decrease in the scale of the important values of the wave vector $\mathbf{k}$ and frequencies ω in the Fourier representation of the function f. The excess energy per particle $\sim n_0 U$ goes off into the kinetic region. It turns out that the condition $\bar{t} \gg 1$ is by itself a sufficient condition for substantial suppression of the fluctuations in the absolute value of f, that is, the density.

To demonstrate this point, we introduce a deviation δn from the mean spatial density n_0. Using (2.4), we write separate equations for the absolute value and phase of function f. After linearization, the equations for the Fourier components of the density and phase become

$$i\omega_{\mathbf{k}}\tau_c(\delta n_k/n_0) = -2r_c^2k^2\Phi_k,$$

$$i\omega_{\mathbf{k}}\tau_c\Phi_k = (\delta n_k/n_0)(1 + r_c^2k^2/2). \qquad (2.5)$$

This system of equations yields a spectrum which essentially is the Bogoliubov spectrum,

$$\omega_{\mathbf{k}}^2 = (2r_c^2k^2/\tau_c^2)(1 + r_c^2k^2/2). \qquad (2.6)$$

From the first of the equations in (2.5), we find

$$[\delta n_k/n_0]^2 = (4r_c^4k^4/\omega_k^2\tau_c^2) \mid \Phi_k \mid^2 . \qquad (2.7)$$

For $\bar{t} \gg 1$, the characteristic value $\mathbf{k}_*$ of the momentum becomes smaller than $k_c = r_c^{-1}$. For such values of k, the dispersion relation given by (2.6) is acoustic,

$$\omega = ck, \qquad c = 2^{1/2}(r_c/\tau_c) \qquad (2.8)$$

and we have

$$[\delta n_k/n_0]^2 = 2r_c^2k^2 \mid \Phi_k \mid^2 .$$

Hence we obtain

$$< (\delta n/n_0)^2 > = \sum_{\mathbf{k}} \mid \delta n_k/n_0 \mid^2 = 2r_c^2 \sum_{\mathbf{k}} k^2 \mid \Phi_k \mid^2 . \qquad (2.9)$$

Taking account of the relation $E \sim (k_*^2/2m)n_0 V$ for the kinetic energy remaining in the system, and using

$$E \approx (n_0/m) \int d\mathbf{r} (\nabla\Phi)^2 = (n_0 V/m) \sum_{\mathbf{k}} k^2 \mid \Phi_k \mid^2,$$

we find from (2.9)

$$< (\delta n/n_0)^2 > \sim (k_*/k_0)^2 \ll 1 . \tag{2.10}$$

Consequently, over times determined by τ_c, a state forms in which density fluctuations are essentially suppressed (as in a true condensate). It is interesting to show that suppression of density fluctuations is a necessary and sufficient condition for the same change in the probability of inelastic processes to appear, as in the presence of a true condensate. With this goal in mind, we consider the example of three-particle recombination, which is a limiting process in the decay of high density spin-polarized atomic hydrogen. For this process the probability is determined by the correlation function

$$Z = < \hat{\Psi}^+(\mathbf{r},t)\hat{\Psi}^+(\mathbf{r},t)\hat{\Psi}^+(\mathbf{r},t)\hat{\Psi}(\mathbf{r},t)\hat{\Psi}(\mathbf{r},t)\hat{\Psi}(\mathbf{r},t) >, \tag{2.11}$$

as was shown in Ref. [5] This is a typical short-range correlation function. The coincidence of arguments reflects only the fact that the interaction takes place at distances much less than r_c and times much less than τ_c. In the case at hand, this correlation function reduces to

$$Z = n_0^3 < f^{*3}(\mathbf{r},t) f^3(\mathbf{r},t) > = < n^3 > . \tag{2.12}$$

If fluctuations in f are suppressed, then

$$Z = n_0^3 . \tag{2.13}$$

If fluctuations are present, then

$$Z = n_0^3 + 3n_0 < (\delta n_0)^2) > + < (\delta n_0)^3) > . \tag{2.14}$$

The expression for δn can be written

$$\delta n/n_0 = (\mid f \mid^2 - 1) . \tag{2.15}$$

Consequently,

$$< (\delta n/n_0)^2 > = n_0^2 [<\mid f \mid^4> -2 <\mid f \mid^2> +1],$$

$$< (\delta n/n_0)^3 > = n_0^3 [<\mid f \mid^6> -3 <\mid f \mid^4> +3 <\mid f \mid^2> -1]. \tag{2.16}$$

Near the boundary between the coherent and kinetic regions, where (2.1) continues to be correct, all the field modes corresponding to different wave vectors **k** can be regarded as independent, i.e.

$$< f^*_{\mathbf{k}} f_{\mathbf{k}'} > = | f_{\mathbf{k}} |^2 \, \delta_{\mathbf{k},\mathbf{k}'} \, . \tag{2.17}$$

Expanding f in a Fourier series, and using an analog of Wick's theorem, we then find

$$<| f |^2> = \sum_{\mathbf{k}} | f_k |^2 = 1, \qquad <| f |^4> = 2, \qquad <| f |^6> = 6 \, .$$

Using these in conjunction with (2.14)-(2.16), we obtain

$$Z = 6 n_0^3 \, . \tag{2.18}$$

We conclude that when density fluctuations are substantially suppressed, the probability for three-particle recombination is reduced from (2.18) by a factor of six (see (2.13)). This result is precisely the same as that found for the case of a true condensate [5,6]. From this point of view, a state with narrow "precondensate" peak ($k_* \ll k_0$) in the particle distribution has all the properties of a true condensate for all processes defined by short distance and short time interval correlations. Such a state can be considered as a quasicondensate. These results demonstrate that the formation of the quasicondensate starts only when the future condensate particles turn out to be in a coherent region. After that, it takes time of the order of τ_c given by (1.2) for a proper transformation to evolve.

3 Kinetic Region. Linear Regime

We now consider the kinetic region, assuming self-evolution of a nonequilibrium system. We suppose that the disturbance of the system after fast cooling, as well as the location of extra particles which should later set up the condensate, are related mainly to high energy degrees of freedom at the initial time. Actually, this is rather typical for all conceivable systems.

To get away from the fluctuation region near T_c, we assume that the condensate density n_0 in the equilibrium will be restricted by the inequality

$$n_0/n \gg (na^3)^{1/3} \, , \tag{3.1}$$

in accordance with the Ginsburg criterion (see, e.g. Ref. [24]). Here n is the total density of particles, and a is the s-wave scattering length. The latter is connected to $\tilde{U}$ via the well-known relation

$$\tilde{U} = \frac{4\pi a}{m}. \tag{3.2}$$

The small value of the gas parameter na^3 allows (3.1) to be compatible with the conditions

$$n_0/n \ll 1 \tag{3.3}$$

and

$$\frac{T_c - T}{T_c} \ll 1, \tag{3.4}$$

where T is the equilibrium temperature of the gas. We take advantage of this possibility to make the discussion more transparent.

For the energy ϵ_* which defines the boundary of the nonlinear region, we have in this case

$$\epsilon_* \ll T. \tag{3.5}$$

This means that in the first stage the excess particle flux in the energy space covers a rather wide interval (T, ϵ_*), where the deviation of the distribution function from its equilibrium value is relatively small.

The general kinetic equation for a Bose gas in a spatially uniform case, for a distribution which is isotropic in terms of momentum, has the well-known form $(n_i \equiv n_{\epsilon_i})$

$$\dot{n}_1 = -W \int d\epsilon_2 d\epsilon_3 d\epsilon_4 \chi(\frac{\epsilon_2}{\epsilon_1}, \frac{\epsilon_3}{\epsilon_1}, \frac{\epsilon_4}{\epsilon_1})\delta(\epsilon_1 + \epsilon_2 - \epsilon_3 - \epsilon_4)$$

$$\times [n_1 n_2 (1 + n_3)(1 + n_4) - (1 + n_1)(1 + n_2) n_3 n_4], \tag{3.6}$$

where

$$W = \tilde{U}^2 m^3 / 4\pi^3 \hbar^3 = \frac{4ma^2}{\pi \hbar^3} \tag{3.7}$$

and

$$\chi = \frac{4}{\pi} \int \frac{dy}{y^2} \sin(y) \sin[y(\frac{\epsilon_2}{\epsilon_1})^{1/2}] \sin[y(\frac{\epsilon_3}{\epsilon_1})^{1/2}] \sin[y(\frac{\epsilon_4}{\epsilon_1})^{1/2}]. \tag{3.8}$$

We have restored $\hbar$ in (3.7) to exhibit the dimensionality. The function χ for the entire set of ϵ_i satisfying the energy conservation condition can be reduced to the simple form (see Ref. [14])

$$\chi = [\min(\epsilon_1, \epsilon_2, \epsilon_3, \epsilon_4)/\epsilon_1]^{1/2}. \tag{3.9}$$

The temporal evolution in this region can be found by means of the solution of the linearized equation (3.6). The following simple considerations enable one to reveal the main features of this stage.

We assume that a local quasiequilibrium is reached in the energy interval where the bulk of excess particles is concentrated. Then the particle distribution can be written as

$$n_\epsilon \approx [\exp[(\epsilon - \mu(t))/T] - 1]^{-1} \approx T/(\epsilon - \mu(t)), \tag{3.10}$$

with a time-dependent chemical potential $\mu(t) > 0$. In general, the temperature in (3.10) may also depend on the time. However, it is easy to verify that this time dependence can be ignored by virtue of (3.1). Relation (3.5) of course holds only for $\epsilon > \mu(t)$. Over most of the linear interval, the condition $\mu \ll \epsilon$ holds, as we will see below. We can thus write

$$\Delta n_\epsilon = n_\epsilon - \overline{n}_\epsilon \approx \mu(t)T/\epsilon^2\,. \tag{3.11}$$

The ϵ dependence corresponds to motion of the front of the distribution $\epsilon_0(t)$, with a width determined by the same parameter $\epsilon_0(t)$. The value $\epsilon_0(t)$ itself is found from the condition

$$\alpha \int_{\epsilon_0}^{T} \epsilon^{1/2} \Delta n_\epsilon d\epsilon = n_0, \tag{3.12}$$

where $\alpha = m^{3/2}/2^{1/2}\pi^2$. The integral is determined by its lower limit and has the value $2T\mu(t)/\epsilon_0^{1/2}(t)$. Hence

$$\mu(t) = n_0 \epsilon_0^{1/2}(t)/2\alpha T. \tag{3.13}$$

The boundary of the linear regime, ϵ_*, is determined by the condition $\delta n_\epsilon/\overline{n}_\epsilon \sim 1$. It can be found directly:

$$\epsilon_* \approx T(n_0/n)^2 \tag{3.14}$$

(here we have made use of the circumstance that T is close to T_c, and we have used the relationship between T_c and n). In this case we have

$$\mu(t) = [\epsilon_* \epsilon_0(t)]^{1/2} < \epsilon_0(t)\,. \tag{3.15}$$

The parameter $\mu(t)$ lags behind $\epsilon_0(t)$ at all times and is comparable to this quantity when the boundary of the linear regime, ϵ_*, is reached. This circumstance demonstrates that it is legitimate to assume a quasiequilibrium distribution of the form given in (3.10) and (3.11).

Let us find an expression for the particle flux in energy space. This flux is related to the motion of the front $\epsilon_0(t)$. Using (3.11) and (3.13) we find

$$Q = -\alpha \epsilon_0^{1/2} \epsilon_0 n_{\epsilon_0} \approx -\frac{1}{2} n_0 (\dot{\epsilon}_0/\epsilon_0) . \tag{3.16}$$

The ratio $\dot{\epsilon}_0/\epsilon_0$ can be found from the kinetic equation (3.6). Analysis of the collision integral shows that the main contribution is provided by processes in which all particles have an energy of the same order of magnitude. Besides, after linearization, the collision-integral operator is a homogeneous functional of the energy to zeroth order. In other words, it does not change under the replacement $\epsilon_i \rightarrow \lambda \epsilon_i$. As a result, the collision integral is characterized by an energy-independent relaxation time and we have approximately

$$\dot{\epsilon}_0/\epsilon_0 \approx -\gamma/\tau_{\text{kin}} . \tag{3.17}$$

Explicit calculations using the kinetic equation (3.6) result in the estimate $\gamma \approx 4$ (τ_{kin} is equal to (1.6) with $\sigma = 8\pi a^2$). The particle flux in energy space thus takes the simple form

$$Q \approx \gamma n_0 / 2\tau_{\text{kin}} . \tag{3.18}$$

It is interesting to note that the flux has a wave-shape with an increasing front amplitude and a decreasing particle-distribution width during the passage of the considered energy interval (see (3.11)). Using (3.17) and (3.14), we can estimate the time which takes the front of the distribution to reach the boundary of the linear region ϵ_* as

$$\tau_L \approx \frac{2\tau_{\text{kin}}}{\gamma} \ln\left(\frac{n}{n_0}\right) . \tag{3.19}$$

The time required to cross the linear region is thus determined by the ordinary collision time scale τ_{kin}, enhanced by a factor $\ln(n/n_0)$ at small n_0.

4 Nonlinear Kinetic Regime

Consider the nonlinear region $n_0 \tilde{U} < \epsilon < \epsilon_*$, where we have

$$n_\epsilon \gg \overline{n}_\epsilon \gg 1 . \tag{4.1}$$

Under condition (3.1) the inequality

$$\epsilon_* \gg n_0 \tilde{U}$$

is always satisfied, so the system necessarily passes through the nonlinear kinetic regime. The evolution in this region is described by the kinetic equation (1.4). After integrating over the directions of momenta, this equation takes on the form (3.6) if the expression in square brackets is replaced by

$$n_1 n_2 (n_3 + n_4) - (n_1 + n_2) n_3 n_4 \,. \tag{4.2}$$

It can be concluded from the structure of the collision integral that the effective kinetic time is determined by

$$\tau_{\text{eff}}^{-1} \sim W n_\epsilon^2 \epsilon^2 \,. \tag{4.3}$$

Analyzing a nonlinear kinetic equation of such a type, Zakharov [22] found that the equation $J_{\text{coll}} = 0$ has two extra solutions, corresponding to a steady state particle flux and to a steady state energy flux in the energy space. In first case, the solution has the form

$$n_\epsilon = A/\epsilon^{7/6} \,. \tag{4.4}$$

Substituting this solution into (4.2), we find

$$\tau_{\text{eff}}^{-1}(\epsilon) \sim W A^2/\epsilon^{1/3} \,. \tag{4.5}$$

Thus, the solution (4.4) leads to the enhancement of the collision processes with decreasing energy in both the incoming and the outgoing terms.

Considering the particle flux moving toward $\epsilon = 0$, there are good reasons to suppose that behind the front the particle distribution should be close to (4.4). Direct numerical calculations performed by Svistunov [23] have corroborated this conjecture.

Thus, the effective transformation from the distribution in (3.11) to that in (4.4) starts at the energy $\epsilon \sim \epsilon_*$. Once the distribution (4.4) has been established behind the front $\tilde{\epsilon}_0(t)$ of a particle-flux wave, the parameter A can be found from the condition

$$\alpha \int_{\tilde{\epsilon}_0}^{\epsilon_*} d\epsilon \; \epsilon^{1/2} A/\epsilon^{7/6} \approx n_0 \,.$$

For $\tilde{\epsilon}_0(t)$ much lower than ϵ_* we have

$$A \approx \frac{n_0}{3\alpha} \frac{1}{\epsilon_*^{1/2}} \,. \tag{4.6}$$

Note that the restructuring of the distribution from (3.11) to (4.4) leaves a substantial fraction of the particles in the tail of the distribution ($\epsilon \sim \epsilon_*$)

Making use of (4.6), one can rewrite (4.5) in the form

$$\tau_{\text{eff}}^{-1}(\epsilon) \sim \tau_{\text{kin}}^{-1} \left(\frac{\epsilon_*}{\epsilon} \right)^{1/3} \,. \tag{4.7}$$

Writing the kinetic equation as $\dot{n}_\epsilon = J_+ - J_-$, we have $J_\pm \sim n_\epsilon/\tau_{\text{eff}}(\epsilon)$, while for the left-hand side we have $\dot{n}_\epsilon \sim n_\epsilon/\tau_{\text{kin}}$. It means that in the first approximation $J_{\text{coll}} \approx 0$ at $\epsilon \ll \epsilon_*$ and, consequently, the solution (4.4) is self-consistent.

The value of the flux associated with the motion of the front is determined by a relation such as (3.16). Taking into account (4.4), we obtain

$$Q = -3\alpha A \frac{(d\tilde{\epsilon}_0^{1/3})}{dt}. \tag{4.8}$$

Under the assumption of a constant flux, this relation leads to

$$\tilde{\epsilon}_0(t) \approx \epsilon_*(1 - t/t_0)^3, \tag{4.9}$$

where

$$t_0 = n_0/Q \tag{4.10}$$

(t_0 is reckoned from the beginning of the nonlinear regime). The quantity t_0 determines the time at which the front of the particle-flux wave arrives in the coherent region. Using the value of the flux Q entering from the linear region (see (3.18)), we immediately conclude $t_0 \sim \tau_{\text{kin}}$.

In the initial stage, the buildup of particles in the coherent region is described by

$$n_c \approx Q(t - t_0). \tag{4.11}$$

Considering the subsequent evolution, we can easily understand that the particles in the kinetic region $\epsilon > n_0\tilde{U}$ interact with the particles in the coherent region just as in the case of a real condensate. Without paying attention to the evolution in the coherent region by itself, one can introduce as an approximation the following system of coupled equations (see Ref. [23]),

$$\dot{n}_\epsilon = -J_{\text{coll}}([n], \epsilon) - n_c J'_{\text{coll}}([n], \epsilon), \tag{4.12}$$

$$\dot{n}_c = n_c J''_{\text{coll}}([n]), \tag{4.13}$$

where

$$J'_{\text{coll}}([n], \epsilon_1) = \frac{W}{\alpha}\frac{1}{\epsilon_1^{1/2}}\Big\{\int_{n_0\tilde{U}}^{\epsilon_1} d\epsilon_2[n_1(n_2 + n_{1-2}) - n_2 n_{1-2}]$$

$$+2\int_{\epsilon_1}^{\infty} d\epsilon_2[n_1 n_{2-1} - n_2(n_1 + n_{2-1})]\Big\}, \tag{4.14}$$

$$J''_{\text{coll}}([n]) = W \int d\epsilon_2 d\epsilon_3 [n_2 n_3 - n_{2+3}(n_2 + n_3)]. \tag{4.15}$$

The first term on the right-hand side of (4.12) describes the interaction of particles belonging to the kinetic region before and after a collision. The distribution (4.4) causes the divergence of the integral in (4.15) at low energies. If one defines the lower bound of the energy as ϵ_c, one obtains for the collision integral

$$J''_{\text{coll}} \sim \frac{1}{\epsilon_c^{1/3}}. \tag{4.16}$$

This result helps one to understand that the transition to the coherent region is evolving from intervals with successively increasing values of energy. As we have already mentioned, the majority of the particles in the nonlinear kinetic region are concentrated in the tail of the particle distribution. This is why the transition will effectively be over when ϵ_c increases up to energy of the order of ϵ_{*}. It is easy to estimate from (4.13) and (4.15) that this time has the same order of magnitude as τ_{kin}.

Thus, the crossing of the nonlinear kinetic region and the arrival of the majority of the excess particles into the coherent region occurs on the time scale determined by τ_{kin}. Since $\tau_{\text{kin}} \gg \tau_c$ this means that the overall time of the quasicondensate formation has the same time scale τ_{kin}.

The results found above are not dependent on the inequality $n_0 \ll n$. As n_0 increases, the boundary ϵ_* moves into the high energy region, and the flux Q increases. However, the time scale stays the same. If $n_0 \sim n$ holds, then $\epsilon_* \sim T$ and the nonlinear regime covers practically the entire kinetic region while retaining τ_{kin} as the time scale.

5 Formation of Topological Order

As we have discussed, the formation of the quasicondensate, related to the suppression of density fluctuations, leaves out large long-range phase fluctuations. The relaxation of these fluctuations on one hand and the appearance, as a result, of true superfluidity and a genuine condensate, require quite different time scales.

Suppose that the quasicondensate correlation properties have already been set up in a region with size scale $r_0 \gg r_c$. At distances greater than r_0 any phase correlations are absent. Let us mentally divide the space into boxes with linear size $l > r_0(t)$. The absence of correlations between the phase in the different boxes easily leads to the appearance of a vortex line or a vortex ring of the size of the l scale. If we increase l

but simultaneously remove a part of the phase field created by the rings with size less than l, then the absence of the correlation between phases in bigger boxes leads to the same conclusion. As a result, for the number of rings $W(R)$ with radius R in the volume unity we have

$$W(R) \sim \frac{1}{R^3}. \tag{5.1}$$

In the general case, the phase Φ of the function Ψ can be represented as the sum

$$\Phi = \Phi_0 + \phi, \tag{5.2}$$

where Φ_0 is the phase field, determined purely by the configuration of the vortex lines. To remove the ambiguity, the relation

$$\oint \nabla\Phi_0 d\mathbf{l} = \pm 2\pi$$

for any contour enclosing a vortex line should be supplemented by the equation $\Delta\Phi_0 = 0$ for the entire region out of the vortex lines.

If we make use of the same picture of uncorrelated boxes, one can ascertain that the difference between the values of the regular part of phase ϕ in different boxes is of the order of magnitude π. Owing to the arbitrariness of the box size this leads to the following relation for the Fourier harmonics of the field ϕ:

$$|\phi_k|^2 k^3 \sim 1, \qquad k \ll r_0^{-1}(t). \tag{5.3}$$

Making use of the well-known relation between $|\phi_k|^2$ and the phonon number occupations (see, e.g. Ref. [21]) allows us to proceed from the relaxation of phase fluctuations to the evolution of the nonequilibrium phonon distribution. Since in the process of the quasicondensate formation, the excitation spectrum becomes acoustic (see Section 2), we find, with allowance for (5.3),

$$n_k \approx \frac{n_0}{mc}\frac{1}{k^2}, \qquad k \ll r_0^{-1}(t). \tag{5.4}$$

Here c is the sound velocity (2.8). Let us first consider the relaxation of topological defects. The main channel of the relaxation is self-annihilation of the vortex rings caused by interaction with the elementary excitations which provide the drag force and thus the dissipation of the energy. The small-radius rings naturally disappear first. We define the minimum radius in the distribution (5.1) at a moment t as $\overline{R}(t)$. Then it is easy to make sure that the average radius of the rings is equal to $\overline{R}(t)$ and the total length L of the vortex line in a unity volume equals

$$L(t) \approx \frac{1}{\overline{R}^2(t)}. \tag{5.5}$$

This means that the relaxation at any moment t is related to the rings of one size scale $\overline{R}(t)$.

In reality, the emerging vortex structure may resemble a tangle rather than an ensemble of the rings. Usually, when studying the kinetics of such a tangle, the average distance between the vortex lines is considered as a basic parameter of a system. The same scale defines the characteristic radius of the curvature. Since this parameter and L are linked by the relation (5.5), the kinetics of the two models prove to be equivalent. For simplicity we shall only refer to the first one.

The characteristic time of the self-annihilation of a ring of radius R is (see, e.g. Ref. [19])

$$\tau_v(R) \approx \frac{2\pi\rho_s}{\gamma}\frac{R^2}{\ln R/r_c}. \tag{5.6}$$

Here γ is the drag force coefficient per unit length of a vortex line. We have taken into account that the core radius is of the order of r_c and $\overline{R}(t) \gg r_c$. It can be shown [18] that in an interacting Bose gas in the temperature interval

$$n_0\tilde{U} \ll T \ll T_c, \tag{5.7}$$

the coefficient γ is equal to

$$\gamma = \rho_s\Gamma_0\alpha, \tag{5.8}$$

where

$$\alpha \approx \left(\frac{n_0\tilde{U}}{T_c}\right)^{3/2} \tag{5.9}$$

and Γ_0 is the quantum of circulation.

It is a specific feature of the gas that α has a constant value in the interval (5.7), where the elementary excitations practically coincide with free particles. Any ring has a self-induced velocity related to the radius value (the local radius of curvature for an element of the tangle structure). One can estimate that in the time τ_v given by (5.6), the ring passes over a distance

$$l_R \approx \frac{1}{\alpha}R. \tag{5.10}$$

Since $\alpha \ll 1$, this signifies that the ring experiences many crossing encounters before disappearing. If these crossings are not accompanied by reconnection, an estimate of the time for removal of topological defects can be obtained from (5.6), (5.8) and (5.9). For a system of size D_0, we find

$$\tau_v(D_0) \sim \frac{1}{\Gamma_0 \alpha} \frac{D_0^2}{\ln D_0/r_c} . \tag{5.11}$$

Apparently, it is more realistic to think that the reconnection happens efficiently (see, e.g. the discussion in Refs. [25] and [26]). Analysis of the relaxation in this more complicated case [18] leads to the result

$$\tau_v(D_0) \sim \frac{1}{\Gamma_0 \alpha^{1/3}} \frac{D_0^2}{\ln D_0/r_c} . \tag{5.12}$$

Note that this expression differs from (5.11) only in a substantial decrease of the dependence upon the small interaction parameter α. (In the region $T < n_0 \tilde{U}$ one finds that $\alpha \sim T^5$ [27].)

Thus the time of the formation of topological long-range order, which is determined by (5.12) (or (5.11)), turns out to be dependent on the size of the system. In a time of the order of $\tau_v(D_0)$, the Bose gas becomes truly superfluid.

6 Formation of the Off-diagonal Long-range Order

The complete relaxation of topological defects does not in the general case imply the appearance of off-diagonal, long-range order together with a strictly δ-function momentum distribution of particles. Accomplishing the crossover from the quasicondensate to the genuine condensate also requires the attenuation of the long-wave fluctuations of the phase ϕ or the proper nonequilibrium distribution of phonons. It is interesting to mention that the phonon distribution (5.4) leads to a logarithmic divergence of the correlator $< \phi(r)\phi(0) >$ at large distance, quite similar to the two-dimensional case.

Considering the relaxation of the nonequilibrium distribution of phonons (5.4), we deal with the problem of the attenuation of the sound using the Bogoliubov dispersion law (2.8). At a moment t, when a region free from vortices of size $\overline{R}(t)$ is formed, the dissipation of sound with $k \geq \overline{R}(t)^{-1}$ proves to be much the same as in a true superfluid system. This makes it possible to make use of some results obtained for a superfluid liquid.

In a uniform system, in which the relaxation time is connected with k stipulated by large distances, it is natural to assume the inequality

$$kl \ll 1. \tag{6.1}$$

Here l is a free path length of particles. This means that we have to analyze the attenuation of a sound in the hydrodynamic regime. As is well known, condition (6.1) allows one to start with a nondissipative system of equations and then to find the attenuation as a perturbation.

From the general system of the linear hydrodynamic equations for the superfluid liquid, we have (see e.g. Ref. [28])

$$\frac{\partial^2 \rho}{\partial t^2} = -\Delta p, \qquad \frac{\partial^2 s}{\partial t^2} = \frac{\rho_s s^2}{\rho_n}\Delta T\,, \tag{6.2}$$

where ρ_n is the density of normal liquid, and s is the entropy per unit mass. Considering ρ and T as independent thermodynamic variables, we can write for a small variation of parameters in a sound wave

$$p' = \left(\frac{\partial p}{\partial \rho}\right)_T \rho' + \left(\frac{\partial p}{\partial T}\right)_\rho T', \quad s' = \left(\frac{\partial s}{\partial \rho}\right)_T \rho' + \left(\frac{\partial s}{\partial T}\right)_\rho T'. \tag{6.3}$$

Introducing (6.3) into (6.2), we find the system of equations for ρ' and T' . The equations (6.2)–(6.3) are correct in the general case. Applying these equations to the interacting Bose gas, we find that the results differ noticeably from the usual results for superfluid helium.

In the temperature interval (5.7) the normal excitations practically have the dispersion law of free particles. Consequently, the normal component has the properties of an ideal Bose gas. In this case, for the temperature-dependent part of the pressure p_T, we have

$$\left(\frac{\partial p_T}{\partial \rho}\right)_T \approx 0, \tag{6.4}$$

and as a result

$$\left(\frac{\partial p}{\partial \rho}\right)_T = c^2, \tag{6.5}$$

where c is the condensate sound velocity (2.8). On the other hand, we have

$$c_v, s, \rho_n \sim (\frac{T}{T_c})^{3/2}, \qquad \frac{c_p}{c_v} = \frac{5}{3}, \tag{6.6}$$

where c_v and c_p are the specific heats at constant volume and pressure respectively. The solution of the system of equations (6.2) and (6.3), with allowance for (6.5) and (6.6), leads to the following result:

$$\frac{T'}{T} \approx \frac{n_0 \tilde{U}}{T} \frac{\rho'}{\rho}. \tag{6.7}$$

The normal velocity $\mathbf{v}_n$ can be found from the equation

$$\frac{\partial(\rho_s)}{\partial t} + \rho_s \nabla \cdot \mathbf{v}_n = 0 .$$

Taking into account the fact that $\rho_s \neq f(\rho)$ we find, with allowance for (6.6), that

$$v_n \approx \frac{3}{2} c \frac{T'}{T} . \tag{6.8}$$

Using (6.7) and (6.8), we can find the sound dissipation by employing the standard approach (see, e.g. [28]). Since $n_0 \tilde{U} \ll T$ and the specific heat has temperature dependence given by (6.6), it is easy to show that the role of the viscosity is negligible. As a result, the attenuation of sound is determined by the thermal conductivity. Direct evaluation gives for the phonon relaxation time

$$\tau_{ph}^{-1} \approx \frac{\kappa}{n} \frac{n\tilde{U}}{T} k^2 \sim \frac{T}{\hbar} \frac{1}{na} k^2, \tag{6.9}$$

where κ is the coefficient of the thermal conductivity. It is interesting to compare $\tau_v(\overline{R})$ from (5.12) and $\tau_{ph}(k = \overline{R}^{-1})$:

$$\frac{\tau_v(\overline{R})}{\tau_{ph}(k = \overline{R}^{-1})} \approx \frac{1}{(na^3)^{1/2}} \frac{T}{T_c} \frac{1}{\ln \overline{R}/r_c} . \tag{6.10}$$

For a realistic value of the system size, this ratio is always much greater than unity. This assertion has an additional basis. As is well known, sound reflection from a wall is accompanied by strong absorption (see, e.g. Ref. [28]). At the same time, the phonon free path length approaches the value D_0 much earlier than $\overline{R}(t)$ approaches this size.

Since the phase relaxation time τ_ϕ actually coincides with τ_{ph}, we can conclude that $\tau_\phi < \tau_v$. This signifies that the phase relaxation happens faster than the vortex annihilation. Therefore, for the conditions under consideration, complete long-range order should become established when the topological order appears.

In a system of limited size, if the inequality $l > D_0$ is justified, the hydrodynamic condition (6.1) is violated. In this case, the sound attenuation is determined by the direct scattering of phonons with the normal fluid excitations. The calculation of this process leads, in the temperature interval (5.7), to the relation [18]

$$\frac{1}{\tau_{ph}} \sim kaT \,. \tag{6.11}$$

Comparing this result with (5.12), we find

$$\frac{\tau_v(D_0)}{\tau_{ph}(k = D_0^{-1})} \sim \frac{D_0}{r_c}\frac{T}{T_c} \,. \tag{6.12}$$

Provided $D_0 \gg r_c$, (6.12) leads in many cases to the same conclusion as before. However, we cannot exclude the case when the ratio τ_v/τ_{ph} turns out less than unity. In this case, τ_ϕ proves to be larger than τ_v and the formation of the off-diagonal order is characterized by the time

$$\tau_\phi^{-1} \sim \frac{aT}{\hbar D_0} \tag{6.13}$$

(we have again restored $\hbar$ in the explicit form).

7 Final Comments

In a number of experimental situations, nonuniform conditions are involved and the condensate should form in a small volume. This takes place, for example, in different kinds of magnetic traps or during the creation of excitons by a laser pulse. We make some remarks in this connection.

The formation of the quasicondensate in this case does not differ noticeably from the uniform case characterized by the same scale of relaxation time. The number of vortices is now limited, especially at low densities. The time of their self-annihilation drastically shortens due to the small size of the condensate location region. The free expansion of the system, which is characteristic in particular of the excitonic gas in a semiconductor, causes further decrease of the time τ_v. This is important for the appearance of the superfluid properties of such a system.

The quasicondensate formation in a nonuniform case is accompanied by a sharp increase of density. This leads inevitably to the strong increase of the rate of any inelastic process which depends nonlinearly on the density [29,7]. In a recent paper[30], it was shown that this fact can prevent the nonequilibrium system from penetrating deeply into the Bose condensation region. (It is useful to mention that all the results of the preceding sections are correct at $T \sim T_c$.) One can conjecture that the experimental observation [2] that the excitonic gas does not cross the

BEC line while cooling is a consequence of just this effect. The motion of the system along the BEC line during the evolution, which was also revealed, can be explained by the adiabatic expansion of the Bose gas which has the ratio $c_p/c_v = 5/3$.

It is clear that the general trends for nonuniform systems can be understood using the results obtained for the uniform case. However, the precise values of the relaxation times are very sensitive to the details of the experimental situation.

Acknowledgements. This study was supported by the Russian fund of fundamental research (#93-02-2540) and by the Netherlands Organization for Scientific Research (NWO).

References

[1] T.J. Greytak and D. Kleppner, in: *New Trends in Atomic Physics*, G. Grynberg and R. Stora, eds. (North-Holland, Amsterdam, 1984) v.2, p.1127; I.F. Silvera and J.T.M. Walraven, *Progress in Low Temperature Physics*, D.F. Brewer, ed. (North-Holland, Amsterdam, 1986), v.10, p.39; I.F. Silvera and M. Reynolds, Journ. Low Temp. Phys. **87**, 343 (1992); see articles by Silvera and by Greytak, this volume.

[2] D.W. Snoke, J.P. Wolfe, and A. Mysyrowicz, Phys. Rev B **39**, 11171 (1990); A.Mysyrowicz, D.W. Snoke, and J.P. Wolfe, Phys. Stat. Sol. **159**, 387 (1990); see article by Wolfe *et al.*, this volume.

[3] M. Hasuo, N. Nagasawa, T. Itoh, and A. Mysyrowicz, Phys. Rev. Lett. **70**, 1303 (1993).

[4] C. Monroe, W. Swann, H. Robinson, and C. Wieman, Phys. Rev. Lett. **65**, 1571 (1990); see article by Castin *et al.*, this volume.

[5] Yu. Kagan, B.V. Svistunov, and G.V. Shlyapnikov, Pis'ma Zh. Eksp. Teor. Fiz. **42**, 169 (1985) [JETP Lett. **42**, 209 (1985)].

[6] Yu. Kagan, B.V. Svistunov, and G.V. Shlyapnikov, Zh. Eksp. Teor. Fiz. **93**, 552 (1987) [Sov. Phys. JETP **66**, 314 (1987)].

[7] Yu. Kagan and G.V. Shlyapnikov, Phys. Lett. A**130**, 483 (1988).

[8] H.E. Hess, G.P. Kochanski, J.M. Doyle, N. Masuhara, D. Kleppner and T.J. Greytak. Phys. Rev. Lett. **59**, 672 (1987); R. van Roijen, J.J. Berkhout, S. Jakkola, and J.T.M. Walraven, Phys. Rev. Lett. **61**, 931 (1988).

[9] E. Fortin and A. Mysyrowicz, Phys. Rev. Lett. **70**, 3951 (1993).

[10] E. Levich and V. Yakhot, Phys. Rev. B **15**, 243 (1977).

[11] P. Nozières (unpublished).

[12] S.G. Tikhodeev, Zh. Eksp. Teor. Fiz. **97**, 681 (1990).
[13] E. Levich and V. Yakhot, J. Phys. A **11**, 2237 (1978).
[14] D. Snoke and J.P. Wolfe, Phys. Rev. B **39**, 4030 (1989).
[15] H.T.C. Stoof, Phys. Rev. Lett. **66**, 3148 (1991).
[16] Yu. Kagan, B.V. Svistunov, and G.V. Shlyapnikov, Zh. Eksp. Teor. Fiz. **101**, 528 (1992)[Sov. Phys. JETP **75**, 387 (1992)].
[17] J.S. Langer, Phys. Rev. **167**, 183 (1968).
[18] Yu. Kagan and B.V. Svistunov, Zh. Eksp. Teor. Fiz. **105**, No. 2 (1994).
[19] C.P. Barenhi, R.J. Donnelly, and W.F. Vinen, Journ. Low Temp. Phys. **52**, 189 (1983).
[20] J.T. Tough, in *Progress in Low Temperature Physics*, D.F. Brewer, ed. (North-Holland, Amsterdam, 1982), v.8, p.133.
[21] E.M. Lifshitz and L.P. Pitaevskii, *Statistical Physics*, Part 2 (Pergamon, Oxford, 1980).
[22] V.E. Zakharov, in *Basic Plasma Physics Vol. 2*, A.A. Galeev and R.N. Sudan, eds. (North-Holland, Amsterdam, 1984).
[23] B.V. Svistunov, J. Mosccow Phys. Soc. **1**, 363 (1991).
[24] L.D. Landau and E.M. Lifshitz, *Statistical Physics*, Part 1 (Pergamon, Oxford, 1980), p. 476.
[25] K.W. Schwarz, Phys. Rev. B **18**, 245 (1978).
[26] K.W. Schwarz, Phys. Rev. B **38**, 2398 (1988).
[27] S.V. Iordanskii, Zh. Eksp. Teor. Fiz. **49**, 225 (1966)[Sov. Phys. JETP **22**, 160, (1966)].
[28] L.D. Landau and E.M. Lifshitz, *Fluid Mechanics* (Pergamon, Oxford, 1959).
[29] V.V. Goldman, I.F. Silvera, and A.J. Leggett, Phys. Rev. B **24**, 2870 (1981).
[30] T.W. Hijmans, Yu. Kagan, G.V. Shlyapnikov, and J.T.M. Walraven, Phys. Rev. B **48**, 12886 (1993); see also this volume.

11

Condensate Formation in a Bose Gas

H. T. C. Stoof†

Department of Physics, University of Illinois at Urbana-Champaign
Urbana, Illinois 61801, USA
and
Department of Theoretical Physics, Eindhoven University of Technology
5600 MB Eindhoven, The Netherlands

Abstract

Using magnetically trapped atomic hydrogen as an example, we investigate the prospects of achieving Bose–Einstein condensation in a dilute Bose gas. We show that, if the gas is quenched sufficiently far into the critical region of the phase transition, the typical time scale for the nucleation of the condensate density is short and of $O(\hbar/k_B T_c)$. *As a result we find that thermalizing elastic collisions act as a bottleneck for the formation of the condensate. In the case of doubly polarized atomic hydrogen these occur much more frequently than the inelastic collisions leading to decay and we are led to the conclusion that Bose–Einstein condensation can indeed be achieved within the lifetime of the gas.*

1 Introduction

In the last few years it has been clearly demonstrated that not only charged ions but also neutral atoms can be conveniently trapped and cooled by means of electro-magnetic fields. Although the physics of the various ingenious scenarios developed to accomplish this is already interesting in itself [1], the opportunities offered by an atomic gas sample at very low temperatures are exciting in their own right. Examples in this respect are the performance of high-precision spectroscopy, the search for a violation of CP invariance by measuring the electric dipole moment of atomic cesium [2], the construction of an improved time standard based on an atomic fountain [3] and the achievement of Bose–Einstein condensation in a weakly interacting gas.

In particular the last objective has been pursued mainly with atomic

† Present address: Institute for Theoretical Physics, University of Utrecht, Utrecht, The Netherlands.

hydrogen [4, 5]. It has been recently proposed that the alkali-metal vapors cesium [6] and lithium [7] are also suitable candidates for Bose–Einstein condensation. We will nevertheless concentrate here on atomic hydrogen, because it still seems to be the most promising system for the observation of the phase transition in the near future. (See the review articles by Greytak and Silvera in this volume.) Moreover, it has the advantage that the atomic interaction potential is known to a high degree of accuracy. As a result we can have confidence in the fact that the scattering length is positive, which is required for the condensation to take place in the gaseous phase [8], and small enough to rigorously justify the approximations made in the following for the typical temperatures ($T \simeq 10\,\mu$K) and densities ($n \simeq 10^{14}\,\text{cm}^{-3}$) envisaged in the experiments.

Due to the spin of the electron and the proton, the 1s-hyperfine manifold of atomic hydrogen consists of four states which are in order of increasing energy denoted by $|a\rangle$, $|b\rangle$, $|c\rangle$, and $|d\rangle$, respectively. Only the $|c\rangle$ and $|d\rangle$ states can be trapped in a static magnetic trap, because in a magnetic field they have predominantly an electron spin-up component and are therefore low-field seeking [9]. Furthermore, if we load a trap with atoms in these two hyperfine states, the $|c\rangle$ state is rapidly depopulated as a result of the much larger probability for collisional relaxation to the high-field seeking $|a\rangle$ and $|b\rangle$ states which are expelled from the trap. In this manner the system polarizes spontaneously and we obtain a gas of $|d\rangle$-state atoms, known as doubly spin-polarized atomic hydrogen since both the electron and the proton spin are directed along the magnetic field. Unfortunately, such a doubly polarized hydrogen gas still decays due to the dipole interaction between the magnetic moments of the atoms. Although the time scale τ_{inel} for this decay is much longer than the time scale for the depopulation of the $|c\rangle$ state mentioned above, it nevertheless limits the lifetime of the gas sample to the order of seconds for the densities of interest [10].

Having filled the trap with doubly polarized atoms, we must subsequently lower the temperature of the gas to accomplish Bose–Einstein condensation. At present it is believed that the most convenient way to achieve this is by means of conventional [11, 12] or light-induced [13] evaporative cooling. In both cases the idea is to remove, by lowering the well-depth or by photon absorption in the perimeter, the most energetic particles from the trap and thus to create momentarily a highly nonequilibrium energy distribution that will evolve into a new equilibrium distribution at a lower temperature. According to the quantum Boltzmann equation describing this process, a typical time scale for the evolution is the average time between two elastic collisions $\tau_{el} = 1/n\langle v\sigma \rangle$, with $\langle v\sigma \rangle$ the thermal average of the relative velocity v of two colliding atoms times their elastic cross section σ. Clearly, τ_{el} must be small compared to τ_{inel} to ensure that thermal equilibrium is achieved within the lifetime of the system. As a result, the minimum temperature that can be reached by evaporative cooling is about 1 μK and indeed below the critical temperature of atomic hydrogen at a density of 10^{14} cm^{-3}.

The previous discussion appears to indicate that a typical time scale for the formation of the condensate is given by τ_{el}. However, this is not correct because simple phase-space arguments show that a kinetic equation cannot lead to a macroscopic occupation of the one-particle ground state: considering a homogeneous system of N bosons in a volume V, we find from the Boltzmann equation that the production rate of the condensate fraction is

$$\left. \frac{d}{dt} \frac{N_0}{N} \right|_{in} = C \frac{\langle v\sigma \rangle}{V} (1 + N_0) , \quad (1)$$

where N_0 is the number of particles in the zero-momentum state and C is a constant of $O(1)$. Hence, in the thermodynamic limit ($N, V \to \infty$ in such a way that their ratio $n = N/V$ remains fixed) a nonzero production rate is only possible if a condensate already exists [14] and we are forced to conclude that Bose–Einstein condensation cannot be achieved by evaporative cooling of the gas.

2 Nucleation

In the above argument we have only considered the effect of two-body collisions. It is therefore legitimate to suggest that perhaps three- or more body collisions are required for the formation of the condensate, even though they are very improbable in a dilute gas [15]. However, we can easily show that the same argument also applies to these processes: in a m-body collision that produces one particle with zero momentum, we have $2m-2$ independent momentum summations, leading to a factor of V^{2m-2}. Moreover, the transition matrix element is proportional to $V \cdot V^{-m}$ due to the integration over the center-of-mass coordinate and the normalization of the initial and final state wave functions, respectively. In total, the production rate for the condensate fraction is thus proportional to $V^{2m-2}(V^{1-m})^2 V^{-1}(1+N_0)$ or $V^{-1}(1+N_0)$, which again vanishes in the thermodynamic limit if there is no nucleus of the condensed phase. As expected, the contributions from collisions that produce more than one zero-momentum particle have additional factors of V^{-1} and vanish even more rapidly if $V \to \infty$.

Clearly, we have arrived at a nucleation problem for the achievement of Bose–Einstein condensation which seriously endangers the success of future experiments. Fortunately, we suspect that the line of reasoning presented above is not completely rigorous because otherwise it implies that liquid helium also cannot become superfluid, in evident disagreement with experiment. Indeed, by using a kinetic equation to discuss the time evolution of the gas we have in effect neglected the buildup of coherence which is crucial for the formation of the condensate. Our previous argument therefore only shows that by means of evaporative cooling the gas is quenched into the critical region on a time scale τ_{el}, not that Bose–Einstein condensation is impossible. To discuss that point we need a different approach that accurately describes the time evolution of the system after the quenching by taking the buildup of coherence into account exactly. Such a nonequilibrium approach was recently developed on the basis of the Keldysh formalism and can, in the case of a dilute Bose gas, be seen as a generalization of the Landau theory of second-order phase transitions [16, 17]. As a consequence it is useful to consider the Landau theory first. This leads to a better understanding of the more complicated nonequilibrium theory and ultimately of the physics involved in the nucleation of Bose–Einstein condensation.

2.1 Landau Theory

As an introduction to the Landau theory of second-order phase transitions we use the example of a ferromagnetic material [17]. To be more specific we consider a cubic lattice with spins $\mathbf{S}_i$ at the sites $\{i\}$. The Hamiltonian is taken to be

$$H = -J \sum_{\langle i,j \rangle} \mathbf{S}_i \cdot \mathbf{S}_j \, , \tag{2}$$

where J is the exchange energy and the sum is only over nearest neighbors. For further convenience we also introduce the magnetization

$$\mathbf{M} = \frac{1}{V} \sum_i \mathbf{S}_i \, . \tag{3}$$

Physically it is clear that this model has a phase transition at a critical temperature T_c of $O(J/k_B)$. Above the critical temperature the thermal fluctuations randomize the direction of the spins and the system is in a disordered (paramagnetic) state having a vanishing average magnetization $\langle \mathbf{M} \rangle_{\text{eq}}$. However, below the critical temperature the thermal fluctuations are not large enough to overcome the directional effect of the Hamiltonian and the spins favor an ordered (ferromagnetic) state with $\langle \mathbf{M} \rangle_{\text{eq}} \neq 0$. The different phases of the material are thus conveniently characterized by the average magnetization, which for this reason is known as the order parameter of the ferromagnetic phase transition.

In the phenomenological approach put forward by Landau the above mentioned temperature dependence of the equilibrium order parameter $\langle \mathbf{M} \rangle_{\text{eq}}$ is reproduced by anticipating that the free-energy density of the system at a fixed but not necessarily equilibrium value of the average magnetization has the following expansion

$$f(\langle \mathbf{M} \rangle, T) \simeq f(0, T) + \alpha(T)\langle \mathbf{M} \rangle^2 + \frac{\beta(T)}{2}\langle \mathbf{M} \rangle^4 \tag{4}$$

for small values of $\langle \mathbf{M} \rangle$, and that the coefficients of this expansion behave near the critical temperature as

$$\alpha(T) \simeq \alpha_0 \left(\frac{T}{T_c} - 1 \right) \tag{5}$$

and

$$\beta(T) \simeq \beta_0 \,, \tag{6}$$

respectively, with α_0 and β_0 positive constants.

Hence, above the critical temperature $\alpha(T)$ and $\beta(T)$ are both positive. As a result the free energy is minimal for $\langle \mathbf{M} \rangle = 0$, which corresponds exactly to the paramagnetic phase. Moreover, for temperatures below the critical one $\alpha(T)$ is negative and the free energy is indeed minimized by a nonzero average magnetization with magnitude $\sqrt{-\alpha(T)/\beta(T)}$. Just below the critical temperature the latter equals

$$\langle M \rangle_{\text{eq}} \simeq \sqrt{\frac{\alpha_0}{\beta_0}\left(1 - \frac{T}{T_c}\right)} \,, \tag{7}$$

which after substitution in Eq. (4) gives rise to an equilibrium free-energy density of

$$f(\langle \mathbf{M} \rangle_{\text{eq}}, T) \simeq f(0, T) - \frac{\alpha_0^2}{2\beta_0}\left(1 - \frac{T}{T_c}\right)^2 . \tag{8}$$

Therefore, the second derivative d^2f/dT^2 is discontinuous at the critical temperature and the phase transition is of second order according to the Ehrenfest nomenclature.

Note that minimizing the free energy only fixes the magnitude and not the direction of $\langle \mathbf{M} \rangle_{eq}$. This degeneracy is caused by the fact that the Hamiltonian in Eq. (2) is symmetric under an arbitrary rotation of all the spins $\mathbf{S}_i$. Consequently, the free energy must be symmetric under a rotation of the average magnetization and only even powers of $\langle \mathbf{M} \rangle$ can appear in its expansion (cf. Eq. (4)). Due to this behavior the ferromagnet is a good example of a system with a spontaneously broken symmetry, i.e. although the Hamiltonian is invariant under the operations of a group, its ground state is not. In the case of a ferromagnet the symmetry group is $SO(3)$, which is broken spontaneously below the critical temperature because the average magnetization points in a certain direction. Which direction is chosen in practice depends on the surroundings of the system and in particular on (arbitrary small) external magnetic fields that favor a specific direction.

After this summary of the Landau theory we are now in a position to introduce two time scales which turn out to be of great importance for the nucleation of Bose–Einstein condensation. To do so we consider the following experiment. Imagine that we have a piece of ferromagnetic material at some temperature T_1 above the critical temperature. Being in thermal equilibrium the material is in the paramagnetic phase with $\langle \mathbf{M} \rangle_{eq} = 0$. We then quickly cool the material to a new temperature T_2 below the critical temperature. If this is done sufficiently fast, the spins will have no time to react and we obtain a nonequilibrium situation in which the free energy has developed a 'double-well' structure but the average magnetization is still zero. This is depicted in Fig. 1(a). In such a situation there is a typical time scale for the relaxation of the average magnetization to its new equilibrium value $\sqrt{-\alpha(T_2)/\beta(T_2)}$, which we denote τ_{coh}.

However, in the case of magnetically trapped atomic hydrogen, the gas is isolated from its surroundings and it is not possible to perform the cooling stage mentioned above. As a result the gas has to develop the instability associated with the phase transition by itself. The time scale corresponding to this process is called τ_{nucl} and is schematically shown in Fig. 1(b). Combining the two processes we are led to the following physical picture for the nucleation of Bose–Einstein condensation. After the quench into the critical region the gas develops an instability on the time scale τ_{nucl}. On this time scale the actual nucleation takes place and a small nucleus of the condensate is formed, which then grows on the time scale τ_{coh} as shown in Fig. 2. To solve the nucleation problem we are thus left with the actual determination of these two time scales. Clearly,

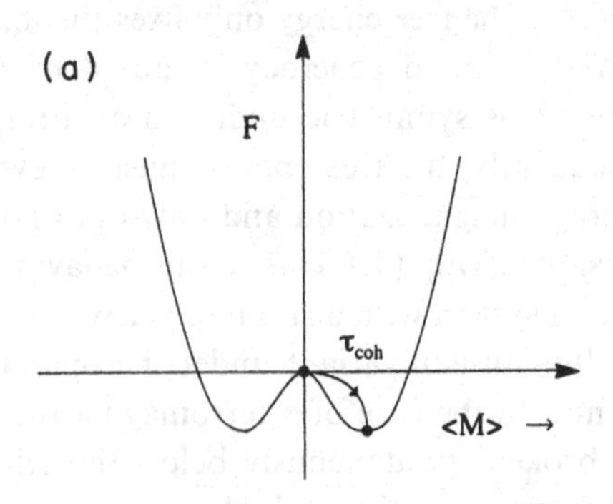

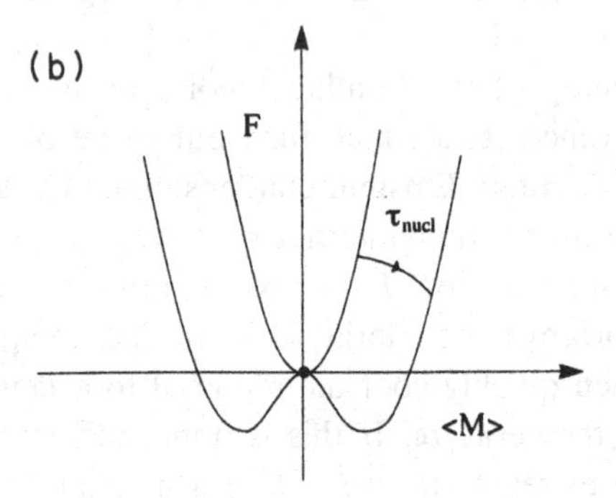

Fig. 1. Visualization of (a) the time scale τ_{coh} for the relaxation of the order parameter to its equilibrium value and (b) the time scale τ_{nucl} associated with the appearance of the instability.

before this can be done we need to know the correct order parameter of the phase transition.

2.2 Order Parameter

Starting with the pioneering work of Bogoliubov [18], it is well known that the order parameter for Bose–Einstein condensation in a weakly interacting Bose gas is conveniently treated by using the method of second quantization. In this method, all many-body observables are expressed in terms of the creation and annihilation operators of a particle at position **x** denoted by $\psi^{\dagger}(\mathbf{x})$ and $\psi(\mathbf{x})$, respectively [19]. For example, for a gas of particles with mass m and a two-body interaction potential

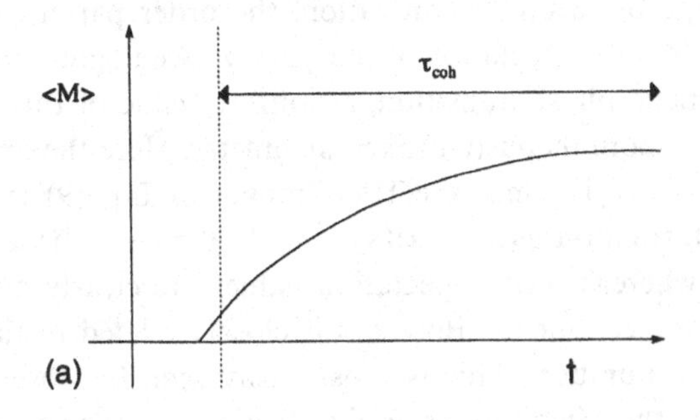

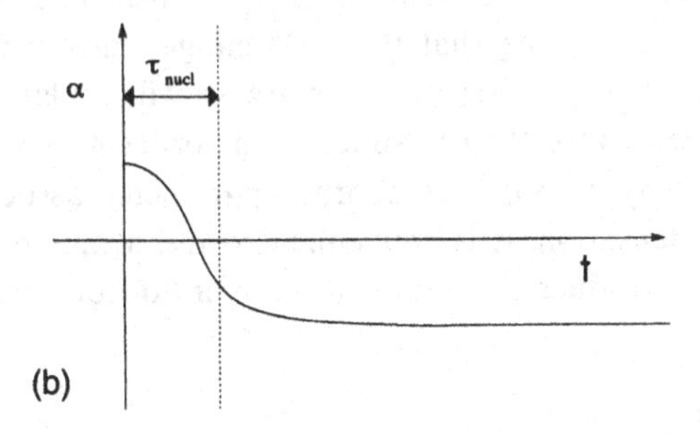

Fig. 2. Visualization of the time scales $\tau_{\rm coh}$ and $\tau_{\rm nucl}$, using the time dependence of (a) the order parameter and (b) the coefficient α of the quadratic term in the free energy.

$V(\mathbf{x}-\mathbf{x}')$, the Hamiltonian equals

$$\begin{aligned} H &= \int d\mathbf{x}\, \psi^\dagger(\mathbf{x}) \left(-\frac{\hbar^2\nabla^2}{2m} \right) \psi(\mathbf{x}) \\ &+ \frac{1}{2} \int d\mathbf{x} \int d\mathbf{x}'\, \psi^\dagger(\mathbf{x})\psi^\dagger(\mathbf{x}') V(\mathbf{x}-\mathbf{x}')\psi(\mathbf{x}')\psi(\mathbf{x}) \end{aligned} \tag{9}$$

and the total number of particles is given by

$$N = \int d\mathbf{x}\, \psi^\dagger(\mathbf{x})\psi(\mathbf{x}) \,. \tag{10}$$

The method is also particularly useful for a Bose system because the permutation symmetry of the many-body wave function is automatically accounted for by assuming the Bose commutation relations $[\psi(\mathbf{x}), \psi(\mathbf{x}')] = [\psi^\dagger(\mathbf{x}), \psi^\dagger(\mathbf{x}')] = 0$ and $[\psi(\mathbf{x}), \psi^\dagger(\mathbf{x}')] = \delta(\mathbf{x}-\mathbf{x}')$ between the creation and annihilation operators.

In the language of second quantization, the order parameter for the dilute Bose gas is the expectation value $\langle\psi(\mathbf{x})\rangle$. Analogous to the case of the ferromagnetic phase transition, a nonzero value of this order parameter signals a spontaneously broken symmetry. Here the appropriate symmetry group is $U(1)$, since the Hamiltonian of Eq. (9) is invariant under the transformation $\psi(\mathbf{x}) \rightarrow \psi(\mathbf{x})e^{i\vartheta}$ and $\psi^\dagger(\mathbf{x}) \rightarrow \psi^\dagger(\mathbf{x})e^{-i\vartheta}$ of the field operators, whereas their expectation values are clearly not. Notice that the $U(1)$ symmetry of the Bose gas is closely related to the conservation of particle number. This is most easily seen by observing that the invariance of the Hamiltonian is due to the fact that each term in the right-hand side of Eq. (9) contains an equal number of creation and annihilation operators. The relationship can also be established in a more formal way by noting that the $U(1)$ gauge transformations are generated by the particle number operator. As we will see later on, it has important consequences for the dynamics of the order parameter.

To understand why $\langle\psi(\mathbf{x})\rangle$ is the order parameter associated with Bose–Einstein condensation, it is convenient to use a momentum-space description and to introduce the annihilation operator for a particle with momentum $\hbar\mathbf{k}$

$$a_{\mathbf{k}} = \int d\mathbf{x}\, \psi(\mathbf{x}) \frac{e^{-i\mathbf{k}\cdot\mathbf{x}}}{\sqrt{V}} \tag{11}$$

and the corresponding creation operator $a_{\mathbf{k}}^\dagger$ by Hermitian conjugation. The basis of states for the gas is then characterized by the occupation numbers $\{N_{\mathbf{k}}\}$. If the gas is Bose-condensed, there is a macroscopic occupation of the zero-momentum state and the relevant states are $|N_0, \{N_{\mathbf{k}}\}_{\mathbf{k}\neq 0}\rangle$ with only N_0 proportional to N. Within this subspace of states we have

$$\langle a_0{}^\dagger a_0\rangle = \langle N_0\rangle \simeq \langle N_0\rangle + 1 = \langle a_0 a_0{}^\dagger\rangle \tag{12}$$

and we can neglect the fact that a_0 and $a_0{}^\dagger$ do not commute. As a result we can treat these operators as complex numbers [18, 19] and say that $\langle a_0{}^\dagger a_0\rangle = \langle a_0{}^\dagger\rangle\langle a_0\rangle$ or equivalently that $\langle a_0\rangle = \sqrt{N_0}$. In coordinate space, the latter reads $\langle\psi(\mathbf{x})\rangle = \sqrt{n_0}$, with $n_0 = N_0/V$ the condensate density.

The above argument essentially tells us that a sufficient condition for a nonzero value of $\langle\psi(\mathbf{x})\rangle$ is $\langle N_0\rangle \gg 1$. Although this is intuitively appealing, it is important to point out that it is not generally true. Consider, for example, the ideal Bose gas [20]. In the grand canonical ensemble, the total number of particles in the gas is given by

$$N = \sum_{\mathbf{k}} \langle N_{\mathbf{k}} \rangle = \sum_{\mathbf{k}} \frac{1}{\zeta^{-1} e^{\beta \epsilon_{\mathbf{k}}} - 1}, \tag{13}$$

where β is $1/k_B T$, $\epsilon_{\mathbf{k}}$ is the kinetic energy $\hbar^2 \mathbf{k}^2/2m$, ζ is the fugacity $e^{\beta\mu}$ and μ is the chemical potential.

At high temperatures, the fugacity is small and we are allowed to take the continuum limit of Eq. (13), which results in the equation of state

$$n = \frac{1}{\Lambda^3} g_{3/2}(\zeta), \tag{14}$$

where we have introduced the de Broglie wavelength $\Lambda = \sqrt{2\pi\hbar^2/mk_B T}$ and the Bose functions $g_n(\zeta)$ defined by

$$g_n(\zeta) = \frac{1}{\Gamma(n)} \int_0^\infty dx \, \frac{x^{n-1}}{\zeta^{-1} e^x - 1}. \tag{15}$$

Lowering the temperature while keeping the density fixed, the fugacity increases until it ultimately reaches the value one at the critical temperature

$$T_0 = \frac{2\pi\hbar^2}{mk_B} \left(\frac{n}{g_{3/2}(1)} \right)^{2/3} \simeq \frac{2\pi\hbar^2}{mk_B} \left(\frac{n}{2.612} \right)^{2/3}. \tag{16}$$

At this point, Eq. (14) ceases to be valid because the occupation number of the zero-momentum state, which is equal to $\zeta/(1-\zeta)$, diverges and must be taken out of the discrete sum in Eq. (13) before we take the continuum limit. Moreover, we only need to treat the zero-momentum term separately because in the thermodynamic limit the chemical potential goes to zero as V^{-1}, whereas the kinetic energy for the smallest nonzero momentum decreases only as $V^{-2/3}$. Consequently, below the critical temperature the equation of state becomes

$$n = n_0 + \frac{1}{\Lambda^3} g_{3/2}(1) \tag{17}$$

and leads to a condensate density given by the well-known formula

$$n_0 = n \left(1 - \left(\frac{T}{T_0} \right)^{3/2} \right). \tag{18}$$

We thus find that the average occupation number $\langle N_0 \rangle$ is at all temperatures given by $\zeta/(1-\zeta)$, that is, its value in the grand canonical ensemble with the density matrix $e^{-\beta(H-\mu N)}$. Since this density matrix commutes with the particle number operator, we conclude that in the case of an ideal Bose gas there is a macroscopic occupation of the zero-momentum state without a spontaneous breaking of the $U(1)$ symmetry. To show more rigorously that $\langle\psi(\mathbf{x})\rangle_{eq} = 0$ at all temperatures, we determine the free-energy density of the gas as a function of the order parameter $\langle\psi(\mathbf{x})\rangle$. Dealing with a noninteracting system, it is not difficult to obtain

$$f(\langle\psi(\mathbf{x})\rangle, T) = -\mu(T)|\langle\psi(\mathbf{x})\rangle|^2 \tag{19}$$

for a homogeneous value of the order parameter. Because $\mu \leq 0$ the minimum is indeed always at $\langle\psi(\mathbf{x})\rangle = 0$ and it is necessary to identify the condensate density n_0 with the order parameter of the ideal Bose gas (cf. Eq. (18)).

Notwithstanding the previous remarks, the order parameter for Bose–Einstein condensation in a weakly interacting Bose gas is given by $\langle\psi(\mathbf{x})\rangle$. This was put on a firm theoretical basis by Hugenholtz and Pines [21], who calculated the free energy as a function of the above order parameter and showed that at sufficiently low temperatures the system develops an instability that is removed by a nonzero value of $\langle\psi(\mathbf{x})\rangle$. In addition, they derived an exact relationship between the chemical potential and the condensate density, which turns out to be valid also in the nonequilibrium problem of interest here and is important for an understanding of how the $U(1)$ symmetry is broken dynamically.

2.3 Condensation Time

We have argued that by means of evaporative cooling a doubly polarized atomic hydrogen gas can be quenched into the critical region of the phase transition and that this kinetic part of the condensation process is described by a quantum Boltzmann equation. As a result, the gas acquires on the time scale τ_{el} an equilibrium distribution with some temperature T, which is slightly above the critical temperature T_0 of the ideal Bose gas because a condensate cannot be formed at this stage.

For the study of the subsequent coherent part of the condensation process, it is therefore physically reasonable to assume that at a time t_0 the density matrix $\rho(t_0)$ of the gas is well approximated by the density matrix of an ideal Bose gas with temperature T. The evolution of the order parameter $\langle\psi(\mathbf{x})\rangle$ for times larger than t_0 is then completely determined by the Heisenberg equation of motion

$$i\hbar\frac{d\psi(\mathbf{x}, t)}{dt} = [\psi(\mathbf{x}, t), H] \tag{20}$$

for the field operator. Substituting the Hamiltonian in Eq. (9), and taking the expectation value with respect to $\rho(t_0)$, we find

$$i\hbar\frac{d\langle\psi(\mathbf{x},t)\rangle}{dt} = \frac{-\hbar^2\nabla^2}{2m}\langle\psi(\mathbf{x},t)\rangle + \int d\mathbf{x}'\, V(\mathbf{x}-\mathbf{x}')\langle\psi^\dagger(\mathbf{x}',t)\psi(\mathbf{x}',t)\psi(\mathbf{x},t)\rangle\,, \tag{21}$$

where the complicated part is the evaluation of $\langle\psi^\dagger(\mathbf{x}',t)\psi(\mathbf{x}',t)\psi(\mathbf{x},t)\rangle$. In lowest order, we simply have

$$\begin{aligned}\langle\psi^\dagger(\mathbf{x}',t)\psi(\mathbf{x}',t)\psi(\mathbf{x},t)\rangle &\simeq \langle\psi^\dagger(\mathbf{x}',t)\rangle\langle\psi(\mathbf{x}',t)\rangle\langle\psi(\mathbf{x},t)\rangle \\ +\langle\psi^\dagger(\mathbf{x}',t)\psi(\mathbf{x}',t)\rangle\langle\psi(\mathbf{x},t)\rangle &+ \langle\psi^\dagger(\mathbf{x}',t)\psi(\mathbf{x},t)\rangle\langle\psi(\mathbf{x}',t)\rangle\,,\end{aligned} \tag{22}$$

which, after substitution into Eq. (21), leads to

$$\begin{aligned}\left(i\hbar\tfrac{d}{dt} + \tfrac{\hbar^2\nabla^2}{2m}\right)\langle\psi(\mathbf{x},t)\rangle &= \textstyle\int d\mathbf{x}'\, V(\mathbf{x}-\mathbf{x}')\langle\psi^\dagger(\mathbf{x}',t)\psi(\mathbf{x}',t)\rangle\langle\psi(\mathbf{x},t)\rangle \\ &+ \textstyle\int d\mathbf{x}'\, V(\mathbf{x}-\mathbf{x}')\langle\psi^\dagger(\mathbf{x}',t)\psi(\mathbf{x},t)\rangle\langle\psi(\mathbf{x}',t)\rangle \\ &+ \textstyle\int d\mathbf{x}'\, V(\mathbf{x}-\mathbf{x}')\langle\psi^\dagger(\mathbf{x}',t)\rangle\langle\psi(\mathbf{x}',t)\rangle\langle\psi(\mathbf{x},t)\rangle,\end{aligned} \tag{23}$$

and thus corresponds exactly to the Hartree–Fock approximation.

To proceed, we must restrict ourselves to the case of a dilute Bose gas in the quantum regime. Introducing the scattering length a, which is of the order of the range of the interaction, the quantum regime is characterized by $a/\Lambda \ll 1$. We therefore only need to consider s-wave scattering and can neglect the momentum dependence of various collisional quantities. In particular, we can replace the potential $V(\mathbf{x}-\mathbf{x}')$ by the contact interaction $V_0\,\delta(\mathbf{x}-\mathbf{x}')$ with $V_0 = \int d\mathbf{x}\, V(\mathbf{x})$. Hence, in the Hartree–Fock approximation we obtain

$$\left(i\hbar\frac{d}{dt} + \frac{\hbar^2\nabla^2}{2m}\right)\langle\psi(\mathbf{x},t)\rangle = \left(2nV_0 + V_0|\langle\psi(\mathbf{x},t)\rangle|^2\right)\langle\psi(\mathbf{x},t)\rangle\,, \tag{24}$$

having only the trivial solution $\langle\psi(\mathbf{x},t)\rangle = 0$ for a space and time-independent order parameter. Within this lowest order approximation we thus conclude that $\tau_{\text{nucl}} = \infty$ and that the formation of a condensate will not take place.

Fortunately, it is well known that the Hartree–Fock approximation is not sufficiently accurate for a dilute Bose gas because the diluteness condition $na^3 \ll 1$ implies that we need to consider all two-body processes, i.e., two particles must be allowed to interact more than once. The appropriate approximation is therefore the ladder or T-matrix approximation, which is displayed in terms of diagrams in Fig. 3. Moreover, in the degenerate regime, where the temperature T is slightly larger than T_0 and the degeneracy parameter $n\Lambda^3$ is of $O(1)$, the condition

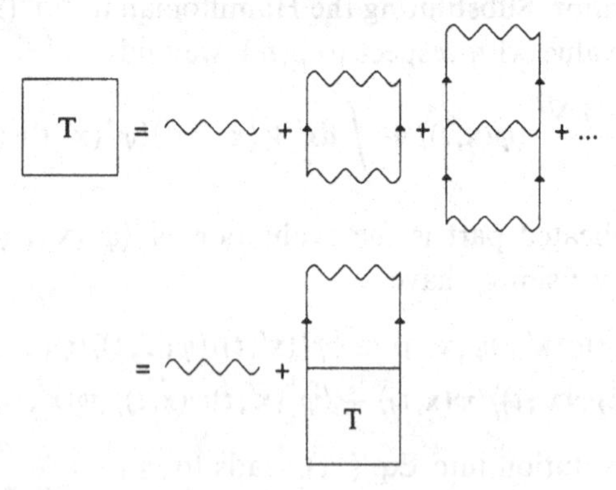

Fig. 3. Diagrammatic representation of the T-matrix equation. The wavy line corresponds to the interaction and the straight line to the noninteracting one-particle Green's function.

$a/\Lambda \ll 1$ implies that also $na\Lambda^2 \ll 1$ or physically that the average kinetic energy of the gas is much larger than the typical interaction energy. Consequently, an accurate discussion of the nucleation of Bose–Einstein condensation in a weakly interacting Bose gas requires an evaluation of $\langle \psi^\dagger(\mathbf{x}',t)\psi(\mathbf{x}',t)\psi(\mathbf{x},t)\rangle$ within the T-matrix approximation and to zeroth order in the gas parameters a/Λ and $na\Lambda^2$.

Although it is easy to formulate this objective, to actually perform the calculation is considerably more difficult. It is most conveniently accomplished by making use of the Keldysh formalism [22] which has been reviewed by Danielewicz [23] using operator methods. For a functional formulation of this nonequilibrium theory and for the technical details of the somewhat tedious mathematics, we refer to our previous papers [16]. Here we only present the final results and concentrate on the physics involved.

Due to the fact that we are allowed to neglect the (relative) momentum dependence of the T matrix, the equation of motion for the order parameter $\langle \psi(\mathbf{x},t)\rangle$ acquires the local form of a time-dependent Landau–Ginzburg theory

$$\left(i\hbar\frac{d}{dt} + \frac{\hbar^2\nabla^2}{2m}\right)\langle \psi(\mathbf{x},t)\rangle = \left(S^{(+)}(t) + T^{(+)}|\langle \psi(\mathbf{x},t)\rangle|^2\right)\langle \psi(\mathbf{x},t)\rangle\,, \quad (25)$$

which is recovered from a variational principle if we use the action

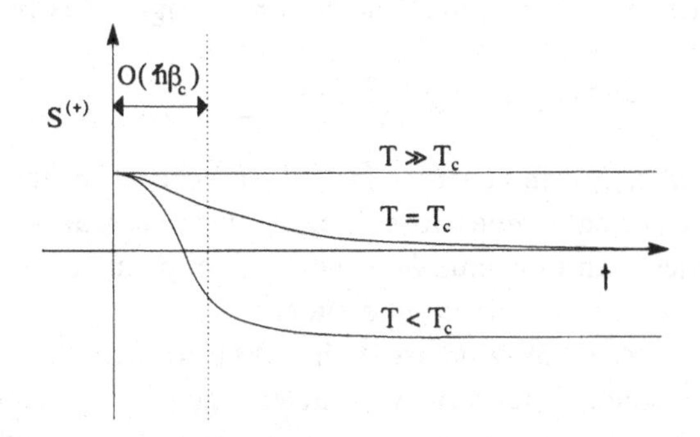

Fig. 4. Time dependence of the coefficient $S^{(+)}$ for three different initial temperatures of the Bose gas.

$$S(\langle\psi(\mathbf{x},t)\rangle, T) = \int dt \int d\mathbf{x}\, \langle\psi(\mathbf{x},t)\rangle^{*}$$
$$\times \left(i\hbar\frac{d}{dt} + \frac{\hbar^2\nabla^2}{2m} - S^{(+)}(t) - \frac{T^{(+)}}{2}|\langle\psi(\mathbf{x},t)\rangle|^2 \right) \langle\psi(\mathbf{x},t)\rangle\,. \quad (26)$$

Here $S^{(+)}(t)\delta(t-t')$ is a simple approximation for the retarded self-energy $\hbar\Sigma^{(+)}(0;t,t')$ of a hydrogen atom with zero momentum and $T^{(+)} \simeq 4\pi\hbar^2 a/m$ is the effective interaction between two such atoms. Clearly, the action in Eq. (26) is the desired generalization of the Landau free energy and corresponds precisely to the physical picture presented previously in Figs. 1(b) and 2(b).

Therefore, τ_{nucl} is determined by the time dependence of the coefficient $S^{(+)}$ which is shown in Fig. 4 for three different initial temperatures. If the temperature T is much larger than T_0, $S^{(+)}(t)$ is constant and equal to $8\pi\hbar^2 an/m$. In this region of the phase diagram, coherent processes are negligible and the evolution of the gas is described by a Boltzmann equation. Lowering the temperature, the occupation numbers for momenta $\hbar k < O(\hbar/\Lambda)$ increase and lead to an enhancement of the coherent population of states with momenta $\hbar k < O(\hbar\sqrt{na}) \ll O(\hbar/\Lambda)$. This is signaled by the increasing correlation length $\xi = \hbar/\sqrt{2mS^{(+)}(\infty)}$. At the critical temperature $T_c = T_0(1 + O(a/\Lambda_0))$, we have $S^{(+)}(\infty) = 0$ and this correlation length diverges. Below that temperature, but still above T_0 so as not to have a condensate already in the initial state, $S^{(+)}(t)$ actually changes sign and the gas develops the required instability for a Bose–Einstein condensation. The change of sign takes place at

$$t \equiv t_c = t_0 + O\left(\frac{a}{\Lambda_c}\frac{\hbar}{k_B(T_c - T)}\right), \tag{27}$$

which shows that τ_{nucl} is in general of $O(\hbar/k_B T_c)$ except for temperatures very close to the critical temperature. Clearly, this time scale is due to the fact that all states with momenta $\hbar k < O(\hbar/\Lambda)$ cooperate in the coherent population of the one-particle ground state.

After a small nucleus of $O(n(a/\Lambda_c)^2)$ has been formed, the subsequent buildup of the condensate density is determined by the equation of motion Eq. (25). Looking at the right-hand side, we immediately see that the time scale τ_{coh} involved in this process is typically of order $\hbar/n_0 T^{(+)}$ or, equivalently, of $O((\hbar/k_B T_c)(1/n_0 a\Lambda_c^2))$. Therefore, we find $\tau_{\text{coh}} \gg \tau_{\text{nucl}}$, as anticipated in Fig. 2. The physical reason for this time scale is that after the nucleation of the phase transition, the buildup of the condensate density is accompanied by a depopulation of the momentum states with $\hbar k < O(\hbar\sqrt{n_0 a})$. As a result, it is not difficult to show that in the limit $t \to \infty$, the condensate density is of $O(na/\Lambda_c)$ and thus we have $\tau_{\text{coh}} = O((\Lambda_c/a)^2\hbar/k_B T_c)$.

Finally, it is interesting to point out how the gas can conserve the total number of particles and, apparently at the same time, break the $U(1)$ gauge symmetry that is responsible for this conservation law. To that end, we write the field operator $\psi(\mathbf{x}, t)$ as the sum of its expectation value $\langle\psi(\mathbf{x}, t)\rangle$ and the fluctuation $\psi'(\mathbf{x}, t)$, and introduce a time-dependent chemical potential $\mu(t)$ by means of

$$\langle\psi(\mathbf{x}, t)\rangle = \sqrt{n_0(t)}\, exp\left(-\frac{i}{\hbar}\int_{t_0}^{t} dt'\, \mu(t')\right). \tag{28}$$

Substituting the latter into the action of Eq. (26) and minimizing with respect to $\sqrt{n_0(t)}$ gives, for $t > t_c$,

$$n_0(t) = \frac{-S^{(+)}(t) + \mu(t)}{T^{(+)}}, \tag{29}$$

which determines the growth of the condensate density and is in effect a nonequilibrium version of the Hugenholtz–Pines theorem [21]. Furthermore, by considering the fluctuations around $\sqrt{n_0(t)}$, we can show that the chemical potential is determined by the constraint

$$n = n_0(t) + \frac{1}{V}\int d\mathbf{x}\, \langle\psi'^{\dagger}(\mathbf{x}, t)\psi'(\mathbf{x}, t)\rangle, \tag{30}$$

enforcing the conservation of particle number at all times. In the complex

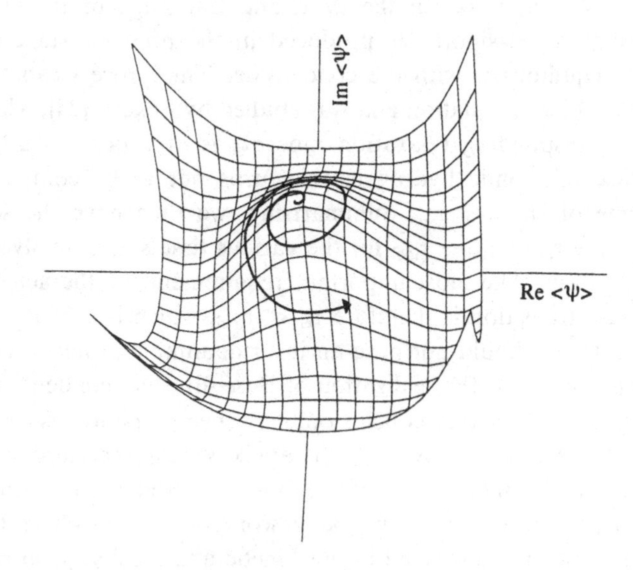

Fig. 5. Evolution of the complex order parameter $\langle\psi(\mathbf{x})\rangle$, which is constrained by the requirement of particle number conservation.

plane, $\langle\psi(\mathbf{x},t)\rangle$ thus moves radially outward along a spiral, as shown in Fig. 5. Consequently, the phase of the order parameter never has a fixed value and the $U(1)$ symmetry is not really broken dynamically. This is of course expected since the system evolves according to a symmetric Hamiltonian.

3 Conclusions and Discussion

We studied the evolution of a doubly polarized atomic hydrogen gas in a magnetic trap and showed that by means of evaporative cooling the gas can accomplish the Bose–Einstein phase transition within its lifetime τ_{inel}. The condensation process proceeds under these conditions in the following three stages. In the first kinetic stage, the gas is quenched into the critical region $T_0 < T \leq T_c$. A typical time scale in this part of the evolution is given by the time between elastic collisions τ_{el}, which for a degenerate gas is of $O((\Lambda_c/a)^2\hbar/k_B T_c)$. In the following coherent stage, the actual nucleation takes place on the time scale $\tau_{\text{nucl}} = O(\hbar/k_B T_c)$ by means of a coherent population of the zero-momentum state. The small nucleus formed in this manner then grows on the much longer time scale $\tau_{\text{coh}} = O((\Lambda_c/a)^2\hbar/k_B T_c)$ by a depopulation of the low-momentum states having $\hbar k < O(\hbar\sqrt{n_0 a})$. In the third and last stage of the evolution, the Bogoliubov quasiparticles produced in the previous stage have to come into equilibrium with the condensate. This process can again be treated by a kinetic equation and was studied by Eckern [24], who found that the corresponding relaxation time $\tau_{\rm rel}$ is of order $(\Lambda_c/a)^3\hbar/k_B T_c$. In the case of atomic hydrogen, this turns out to be comparable to the lifetime of the system. Summarizing, we thus have the sequence $\tau_{\rm nucl} \ll \tau_{\rm coh} \simeq \tau_{\rm el} \ll \tau_{\rm rel} \simeq \tau_{\rm inel}$ for the various time scales involved in the phase transition. The most important requirement for the achievement of the phase transition is therefore $\tau_{\rm el} \ll \tau_{\rm inel}$, which is a relatively mild requirement and should not pose an insurmountable problem for future experiments aimed at the realization of Bose–Einstein condensation.

Having arrived at this conclusion, it is necessary to discuss recent work by Kagan, Svistunov and Shlyapnikov [25] that also considers the evolution of a weakly interacting Bose gas after the removal of the most energetic atoms. Kagan and coworkers also conclude that the evolution of the gas is divided into a kinetic and a subsequent coherent stage. Moreover, their detailed study of the kinetic part of the evolution confirms our conjecture that the gas is quenched into the critical region on the time scale $\tau_{\rm el}$. The investigation of the coherent part, however, leads to the extreme result that complete Bose–Einstein condensation cannot occur in a finite amount of time. To understand why this conclusion is reached we briefly summarize their line of thought. For more details, see the review article by Kagan in this volume.

At the end of the kinetic stage, the gas has acquired large average occupation numbers for the states with momenta $\hbar k < \hbar k_0 = O(\hbar\sqrt{n_0 a})$, where n_0 is the density of particles with these small momenta. Therefore, Kagan, Svistunov and Shlyapnikov argue that for a study of the coherent part of the evolution we must use the initial condition

$$\langle\psi(\mathbf{x}, t_0)\rangle = \sum_{k<k_0} \sqrt{\langle N_{\mathbf{k}}\rangle}\,\frac{e^{i\mathbf{k}\cdot\mathbf{x}}}{\sqrt{V}} \neq 0, \tag{31}$$

together with the nonlinear Schrödinger equation

$$i\hbar\frac{d\langle\psi(\mathbf{x}, t)\rangle}{dt} = \left(\frac{-\hbar^2\nabla^2}{2m} + T^{(+)}|\langle\psi(\mathbf{x}, t)\rangle|^2\right)\langle\psi(\mathbf{x}, t)\rangle\,, \tag{32}$$

which has the equilibrium solution $\langle\psi(\mathbf{x}, t)\rangle = \sqrt{n_0}\exp(-i\mu_0 t)$ and $\mu_0 = n_0 T^{(+)}$. Consequently, all the particles that initially have momenta $\hbar k < \hbar k_0$ are in the limit $t \to \infty$ assumed to be in the condensate.

Linearizing the Hamiltonian (32) around this equilibrium solution, Kagan *et al.* then observe that the energy involved with a magnitude fluctuation of the order parameter is $\epsilon_{\mathbf{k}} + n_0 T^{(+)}$, whereas the energy involved with a phase fluctuation is only $\epsilon_{\mathbf{k}}$. As a result, they assert that on the time scale $\tau_{\text{ampl}} = \tau_{\text{coh}} = O(\hbar / n_0 T^{(+)})$, a state is formed in which the amplitude of $\langle \psi(\mathbf{x}, t) \rangle$ is fixed, but the phase is still strongly fluctuating because the corresponding time scale τ_{ph} is much longer (and even diverges as $V^{2/3}$ in the thermodynamic limit). Hence, at finite time, the gas is in a state with only a quasicondensate [25, 26] and a complete or true condensate is only formed in the limit $t \to \infty$.

Clearly, this physical picture of two different time scales for the amplitude and phase fluctuations of the order parameter is only applicable if these fluctuations exist independently of each other. Looking only at the Hamiltonian this indeed seems to be the case. However, a correct discussion of the fluctuations must be based on the equations of motion or, equivalently, the Lagrangian. The latter contains a first-order time derivative which strongly couples the amplitude and phase fluctuations. Therefore, a dilute Bose gas only has one collective mode (not two) with a single dispersion relation, i.e. the well-known Bogoliubov dispersion $\sqrt{\epsilon_{\mathbf{k}}(\epsilon_{\mathbf{k}} + 2n_0 T^{(+)})}$, and we are lead to $\tau_{ph} = \tau_{\text{ampl}}$. It is interesting to note that in the case of a neutral BCS-type superfluid, we do have two different time scales because the Lagrangian now contains a second-order time derivative and the amplitude and phase fluctuations are indeed independent in lowest order [27, 28].

An even more serious problem with the approach of Kagan, Svistunov and Shlyapnikov is their claim that the use of the initial condition in Eq. (31) is justified because $\langle N_{\mathbf{k}} \rangle \gg 1$. As we have pointed out earlier, this is not true in general. For $\langle \psi(\mathbf{x}, t) \rangle$ to be nonzero, one must show that the system has a corresponding instability. However, within the T-matrix approximation it is not difficult to show that the instability associated with a quasicondensate is always preceded by the instability corresponding to the formation of a condensate. This implies that we always have to take Bose–Einstein condensation into account first. After that has been accomplished by means of the theory reviewed here, it is of course no longer relevant to consider the appearance of a quasicondensate.

Acknowledgments. It is a great pleasure to thank Tony Leggett for various helpful discussions and for giving me the opportunity to visit the University of Illinois at Urbana-Champaign. I also benefited from conversations with Steve Girvin, Daniel Loss, Kieran Mullen and Jook

Walraven. This work was supported by the National Science Foundation through Grant No. DMR-8822688.

References

[1] See the special issue of J. Opt. Soc. Am. B **6**, No. 11 (1989), edited by S. Chu and C. Wieman.
[2] W. Bernreuther and M. Suzuki, Rev. Mod. Phys. **63**, 131 (1991).
[3] A. Clairon, C. Salomon, S. Guellati, and W.D. Phillips, Europhys. Lett. **12**, 683 (1990); K. Gibble and S. Chu, Phys. Rev. Lett. **70**, 1771 (1993).
[4] H.F. Hess, G.P. Kochanski, J.M. Doyle, N. Masuhara, D. Kleppner, and T.J. Greytak, Phys. Rev. Lett. **59**, 672 (1987); N. Masuhara, J.M. Doyle, J.C. Sandberg, D. Kleppner, T.J. Greytak, H.F. Hess, and G.P. Kochanski, Phys. Rev. Lett. **61**, 935 (1988).
[5] R. van Roijen, J.J. Berkhout, S. Jaakkola, and J.T.M. Walraven, Phys. Rev. Lett. **61**, 931 (1988).
[6] C. Monroe, W. Swann, H. Robinson, and C. Wieman, Phys. Rev. Lett. **65**, 1571 (1990).
[7] J.J. Tollett, C.C. Bradley, and R.G. Hulet, Bull. Am. Phys. Soc. **37**, 1126 (1992).
[8] H.T.C. Stoof, to be published.
[9] For a review we refer to T.J. Greytak and D. Kleppner, in *New Trends in Atomic Physics*, edited by C. Grynberg and R. Stora (North-Holland, Amsterdam, 1984), p. 1125 and I.F. Silvera and J.T.M. Walraven, in *Progress in Low Temperature Physics*, edited by D.F. Brewer (North-Holland, Amsterdam, 1986), Vol. 10, p. 139.
[10] A. Lagendijk, I.F. Silvera, and B.J. Verhaar, Phys. Rev. A **33**, 626 (1986); H.T.C. Stoof, J.M.V.A. Koelman, and B.J. Verhaar, Phys. Rev. B **38**, 4688 (1988).
[11] J.M. Doyle, J.C. Sandberg, I.A. Yu, C.L. Cesar, D. Kleppner, and T.J. Greytak, Phys. Rev. Lett. **67**, 603 (1991).
[12] O.J. Luiten, H.G.C. Werij, I.D. Setija, T.W. Hijmans, and J.T.M. Walraven, Phys. Rev. Lett. **70**, 544 (1993).
[13] I.D. Setija, H.G.C. Werij, O.J. Luiten, M.W. Reynolds, T.W. Hijmans, and J.T.M. Walraven, Phys. Rev. Lett. **70**, 2257 (1993).
[14] See also E. Levich and V. Yakhot, Phys. Rev. B **15**, 243 (1977) and S.G. Tikhodeev, Zh. Eksp. Teor. Fiz. **97**, 681 (1990) [Sov. Phys.-JETP **70**, 380 (1990)].
[15] D.W. Snoke and J.P. Wolfe, Phys. Rev. B **39**, 4030 (1989).

[16] H.T.C. Stoof, Phys. Rev. Lett. **66**, 3148 (1991); Phys. Rev. A **45**, 8398 (1992).

[17] For a more detailed discussion see, for instance, S. Ma, *Modern Theory of Critical Phenomena* (Addison-Wesley, New York, 1976) and J.W. Negele and H. Orland, *Quantum Many-Particle Systems* (Addison-Wesley, New York, 1988).

[18] N.N. Bogoliubov, J. Phys. USSR (Moscow) **11**, 23 (1947).

[19] See, for example, A.L. Fetter and J.D. Walecka, *Quantum Theory of Many-Particle Systems* (McGraw-Hill, New York, 1971).

[20] K. Huang, *Statistical Mechanics*, second edn (Wiley, New York, 1987).

[21] N.M. Hugenholtz and D. Pines, Phys. Rev. **116**, 489 (1958).

[22] L.V. Keldysh, Zh. Eksp. Teor. Fiz. **47**, 1515 (1964) [Sov. Phys.-JETP **20**, 235 (1965)].

[23] P. Danielewicz, Ann. Phys. (N.Y.) **152**, 239 (1984).

[24] U. Eckern, J. Low Temp. Phys. **54**, 333 (1984).

[25] Yu. M. Kagan, B.V. Svistunov and G.V. Shlyapnikov, Zh. Eksp. Teor. Fiz. **101**, 528 (1992) [Sov. Phys.-JETP **74**, 279 (1992)]; Sov. Phys.-JETP **75**, 387 (E) (1992); B.V. Svistunov, J. Moscow Phys. Soc. **1**, 373 (1991); Yu. Kagan, this volume.

[26] V.N. Popov, *Functional Integrals in Quantum Field Theory and Statistical Physics* (Reidel, Dordrecht, 1983).

[27] N.N. Bogoliubov, V.V. Tolmachev, and D.N. Shirkov, *New Methods in the Theory of Superconductivity* (Consultants Bureau Enterprises, New York, 1959).

[28] P.W. Anderson, Phys. Rev. **112**, 1900 (1958).

12

Macroscopic Coherent States of Excitons in Semiconductors

L. V. Keldysh

P. N. Lebedev Physics Institute
Leninsky Prospect 53
117924 GSP, Moscow B-333
Russia

Abstract

Initially put forward by Moskalenko [1] *and Blatt et al.* [2], *the idea of a possible Bose–Einstein condensation* (BEC) *of excitons in semiconductors has attracted the attention of both experimentalists and theoreticians for more than three decades. At different stages of this long history, the results of their efforts have been described and discussed in review articles* [3–10]. *A brief introduction and summary of the main qualitative conclusions of this older work is presented here (Sections* 1 *and* 2), *followed by a more detailed discussion of some more recent developments (Sections* 3 *and* 4).

1 Electronic Excitations in Semiconductors

Schematically presented in Fig. 1 is the typical electronic spectrum of a semiconductor: two bands (or two groups of bands) of continuous spectrum – conduction (c) and valence (v) – separated by the energy gap $E_g = E_{c,\min} - E_{v,\max}$. In the ground state, all of the states in the valence band(s) are occupied by valence electrons of the semiconductor, and all states in the conduction band are empty.

The lowest single-particle electronic excitations are an additional electron (e) in the conduction band or a single empty state – a hole (h) – in the valence band. Both of these excitation types are mobile fermions (spin = 1/2) characterized by effective masses, m_e and m_h and effective charges $e_e = e$ and $e_h = -e$, respectively. Here e is the usual (negative) elementary charge. Effective charges of electrons and holes are universal, unlike the effective masses, which are different not only in different semiconductors but also in different bands of the same semiconductor and for different directions in the same band (anisotropy).

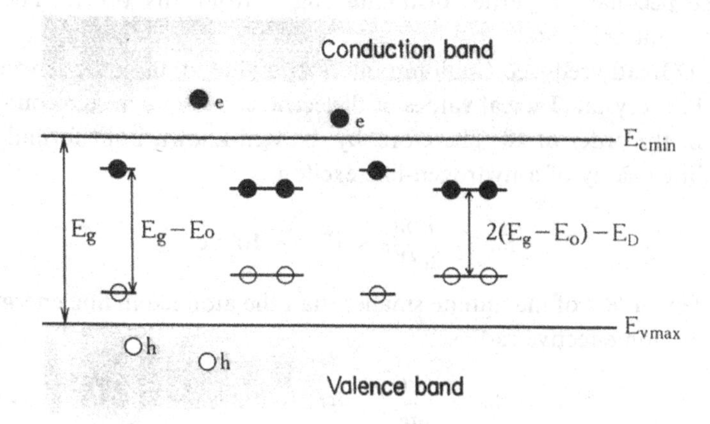

Fig. 1. Schematic picture of the main components of the nonequilibrium electron–hole plasma – free electrons and holes, excitons and excitonic molecules – with their characteristic energies. The top and bottom represent the continuous spectrum of single-particle excitations – electrons and holes. The energy levels of electrons and holes bound in excitons or excitonic molecules have no exact meaning and are depicted only for the sake of illustration. Only the total energy of a complex as a whole is well defined.

Being oppositely charged (quasi)free particles, the electron and hole attract one another by Coulomb forces just as the electron and proton do. And like an electron and proton which can be bound into a hydrogen atom, the electron and hole can be bound via this interaction into the two-particle excitation called the Wannier–Mott exciton [11, 12]. Being the combination of two fermions, an exciton is expected to be a boson, and so the problem of the possibility of BEC of excitons arises. But to discuss this problem, we will need more information about the possible states and transformations in the system of many excitons, or, more correctly, in the many electron–hole system. The above-mentioned similarity to the electron–proton system can be extended to the many-particle system and so suggests the existence of the excitonic molecule (biexciton), the bound state of two excitons similar to a hydrogen molecule [13, 14] and, moreover, the gas–liquid-type phase transition of exciton and biexciton "gas" into an electron–hole Fermi liquid (EHL) – the bound state of a macroscopically large number of electrons and holes. [15, 6, 9, 10]

But this similarity is not complete. The electron–hole system possesses some peculiar properties distinguishing it from any other. The most important are

(1) Greatly reduced Coulomb interaction due to dielectric screening in the host crystal. Typical values of dielectric constant ϵ in semiconductors are of the order of 10. Therefore, by the well-known Bohr formulas, the binding energy of a hydrogen-like exciton

$$E_0 = \frac{e^4 m}{2\epsilon^2 \hbar^2} \sim 10^{-1} - 10^{-3}\,\mathrm{eV}, \tag{1}$$

is a few orders of magnitude smaller than the atomic binding energy, and the exciton effective radius

$$a_0 = \frac{\epsilon \hbar^2}{me^2} \sim 10^{-7} - 10^{-6}\mathrm{cm} \tag{2}$$

is macroscopically large compared to interatomic distances in the host crystal. Formulas (1) and (2) represent the natural units of quantum scales of energy and length in the interacting electron–hole system, in just the same sense as the Rydberg and Bohr radius are the natural scales in atomic, molecular and solid state physics. Therefore in what follows we will use these units. Instead of the temperature, a dimensionless ratio kT/E_0, designated by T, will be used, and instead of particle concentration the product na_0^3, designated by n. In this notation, the usual formula for the critical temperature of an excitonic gas BEC, neglecting any interaction corrections, is given by

$$T_c = 6.62 \frac{m}{M} (\frac{n}{g})^{2/3}. \tag{3}$$

Here $M = m_e + m_h$ is the effective mass of the exciton, equal to the sum of electron m_e and hole m_h effective masses; $m = m_e \cdot m_h / M$ is their reduced mass, entering (1) and (2); and g is the exciton ground state statistical weight (degeneracy).

(2) The effective masses m_e and m_h in any particular semiconductor, though different, are usually of the same order of magnitude (or differ at most by one order of magnitude). The absence in the electron–hole system of really heavy particles such as nuclei results in a dominant role of quantum effects at all temperatures $T \leq 1$. In particular as formula (3) shows, $T_c \sim 1$ for $n \sim 1$, unlike, for example, in ^{4}He where $T_c \sim 10^{-5}$ in corresponding units. Other important manifestations of the absence of heavy particles are very large zero vibrations of excitons in the excitonic molecule. The amplitude of this vibration is of the order of, or even larger than, the excitonic radius a_0 itself. So biexcitons appear to be

very loosely bound complexes with binding energy E_D not exeeding 0.1, compared to 0.35 in the hydrogen molecule [16, 17].

For the same reason, nothing like an "excitonic crystal" can exist. The weak van der Waals attraction dominating at large intermolecular distances is not able to confine light particles such as excitons or biexcitons. According to Ref. [18], the s-wave scattering length of two excitonic molecules is large ($\simeq 7$) and positive (repulsive). Therefore a condensed phase of excitonic molecules – a molecular "liquid" similar to liquid hydrogen – also cannot exist. But an electron–hole liquid (EHL) similar to metallic hydrogen or alkali metals does exist [15, 6, 9, 10]. Unlike common metals, in the EHL not only electrons but also holes ("nuclei") are Fermi degenerate. If the effective masses of both electrons and holes are more or less isotropic, the EHL at low temperatures transforms to an "excitonic insulator" phase [19–21]. In this two-Fermi-liquid state the collective pairing of electrons and holes in the vicinity of Fermi surfaces arises, quite similar to BCS pairing in superconductors. This pairing manifests itself in the appearance of the energy gaps Δ around the Fermi surfaces, as shown in Fig. 2. These gaps may be considered a remnant of the binding energy of a single exciton, transformed by collective many-body interactions in the condensed phase. The gap diminishes fast with increasing particle density or m_e–m_h difference, and it does not exist at all if the anisotropy is large. In that sense, the excitonic insulator state in the nonequilibrium electron–hole (e–h) system is a coherent BEC state of high density ($n \geq 1$) excitons, as the superconducting state is a coherent BEC state of Cooper pairs. Like Cooper pairs, e–h pairs in an excitonic insulator have very large radii $\sim \Delta^{-1/2}$, much larger than interparticle distance. So they are true collective phenomenon. But, unlike Cooper pairs in superconductors which are charged ($-2e$), excitons are electrically neutral.

The above-described phenomena can be schematically represented by the phase diagrams depicted in Figs. 3-4. They show possible states of the e–h system in the plane $(\bar{n}, T)$. Here $\bar{n}$ is the average particle density.

The phase diagram type of any particular semiconductor is defined by the relationship of the binding energy per e–h pair in the biexciton gas, given by $(E_0 + E_D/2)$, and in the EHL, indicated by $-\mu_l$. Strictly speaking, μ_l is the difference between the chemical potential per e–h pair in the EHL and E_g, where E_g is the minimum energy of the free e–h pair. Different phases presented in Figs. 3 and 4 are: G, the gas (plasma) of weakly interacting electrons, holes, excitons and excitonic molecules; BG, the degenerate gas of excitons and(or) excitonic molecules; ML, the

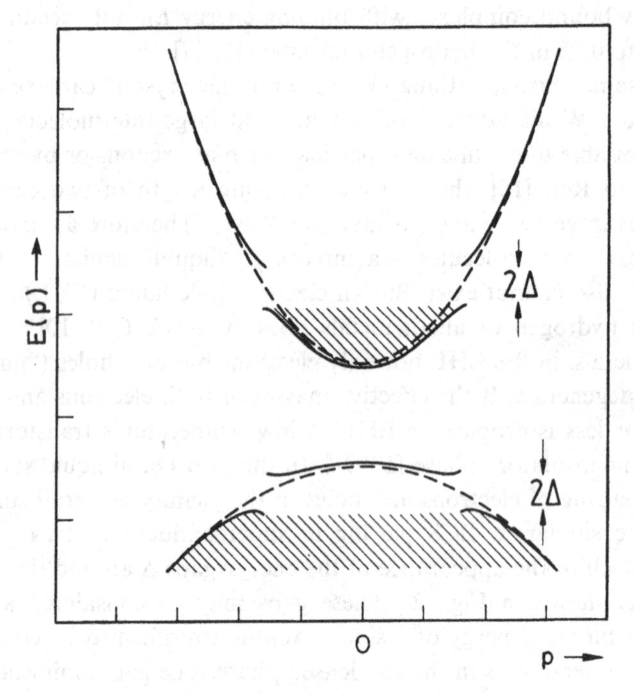

Fig. 2. Energy spectrum in the nonequilibrium "excitonic insulator" phase. The striped regions are energy ranges occupied by electrons.

metallic liquid (EHL); and IL, the insulating liquid (excitonic insulator). The striped area is the region of phase coexistance: droplets of liquid phase surrounded by gas, degenerate or nondegenerate. Coherent states BG or IL are present in the phase diagrams of the types shown in Figs. 3(b) and 4(a),(b). But up to now only the simplest type, shown in Fig. 3(a), is firmly established and studied experimentally, in germanium and silicon.

Both phases in Fig. 3 correspond to the case $| \mu_l | > (E_0 + E_D/2)$; that is, the EHL is more tightly bound than excitonic molecules. It is theoretically proven and experimentally confirmed that the fulfilment of this condition is strongly favored by multivalley band structure, i.e. the presence in the electronic spectrum of a semiconductor of a few equivalent groups of electrons or holes (or both). Just such a band structure is characteristic of both germanium and silicon. No BEC exists in the gas phase in this case, exactly as in the case of helium. It does exist in the phase diagrams depicted in Fig. 4, corresponding to the fulfillment

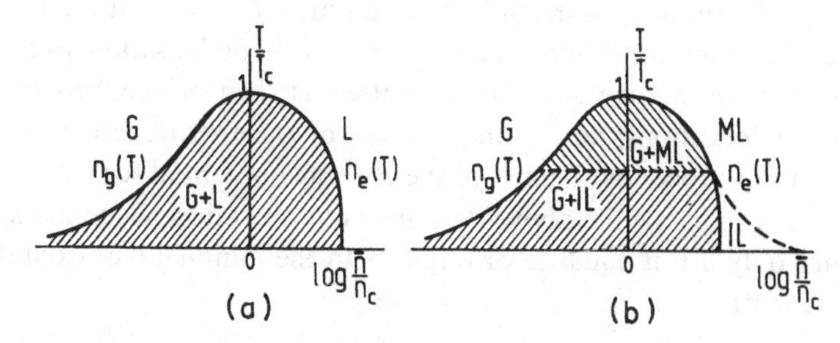

Fig. 3. Possible types of nonequilibrium electron–hole system phase diagrams in semiconductors with electron–hole liquid more tightly bound than the excitonic molecule. The striped regions are the coexistence domains of the liquid and gas phases.

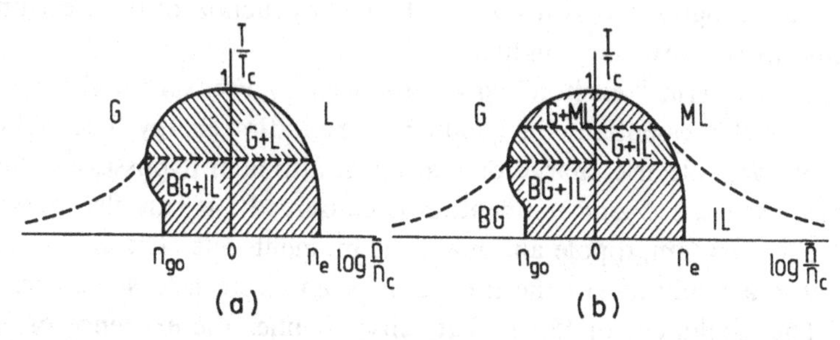

Fig. 4. The same as in Fig. 3, but for semiconductors with excitonic molecules more tightly bound than electron–hole liquid.

of the opposite condition,

$$|\mu_l| < E_0 + 0.5E_D\,. \qquad (4)$$

Theoretically, such a case may be expected in semiconductors with single-valley band structure for both electrons and holes, Fig. 4(a) corresponding to semiconductors with a very large difference of the electron and hole effective mass and moderate anisotropy, and Fig. 4(b) to comparable effective masses and small anisotropy. Experimentally, the fulfillment of (4) is not firmly established in any semiconductor. And it should be noted also that the difference of binding energies per *e–h* pair in the EHL and the biexciton in the case corresponding to Fig. 4(b) is very small and comparable to the accuracy of theoretical computation. So even the theoretical possibility of the phase diagram shown in Fig. 4(b) is, strictly speaking, not proven.

(3) Another important difference of the *e–h* system from any ordinary

matter is its essentially nonequilibrium nature. For sure, it also exists under equilibrium conditions, but without any condensation phenomena, because the number of *e–h* pairs itself is then determined by the equilibrium conditions, i.e. by the balance of the thermal creation and recombination processes. Therefore the chemical potential for pairs must be equal to zero, while the above-described condensation phenomena can take place only if it is equal to or larger than the minimal pair excitation energy. It implies

$$\min(E_g - (E_0 + 0.5E_D), E_g + \mu_l) \leq 0, \tag{5}$$

which means that the binding energy per pair in the excitonic molecule or in the EHL becomes larger than the minimal energy for a free *e–h* pair creation, E_g. Such a possibility was also considered theoretically [19–22], and it was recognized that it implies the reconstruction of the electronic structure of the host semiconductor.

Indeed, the term " exciton" was introduced by Frenkel for the quasiparticle which is essentially the quantum of excitation energy. The exciton does not carry electric charge or real mass, although it possesses effective mass (exciton *crystal* momentum is conserved). It can also possess angular momentum, dipole and magnetic moments, etc. The creation of an exciton actually means the transition of an electron to some excited level. The fulfillment of (5) in that sense signifies the existence of another ground state electronic configuration than is supposed in Fig. 1, with lower total energy. The equilibrium condensation of excitons with zero chemical potential is, then, another formal language to describe an electronic phase transition, i.e. reconstruction of electronic states in the valence and conduction bands, accompanied by the formation of a charge or spin density wave, magnetic ordering, or something like that. Nothing like superfluidity can arise because the Hamiltonian does not conserve the total number of excitons (*e–h* pairs). This inevitably results in pinning of the "condensate" by interband scattering matrix elements of the Coulomb interaction [23], which in this language appears as the source of the spontaneous creation of two *e–h* pairs lifting the gauge invariance of the Hamiltonian to yield a phase transformation, different in different bands, of the kind

$$\psi_j \rightarrow \psi_j \exp(i\phi_j). \tag{6}$$

Here $j = (c, v)$ is the band index. Also, interband scattering of electrons by static lattice distortions – uniform or periodic – can serve as a single pair creation source and thus a pinning mechanism.

Quite different is the problem of the collective properties of the nonequilibrium dense *e–h* system produced in a semiconductor by some external action, usually illumination. Then the total number of *e–h* pairs N (or as used in Figs. 3 and 4, their average density $\bar{n} = N/V$, where V is the volume of the specimen) really becomes an independent variable with a value controlled by an external source. The conservation of N is broken by recombination processes. But in some semiconductors the direct radiative recombination is forbidden by a parity selection rule (as in Cu_2O) or by quasimomentum conservation, as in germanium and silicon. The lifetime of the nonequilibrium *e–h* system in such a case may be much longer than the thermalization time and the system appears to be in quasiequilibrium, the only nonequilibrium parameter being the total number of particles N itself. Under this condition, the phase diagrams of Figs. 3 and 4 have an exact meaning. On the other hand, in semiconductors with a dipole-allowed excitonic radiative transition, the lifetimes are of the order $10^{-9} - 10^{-10}$ s, and a dense enough *e–h* system is always far from equilibrium, so that these phase diagrams can be considered only as qualitative indications of trends in its evolution.

One more question of importance is whether excitons are really bosons [24]. The reason for asking this is that the exciton is a compound system composed of two fermions, and increasing the density of excitons and their overlap also means increasing the local density of fermions, perhaps leading to problems with the Pauli principle. Also, from a formal point of view, the exciton annihilation operator $\hat{A}_{\mathbf{P}}$ is composed of electron and hole annihilation operators $\hat{a}_{e\mathbf{p}}$ and $\hat{a}_{h\mathbf{p}}$ ($\mathbf{P}, \mathbf{p}$ are momenta)

$$\hat{A}_{\mathbf{P}} = \sum_{\mathbf{p}} \varphi_0(\mathbf{p}) \hat{a}_{e\frac{\mathbf{P}}{2}+\mathbf{p}} \hat{a}_{h\frac{\mathbf{P}}{2}-\mathbf{p}} , \qquad (7)$$

where φ_0 is the wave function describing the relative *e–h* motion in the exciton. The commutator of the two exciton operators, $\hat{A}^{\dagger}_{\mathbf{P}}$ and $\hat{A}_{\mathbf{P}'}$, is then

$$[\hat{A}_{\mathbf{P}}, \hat{A}^{\dagger}_{\mathbf{P}'}] = \delta_{\mathbf{P},\mathbf{P}'} + \sum_{\mathbf{q}} \left\{ \varphi_0^*(\mathbf{q} - \frac{\mathbf{P}'}{2}) \varphi_0(\mathbf{q} - \frac{\mathbf{P}}{2}) \hat{a}^{\dagger}_{e\mathbf{q}-\mathbf{Q}} \hat{a}_{e\mathbf{q}+\mathbf{Q}} \right.$$
$$\left. + \varphi_0^*(\frac{\mathbf{P}'}{2} - \mathbf{q}) \varphi_0(\frac{\mathbf{P}}{2} - \mathbf{q}) \hat{a}^{\dagger}_{h\mathbf{q}-\mathbf{Q}} \hat{a}_{h\mathbf{q}+\mathbf{Q}} \right\} , \qquad (8)$$

where $\mathbf{Q} = (\mathbf{P}' - \mathbf{P})/2$. The righthand side of (8) differs from the usual one for Bose operators $\delta_{\mathbf{P},\mathbf{P}'}$ by operators that are of the order of n, the exciton density. So for $n \sim 1$, the operator (7) has nothing to do with usual

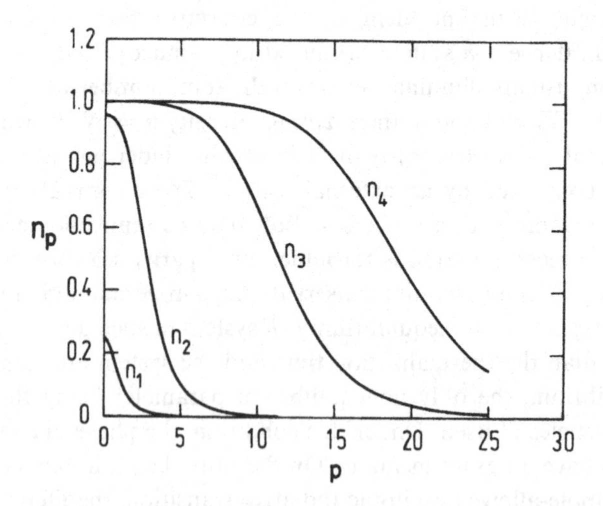

Fig. 5. Momentum distribution function for electrons (or holes) in a system of Bose-condensed excitons. Different curves correspond to different exciton densities ($n_1 < n_2 < n_3 < n_4$).

Bose operators. What it really means and what happens to excitons at high densities is illustrated by Fig. 5, in which the momentum distribution function of electrons (or holes) for the system of Bose-condensed excitons of different densities is presented. For a single exciton, it is $|\varphi_0(\mathbf{p})|^2$, the probability of finding an electron with momentum $\mathbf{p}$ in the exciton. For low concentration it increases proportional to density, since all excitons are identical: $f(\mathbf{p}) = n\,|\varphi_0(\mathbf{p})|^2$. But as $f(\mathbf{p} = 0)$ approaches 1 with increasing n, it cannot continue to increase because of the Pauli principle – $f(\mathbf{p})$ is the fermion distribution function. So at larger densities, the electrons and holes bound into excitons are forced to occupy states with larger momenta until their $f(\mathbf{p})$ also aproaches 1, and so the distribution smoothly transforms to the Fermi distribution function. The binding energy of the exciton transforms to the narrow gap around the Fermi level which diminishes fast as the kinetic Fermi energy becomes large compared to the Coulomb interaction. Thus Bose-condensed excitons continuously transform to the "excitonic insulator" described above, and finally to a degenerate EHL.

Such transformations are not unique for excitons. They must hold also for ^{4}He or any other atomic or molecular system. But it is very difficult experimentally to compress, for example, helium to densities corresponding to $n \sim 1$. The peculiarities of excitons are their large radii and the possibility of creating them by external excitation, so that high compression can be easily realized.

2 Polaritons

A very peculiar, but undoubtedly also the most important, case is presented by BEC of dipole-active excitons. Not only is it always essentially a nonequilibrium problem, as explained above, but the nature of the quasiparticles themselves is very special. They are the well-known polaritons – the quantum superposition of photons and electronic excitation (excitons) [25]. In other words, a polariton is a real photon in a medium: an exciton representing a polarization cloud accompanying propagation of electromagnetic field through the medium [26]. These polarization effects are especially strong under resonant conditions, that is, for photon energies $\hbar\omega$ close to the energy of exciton creation $\hbar\omega_0 = (E_g - E_0)$. Schematically depicted in the Fig. 6 is the well-known polariton dispersion law. Strong mixing arises close to the intersection point of the undisturbed photon and exciton branches $(\mathbf{k}_0, \omega_0)$, and the typical picture of level repulsion appears. So the problem of BEC of dipole-active excitons is essentially that of photons, or, in other words, the problem of a coherent electromagnetic wave of finite amplitude in the polariton frequency range. It is a very familiar problem but is still far from trivial. It includes all the many-body phenomena typical of BEC, but they acquire the meaning of nonlinear optics phenomena. Two different problems are usually considered in this context.

The accumulation of a macroscopic number of initially incoherent excitation quanta in a single-photon mode is lasing. It was recognized long ago [27–31] that the appearance of a coherent mode from multimode intense noise in a pumped system is a very typical example of a nonequilibrium phase transition. The approaches used for its general theoretical treatment are based on mean-field approximations, the Langevin equation, the Fokker–Planck equation, etc. Their detailed review is beyond the scope of this article. As applied to the particular case of dipole-active excitons in semiconductors, the qualitative picture of the lasing process should look like the following [32]. Initially excited electrons and holes combine into excitons, which dissipate their kinetic energy by emission of phonons – quanta of the crystal lattice vibrations – and so move downwards in energy along the lower polariton branch (LPB) in Fig. 6. Approaching the exciton–photon intersection point, which corresponds to very small momenta $\sim E_g/c$ (c the light velocity), they smoothly

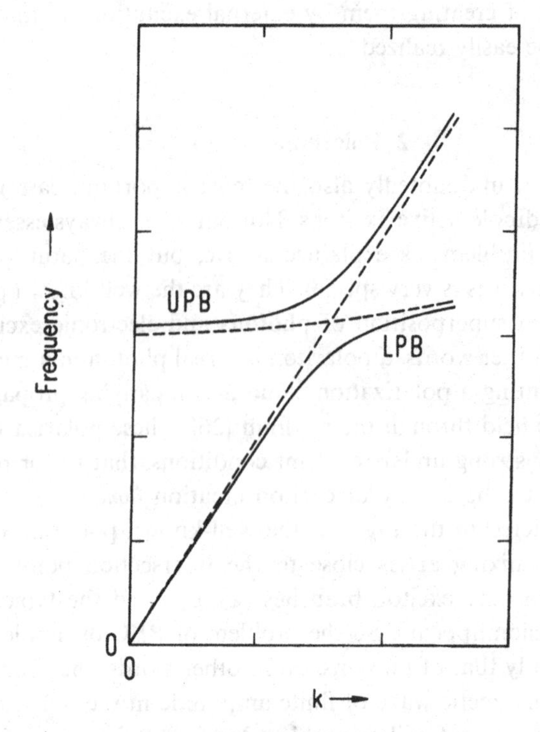

Fig. 6. Polariton dispersion law.

transform into polaritons, with the photonic component increasing at the expense of excitonic. But this transformation is accompanied by a decrease of energy dissipation rate, since just the excitonic component interacts with lattice vibrations, and also because of the dramatic decrease in the density of states due to increase of the polariton velocity below the intersection point. Thus a "bottleneck" arises in this momentum range where polaritons accumulate. Because of the small density of states in the bottleneck region, polariton occupation numbers there easily reach values of the order of unity even if the total exciton density is small, $n \ll 1$. Then the mechanism, common in laser physics, of line narrowing by stimulated emission starts. As applied to the polariton kinetics, it looks like stimulation of transitions to the bottleneck region. Under the condition of sufficiently strong stationary pumping, this process results in the imaginary part of the optical susceptibility becoming negative for one or a few polariton modes. This denotes an instability of the stationary incoherent radiation regime and the appearance of a finite-amplitude coherent polariton mode, i.e. BEC of polaritons. In the transient regime, as a peak in polariton frequency distribution becomes large enough and narrow, the exciton interaction synchronizes all of the modes inside this peak. So in a finite time interval, the coherent state of a macroscopically large number of polaritons arises. It should be noted that existing semiconductor lasers usually operate either at excitation densities $n \gg 1$ or with participation of optical phonons, impurities or lattice defects, and do not correspond to this idealized picture.

Another direction of research is the nonlinear optics proper, i.e. propagation of the intense electromagnetic wave or pulse and its interaction with the medium under conditions of polariton formation [31, 33]. In such a case the coherence is induced by some external source of radiation. In what follows, mainly this type of problem will be discussed. The main difference in the formal description of these phenomena corresponding to the above physical picture, compared to common "excitonic insulator" theory, is the presence in the Hamiltonian of the source of e–h pairs – the electromagnetic field f. The additional term is

$$H_{eh} = \int [\psi_e(\mathbf{r})f^*(\mathbf{r},t)\psi_h(\mathbf{r}) + \psi_h^\dagger(\mathbf{r})f(\mathbf{r},t)\psi_e^\dagger(\mathbf{r})]d^3r, \tag{9}$$

where

$$f = ie\hbar\mathscr{E}\frac{\mathbf{p}_{cv}}{mE_g E_0} \tag{10}$$

is the matrix element of the electronic dipole transition between the conduction and valence bands induced by the electric field $\mathscr{E}$; $\mathbf{p}_0$ is the momentum of the photons; and $\mathbf{p}_{cv}$ is the interband momentum matrix element. In its physical interpretation, f is the interband Rabi frequency in terms of $E_0/\hbar$.

It was indicated in Ref. [34] that for photon energies above the absorption threshold $\hbar\omega > E_g$, a coherent field produces a collective excited state of semiconductors quite similar to the excitonic insulator state, even neglecting the Coulomb interaction. Creating electrons and holes coherently in correlated pairs, the field itself acts as the pairing force. Also, an energy gap arises in the electronic spectrum at the Fermi level with a magnitude $| f |$ (Fig. 2). The obvious condition for the existence of such a state is $f\tau \gg 1$, where τ is electron (hole) relaxation time. The complete treatment of this problem needs inclusion of many-body effects. This was done in Refs. [35–36] (see also the reviews and monographs [31, 33], [38–40] and especially [37], where the problem has received its most complete formulation, in terms of nonequilibrium Green functions [41]. (The details of this technique are presented in Refs. [42–51], [31] and many other reviews and monographs.) Here we summarize only a few general definitions and basic relations and also some features specific to the nonequilibrium e–h system.

The nonequilibrium Green's function is a 2×2 matrix, which in the so-called "triangular representation" looks as follows:

$$\hat{G} = \begin{pmatrix} 0 & G^{(a)} \\ G^{(r)} & F \end{pmatrix}. \tag{11}$$

$G^{(a,r)}$ are advanced and retarded functions, respectively, defined in the usual way

$$G^{(r)}(x,x') = -i\Theta(t-t') < [\psi(x), \psi^{\dagger}(x')]_{\pm} >, \tag{12}$$

$$G^{(a)}(x,x') = [G^{(r)}(x',x)]^{*}, \tag{13}$$

$$F(x,x') = -i < [\psi(x), \psi^{\dagger}(x')]_{\mp} >. \tag{14}$$

Here $x = (\mathbf{r}, t)$; indices $+(-)$ in (12) and $-(+)$ in (14) refer to the case of Fermi (Bose) statistics. In the electron–hole system all of these functions are 2×2 matrices in band indices (c, v). In this notation, the interband (electron–hole) Green's function

$$\hat{G}_{vc}(x,x') \equiv \hat{G}_{he}(x,x') = \begin{pmatrix} 0 & G_{he}^{(a)} \\ G_{he}^{(r)} & F_{he} \end{pmatrix} \tag{15}$$

describes the exciton (electron–hole pair) condensate just as a Gor'kov function in the theory of superconductivity describes a condensate of Cooper pairs:

$$\begin{aligned} G_{he}^{(r)}(x,x') &= -i\Theta(t-t') < [\psi_v(x), \psi_c^{\dagger}(x')]_{+} > \\ &= -i\Theta(t-t') < [\psi_h^{\dagger}(x), \psi_e^{\dagger}(x')]_{+} >, \end{aligned} \tag{16}$$

$$\begin{aligned} F_{he}(x,x') &= -i < [\psi_v(x), \psi_c^{\dagger}(x')]_{-} > \\ &= -i < [\psi_h^{\dagger}(x), \psi_e^{\dagger}(x')]_{-} >. \end{aligned} \tag{17}$$

Diagonal in both types of matrix indices, the functions F_{ee} and F_{hh} at coincident times $t = t'$ are closely related to the single-particle density matrices and in the momentum representation to the distribution functions (occupation numbers)

$$F_{ee}(\mathbf{p}; t, t) = -i[1 - 2n_{e\mathbf{p}}(t)]. \tag{18}$$

The interband function $F_{he}(\mathbf{p};t,t)$ can be considered as an effective exciton wave function. This interpretation follows from the comparison of definitions (17) and (16) and is exact, at least for $n \ll 1$. In a dense many-body system it should not be treated too literally.

In general, Green's functions can be found from the Dyson equation

$$\left(i\frac{\partial}{\partial t} - \frac{\hbar^2}{2m_i}\nabla^2\right)\hat{G}_{ij}(x,x') - \int \hat{\Sigma}_{ik}(x,x'')\hat{G}_{kj}(x'',x')d^4x'' = \delta_{ij}\delta(x-x'), \tag{19}$$

where "self-energy" matrices $\hat{\Sigma}_{ij}$ also have "triangular" form

$$\hat{\Sigma}_{ij} = \begin{pmatrix} \Omega_{ij} & \Sigma_{ij}^{(r)} \\ \Sigma_{ij}^{(a)} & 0 \end{pmatrix} \tag{20}$$

and can be expressed in terms of Green's functions and the interaction potential in any approximation by a set of Feynman graphs. The specific feature of the system under consideration is that, apart from the infinite set of diagrams accounting for different order interaction processes, the interband self-energy Σ_{eh} contains, according to (9), the driving force contribution

$$\hat{\Sigma}_{ij}^{(0)}(x,x') = \begin{pmatrix} 0 & f^*(x) \\ f(x) & 0 \end{pmatrix} \cdot \delta(x-x'). \tag{21}$$

The electric field $\mathscr{E}$ entering into Eq. (10) has to be found self-consistently from Maxwell's equation,

$$\nabla \times \nabla \mathscr{E} + \frac{1}{c^2}\frac{\partial^2(\epsilon\mathscr{E})}{\partial t^2} = -\frac{4\pi}{c^2}\frac{\partial^2 \mathscr{P}_{ex}}{\partial t^2}, \tag{22}$$

where the resonant excitonic contribution to the polarization [4] is

$$\mathscr{P}_{ex} = e < \psi_e(x)\mathbf{r}\psi_h(x) > = -\frac{e\hbar}{2mE_g}\left[\mathbf{p}_{cv} \cdot F_{he}(x,x)\right]^* \tag{23}$$

and ϵ in (22) accounts for all the nonresonant contributions.

Equations (19)–(23) are the complete set describing formally all nonlinear polarization problems. The interband matrix function $\hat{G}_{eh}$ represents the macroscopic coherent state of electron–hole pairs (excitons). In general, the theory under consideration is that of a driven two-component Fermi liquid with all its many-body aspects, with the additional complications due to being far from equilibrium. Therefore, the real problem in any particular case is the choice of an appropriate approximation.

If the retardation effects in the self-energies may be neglected, i.e. $\hat{\Sigma}_{ij} \sim \delta(t-t')$, which is the case at least in the mean-field approximation,

(19) can be reduced to the equation for the single-time ($t = t'$) density matrix [31, 33, 36, 37, 51]

$$f_{ij}(\mathbf{r},\mathbf{r}';t) = F_{ij}(\mathbf{r},\mathbf{r}';t = t'). \tag{24}$$

Symbolically written, this equation is similar to the well-known Bloch equation for two-level systems,

$$i\frac{\partial \hat{f}}{\partial t} = \left(\hat{h}_0 + \hat{\Sigma}^{(r)}\right)\hat{f} - \hat{f}\left(\hat{h}_0 + \hat{\Sigma}^{(a)}\right). \tag{25}$$

Here the two-band Hamiltonian matrix is introduced by

$$(h_0)_{ij} = \left(E_{0i} - \frac{\hbar^2}{2m_i}\nabla^2\right)\cdot\delta_{ij}. \tag{26}$$

Matrices $(\hat{\Sigma}^{(r)})_{ij}$ and $(\hat{\Sigma}^{(a)})_{ij}$ are defined as $(\hat{\Sigma}^{(r)})_{ij} = \Sigma_{ij}^{(r)}$, $(\hat{\Sigma}^{(a)})_{ij} = \Sigma_{ij}^{(a)}$; E_{0i} and m_i are the corresponding band-edge energy and the effective mass (negative in the valence band).

In the simplest case of stationary and spatially homogeneous pumping, the solution of (25) is

$$f_{ee,hh} = f_{e,h}^{(0)} - \frac{\gamma_{e,h}}{\gamma_e + \gamma_h}\frac{\lambda^2|\,\Sigma_{eh}\,|^2}{|\,E(\mathbf{p}) + \Sigma^{(r)}(\mathbf{p})\,|^2 + \lambda^2|\,\Sigma_{eh}\,|^2}(f_e^{(0)} + f_h^{(0)}), \tag{27}$$

$$f_{eh}(\mathbf{p}) = \Sigma_{eh}(\mathbf{p})^2\frac{E(\mathbf{p}) + \Sigma^{(r)}(\mathbf{p})}{|\,E(\mathbf{p}) + \Sigma^{(r)}(\mathbf{p})\,|^2 + \lambda^2|\,\Sigma_{eh}\,|^2}(f_e^{(0)} + f_h^{(0)}), \tag{28}$$

where

$$\lambda^2 = \frac{(\gamma_e + \gamma_h)^2}{\gamma_e\gamma_h}, \qquad \Sigma^{(r)}(\mathbf{p}) = \Sigma_e^{(r)}(\mathbf{p}) + \Sigma_h^{(r)}(\mathbf{p}),$$

$$E(\mathbf{p}) = E_g - \hbar\omega_0 + \frac{p^2}{2m}. \tag{29}$$

Here, $\mathbf{p}_0$ and ω_0 are the momentum and frequency of the pumping wave, respectively, and

$$\gamma_{e,h} = Im\Sigma_{e,h}^{(a)}, \quad f_{e,h}^{(0)} \equiv -\frac{\Omega_{ee,hh}}{2\gamma_{e,h}}. \tag{30}$$

In the mean-field approximation, the self-energies themselves can be expressed in terms of f_{ij},

$$\Sigma_{ij}^{(r)}(x,x';t) = \frac{1}{2}V(\mathbf{r}-\mathbf{r}')\left[if_{ij}(\mathbf{r},\mathbf{r}';t) - \delta_{ij}\cdot\delta(\mathbf{r}-\mathbf{r}')\right] + \Sigma_{ij}^{(0)}\cdot\delta(\mathbf{r}-\mathbf{r}') \tag{31}$$

and so the set of equations (25)–(30) is complete. Under stationary conditions they reduce to two equations for the self-energies,

$$\frac{1}{2}\int V_{\mathbf{p},\mathbf{p}'} \frac{\left(|f_e^{(0)}|+|f_h^{(0)}|\right)\Sigma_{eh}(\mathbf{p}')\cdot\left(E(\mathbf{p}')+\Sigma^{(r)}(\mathbf{p}')\right)}{|E(\mathbf{p}')+\Sigma^{(r)}(\mathbf{p}')|^2+\lambda^2|\Sigma_{eh}(\mathbf{p}')|^2}\frac{d^3p'}{(2\pi)^3} + \Sigma_{eh}(\mathbf{p}) = f, \tag{32}$$

$$\Sigma^{(r)}(\mathbf{p}) = \frac{1}{2}\int V_{\mathbf{p},\mathbf{p}'} \frac{\lambda^2\left(|f_e^{(0)}|+|f_h^{(0)}|\right)|\Sigma_{eh}(\mathbf{p}')|^2}{|E(\mathbf{p}')+\Sigma^{(r)}(\mathbf{p}')|^2+\lambda^2|\Sigma_{eh}(\mathbf{p}')|^2}\frac{d^3p'}{(2\pi)^3}. \tag{33}$$

The most important equation here is (32). It defines the gap in the renormalized excitation spectrum,

$$\epsilon(\mathbf{p}) = \sqrt{|E(\mathbf{p})+\Sigma^{(r)}(\mathbf{p})|^2+|\Sigma_{eh}(\mathbf{p})|^2} + \frac{p^2}{2}\left(\frac{1}{m_e}-\frac{1}{m_h}\right). \tag{34}$$

For $\hbar\omega_0 > |E_g+\Sigma^{(r)}(0)|$ it is just equal to $|\Sigma_{eh}|$. So (32) is similar to a BCS equation, but inhomogeneous and with a slightly different integral kernel. The last difference is due to the nonequilibrium nature of our problem. At moderate field $|f|\leq 1$, and not too close to the resonance $|\Sigma_{eh}| \ll E_g - \hbar\omega_0$, Eq. (32) can be transformed to an inhomogeneous Schrodinger equation for the exciton wave function. At very large intensities $|f| \gg 1$, it gives the gap $|\Sigma_{eh}| \approx f$, as was obtained in Ref. [34]. The functions $f_{e,h}^{(0)}$ entering (32)–(33), defined by Eq. (30), are connected to spontaneous fluctuations of occupation numbers. In thermodynamic equilibrium, the ratios on the right-hand side of (31) are found to be exactly

$$-\frac{\Omega_{e,h}(\mathbf{p})}{2\gamma_{e,h}(\mathbf{p})} = i\left[1-2n_{\mathbf{p}}(T)\right] \equiv i\tanh\frac{\epsilon_{e,h}(\mathbf{p})-\mu}{2kT}, \tag{35}$$

where $n_{\mathbf{p}}(T)$ is the equilibrium Fermi distribution function, μ is the chemical potential and $\epsilon_{e,h}(\mathbf{p})$ are the energies of corresponding single-particle excitations. Correct calculation of both the diagonal self-energy matrix element $\Omega(\mathbf{p})$ and the damping $\gamma(\mathbf{p})$ demands consideration of relaxation processes at least to lowest order. But this is out of the scope of our mean-field approximation. For our present considerations, it does not seem to be crucial, since in an intrinsic semiconductor, $n_{\mathbf{p}}(T)$ is very small and it seems reasonable to suppose that the difference of $|f_{e,h}^{(0)}|$ from 1 may also be neglected in a pumped system, at least for fast processes.

Nonlinearites in (31) account for the exchange interaction of electrons and holes and so-called phase-space filling, which means reduction of the interaction due to the blocking of some virtual scattering processes because of occupation of corresponding final states.

The only shortcoming of this approach is the difficulty of including the correlation effects, the most important of them being the formation of excitonic molecules. Because of the polaritonic nature of excitons, the existence of biexcitons implies the resonant mutual scattering of photons. In the following sections, we will show that at $|f| \sim 1$ it can completely dominate the optical properties in the vicinity of the exciton resonance.

3 Excitonic Molecules in a Coherent Cloud of Virtual Excitons

The existence of excitonic molecules – biexcitons (EM) – in semiconductors is well-established theoretically [13, 14, 16, 17] and experimentally [52–62]. It is also well recognized [62–67] that, as the excitons themselves are the essential characteristic feature of semiconductor optics, the biexciton contribution may be very important in the nonlinear optical phenomena in the frequency range close to the intrinsic absorption threshold. As a rule, two different approaches have been used in the theoretical description of EMs and their contribution to physical phenomena: the direct microscopic treatment [68, 50], considering the EM as a four-particle bound state – two electrons and two holes – and a phenomenological one [63–66], introducing the EM as an independent boson excitation capable of decaying into an exciton and a photon. Both of these approaches, in many cases very successful, have essential disadvantages for application to the problem under consideration. It seems that the first approach is too complicated to be used as a starting point for a true many-body problem. The second one is inadequate to account for the changes of EM properties and parameters, similar to those described above for excitons, due to interactions in a dense exciton–biexciton system. Nevertheless, according to modern ideas [24, 37] about many-body exciton physics, just these transformations, including an exciton's and other complexes' continuous destruction, constitute the essence of the high-excitation problem. Moreover, the assignment of definite statistics (e.g. Bose–Einstein) to compound particles at such high densities becomes inadequate, as explained in Section 2. At moderate exciton densities both of these effects contribute corrections of the order n, where n is the exciton dimensionless density. For the EM it will evidently be proportional to na_{bx}^3. Here, a_{bx} is the effective EM radius (in terms of the exciton radius). As EM binding energies E_{bx} are typically an order of magnitude smaller than the corresponding exciton Rydberg $E_0 = 1$, usually a_{bx} must be a few times larger than unity. So there is a relatively wide interval of densities in which excitons are virtually unchanged, whereas the EM may transform drastically. Nonlinear optical phenomena due to excitons coherently driven by an intense electromagnetic pump, with the densities in the interval described above, are the main subjects of this section.

For this purpose the following model seems to be adequate: excitons will be considered as true boson particles, described in terms of the Bose field operator $\psi(r,t)$ and the dispersion law $\epsilon_p^{(0)} = E_g - 1 + p^2/2m$. So defined, $\psi(x)$ is essentially $F_{eh}(x,x)$ of Section 2. In contrast with that section, an EM will be treated as composite particles – the bound state of two excitons arising due to the interaction potential $V(\mathbf{r},\mathbf{r}')$. Obviously such an interaction has to be attractive. But then the well-known problem in quantum liquid theory arises: at low temperatures, a system of Bose particles with an attractive interaction is unstable against spontaneous contraction. In order to avoid this difficulty, two equivalent types of excitons $\psi_\alpha(\alpha = 1,2)$ with the same parameters may be introduced. The only distinction is that the excitons of different types attract each other ($U_{12} = U_{21} = -V < 0$), whereas particles of the same type have a repulsive interaction ($U_{11} = U_{22} \equiv U > V > 0$). Such an appoach leads simultaneously to the existence of bound complexes of two excitons of different types and to the stability of the many-particle system as a whole, i.e. the stability of the "molecular gas" of excitons. Moreover, the proposed model is not too artificial, since two types of exciton may be considered as corresponding to singlet excitons with mutually opposite directions of electron (and also hole) spins. The main difference of this picture from reality is the absence of triplet excitons. Nevertheless, it seems that such an assumption is not crucial because a triplet exciton in the dipole approximation does not interact with photons (though it must in a real system, due to electron exchange.) A similar model has been used in Ref. [50]. Thus the exciton–exciton interaction is introduced as a 2×2 matrix:

$$\hat{U} = U\hat{I} - V\hat{\tau}_1, \tag{36}$$

where

$$\hat{I} = \begin{pmatrix} 1 & 0 \\ 0 & 1 \end{pmatrix}, \tag{37}$$

$$\hat{\tau} = \begin{pmatrix} 0 & 1 \\ 1 & 0 \end{pmatrix}.$$

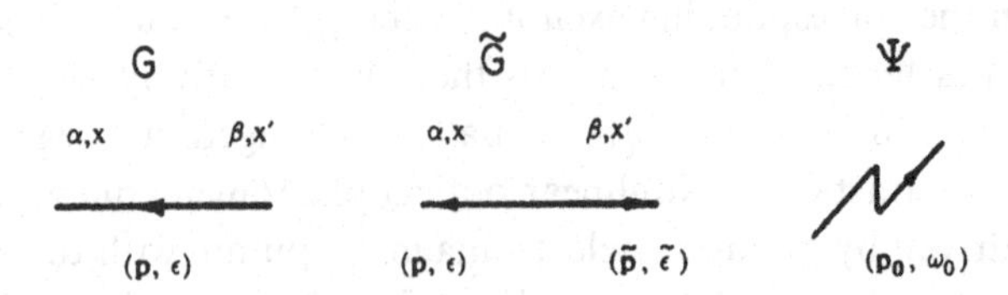

Fig. 7. Schematic illustration of elements in the Feynman graphs: Green's functions and coherent exciton field amplitude.

Qualitatively this corresponds to the exchange interaction of two hydrogen-like atoms (excitons).

From the theoretical point of view, the system under consideration is a "driven Bose liquid". It arises from a macroscopically occupied state (mode) $\psi_\alpha(\mathbf{r}, t)$ of the excitons coherently created by an external electromagnetic field $\mathscr{E}(\mathbf{r}, t)$. As the result of mutual scattering of excitons of this mode, the other modes become populated and also the EM may arise. A consistent description of this system may be given using the nonequilibrium Green's functions technique [41], [46]–[49] and [31]. The following functions are involved in such a treatment:

(1) Normal Green's functions $\hat{g}_{\alpha\beta}(x, x')$, which describe the propagation of a particle from space-time point $x = (\mathbf{r}, t)$ to x' including the possible process of the change of the particle type $(\alpha \to \beta)$.
(2) Anomalous Green's functions [46]–[49] $\hat{\tilde{g}}_{\alpha\beta}\ (x, x')$ which characterize the macroscopically coherent states and describe the correlated appearance (or disappearance – $\hat{\tilde{g}}^{+}_{\alpha\beta}(x, x')$) of two particles α and β at points x and x', respectively. Being the manifestation of the presence of a macroscopically large number of coherently correlated pairs, these functions contain complete information about the EM. In essence, $\tilde{g}_{12}$ at equal times $t = t'$ is the effective EM wave function. Throughout the following discussion we will denote matrices (operators) by the 'hat' $(\hat{o})$ symbol and anomalous Green's functions and self energies by a 'tilde' $(\tilde{o})$.

Graphical representation of all these functions (in real space the same as in momentum representation) is shown in the Fig. 7. Here, $\mathbf{p}_0$ and ω_0 are the momentum and frequency (energy) of the pump wave quanta, respectively, $\tilde{\mathbf{p}} = 2\mathbf{p}_0 - \mathbf{p}, \tilde{\epsilon} = 2\omega_0 - \epsilon$. It is easy to understand the origin of the correlation of pairs of particles with momenta $\mathbf{p}$ and $\tilde{\mathbf{p}}$. Such a pair correlation arises as the result of the mutual scattering of two initial excitons with the momenta $\mathbf{p}_0$. Similar correlations are well-known in the phenomenological description of the nonlinear optical processes such as four-wave mixing, self-phase modulation, etc. In what follows, in order to consider the resonant phenomena in the vicinity of the exciton and biexciton (two-photon) resonances, the "rotating coordinate frame" method will be used. In this case, the frequency ω_0 of the pump wave becomes the origin of the frequency (energy) axis, so that $\omega \to \omega - \omega^0, \epsilon_{\mathbf{p}}^0 = E_g - 1 + p^2/M - \omega^0$ and $\tilde{\omega} = -\omega$. As the photon wave vector $\mathbf{p}_0$ is very small in comparison to typical momenta of scattered excitons, one can neglect it in the process of biexciton creation and put $\tilde{\mathbf{p}} \simeq -\mathbf{p}$. For example, $\epsilon_{\tilde{\mathbf{p}}}^{(0)} \simeq \epsilon_{-\mathbf{p}}^{(0)} = \epsilon_{\mathbf{p}}^{(0)}$. Only for an analysis of the optical manifestations of the renormalization phenomena in the exciton–EM system is it necessary to treat the difference between $\tilde{\mathbf{p}}$ and $-\mathbf{p}$ explicitly. But in what follows we will assume the special arrangement permitting realization of BEC of excitonic molecules with exactly zero momentum: pumping by two counterpropagating, circularly polarized waves [69, 70], which are in our model independent sources of excitons of different species α. Then obviously $\tilde{\mathbf{p}} = -\mathbf{p}$.

As explained above, according to the nonequilibrium Green's function formalism, each matrix element of $\hat{g}_{\alpha\beta}$ or $\hat{\tilde{g}}_{\alpha\beta}$ matrices is itself a 2×2 matrix,

$$\hat{g}_{\alpha\beta} = \begin{pmatrix} 0 & g_{\alpha\beta}^{(a)} \\ g_{\alpha\beta}^{(r)} & f_{\alpha\beta}^{ex} \end{pmatrix}. \tag{38}$$

Here again, $g_{\alpha\beta}^{(r,a)}$ denote retarded and advanced Green's functions, and the diagonal element $f_{\alpha\beta}^{ex}$ is closely related to the distribution function. But unlike the treatment of excitons in the preceding section, here we are dealing with bosonic Green's functions. Therefore for $t = t'$,

$$f_{\alpha\beta}^{ex}(\mathbf{p}; t, t) = -i[1 + 2n_{\mathbf{p}}(t)]. \tag{39}$$

The $+$ sign in the above formula corresponds to Bose particles, and $n_{\mathbf{p}}$ are the occupation numbers of the excitons.

The Green's functions have to be found from the Dyson equations, which are in this case similar to the corresponding equations for the Bose liquid [46, 51]:

$$\left(i\frac{d}{dt} - \epsilon_{\mathbf{p}}^{(0)}\right) \hat{g}_{\alpha\beta}(\mathbf{p}; t, t') - \int \hat{\sigma}_{\alpha\gamma}(\mathbf{p}; t, t_1)\hat{g}_{\gamma\beta}(\mathbf{p}; t_1, t')dt_1$$
$$\int \hat{\tilde{\sigma}}_{\alpha\gamma}^{+}(\mathbf{p}; t, t_1)\hat{\tilde{g}}_{\gamma\beta}(\mathbf{p}; t_1, t')dt_1 = \delta_{\alpha\beta}\delta(t - t'), \tag{40}$$

$$\left(-i\frac{d}{dt}-\epsilon^{(0)}_{\tilde{\mathbf{p}}}\right)\hat{\tilde{g}}_{\alpha\beta}(\tilde{\mathbf{p}};t,t')-\int\hat{\sigma}_{\alpha\gamma}(\tilde{\mathbf{p}};t,t_1)\hat{\tilde{g}}_{\gamma\beta}(\tilde{\mathbf{p}};t_1,t')dt_1$$

$$\int\hat{\tilde{\sigma}}_{\alpha\gamma}(\tilde{\mathbf{p}};t,t_1)\hat{g}_{\gamma\beta}(\tilde{\mathbf{p}};t_1,t')dt_1=0\,. \tag{41}$$

Here, the normal and anomalous self-energies $\hat{\sigma}_{\alpha\beta}$ and $\hat{\tilde{\sigma}}_{\alpha\beta}$ include all interaction processes in the system and are represented as usual by an infinite series of Feynman graphs. The simplest class, which corresponds to the self-consistent field approximation (SCF), is shown in Fig. 8. The dashed lines in these graphs denote the interaction matrix $\hat{U}$. Nonequilibrium self-energies are also 2×2 matrices of the type

$$\sigma_{\alpha\beta}=\begin{pmatrix}\Omega^{ex}_{\alpha\beta} & \sigma^{(r)}_{\alpha\beta}\\ \sigma^{(a)}_{\alpha\beta} & 0\end{pmatrix}. \tag{42}$$

Here, $\Omega^{ex}_{\alpha\beta}$ is the noise correlator in the system, closely related to the probability of incoherent creation and annihilation of excitons. The macroscopic wave function $\psi_\alpha(\mathbf{p}_0,t)$ of coherent excitons satisfies the following field equation:

$$\left(i\frac{d}{dt}-\epsilon^{(0)}_{\mathbf{p}_0}\right)\psi_\alpha(\mathbf{p}_0,t)-\int\sigma^{(r)}_{\alpha\beta}(\mathbf{p}_0;t,t')\psi_\beta(\mathbf{p}_0;t')dt'$$

$$-\int\tilde{\sigma}^{+}_{\alpha\beta}(\mathbf{p}_0;t,t')\psi^{*}_{\beta}(\mathbf{p}_0;t')dt'=f_\alpha(t), \tag{43}$$

or, in a general spatially inhomogeneous case,

$$\left(i\hbar\frac{\partial}{\partial t}+\frac{\hbar^2}{2m}\nabla^2-\epsilon_0\right)\psi_\alpha(\mathbf{r},t)-\int\sigma^{(r)}_{\alpha\beta}(\mathbf{r},\mathbf{r}';t,t')\psi_\beta(\mathbf{r}';t')dr'dt'$$

$$-\int\tilde{\sigma}^{+}_{\alpha\beta}(\mathbf{r},\mathbf{r}';t,t')\psi^{*}_{\beta}(\mathbf{r}';t')dr'dt'=f_\alpha(\mathbf{r},t), \tag{44}$$

where $\epsilon_0=\epsilon^{(0)}_{\mathbf{p}=0}$ is the detuning of the pump frequency from the exciton resonance, and the slowly varying electromagnetic field amplitude $f^{ex}_\alpha(\mathbf{r},t)$ is the source of the dipole-active coherent excitons. This function is determined by the dipole matrix element $\mathbf{m}_\alpha$ of the α-exciton–photon interaction:

$$f^{ex}_\alpha(t)=m_{\alpha i}\mathscr{E}_i(t)=\frac{ie\hbar}{m_0E_g}(p_{cv})_{\alpha i}\phi(0)\mathscr{E}_i(t). \tag{45}$$

Here, $(p_{cv})_{\alpha i}$ is the matrix element of the interband transition, m_0 is the electron mass and $\phi(0)$ is the value of the exciton ground state wave function for the coincident electron and hole coordinates.

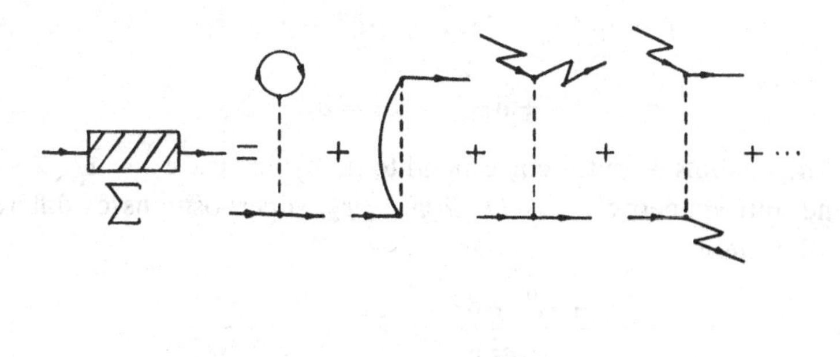

Fig. 8. Diagonal (Σ) and anormalous ($\tilde{\Sigma}$) self-energy graphs in the mean-field approximation.

The equations (40)–(41) are the matrix equations in the sense of the definitions (38) and (42). In fact, the equations for $g^{(r)}$ and $\tilde{g}^{(r)}$ look exactly like (40) and (41) with the substitution of the proper self-energy functions. The same statement holds for the advanced Green's functions. As to the function $f^{ex}_{\alpha\beta}(\mathbf{p}; t, t')$, it can be expressed in terms of the self-energies $\Omega^{ex}_{\alpha\beta}$, and the retarded and advanced functions [41]:

$$f^{ex}_{\alpha\beta}(\mathbf{p}; t, t') = \int g^{(r)}_{\alpha\gamma_1}(t, t_1)\Omega^{ex}_{\gamma_1\gamma_2}(t_1, t_2) g^{(a)}_{\gamma_2,\beta}(t_2, t') dt_1 dt_2 \,. \tag{46}$$

In what follows, the assumption that the self-energies $\Omega^{ex}_{\alpha\beta}$ do not considerably deviate from their quasiequilibrium values will again be used. Also, all the retardation effects in the self-energies will be neglected. Some results of such a treatment are presented in Section 4.

If the duration of the pumping pulse exceeds the relaxation time of excitons, the quasi-stationary state arises. In this case, all self-energies become time-independent functions and the following formulas can be used:

$$\hat{g} = \frac{1}{2}(g^+ + g^-) \cdot \hat{I} + \frac{1}{2}(g^+ - g^-) \cdot \hat{\tau}_1 \,, \tag{47}$$

$$\hat{\tilde{g}} = \frac{1}{2}(\tilde{g}^+ + \tilde{g}^-) \cdot \hat{I} + \frac{1}{2}(\tilde{g}^+ - \tilde{g}^-) \cdot \hat{\tau}_1, \tag{48}$$

$$\sigma^{\pm} = \sigma_{11} \pm \sigma_{12}, \qquad \tilde{\sigma}^{\pm} = \tilde{\sigma}_{11} \pm \tilde{\sigma}_{12}. \tag{49}$$

Here, the symbols + and − correspond to the symmetric $\psi_+ = 1/\sqrt{2}(\psi_1 + \psi_2)$ and antisymmetric $\psi_- = 1/\sqrt{2}(\psi_1 - \psi_2)$ superpositions of different types of excitons,

$$g^{\pm}(p) = \frac{\epsilon + \epsilon_{\mathbf{p}}^{(0)} + \sigma^{\pm}}{\epsilon^2 - (\epsilon_{\mathbf{p}}^{\pm})^2}; \quad \tilde{g}^{\pm} = -\frac{\tilde{\sigma}^{\pm}}{\epsilon^2 - (\epsilon_{\mathbf{p}}^{\pm})^2}; \tag{50}$$

$$\epsilon_{\mathbf{p}}^{\pm} = \sqrt{(\epsilon_{\mathbf{p}}^{(0)} + \sigma^{\pm})^2 - |\,\tilde{\sigma}^{\pm}\,|^2}\,; \tag{51}$$

$$\psi_+ = \frac{f}{\epsilon_0 + \sigma^+ + \tilde{\sigma}^+}. \tag{52}$$

The values $\epsilon_{\mathbf{p}}^{\pm}$ represent the energy difference between the excitonic levels and the carrier frequency of the pump. In a sense, the last formula (52) is an odd result. Instead of the usual resonance relation of the exciton polarization ψ to the field f, the denominator in Eq. (52) is quite different from $\epsilon_{\mathbf{p}}^{\pm}$. This result is one of the manifestations of the presence of pair coherence in the system under consideration.

The polarizability $\chi(\mathbf{k}, \omega)$ which describes the propagation of a weak probe signal $(\mathbf{k}, \omega)$ in a semiconductor in the presence of a pump is given by the following expression:

$$\chi(\mathbf{k}, \omega) = \left|\mathbf{m}_+ \frac{c}{\omega}\right|^2 \left(g(k) + \frac{4\pi\,|\,\mathbf{m}_+\,|^2|\,\tilde{g}(k)\,|^2}{\tilde{\omega}^2 - c^2\tilde{k}^2 - 4\pi\,|\,\mathbf{m}_+\,|^2\,\tilde{g}(\tilde{k})} \right). \tag{53}$$

This formula directly manifests the interconnection of the polariton branches $\mathbf{k}$ and $\tilde{\mathbf{k}}$. The proper Green's function $D(k)$ of the probe electromagnetic field is given by

$$D(k) = \frac{(\omega - \epsilon_{\mathbf{k}}^+)A^+ + \nu_{\mathbf{p}}\,|\,\mathbf{m}_+\,|^2\,\epsilon_{\mathbf{k}}^+}{A^- A^+ - 2\nu_{\mathbf{k}}\,|\,\mathbf{m}_+\,|^2\,\omega_{\mathbf{k}}\epsilon_{\mathbf{k}}^+}, \tag{54}$$

with

$$A^- = (\omega - \omega_{\mathbf{k}})(\omega - \epsilon_k^+) - |\,\mathbf{m}_+\,|^2, \tag{55}$$

$$A^+ = (\omega + \omega_{\mathbf{k}})(\omega + \epsilon_k^+) - |\,\mathbf{m}_+\,|^2, \tag{56}$$

and

$$\nu_{\mathbf{p}} = \frac{\epsilon_{\mathbf{p}}^{(0)} + \sigma^+}{\epsilon_{\mathbf{p}}^+} - 1 = \sqrt{1 + 2n_{\mathbf{p}}} - 1. \tag{57}$$

The poles of $D(k)$ describe the four new polariton-like branches. Each of these branches reflects the joint propagation of the four particles – two excitons and two photons with wave vectors $\mathbf{k}$ and $\tilde{\mathbf{k}}$. Their dispersion laws are

$$(\omega_{\mathbf{p}}^{\pm})^2 = \frac{1}{2}\Bigg(2 \mid \mathbf{m} \mid^2 + (\epsilon_{\mathbf{p}}^{+})^2 + (\omega_{\mathbf{p}}^{(0)})^2$$

$$\pm\sqrt{\left[(\epsilon_{\mathbf{p}}^{+})^2 - (\omega_{\mathbf{p}}^{(0)})^2\right]^2 + 4 \mid \mathbf{m} \mid^2 \left[(\epsilon_{\mathbf{p}}^{+} + \omega_{\mathbf{p}}^{(0)})^2 + 2\nu_{\mathbf{p}}\epsilon_{\mathbf{p}}^{+}\omega_{\mathbf{p}}^{(0)}\right]}\Bigg), \quad (58)$$

Here $\omega_{\mathbf{p}}^{(0)} = c|\mathbf{p}|$ and $n_{\mathbf{p}} = -0.5[1 + if^{ex}(\mathbf{p}; t, t)]$ are the exciton occupation numbers. It is easy to see from (54), and especially (58), that the coupling of two polaritons $\mathbf{p}$ and $\tilde{\mathbf{p}}$ is determined by $\nu_{\mathbf{p}}$, which in its turn is governed essentially by the $\mid \tilde{\sigma}^{+} \mid^2$, that is, by the EM density.

4 Results

In the framework of the model introduced, the considerations in the previous section are quite general. In order to obtain a more detailed description accounting for explicit dependences of self-energies on pumping frequency and intensity, the mean-field approximation will be used below. In this case a matrix equation similar to (25) can be used for the density matrix $\hat{f}_{\alpha\beta}^{ex}(\mathbf{r}, \mathbf{r}'; t, t)$:

$$i\hbar\frac{\partial}{\partial t}\hat{f}^{ex} = \left(\hat{h}_0 + \hat{\sigma}^{(r)}\right)\hat{f}^{ex} - \hat{f}^{ex}\left(\hat{h}_0 + \hat{\sigma}^{(a)}\right) \quad (59)$$

and retarded self-energies can be calculated in terms of the density matrix

$$\sigma_{\alpha\beta}(\mathbf{r}, \mathbf{r}';\, t) = \frac{1}{2}U_{\alpha\beta}(\mathbf{r}, \mathbf{r}')\left[if_{\alpha\beta}^{ex}(\mathbf{r}, \mathbf{r}';\, t) + 2\psi_{\alpha}(\mathbf{r}, t)\psi_{\beta}(\mathbf{r}', t)\right.$$

$$\left. - \delta_{\alpha\beta} \cdot \delta(\mathbf{r} - \mathbf{r}')\right] \quad (60)$$

and

$$\hat{h} = \hat{I} \cdot \left(E_g - E_0 - \hbar\omega_0 - \frac{\hbar^2}{2M}\nabla^2\right). \quad (61)$$

It should be noted that, in its physical content, the mean-field approximation used here is essentially different from that used in Section 2. Having introduced the exciton as a basic "elementary particle", we neglect the possibility of treating its internal structure but include in the mean-field treatment the exciton–exciton iteraction and even biexciton formation. As explained above, the accuracy of this approach is of the order of n or

$a_{bx}^{-1} \ll 1$. Within the same accuracy, we can neglect the effective radius of the exciton–exciton iteraction potential, which is obviously of the order of 1. Then considering Fourier components of matrix elements of $\hat{U}$ as two different constants, we can adjust their values using the two most important experimental data – the dissociation energy of the excitonic molecule E_d and the low-energy mutual scattering length of two biexcitons. To be specific in what follows, the results for a two-dimensional problem will be presented, i.e. excitons in a quantum well (QW) structure and a pump wave propagating normal to the QW plane direction will be treated. Many current experiments are done in such a geometry. Its obvious advantages are that the exciting field amplitude is independent of coordinates and undistorted by propagation phenomena, the excitons are coherently created with zero momentum, etc. The well-known problem of the destruction of the long-range order in two-dimensional systems by fluctuations seems in our case to be unimportant because the coherence is induced by an external source and so no phase degeneracy exists, and also because the finite lifetime and excitation pulse duration essentially restrict the logarithmic divergence of long-wave fluctuations.

Being no longer interested in the exciton internal structure, we will now introduce new ("biexcitonic") scales of variables more convenient for the following discussion, by putting the biexciton dissociation energy E_d and its effective radius

$$a_{bx} = \frac{\hbar}{\sqrt{ME_d}} \tag{62}$$

equal to unity. As a result, time intervals in what follows will be measured in terms of $\hbar/E_d$ and two-dimensional exciton densities in units of a_{bx}^{-2}. Dimensionless coupling constants will be introduced as

$$\alpha = \frac{mU}{4\pi\hbar^2}; \quad \lambda = \frac{mV}{4\pi\hbar^2}; \quad \gamma = \frac{\alpha\lambda}{\alpha+\lambda}. \tag{63}$$

In order to avoid explicit treatment of relaxation processes only two limiting regimes will be considered here: (1) stationary and (2) transient, for time intervals short compared to any relaxation process.

4.1 Stationary Regime

This can be achieved if the excitation-pulse duration is large compared to either the exciton phase-relaxation time or the coherence interval for the pump field itself. Then formulas (47)–(51) within the above-described approximations give the possibility of expressing all the interesting variables in terms of two of them – the anomalous self-energies $\tilde{\sigma}^{\pm}$:

$$\sigma^{\pm} = 4\pi\left[\left(\alpha - \frac{1}{2}\lambda\right)\frac{n_{+} + n_{-}}{2} - \frac{1}{2}\lambda n^{\pm} + \left(\alpha - \frac{1 \pm 1}{2}\lambda\right)n_0\right], \tag{64}$$

where

$$n^{\pm} = \frac{1}{4\pi}\cdot\tilde{\sigma}^{\pm}\cdot \mathrm{Arsh}\left(\frac{\tilde{\sigma}^{\pm}}{\epsilon^{\pm}_{\mathrm{min}}}\right), \tag{65}$$

and

$$\epsilon^{\pm}_{\mathrm{min}} = \sqrt{(\epsilon_0 + \sigma^{\pm})^2 - |\,\tilde{\sigma}^{\pm}\,|^2}. \tag{66}$$

Here, $n_0 = |\,\psi_+\,|^2$ is the density of coherent excitons in the single-particle condensate (polarization) and $\epsilon_0 = \epsilon^{(0)}_{\mathbf{p}=0}$ is the detuning of the pumping wave from the unperturbed excitonic level. Two more equations define the anomalous self-energies themselves:

$$\tilde{\sigma}^{+}\cdot\ln\epsilon^{+}_{\mathrm{min}} - \tilde{\sigma}^{-}\cdot\ln\epsilon^{-}_{\mathrm{min}} = -2n_0\,; \tag{67}$$

$$\tilde{\sigma}^{+}\left(1 - \gamma\ln\epsilon^{+}_{\mathrm{min}}\right) + \tilde{\sigma}^{-}\left(1 - \gamma\ln\epsilon^{-}_{\mathrm{min}}\right) = 2\gamma n_0\,. \tag{68}$$

Together with formula (52), equations (64)–(68) are a complete set, describing in the mean-field approximation the stationary states in the model under consideration. They can be further simplified if $\lambda \ll \alpha$. Then it follows immediately from (68) that $\tilde{\sigma}^{-} \approx -\tilde{\sigma}^{+}$ and then from (64)–(66) it follows that $\sigma^{-} \approx \sigma^{+}$ and $\epsilon^{-}_{\mathrm{min}} \approx \epsilon^{+}_{\mathrm{min}}$. The equations now reduce to

$$\tilde{\sigma}\cdot\ln\epsilon_{\mathrm{min}} = -n_0; \tag{69}$$

$$\sigma = 4\pi\,(n + n_0) = \alpha\left[\tilde{\sigma}\,\mathrm{Arsh}\left(\frac{\tilde{\sigma}}{\epsilon_{\mathrm{min}}}\right) + n_0\right]. \tag{70}$$

Some results of the numerical solution of these equations are depicted in Figs. 9–12. The first three of them show "resonant curves", i.e. dependences of response – ψ_+ and $\tilde{\sigma}$ – on the detuning of pumping frequency from the exciton resonance at a fixed value of a pumping field. The zero of the frequency axis corresponds in these plots to the position of the exciton level, and -1 to the biexciton resonance.

In these diagrams, the anomalous self-energy $\tilde{\sigma}$ is represented by full lines and polarization ψ_+, designated as c_0, by broken lines. Strictly speaking, $\tilde{\sigma}$ cannot be considered as the "biexciton amplitude," in analogy with the exciton amplitude ψ_+. But at low densities this interpretation is exact, as can readily be seen from (65). It can be seen from these plots that in the frequency range $-1 < \omega_0 < 1$ and for fields ≥ 0.2 the response is dominated by excitonic molecules. No resonant structure

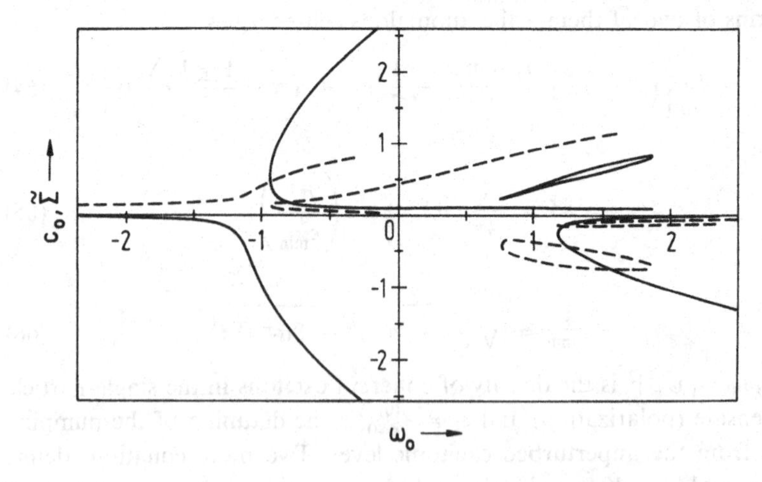

Fig. 9. Coherent exciton amplitude c_0 (broken line) and anomalous self-energy $\tilde{\Sigma}$ (full lines) dependences on the pumping field frequency ω_0 in a stationary regime for pumping field strength $f = 0.2$ and effective repulsion coupling constant $\alpha = 0.6$.

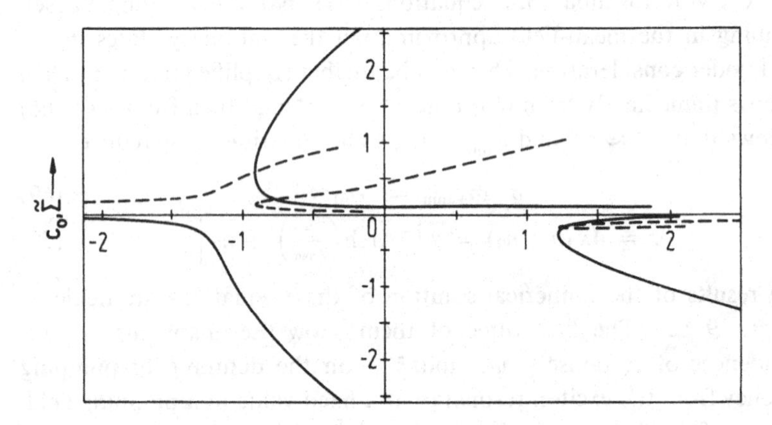

Fig. 10. The same as in Fig. 9, but for $f = 0.25$ and $\alpha = 0.6$.

exists around $\omega_0 = 0$. Only a loop in the interval $0.5 < \omega_0 < 2$ in Fig. 9, also disappearing at higher field, is a small remnant of a "pure" excitonic resonance observed at low fields. Instead of that, a typical resonant structure for excitonic molecules is observed, accompanied by a strongly nonlinear resonance of the excitonic polarization in the vicinity of $\omega_0 = -1$. For frequencies above this stationary resonance the response

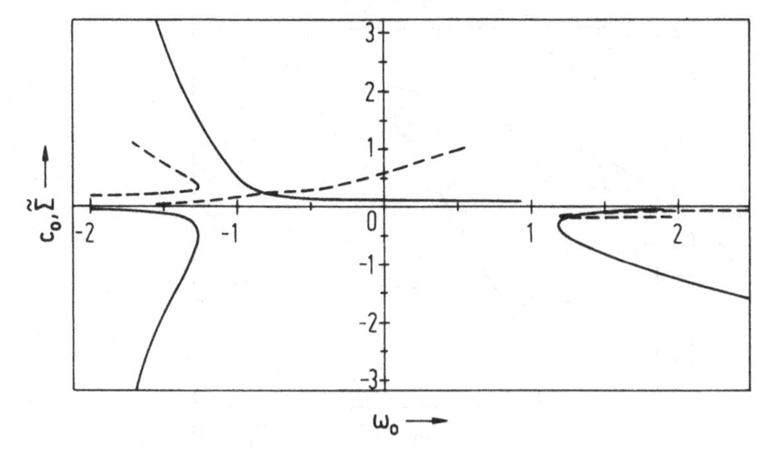

Fig. 11. The same as in Fig. 9, but for $f = 0.25$ and $\alpha = 0.3$.

becomes multistable. Two new states arise. The one with the smaller value of excitonic amplitude corresponds to the one with larger biexciton amplitude, and vice versa. Strictly speaking, solutions with $\omega_0 > 0, c_0 < 0$ are unstable because of stimulated scattering into some states in the lower polariton branch. Therefore, they will not be discussed in what follows. Among the three stable ($c_0 > 0$) solutions, one is "exciton dominated", i.e. beyond the immediate vicinity of the resonance, the majority of excitons are in the single-particle condensate-coherent polarization cloud. Two other solutions are "biexciton dominated" – the majority of excitons are bound into molecules at the expense of coherent polarization. But nonlinear processes become greatly enhanced, such as four-wave mixing (FWM) which increases with $\tilde{\sigma}$, as can readily be seen from (54)–(57) and the following formula, relating the exciton distribution function to the anomalous self-energy

$$n_{\mathbf{p}} = \frac{1}{2} \frac{|\tilde{\sigma}|^2}{|\epsilon_{\mathbf{p}}^{(0)} + \sigma|^2 - |\tilde{\sigma}|^2}. \tag{71}$$

As $\tilde{\sigma}$ increases, this distribution transforms to a narrow peak with the maximum value increasing as $\tilde{\sigma}^2$ and the halfwidth decreasing inversely proportional to $|\tilde{\sigma}|^{1/2}$. This means an increase of the effective biexciton radius proportional to $|\tilde{\sigma}|^{1/2}$.

As the pumping field strength increases, the resonance broadens, different branches shifting in opposite directions from the resonant frequency. So for any frequency $\omega_0 > -1$, at some critical field value dependent

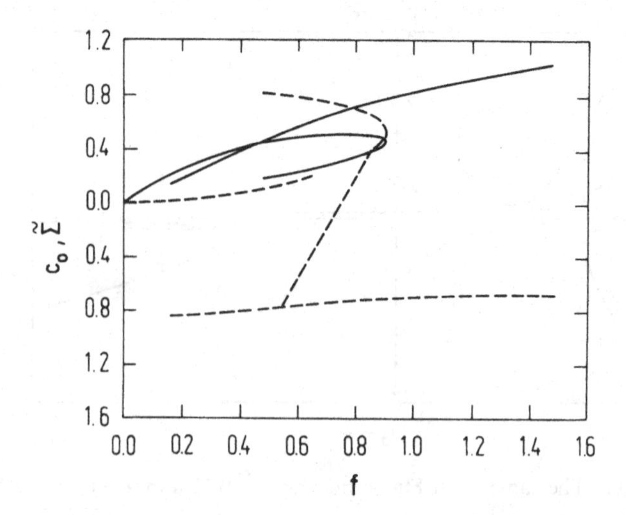

Fig. 12. Field dependence of the coherent exciton amplitude c_0 (full lines; note the difference with Figs. 9–11!) and anomalous self-energy $\tilde{\Sigma}$ (broken lines) for pumping frequency $\omega_0 = -0.75$.

on frequency, two solutions (both with $\tilde{\sigma}$ positive) – the "exciton dominated" one and one "biexciton dominated"– converge and at still larger fields disappear. Only the other "biexciton dominated" state– that with $\tilde{\sigma}$ negative – persists. Such behavior is clearly seen in Fig. 12 where field dependences of response ψ_+ (full lines) and $\tilde{\sigma}$ (broken lines) for $\omega_0 = -0.75$ and $\alpha = 0.75$ are depicted. The common low-field "exciton dominated" solution (the line coming from zero in Fig. 12) corresponding to relatively small biexciton density ($\tilde{\sigma} \sim f^2$) disappears at the field value $f \sim 1$. The system must jump to the only remaining stationary state, which will manifest itself in an abrupt increase of polarizability and large increase of the FWM signal.

Both Figs. 9 and 10 correspond to a relatively large value of the effective-repulsion coupling constant $\alpha = 0.6$. At $\alpha < 0.5$ the characteristic appearance of the biexciton resonance is essentially different [51], as if the effective nonlinearity changes its sign. It is shown in Fig. 11 for $\alpha = 0.3$. Now the multistable regime arises for frequencies below the unperturbed biexciton resonance, starting at

$$\omega_{\min} = -\left[1 - \frac{\alpha^2 z^2}{1-\alpha}\right] \cosh z \qquad (72)$$

with a finite value of biexciton density $|\tilde{\sigma}_0| \approx \sinh z$. Here z is the root

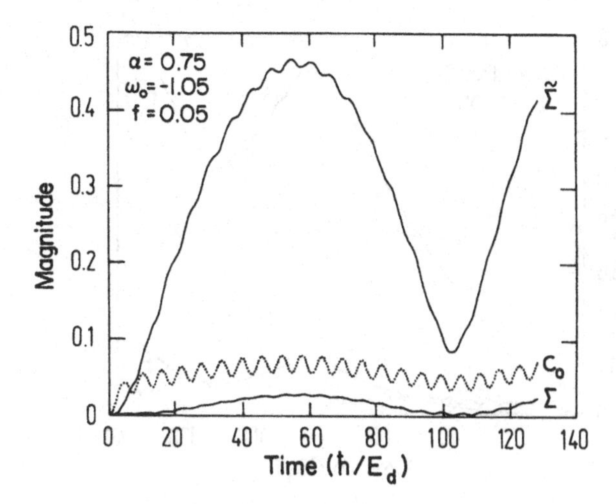

Fig. 13. Time dependences of the coherent exciton amplitude c_0 and normal (Σ) and anomalous ($\tilde{\Sigma}$) self-energies after smooth switching on of a pump field (at $t = 0$) for the pumping field strength $f = 0.05$ and pumping frequency $\omega_0 = -1.05$. Time is in units of $\hbar/E_d$.

of the equation

$$\tanh z = \frac{\alpha}{1-\alpha} z. \tag{73}$$

In such a case, two more branches of stationary states exist, not shown in Fig. 11, with very large biexciton densities $|\tilde{\sigma}| > |\tilde{\sigma}_0|$. It seems that if α is not too close to 0.5, these densities appear to be too large for excitonic molecules to really survive.

4.2 Transient Regime

This regime is shown in Figs. 13–16. No relaxation processes are taken into account. The pump field onset is supposed to have the form $f(t) = [1 - \exp(-t/\tau)]$ with $\tau = 4$. This results in a partial smearing of the Rabi oscillations for excitons. In these diagrams, one can see even more dramatically than in the static regime that, in the vicinity of the biexciton resonance, excitonic molecules completely dominate the system response. Especially spectacular is the time dependence of the normal self-energy, which, according to (70), contains the contributions of both bound and free excitons proportional to their densities. The striking similarity of the time evolution of both self-energies confirms

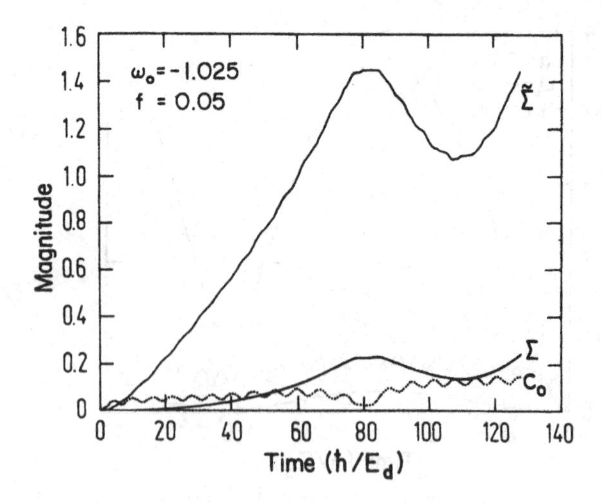

Fig. 14. The same as in Fig. 13, but $f = 0.05, \omega_0 = -1.025$.

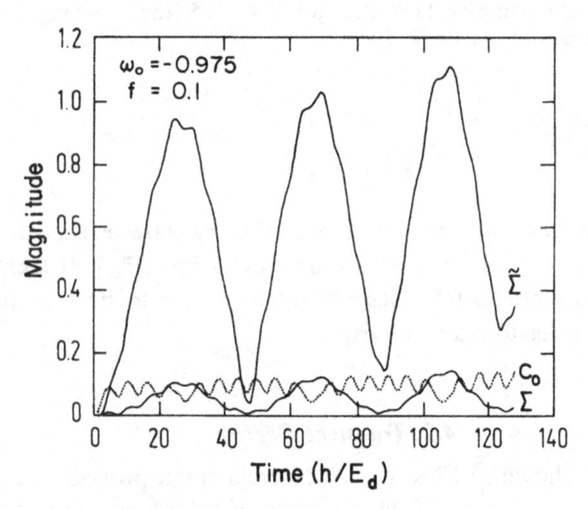

Fig. 15. The same as in Fig. 13, but $f = 0.1, \omega_0 = -1.05$.

the overwhelming contribution of biexcitons. It is evident for both frequencies – 0.05 detuning from the biexciton resonance in Figs. 13 and 15 and 0.025 in Figs. 14 and 16 – and for both (relatively small) pumping fields – 0.05 in Figs. 13–14 and 0.1 in Figs. 15–16.

The Rabi oscillations for excitonic molecules are very slow and extremely nonlinear. At higher exitation and smaller detuning (Fig. 16)

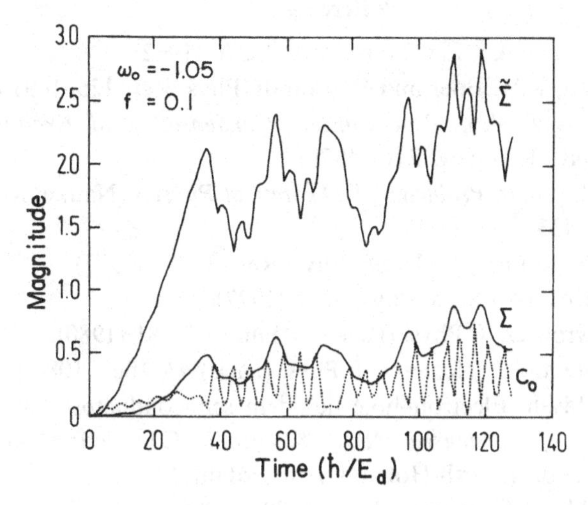

Fig. 16. The same as in Fig. 13, but $f = 0.1, \omega_0 = -1.025$.

they even seem to become chaotic, which does not seem unexpected for such a highly nonlinear system.

While small in reduced units, the field values supposed in Figs. 9–16 are really very large, corresponding to pumping intensities of the order of tens and hundreds of GW/cm^2. These phenomena can be observed only in experiments with ultrashort pulses. Therefore, the details of the relaxation mechanism become both important and interesting, as the standard Boltzmann equation approach becomes inadequate. Different theoretical approaches to this problem are presented in Refs. [71–73]. Being, in a sense, closely related to the problem of the coherence persistance, it remains one of most important topics in the nonlinear optics of semiconductors.

Acknowledgements. This work was partially supported by NATO Collaborative Research Grant CRG 930084. It is also a pleasure to acknowledge here many illuminative and stimulating discussions with H. Haug, A. Mysyrowicz, S. Moskalenko, N. Nagasawa, J. Wolfe, S. Tikhodeev and especially A. Ivanov, whose assistance was very important. I am also grateful to the Institute of Theoretical Physics of Frankfurt University for hospitality while completing this work.

References

[1] S.A. Moskalenko, Fiz. Tverd. Tela **4**, 276 (1962).
[2] I.M. Blatt, K.W. Boer and W. Brandt, Phys. Rev. **126**, 1691 (1962).
[3] S.A. Moskalenko, *Bose–Einstein Condensation of Excitons and Biexcitons*, (Kishinev, RIO, 1970).
[4] L.V. Keldysh, in *Problems of Theoretical Physics* (Nauka, Moskow, 1972), p.433.
[5] E. Hanamura and H. Haug, Phys. Rep. **3**, 209 (1977).
[6] T.M. Rice, Solid State Phys. **32**, 1 (1977).
[7] A. Mysyrowicz, J. Phys. (Paris) **41** Suppl. 7, 281 (1980).
[8] C. Comte and P. Nozieres, J. Phys. (Paris) **43**, 1069, 1083 (1982).
[9] L.V. Keldysh, Electron–hole liquid in semiconductors, in *Modern Problems of Condensed Matter Science* **6**, C.D. Jeffries and L.V. Keldysh, eds. (North-Holland, Amsterdam, 1987).
[10] L.V. Keldysh, Contemp. Phys. **27**, 395 (1986).
[11] G.H. Wannier, Phys. Rev. **52**, 191 (1937).
[12] N.F. Mott, Trans. Farad. Soc. **34**, 500 (1938).
[13] S.A. Moskalenko, Opt. Spectrosk. **5**, 147 (1958).
[14] M.A. Lampert. Phys. Rev. Lett. **1**, 450 (1958).
[15] L.V. Keldysh, *Proc. 9th Int. Conf. on Physics of Semicond.* (Moscow, 1968), p.1303.
[16] O. Akimoto and B. Hanamura, J. Phys. Soc. Jap. **33**, 1537 (1972).
[17] W.F. Brinkman, T.M. Rice, and B. Bell, Phys. Rev. B **8**, 1570 (1973).
[18] W.F. Brinkman and T.M. Rice, Phys. Rev. B **7**, 1508 (1973).
[19] L.V. Keldysh and Yu.V. Kopaev, Sov. Phys. Solid State **6**, 2219 (1965).
[20] J. des Cloizeaux, J. Phys. Chem. Solids **26**, 259 (1965).
[21] B.I. Halperin and T.M. Rice, Solid State Phys. **21**, 115 (1968).
[22] E.A. Andryushin, L. V. Keldysh and A.P. Silin, Sov. Phys. JETP **46**, 616 (1977).
[23] R.R.Guseinov and L.V.Keldysh, Sov. Phys. JETP **36**, 1193, (1973).
[24] L.V. Keldysh and A.N. Kozlov, Sov. Phys. JETP **27**, 521 (1968).
[25] J.J. Hopfield, Phys. Rev. **112**, 1555 (1958)
[26] V.M. Agranovich and V.L. Ginzburg, *Crystal Optics with Spatial Dispersion and Excitons* (Springer, Berlin, 1984).
[27] H.Haug, Z. Phys. **200**, 57 (1967), Phys. Rev. **184**, 338 (1969).
[28] R. Graham and H. Haken, Z. Phys. **237**, 31 (1970).
[29] V. Degiorgio and M.O. Scully, Phys. Rev. A **2**, 1170 (1970).
[30] H. Haken, Rev. Mod. Phys. **47**, 67 (1975).

[31] H. Haug and S.W. Koch, *Quantum Theory of Optical and Electronic Properties of Semiconductors* (World Scientific, London, 1993).

[32] Y. Toyozawa, Progr. Theor. Phys., Supplem.**12**, 11 (1959)

[33] A. Stahl and I. Balslev, *Electrodynamics of the Semiconductor Band Edge*, Springer Tracts in Modern Physics **110** (Springer, Berlin, 1987).

[34] V.M. Galitskii *et al.*, Zh. Eksp. Teor. Phys. **57**, 207 (1969).

[35] V.F. Elesin and Yu.V. Kopaev, Zh. Eksp. Teor. Phys. **63**, 1447 (1972).

[36] S. Schmitt-Rink and D.S. Chemla, Phys. Rev. Lett. **57**, 2752 (1986).

[37] S. Schmitt-Rink, D.S. Chemla and H. Haug, Phys. Rev. B **37**, 941 (1988).

[38] S. Schmitt-Rink, D.S. Chemla and D.A.B. Miller, Adv. Phys. **38**, 89 (1989).

[39] H. Haug and S. Schmitt-Rink, Progress in Quant. Electron. **9**,3 (1984).

[40] R. Zimmerman, *Many Particle Theory of Highly Excited Semiconductors*, (Teubner, Leipzig, 1988).

[41] L.V. Keldysh, Sov. Phys. JETP **20**, 1018 (1965).

[42] E.M. Lifshitz and L.P. Pitaevski, *Statistical Physics*, Part 2 (Pergamon, Oxford, 1980), chapters 10–11.

[43] J. Rammer and H. Smith, Rev. Mod. Phys. **58**, 323 (1986).

[44] G.D. Mahan, Phys. Rev. **145**, 251 (1987).

[45] K. Henneberger, Physica **A150**, 419 (1988).

[46] H. Haug, in *Optical Nonlinearities and Instabilities in Semiconductors*, H. Haug, ed. (Academic, New York, 1988) p. 53.

[47] A.L. Ivanov and L.V. Keldysh, Sov. Phys. JETP **57**, 234 (1983).

[48] L.V. Keldysh and S.G. Tikhodeev, Sov. Phys. JETP **63**, 1086 (1986).

[49] N.A. Gippius, L.V. Keldysh and S.G. Tikhodeev, Sov. Phys. JETP **64**, 1344 (1986).

[50] A.L. Ivanov, L.V. Keldysh and V.V. Panashchenko, Sov. Phys. JETP **72**, 359 (1991).

[51] L.V. Keldysh, Solid State Comm. **84**, 37 (1992)

[52] A. Mysyrowicz, J.B. Grun, R. Levy, A. Bivas, and S. Nikitine, Phys. Lett. A **26**, 615 (1968).

[53] H. Suoma, T. Goto, T. Ohta, and M. Ueta, J. Phys. Soc. Japan, **29**, 697 (1970).

[54] N. Nagasawa, S. Koizumi, T. Mita, and M. Ueta, J. Lumin. **12/13**, 587 (1976).

[55] F. Henneberger, K. Henneberger, and J. Voight, Phys. Stat. Sol. (b) **83**, 439 (1977).

[56] V.D. Phach, A. Bivas, B. Hönerlage, and J.B. Grun, Phys. Status Solidi (b) **84**, 731 (1977).

[57] J.L. Oudar, A. Maruani, E. Batifol, and D.S. Chemla, J. Opt. Soc. Am. **68**, 1638 (1978).

[58] M. Ueta, T. Mito and T. Itoh, Solid State Comm. **32**, 43 (1979).

[59] N. Nagasawa, T. Mito, and M. Ueta, J. Phys. Soc. Japan **41**, 929 (1981).

[60] B. Hönerlage, R. Levy, J.B. Grun, C. Klingshirn, and K. Bohnert, Phys. Rep. **124**, 161 (1985).

[61] V.D. Kulakovski, V.G. Lysenko and V.B. Timofeev, Sov. Phys. Usp. **28**, 735 (1985).

[62] M. Ueta, H. Kanazaki, K. Kobajashi, Y. Tojozava, and E. Hanamura, *Excitonic Processes in Solids* (Springer, Berlin, 1986).

[63] F. Henneberger and J. Voigt, Phys. Status Solidi (b) **76**, 313 (1976).

[64] V. May, K. Henneberger, and F. Henneberger, Phys. Status Solidi (b) **94**, 611 (1979).

[65] P.I. Khadzhi, S.A. Moskalenko, and S.N. Belkin, JETP Lett. **29**, 200 (1979).

[66] H. Haug, R. März, and S. Schmitt-Rink, Phys. Rev. A **77**, 287 (1980).

[67] R. Levy, B. Hönerlage, and J. B. Grun, in *Optical Nonlinearities and Instabilities in Semiconductors*, H. Haug, ed. (Academic, London, 1986), p. 181.

[68] M. Combescot and R. Combescot, Phys. Rev. Lett. **61**, 117 (1988).

[69] M. Hasuo, N. Nagasawa, and A. Mysyrowicz, Phys. Stat. Sol. (b) **173**, 255 (1992).

[70] M. Hasuo, N. Nagasawa, T. Itoh and A. Mysyrowicz, Phys. Rev. Lett. **70**, 1303 (1993).

[71] R. Zimmerman, Phys. Stat. Sol.(b) **173**, 129 (1992).

[72] H. Haug, Phys. Stat. Sol.(b) **173**, 139 (1992).

[73] M. Hartmann and W. Schäfer, Phys. Stat. Sol.(b) **173**, 165 (1992).

13

Bose–Einstein Condensation of a Nearly Ideal Gas: Excitons in Cu_2O

J. P. Wolfe and J. L. Lin

Department of Physics
University of Illinois at Urbana-Champaign
Urbana, IL 61801, USA

D. W. Snoke†

Mechanics and Materials Technology Center
The Aerospace Corporation
Los Angeles, CA 90009-2957, USA

Abstract

Systematic studies of the photoluminescence of free excitons in Cu_2O reveal the temporal and spatial properties of these particles under a variety of conditions. Their relaxation and diffusion are measured by time-resolved spectroscopy and imaging. At low densities the thermalization and diffusivity are well characterized by deformation-potential theory– i.e., electron–phonon coupling. At high densities, the excitonic gas establishes a well-defined temperature as a result of frequent interparticle collisions and displays quantum statistics. The orthoexciton gas tends to saturate at the phase boundary for Bose–Einstein condensation, characterized by the ideal-gas relation $n = CT^{3/2}$. Under appropriate conditions, the paraexciton energy distribution exhibits an extra component which is interpreted as a Bose–Einstein condensate.

1 Excitons as a Nearly Ideal Gas

A gas within a solid. How can that be? That the elementary electronic excitation of a semiconductor – the exciton – can be viewed as a nearly free particle in a box takes quite a leap of the imagination. Yet the basic wave nature of particles leads naturally to this picture. An electron of mass m_0 in a region of volume V with constant potential has the

† Present address: Dept of Physics and Astronomy, University of Pittsburgh, Pittsburgh, PA 15260, USA.

well-known wave function,

$$\psi = \frac{1}{\sqrt{V}} e^{i(\mathbf{k}\cdot\mathbf{r}-\omega t)}, \tag{1}$$

corresponding to the kinetic energy,

$$E = \hbar^2 k^2 / 2m_0, \tag{2}$$

where $\mathbf{k}$ is the wave vector of the particle related to the classical momentum by $\hbar\mathbf{k} = m_0\mathbf{v}$, and ω is the wave frequency related to the kinetic energy by $\hbar\omega = E$. Similarly, an electron in a periodic potential (e.g., that of a crystalline lattice) is described by the famous Bloch wave function,

$$\psi = u_{\mathbf{k}}(\mathbf{r}) e^{i(\mathbf{k}\cdot\mathbf{r}-\omega t)}, \tag{3}$$

where $u_{\mathbf{k}}(\mathbf{r})$ is a periodic function whose period coincides with the lattice structure. In its simplest form, $u_{\mathbf{k}}(\mathbf{r})$ is a sum of single atomic wave functions residing on the nuclei in the crystal. Within some arbitrary constant, the energy associated with a state near $\mathbf{k} = 0$ (at the band extremum, where $u_{\mathbf{k}}(\mathbf{r})$ depends only weakly on $\mathbf{k}$) is given by

$$E = \hbar^2 k^2 / 2m_e, \tag{4}$$

where m_e is the effective mass of the electron reflecting the influence of the atomic lattice. The miraculous implication here is that the electron retains the same wave vector for all time; i.e., it moves *without scattering* through this dense medium. This unique feature of the periodic potential allows one to view the electron in a crystal nearly the same as one that is freely moving in an empty box.

The extension to composite particles is straightforward. An electron and a proton, by virtue of their Coulombic attraction, bind into a hydrogen atom, whose center-of-mass wave vector, $\mathbf{k}_{cm} = (m_0 + m_P)\mathbf{v}_{cm}/\hbar$, is conserved. The hydrogenic wave function consists of an atomic-orbital part multiplied by a plane-wave factor, $\exp[i(\mathbf{k}_{cm} \cdot \mathbf{r} - \omega t)]$. The energy spectrum for the bound pair is the Rydberg energy plus the kinetic energy associated with the center of mass,

$$E_{\text{tot}} = \frac{-e^2}{2a_0 n^2} + \frac{\hbar^2 k_{cm}^2}{2(m_e + m_p)}, \tag{5}$$

where $a_0 = \hbar^2/e^2 m_r$ is the hydrogenic Bohr radius with reduced mass $m_r = m_0 m_P/(m_0 + m_P)$, and $n = 1, 2, ...$ is the principal quantum number. The situation in the semiconductor is quite analogous: an electron in the

Table 1. *Physical parameters for Cu_2O*

Name	Symbol	Value	Reference
Dielectric constant	ϵ	7	[2]
Electron mass	m_e	$1.0m_0$	[3, 4]
Hole mass	m_h	$0.7m_0$	[4]
Orthoexciton Mass	m	$3.0m_0$	[5, 6]
Rydberg (n >1)	R_∞	97 meV	[1]
Binding energy (n=1)	E_x	150 meV	[7]
n=1 Bohr radius	$a_0 \equiv e^2/2E_x\epsilon$	7 Å	
Band gap	E_{gap}	2.17 eV (4K)	[1]
Ortho–Para exchange splitting	E_{ex}	12 meV	[8]

conduction band and a hole in the valence band bind together to form an exciton with a center-of-mass wave vector $\mathbf{k}_{cm} = \mathbf{k}_e + \mathbf{k}_h$ and total mass $m = m_e + m_h$. The factor e^2 in (5) and in a_0 is replaced by e^2/ϵ (where ϵ is the crystalline dielectric constant) and generally the mass of the hole is comparable to that of the electron. In fact, the analogy between excitons and positronium is even better than the analogy between excitons and hydrogen, since the positron can be seen as a "hole" in the negative-energy Dirac sea. The analogy between excitons and positronium is better in Cu_2O than in most semiconductors, since the lowest conduction band and the highest valence band are spherical and well separated from other energy bands, and the effective electron and hole mass are close to the free electron mass. This is illustrated by the fact that the excitonic Rydberg in Cu_2O is almost exactly (13.6 eV)$/2\epsilon^2$, the value for positronium when $e^2 \to e^2/\epsilon$. Values of relevant parameters for the crystal Cu_2O are summarized in Table 1. The Rydberg series of states for excitons in Cu_2O has been clearly observed by optical absorption [1]. In the present review, we deal exclusively with the $n = 1$ ground state because the thermal populations of the excited states are small, since the binding energy of the exciton in Cu_2O is so large. The excitonic binding energy E_x =150 meV is somewhat larger than the Rydberg, $R_\infty = 97$ meV, which describes the excited ($n = 2, 3, ...$) states, due to central-cell corrections for this relatively small Wannier exciton.

Similar to the hydrogen atom, positronium and many atoms, the exciton is a composite boson, having integer total spin. A collection of excitons may be regarded as a gas of bosons provided the average distance between particles is much larger than the Bohr radius, or

$$na_x^3 \ll 1, \tag{6}$$

in terms of the gas density, n. For Cu_2O, a_x equals 7 Å(deduced from the excitonic Rydberg E_x via the above equations), which requires that $n << 5 \cdot 10^{21}$ cm^{-3}, a condition which is well satisfied for the experiments described in this paper. The condition for a gas of hard spheres of radius a_x to be *weakly interacting* is [9]

$$nU(0) \ll \frac{\hbar^2(1/a_x^2)}{2m}, \tag{7}$$

or

$$na_x^3 \ll \frac{1}{4\pi}, \tag{8}$$

for the interaction vertex $U(0) = 4\pi a_x \hbar^2/m$. This condition is also fulfilled for the experiments discussed here.

2 Creation and Observation of Excitons

A pure semiconductor at zero temperature has no mobile charge carriers. The valence band states are completely filled with electrons and the conduction band is completely empty. Photons with energy, $h\nu$, larger than the semiconductor band gap, E_{gap}, create non-equilibrium electrons in the conduction band and holes in the valence band which can bind pairwise into excitons.† Once formed in a relatively pure crystal, the excitons behave as mobile particles with finite lifetime. Unlike the single particle in a box, however, the excitons find a variety of ways to scatter from one wave vector to another. In particular the periodicity of the lattice is easily broken by thermal vibrations, or phonons, so that excitons can lose energy and cool down. The diffusivity and thermal relaxation of excitons will be discussed in detail below. First we must consider how excitons are detected.

Excitons make themselves visible by decaying. The radiative recombination of the electron and the hole composing the ground-state exciton produces a photon with an energy,

† Under typical conditions the density of excitons, n, far exceeds the density of electrons and holes, $n_e = n_h$. Statistical mechanics predicts $n_e^2/n \approx C(T)\exp(E_x/k_BT)$, where $C(T)$ is given approximately by n_Q in (22) below, so that for $E_x = 150$ meV, $k_BT = k_B(50$ K$)$ $= 4$ meV and $n = 10^{14}$ cm^{-3}, we find $n_e/n \approx 10^{-7}$. See Ref. [10].

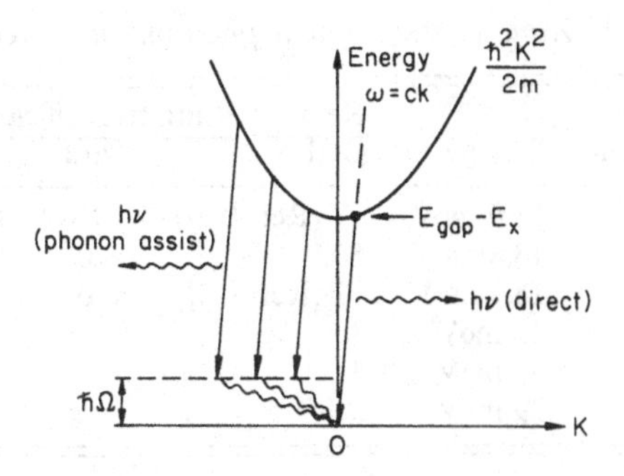

Fig. 1. Recombination processes of excitons in Cu_2O.

$$h\nu = E_{\text{gap}} - E_x + \frac{\hbar^2 k^2}{2m} \qquad \text{(direct transition)}\,, \tag{9}$$

$$h\nu = E_{\text{gap}} - E_x + \frac{\hbar^2 k^2}{2m} \pm \hbar\Omega_i \qquad \text{(phonon-assisted)}\,, \tag{10}$$

where E_{gap} is the semiconductor energy gap, $E_x = e^2/2a_x$ is the excitonic binding energy, † and $\hbar\Omega_i$ is the energy of an optical phonon which is emitted or absorbed simultaneously with the photon emission. Momentum conservation simultaneously requires for the two types of processes,

$$\mathbf{k} = \mathbf{k}_{\text{photon}} \qquad \text{(direct transition)} \tag{11}$$

$$\mathbf{k} = \mathbf{k}_{\text{photon}} + \mathbf{k}_{\text{phonon}} \qquad \text{(phonon-assisted)} \tag{12}$$

These two recombination processes are illustrated in Fig. 1. The direct process can only occur for an exciton with momentum matching that of the emitted photon, yielding a sharp luminescence line at nearly zero kinetic energy ($E \approx E_{\text{gap}}^2/2mc^2 \approx 10^{-6}$ eV). ‡ The phonon-assisted process, on the other hand, samples all of the occupied exciton states because the exciton momentum is easily absorbed by the optical phonon. *Since the energy of the optical phonon (and the transition probability) is*

† $E_x = 150$ meV is the binding energy of the lowest lying exciton, the paraexciton (see below). For the orthoexciton, which lies higher by an exchange energy of 12 meV, E_x in these expressions must be replaced by $E_x - 12$ meV. Applied stress modifies both E_{gap} and E_x and splits the orthoexciton levels.

‡ This one-photon decay process is allowed for excitons, unlike positronium, because the energy gap of the semiconductor is far less than the positronium "gap" of $2mc^2$. Two-photon decay is also allowed for excitons, like positronium, but the rate for this process is much less than the single-photon emission processes.

Table 2. *Zone-center optical phonons in* Cu_2O

Designation	Energy	Relative Radiative Efficiency	
		Ortho	Para
${}^3\Gamma_{25}^-(\Gamma_5^-)$	11.0 meV	17 (Ref. [13])	1 (Ref. [15])
${}^2\Gamma_{12}^-(\Gamma_3^-)$	13.8 meV	500	≈ 0
${}^3\Gamma_{15}^-(\Gamma_4^-)$	18.7 meV	17 (Ref. [14])	≈ 0
${}^1\Gamma_2^-(\Gamma_2^-)$	43 meV	3	
${}^3\Gamma_{25}^+(\Gamma_5^+)$	64 meV	3	
${}^3\Gamma_{15}^-(\Gamma_4^-)$	79 meV	8	

only weakly dependent on the phonon wave vector,† *the phonon-assisted process gives an accurate replica of the energy distribution of the excitons.* This situation is unlike that of liquid helium-4, in which the momentum distribution can only be deduced indirectly from neutron scattering data.

Because there are several optical phonons with appropriate symmetry to participate in this recombination process, each having a different energy $\hbar\Omega_i$, there are several "phonon replicas" of the excitonic kinetic energy distribution observed in the recombination spectrum. Table 2 lists the energies of zone-center optical phonons, from Ref. [12], denoted by their irreducible representation in the cubic group, and gives the relative radiative strengths of the corresponding phonon-assisted luminescence processes.

The symmetries of the electronic and vibrational states are crucial to the relaxation properties of the excitons – in particular, their lifetime and scattering rates. A useful review of these symmetry properties is given in Ref. [7]; also, Birman [16] gives a useful discussion of the effect of symmetry on the phonon-assisted recombination rates. At this point we only need to know the following facts: (1) Cu_2O is a forbidden direct-gap semiconductor; that is, the parity of the conduction-band minimum is the same as that of the valence-band maximum, implying that dipole radiation is forbidden. This property leads to relatively long excitonic lifetimes and correspondingly weak optical transitions. (2) Because the exciton is composed of a spin-1/2 electron and spin-1/2 hole, its ground state is split into an "S = 0 like" paraexciton and an "S = 1 like" orthoexciton by the electron–hole exchange interaction,

† For typical gas temperatures of 100 K or less, the exciton wave vectors are small ($k < 10^7$ cm^{-1}) compared to the Brillouin-zone boundary ($\sim 10^8$ cm^{-1}). The optical-phonon dispersion, as measured by neutron scattering [11], is negligible for these wave vectors.

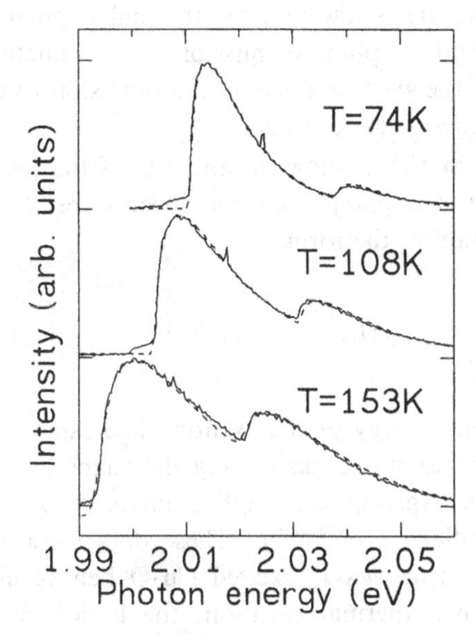

Fig. 2. Luminescence from Cu_20 at various temperatures. (From [18].)

also called the "virtual annihilation" interaction. Only the orthoexciton feels this exchange energy – just as in positronium, the triplet state has higher energy.† (3) The orthoexciton luminescence is quadrupole allowed, whereas direct recombination of the paraexciton is forbidden to all orders. Phonon-assisted recombination of paraexcitons is weakly allowed. d) Formation of other complexes, such as biexcitons or electron–hole liquid, is energetically unfavorable, due to the low band degeneracy and near-unity electron–hole mass ratio [17].

Photoluminescence spectra of orthoexcitons in Cu_2O are shown in Fig. 2. To obtain these, a high-quality natural crystal (from a copper mine) is cooled in a cryostat and weakly excited with a continuous (cw) Ar^+ laser beam (2.4 eV). The sharp spike is due to direct recombination of the orthoexciton. The broad spectral lines flanking the direct emission are phonon-assisted orthoexciton lines, corresponding to the emission or absorption of a Γ_{12}^- optical phonon with an energy of $\hbar\Omega = 14$

† There exists some confusion about whether electron-hole exchange gives a *positive* shift to the *ortho* state, or a *negative* shift to the *para* state. In general, the answer depends on the symmetry of the conduction and valence bands. For Cu_2O, the ortho state is shifted up; as noted in Ref. [6], the mass renormalization of the ortho ($M_{ortho} = 3m_0$ instead of $M_{ortho} = m_e + m_h = 1.7m_0$) implies a k-dependent positive exchange-energy shift of the ortho state, analogous to that of the positronium triplet state.

meV. As the temperature is lowered, the thermal population of optical phonons decreases and the phonon-emission line dominates. The overall temperature shift of the spectrum is due to a blue shift of the band gap, E_{gap}, as the crystal thermally contracts.

The dashed lines in Fig. 2 show an analysis of the low temperature orthoexciton spectra. The phonon-assisted spectra are fit to a Maxwell–Boltzmann distribution of the form,

$$N(E) \propto E^{1/2} e^{-E/k_B T}, \tag{13}$$

where E is the kinetic energy of the exciton. The same fit temperature T is assumed for the excitonic gas (giving the shape of the lines), and for the crystal lattice (giving the relative intensity of the two lines, $I(\text{emission})/I(\text{absorption}) = e^{\hbar\Omega/k_B T}$). These data show unequivocally that the orthoexcitons in weakly excited Cu_2O behave as a "classical gas" in contact with a thermal reservoir, the lattice vibrations. The lifetime of the orthoexcitons at low temperature is measured to be about 30 ns, so we conclude that the energy relaxation of these particles, originating as hot electrons and holes, is sufficiently fast to reach thermal equilibrium with the lattice within this time. As we shall see, the cooling of the gas is accomplished by emission of phonons.

What is the cause of the 30 ns lifetime of the orthoexcitons? Recent measurements indicate that the radiative lifetime of orthoexcitons is much longer – several hundred nanoseconds [18]. It is most likely, therefore, that the measured lifetime in this dilute-gas limit is due to the decay of orthoexcitons to paraexcitons. Snoke *et al.* [19] have suggested that this interconversion process can take place by emission of a single acoustic phonon. This mechanism appears to be consistent with measurements of ortho–para interconversion times by Weiner *et al.* [20] and with earlier measurements of luminescence intensities as a function of temperature.

In contrast, the paraexcitons display lifetimes of many microseconds in high quality naturally grown crystals [21], because their radiative recombination is highly forbidden. † For a given experimental setup, the

† The defect concentration in common synthetic crystals causes non-radiative lifetimes of only nanoseconds or less, even for paraexcitons, which leads to non-Maxwellian exciton distributions even at low density, since the excitons cannot fully thermalize by scattering with phonons within their lifetime [22].

luminescence intensities of the ortho and paraexcitons are given by†

$$I_o(E) = BN_o(E)/\tau_{ro} \qquad \text{(ortho)}\,, \tag{14}$$

$$I_p(E) = BN_p(E)/\tau_{rp} \qquad \text{(para)}\,, \tag{15}$$

where the constant B contains the experimental collection efficiency (which is the same for ortho and para), and $N(E)$ are the respective energy distributions. The most radiatively efficient phonon-replicas are the Γ_{12}^- line for ortho and the Γ_{25}^- line for para, and the ratio of the radiative rates for these two processes is measured to be $\tau_{rp}/\tau_{ro} \approx 500$. (See Table 2.) Consequently, the paraexcitons are difficult to observe in pulsed experiments within time windows less than about 10 ns, and in order to collect enough photon counts one must repeat the experiment at least 10^5 times per second (a 100 kHz pulse-repetition rate).

3 Diffusion of Excitons

The long lifetimes of the paraexcitons have permitted Trauernicht *et al.* [23] to observe their spatial diffusion by time-resolved luminescence imaging. The two images of paraexciton luminescence in Fig. 3 show the expansion of the paraexciton gas into the crystal. The diffusion constant, D, can be determined by a straight-line fit to the square of the cloud radius as a function of time [23]. A temperature dependence of the measured diffusion constant is shown in Fig. 4. With each diffusion measurement, the temperature of the paraexciton gas was obtained by fitting its spectrum to the Maxwell–Boltzmann distribution (13).

The measured diffusion constants are huge, ranging up to 1000 cm^2/s for the lowest experimental temperature of 1.2 K. The strong temperature dependence indicates that this process is mediated by thermal vibrations of the lattice, i.e., exciton–phonon scattering. Taking into account the Planckian distribution of phonons at temperature T, and making the high-temperature approximation $k_B T \gg mv_L^2$, where v_L is the lattice

† Broadening of the spectrum associated with the lifetime of the excitons is negligible. The Lorentzian broadening from the uncertainty principle implies a spectral full width at half-maximum of 1.2 meV for a lifetime of 1 ps. Therefore, the broadening of the orthoexciton spectrum due to a 30 ns lifetime is an unmeasureable 0.04 μeV. Also, broadening due to exciton–phonon interactions, with a typical scattering time of 30 ps, leads to a linewidth of only 0.04 meV at low temperature. On the other hand, interparticle collisions can cause a measurable broadening at high gas densities [13].

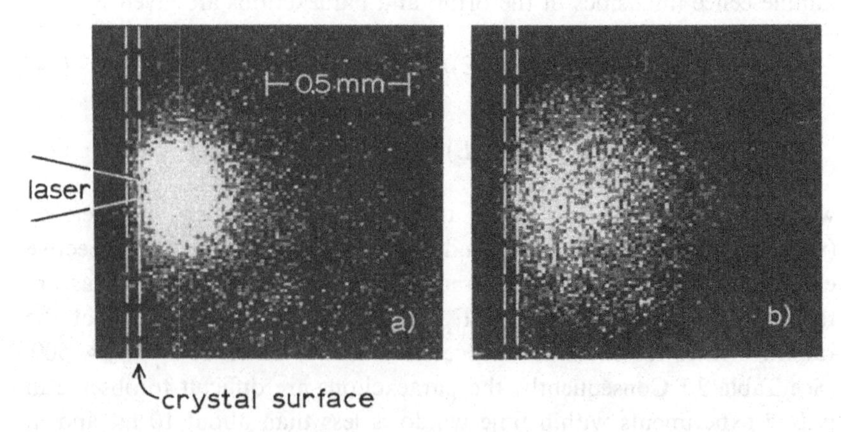

Fig. 3. Time-resolved images of the paraexciton luminescence of Cu_2O at T=2K, obtained by x–y scanning of the crystal image across an entrance aperture of a spectrometer. (a) t=0.2 μs after a 100-ns Ar^+ laser pulse is absorbed at the crystal surface. (b) t=0.6 μs.(From [23].)

phonon velocity, deformation-potential theory gives the result

$$\frac{1}{\tau(\mathbf{k})} = \frac{\Xi^2 m k_B T}{\pi \hbar^3 \rho v_l^2} k, \tag{16}$$

where ρ is the crystal mass density and Ξ is the deformation potential (basically the shift in E_{gap} with strain). If the excitons are distributed according to a Maxwell–Boltzmann distribution at the lattice temperature, then this leads to

$$D = \frac{\tau}{m} k_B T = \frac{k_B}{mR} T^{-1/2}, \tag{17}$$

where R is a constant independent of T (see Ref. [23]). Figure 4 shows that the measured diffusion constant for paraexcitons deviates markedly from this predicted $T^{-1/2}$ dependence for temperatures below about 30 K. Trauernicht *et al.* [24] have shown that this unusually sharp rise in diffusivity at low temperatures can be understood in terms of the breakdown of the high-temperature approximation given above. When the exciton velocity is smaller than the lattice phonon velocity, energy and momentum conservation forbid the exciton from emitting a phonon. Therefore as the gas temperature is lowered, the diffusivity eventually increases exponentially.†

† The extremely high diffusivities of the paraexcitons at low temperatures is a strong testament to the perfection of these natural-growth crystals. At T = 1.2 K, the corresponding mean free path between scattering events is $l = 3D/v_{av} \approx 70\mu$m! The mean scattering

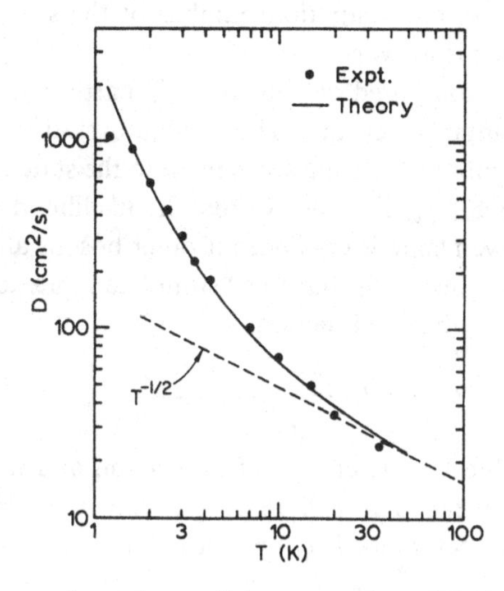

Fig. 4. Temperature dependence of the paraexciton diffusion constant in Cu_2O. (From [23].)

Notice that in (16), the scattering rate varies inversely as the square of the phonon velocity. For spherical electron and hole bands (as in Cu_2O), only longitudinally polarized phonons can couple to the carriers. Trauernicht *et al.* [24] show that if the crystal is stressed, the reduced symmetry allows the excitons to couple also to transverse phonons, which have a velocity about four times smaller than that of the longitudinal phonons. Hence, an externally applied stress acts to reduce the diffusivity of the excitons and may be an effective means of confining the excitons to a smaller volume.

4 Quantum Statistics for the Nearly Ideal Bose Gas

We are all familiar with the profound effects of quantum statistics on electrons in metals. As indistinguishable spin-1/2 particles, electrons are fermions which are unable to share the occupancy of a single state. In terms of scattering theory, this constraint implies that the probability of an electron being scattered into a state with wavevector **k** is proportional

time is $\tau = l/v_{av} \approx 20$ ns, which implies a mobility of $\tau/m = 10^8$ cm^2/eV-s, perhaps the highest measured for any electronic species in a non-superconducting bulk crystal.

to $1 - f_{\mathbf{k}}$, where $f_{\mathbf{k}}$ is the occupation number of the state. This is the Pauli exclusion principle at work.

The exciton, being composed of two spin-1/2 particles, is an integral-spin particle, or composite boson. The quantum statistical rule for a boson is that its probability of being scattered into the state of wavevector $\mathbf{k}$ is proportional to $1 + f_{\mathbf{k}}$. In other words, the likelihood that a boson is scattered into a given state is enhanced if other bosons already occupy that state. It can be shown [25] that this "stimulated" scattering leads to the Bose–Einstein distribution function,

$$f_{\mathbf{k}} = f(E) = \frac{1}{e^{(E-\mu)/k_B T} - 1}, \tag{18}$$

with $E = \hbar^2 k^2/2m$ the kinetic energy of an exciton and μ the chemical potential, which is measured with respect to the total energy of the exciton at $E = 0$ and determined by the condition

$$\sum_{\mathbf{k}} f_{\mathbf{k}} = N, \tag{19}$$

where N is the total number of excitons in the system.

For a macroscopic volume, the number of plane-wave states per unit energy for the ideal (non-interacting) gas has the form $D(E) = C' E^{1/2}$. Therefore, the number of excitons per unit energy is given by the product of this density of states and the occupation number $f(E)$:

$$N(E) = C' \frac{E^{1/2}}{e^{(E-\mu)/k_B T} - 1}. \tag{20}$$

If N excitons occupy a volume V, then $C' = (gV/4\pi^2)(2m/\hbar^2)^{3/2}$ from the usual counting of $\mathbf{k}$ states [26], where g is the spin multiplicity. In an unstrained crystal, $g = 3$ for orthoexcitons and $g = 1$ for paraexcitons. The constraint (19) may be restated as $N = \int N(E) dE$, which in terms of the gas density, $n = N/V$, becomes

$$n = \frac{g}{4\pi^2} \left(\frac{2m}{\hbar^2} \right)^{3/2} \int_0^\infty \frac{E^{1/2}}{e^{(E-\mu)/k_B T} - 1} dE. \tag{21}$$

For a given density and temperature, (21) fixes the chemical potential, μ, of the gas. By defining $\epsilon = E/k_B T$ and the "dimensionless chemical potential" $\alpha = -\mu/k_B T$, Eq. (21) can be written in terms of a dimensionless integral as,

$$n = n_Q \frac{2}{\pi^{1/2}} \int_0^\infty \frac{\epsilon^{1/2}}{e^{\epsilon+\alpha} - 1} d\epsilon. \tag{22}$$

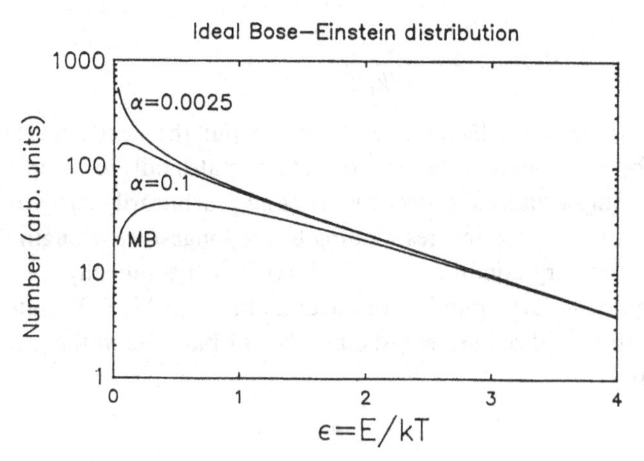

Fig. 5. The Bose–Einstein distribution $N(E) = f(E)D(E)$ for several values of α for an ideal three-dimensional gas at constant potential.

where $n_Q = g(mk_BT/2\pi\hbar^2)^{3/2}$ is the "quantum density", at which the thermal DeBroglie wavelength $\lambda = 2\pi/k_{\rm av}$ of the particles is comparable to the interparticle spacing. In the classical limit, α is much greater than unity and (22) yields $n = n_Q e^{-\alpha}$, or $\mu = -k_BT\ln(n_Q/n)$.

The shape of the kinetic energy distribution depends on the values of μ (or α) and T. In Fig. 5 we plot (20) at a fixed gas temperature as the density of the gas is hypothetically increased. At low densities, the chemical potential is large and negative, which reduces the distribution in (20) to the density-independent shape given by (13). As we increase n, the constraint of Eq. (22) forces α closer to zero. When α approaches 1 (i.e., $n \rightarrow n_Q$), the spectrum becomes more sharply peaked at low energies. In this quantum regime, the $(1 + f_{\mathbf{k}})$ final-state factors become important and scattering into low-energy states is enhanced. *Compared to a classical distribution at temperature T, Bose statistics produce a narrowing of the energy distribution, just the opposite of how Fermi statistics broaden the distribution.*†

As more particles are added to the gas at constant temperature, a surprising thing happens. A density is reached where $\mu = 0$ and the integral in (21) is still finite, as indicated in Fig. 5. This fact can be understood by expanding the integrand about the singularity at $E = 0$

† Fermi broadening of degenerate electron–hole plasmas has been clearly observed, for example, in electron–hole droplets in Ge and Si [27].

to obtain

$$N(E) \sim \frac{E^{1/2}}{1 + E/k_B T - 1} \sim E^{-1/2}, \tag{23}$$

which is integrable. The dilemma of where to put the next particles is surmounted by realizing that the $E^{1/2}$ density of states fails to completely describe a system of discrete states. By bringing μ arbitrarily close to the ground state of the system (corresponding to the longest wavelength that matches the boundary conditions of the box) it is theoretically possible to put an arbitrarily large number of particles into this "$k = 0$" state. In particular, in this idealized system, the number of particles in the ground state is simply,

$$f(0) = \frac{g}{e^{\alpha} - 1} \approx -g/\alpha, \tag{24}$$

which can become an arbitrarily large number as α approaches zero. For a realistic volume, the first excited k-state will also have a large occupation number, but considerably fewer particles than the ground state, due to the sharply peaked distribution function. This effect in which thermodynamic equilibrium of the ideal Bose gas forces a macroscopic number of particles into the lowest state is the idealized concept of Bose–Einstein condensation.

Of course, the quantum states of real gases (atomic or excitonic) can be significantly broadened by interparticle scattering, so the simplistic view of well-defined plane-wave states at low energy is not perfectly valid. Nevertheless, the non-interaction picture is a useful starting point for considering the *weakly* interacting gas and, as we shall see, appears to do a remarkably good job of describing the excitonic energy distributions even at $\alpha \approx 0$. For an excellent review of the thermodynamic properties of a weakly interacting Bose gas, see Ref. [28].

The phase boundary for BEC of the ideal Bose gas is determined by setting $\alpha = 0$ in (21), yielding,

$$n_c = 2.612 n_Q = C T^{3/2}, \tag{25}$$

with $C = 2.612 g (m k_B / 2\pi\hbar^2)^{3/2}$. This relation is plotted for orthoexcitons ($g = 3$) and paraexcitons ($g = 1$) in Fig. 6, using the mass of the orthoexciton in Table 1. We now see why excitons are a prime candidate for observing this quantum effect: their effective mass is approximately that of an electron, rather than that of an atom. At a temperature of 4 K, a density of only about 10^{17} cm^{-3} is required to reach the orthoexciton phase boundary. Alternatively, an experimentally-attainable density of

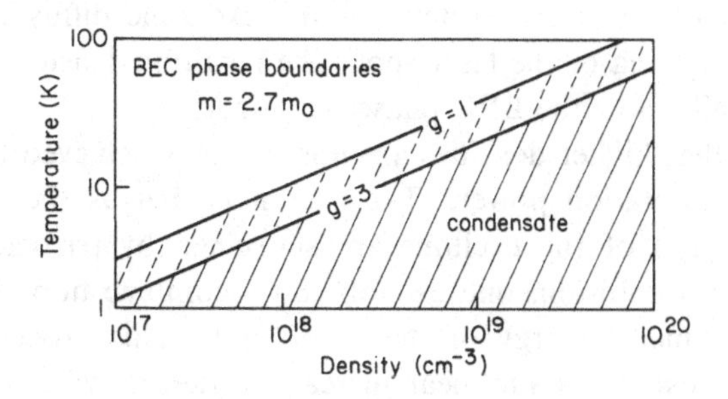

Fig. 6. Phase diagram for excitons in Cu_2O.

10^{19} cm^{-3} requires that the gas be cooled below 30 K. (In contrast, a spin-aligned hydrogen gas at this density must be cooled to millikelvin temperatures.)

Not surprisingly, nature often constructs unexpected and interesting obstacles to lofty goals, as we shall see in the remainder of this article. An interesting complication can be anticipated immediately: we are dealing with a *two-component* gas, composed of the ("S = 0") paraexcitons and ("S = 1") orthoexcitons. We expect that the orthoexcitons and paraexcitons will come into *thermal* equilibrium, or quasiequilibrium (i.e., exhibit the same gas temperature) within a few collision times, which is just a few picoseconds at the relevant densities. However, as we discussed in Section 2, the ortho–para interconversion rate is relatively slow – typically 30 nanoseconds. Therefore on a nanosecond time scale the two components will not attain a *chemical* equilibrium. As they relax from a high temperature and density, the ortho and para components should exhibit distinct chemical potentials, determined by their respective densities.

5 Equilibration and Cooling of the Excitonic Gas

We are confronted with the task of producing an excitonic gas of sufficiently high density and low temperature, as dictated by (25). In the experiment of Fig. 3, orthoexcitons are produced by surface excitation of the crystal. They diffuse away from the surface and convert to paraexcitons within a lifetime of $\tau \approx 30$ ns. With cw excitation, the steady-state number of particles is given by $N = G\tau$, where G is the pair generation rate. For a typical laser power of 100 milliwatts, $G = 3 \cdot 10^{17}$ pairs/s, which implies that $N = 10^{10}$ particles are continuously present. The steady-state volume of the orthoexciton gas is found to be in the range

10^{-5} to 10^{-6} cm^3 from an estimate of the excitonic diffusion length, $l \simeq \sqrt{D\tau}$, and the area of the laser spot. The resulting density of 10^{15} to 10^{16} cm^{-3} is well below the BEC phase boundary.

It turns out that higher densities are not so easily achieved by simply raising the cw excitation power. The reason is that as the density is raised the lifetimes of the excitons are shortened by particle–particle interactions. In a collision, one exciton can recombine non-radiatively by raising the kinetic energy of the other [29]. This process (Auger recombination) leads to a sublinear increase in density with generation rate, namely $n \propto \sqrt{G}$. Consequently, the most effective way of reaching the quantum regime is by using *pulsed surface excitation* and observing the excitonic gas before it diffuses far away from the surface. Instantaneous densities of 10^{19} cm^{-3} can be attained in this way, which may be viewed as a sort of "inertial confinement." †

By "surface excitation" we mean that the incident light is absorbed within a small distance from the crystal surface. A 2.4 eV (green) photon from an Ar^+ laser has an absorption depth of about 1 μm in Cu_2O and produces an "excess energy" of $h\nu - E_{gap} \approx 0.3$ eV per pair, corresponding to an initial carrier temperature of about 3000 K. The "hot" electrons and holes quickly cool towards the lattice temperature by the emission of phonons. As the kinetic energy falls below the excitonic binding energy (150 meV), excitons are formed, which continue to cool by phonon emission.‡

In discussing the approach to "equilibrium," we must consider three separate questions: (1) Under what conditions is the excitonic gas described by a temperature? (2) How quickly does the excitonic gas cool to equilibrium with the crystal via phonon emission? (3) Under what conditions is the exciton gas described by a single chemical potential?

The thermal equilibration time for the excitons to establish a gas temperature has not been measured directly; some reasonable estimates can be made, however. The exciton–exciton scattering rate is given roughly by

$$\frac{1}{\tau_e} = n v_{av} \sigma, \tag{26}$$

where v_{av} is the mean velocity and σ is the exciton–exciton scattering cross section. Although the exciton–exciton scattering cross section has not been measured directly (Ref. [6] gives an upper bound) we can

† An alternative confinement method is considered in Section 11.

‡ There is evidence in some experiments that the excitons form almost immediately, cooling as a unit towards the lattice temperature [30].

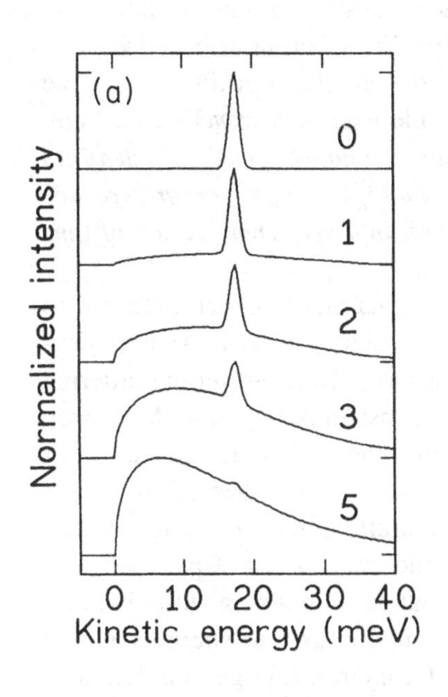

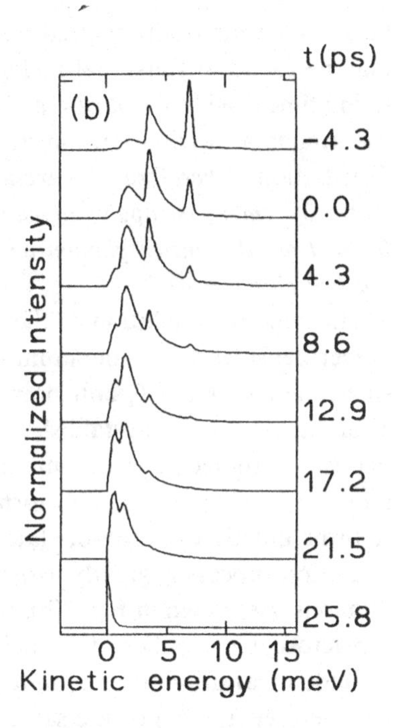

Fig. 7. (a) Theoretical particle distribution for a boson gas undergoing only interparticle scattering, for $n << n_Q$. The label of the curves is the number of scattering events per particle since the initial creation. The last curve is within 1% of a Maxwell–Boltzmann distribution. (From Ref. [35].) (b) Theoretical particle distribution for a boson gas undergoing only inlelastic scattering with phonons, assuming an unchanging phonon bath temperature below the critical temperature for condensation of the gas. The particle mass and particle–phonon scattering matrix element are those of excitons in Cu_2O, but the model does not take into account exciton–exciton interactions important in this density regime. (From Ref. [6].)

assume that σ is close to πa_x^2.† Then for a typical gas temperature of 30 K and density of 10^{18} cm^{-3}, the scattering time τ_e is about 30 picoseconds.

A simulation of hard sphere scattering, shown in Fig. 7(a), tells how many scattering times, τ_e, are required for a hard sphere gas to attain an internal thermal equilibrium. The starting energy distribution is taken to

† For s-wave scattering of hard spheres with radius a, the cross section is $4\pi a^2$. Exciton–exciton scattering almost certainly is not well modeled by infinite-potential hard-sphere scattering, however. The general form for exciton–exciton scattering has not been calculated; Ref. [31] gives a trial form extrapolated from a calculation for hydrogen which does not account for electron–hole exchange and reduced mass; Refs. [17, 32, 33] give estimates of the average interaction including spin effects.

be a narrow distribution at 20 meV. The details of this kinetic simulation may be found in Refs. [34] and [35]. One finds that after only five scattering times, *independent of the shape of the initial energy distribution*, the gas attains a distribution indistinguishable from a Maxwell–Boltzmann distribution. *Therefore, the excitonic gas at relatively moderate densities* ($n > 10^{17}$ cm^{-3}) *is capable of establishing an internal gas temperature, different from the lattice phonon temperature, in a very short period of time – a few picoseconds.*

This type of simulation of the particle dynamics has been extended to higher densities where quantum statistics come into play. At high densities, the factor $1 + f_{\mathbf{k}}$ enhances the scattering rate, but more scattering events are required to establish the Bose–Einstein distribution. Moreover, one is confronted with the important question of how long it takes for a macroscopic condensate to form, a subject of some theoretical debate. It turns out that density-of-states considerations indicate that the condensation process is greatly assisted by the interactions of particles with phonons. As shown in Fig. 7(b), in an ideal gas interacting with phonons, a macroscopic number of particles can enter near-zero energy states in a sudden transition after a finite time. Of course, a large population at near-zero energy is not the same thing as a condensate, and the articles by Kagan and by Stoof in this volume give theoretical calculations of the condensation time from a many-body approach. Nevertheless it seems clear that the exciton gas can reach a very degenerate distribution within its lifetime.

Now let us consider more directly the second question of how quickly the kinetic energy of a hot gas is lost by phonon emission in a cold crystal. Recently this cooling process has been measured by Snoke *et al.* [6] for a *low-density* gas of excitons, and the results are shown in Fig. 8(a). The crystal is excited by a 5 ps dye-laser pulse, and the photoluminescence spectra are recorded by a streak camera with 10 ps time resolution. (Readers who are curious about how one records a photon spectrum in billionths to trillionths of a second will find a useful review in Ref. [36].) The laser photons are tuned to 2.067 eV, which is 34 meV above the 2.033 eV band minimum of the orthoexcitons; in this phonon-assisted process, 14 meV goes towards creating a Γ_{12}^{-} phonon and 20 meV is given as kinetic energy to the exciton. The exciton luminescence is assisted by emission of a Γ_{12}^{-} phonon, so the zero in kinetic energy appears at about 2.018 eV in the data. At $t = 0$, the largest feature in the spectrum corresponds to excitons with 20 meV kinetic energy.

As seen in Table 2, there are several types of optical phonons with

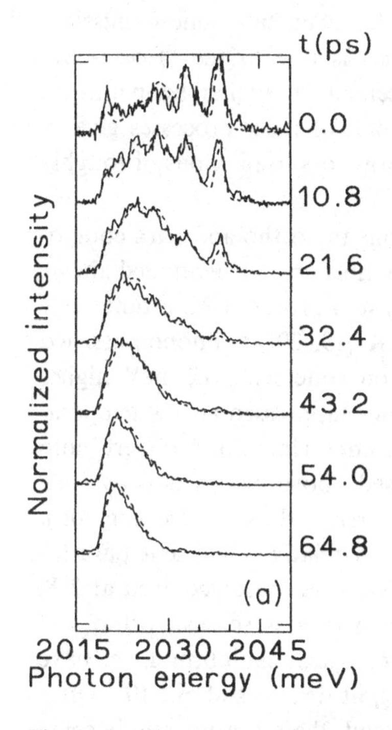

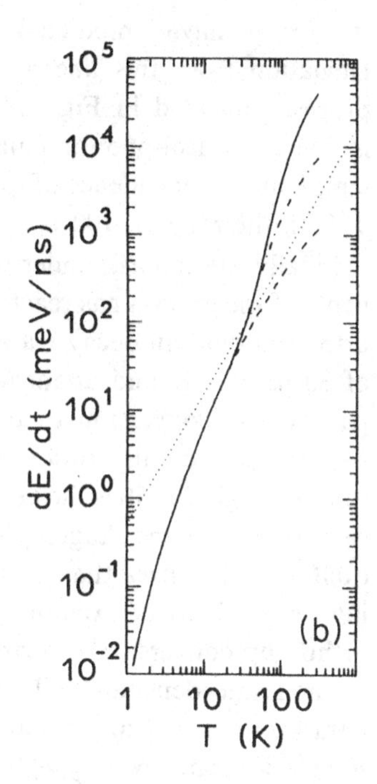

Fig. 8. (a) Time-resolved luminescence from Cu_2O following a very short (5 ps) laser pulse at 2.033 eV, measured with a streak camera. The dashed lines are a fit to the solution of the Boltzmann equation for exciton–phonon scattering based on deformation-potential theory. (From Ref. [6].) (b) The energy loss rate for excitons in a Maxwell–Boltzmann distribution and zero phonon temperature, using the phonon emission rates deduced in Ref. [6].

$\hbar\Omega_i$ < 20 meV. The most favored process is emission of the 18.7-meV TO phonon, apparent in the data of Fig. 8(a) in the peak near zero kinetic energy. The other sharp peaks are longitudinal-acoustic (LA) phonon emission peaks. Phonon emission depletes the 20-meV kinetic energy states in about 30 ps, rather slow compared to semiconductors such as GaAs, but still much less than the exciton lifetime. At 64.8 ps, the distribution resembles a Maxwell–Boltzmann distribution with $T \approx$ 50 K, which is still considerably higher than the 18 K lattice temperature. The exciton gas does not reach the lattice temperature until hundreds of picoseconds later (see Fig. 12). Experiments such as these give the matrix elements for the optical and acoustic phonon emission processes. Fig. 8(b) is a summary of the energy loss rate for an exciton gas at temperature T, with the lattice temperature assumed to be zero. The dotted curve is the $T^{3/2}$ behavior predicted in Section 3 for acoustic-phonon emission. Deviation from this line at lower energies is due to the "freeze-out" process indicated in Fig. 4. The enhancement at higher temperatures is due to optical-phonon emission. The total of these processes gives a temperature dependence of the rate of energy loss to phonons of roughly $T^{5/2}$ in the range 1–100 K.

Finally, we can ask under what conditions the ortho and para components of the exciton gas reach chemical equilibrium. As mentioned above, orthoexcitons can decay via a phonon-assisted process with a time scale of 30 ns at 2 K and about 600 ps at 30 K [19, 20]. † Phonon-assisted paraexciton conversion to the orthoexciton state lying 12 meV higher requires phonon absorption, which is highly suppressed at low temperature. At high densities, however, the two components can "convert" into each other via the Auger process discussed above, which destroys one exciton and ionizes the other. The free carriers thus created can form into either kind of exciton. This leads to a relatively efficient para-to-ortho "up conversion" – orthoexcitons have been observed even at 2 K at moderate densities [37]. Because this process is so fast, ortho and para excitons will in general *never* be in chemical equilibrium, i.e. have population ratio $N_o/N_p = 3\exp(-\Delta/k_B T)$, at densities above 10^{17} cm^{-3} and lattice temperatures below 30 K; instead, their relative number will be determined by a rate balance between conversion processes. The two components will have the same temperature but two different "effective chemical potentials", α_o and α_p.

If the two components *were* in chemical equilibrium, the orthoexcitons would hardly ever have α_o less than $\Delta/k_B T \approx 4$ at 30 K, and they would never deviate from a Maxwell–Boltzmann distribution (see Fig. 5) at low temperature. As we shall see, in fact the orthoexcitons exhibit quantum statistics over a wide range of density and temperature. The ortho-to-para conversion process can be enhanced, however, by raising the lattice temperature, pushing the gas toward chemical equilibrium.

Table 3 gives a summary of the various rates of relaxation processes for excitons in Cu_2O. Relaxation and recombination processes of excitons occur in Cu_2O on time scales which vary from picoseconds to milliseconds, and properly modeling this system requires understanding which processes are important in a given density and temperature regime.

† Orthoexcitons are also expected to convert to paraexcitons by flipping spins in two-body collisions, analogous to the process observed in hydrogen and positronium, but definitive evidence for this process has not yet appeared. See Section 7.

Table 3. *Relaxation processes of excitons in Cu_2O. (For excitons at T=30 K and n=10^{19} cm^{-3}, and lattice at T_L=0)*

Process	Timescale	Proportional to	Reference
exciton formation from free carriers	250 fs		[30]
exciton–exciton scattering	3 ps	$nT^{1/2}$	[6]
exciton–phonon scattering	30 ps	$\sim T^{5/2}$	[6]
exciton–exciton Auger recombination	$\geq$ 200 ps	$nT^{1/2}$	[38]
ortho-to-para phonon-assisted conversion	600 ps	$\sim T^{3/2}$	[19, 20]
para-to-ortho phonon-assisted conversion	30 ns	$\sim T^{3/2}e^{-\Delta/k_BT}$	[19, 20]
ortho radiative recombination	300 ns	constant	[18]
para radiative recombination	$\gtrsim 150\mu s$	constant	[21, 18]

6 The Exciton as a Boson: Spectroscopic Considerations

How do we analyze the luminescence spectra of excitons at high densities? Over a range of conditions, the ortho and paraexciton spectra are expected to have the same temperature but different chemical potentials, so we must consider each component separately. The basic procedure is to extract α and T from a fit of a phonon-assisted spectrum to (20). In the quantum regime, the chemical potential and temperature are quite orthogonal fit parameters, as indicated in Fig. 5. This logarithmic plot shows that the gas temperature manifests itself as a linear high-energy tail, whereas the chemical potential (or $\alpha = -\mu/k_BT$) governs the sharpness of the low-energy portion. Given the spectrally determined α and T, we can then determine the density of this component of the gas by calculating the integral in (22).

An important experimental example of the spectroscopic information gained from the two-component exciton gas is given in Fig. 9. Time-resolved spectra following a 10-ns laser pulse capture the energy distributions of the ortho and paraexcitons at "high" and "low" densities by choosing sampling times of 16 ns and 44 ns, respectively, after the onset of the laser pulse. The ortho and para components are superimposed by shifting their spectral positions. At low densities (t $\geq$ 44 ns), both the para (solid line) and ortho (dashed line) spectra at all times exhibit a Maxwell–Boltzmann distribution with $T = 32$ K, which is taken as the lattice temperature.† The off-scale signal on the high-energy side of the para spectrum is due to a strong ortho luminescence replica.

At high density (t = 16 ns), the energy distribution of the paraexcitons is considerably narrower than that of the orthoexcitons, which fit a Maxwellian distribution at a slightly elevated gas temperature of 38 K (the gas is heated at early times). Assuming the same temperature for the paraexcitons, the paraexciton distribution is described by a Bose–Einstein distribution (circles) with $\alpha = 0.07$. At these relatively high lattice temperatures, the orthoexcitons convert efficiently to paraexcitons via phonon emission, which are consequently the dominant species at nearly all times. Part (c) of this figure plots the temporal evolution of the spectral widths, clearly showing the narrowing of the paraexciton energy distribution associated with Bose–Einstein statistics [40]. At the highest densities, the full width at half maximum of the paraexciton spectrum is about half that of a Maxwell–Boltzmann distribution at the lattice temperature. *No classical mechanism can explain this result*, which does not rely on any model-dependent fit. The exciton truly is a boson!

Now consider how a *condensate* would manifest itself in the luminescence spectrum. If the gas density exceeds the critical density given by (25), we might expect to see a sharp "condensate" peak near $E = 0$ superimposed on the broader "excited-state spectrum" given by (20). Unfortunately, as α approaches zero the peak in the excited-state spectrum becomes arbitrarily intense and very close to $E = 0$. Taking the derivative of $N(E)$ in the quantum regime where $\alpha \ll 1$, one finds that the peak in the excited-state spectrum occurs at $E \approx |\mu|$. It appears that for the ideal Bose gas, the energy distribution will continuously become sharper and sharper without an abrupt change as the phase boundary is crossed at $\alpha = 0$.

Nevertheless even in this idealized case, the presence of condensate should be readily detectable in the spectrum. Recall that the excited-

† Unlike other data presented below, the sample in this case is cooled by helium gas at elevated temperature rather than immersed in liquid helium.

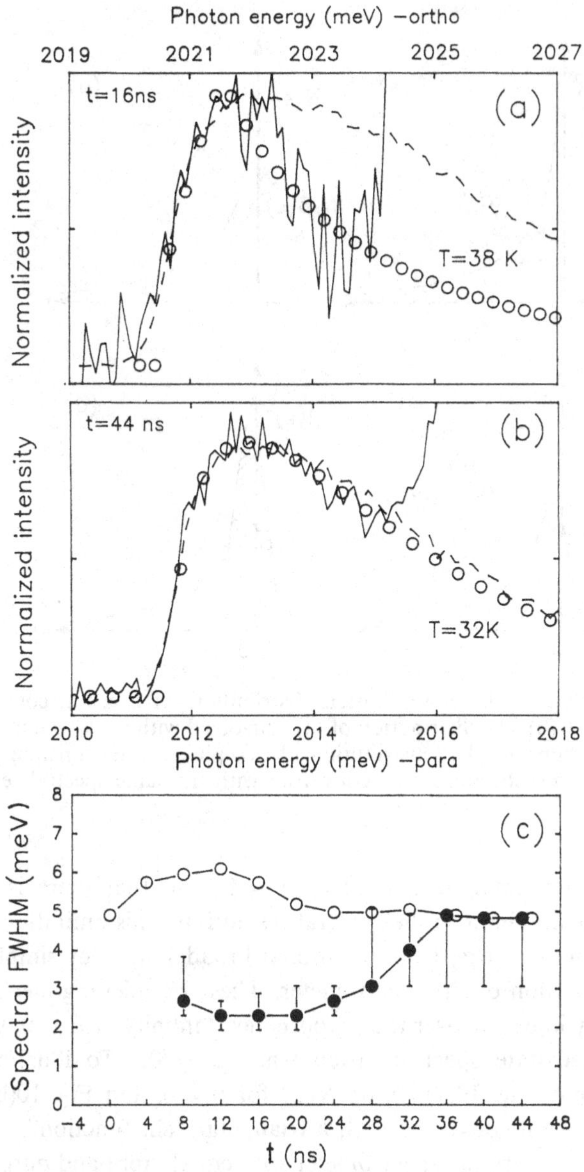

Fig. 9. (a) The orthoexciton (dashed line) and paraexciton (solid line) energy distributions during laser-pulse excitation at photon energy of 2.4 eV. The two spectra are normalized to the same height and shifted in energy for comparison. The orthoexciton distribution fits a Maxwell–Boltzmann with T=38 K. The paraexciton spectrum is obscured by orthoexciton luminescence at energies above 2.015 eV. The open circles are a 1.3 meV-broadened Bose–Einstein distribution with $\alpha = 0.07$ and T=38 K. (b) The distributions well after the laser pulse. Both fit a Maxwell–Boltzmann distribution with T=32 K (open circles). (c) The spectral FWHM of each energy distribution as a function of time (open circles: orthoexcitons; solid circles: paraexcitons). The 10-ns laser pulse is centered at t=10 ns. (From Ref. [39].)

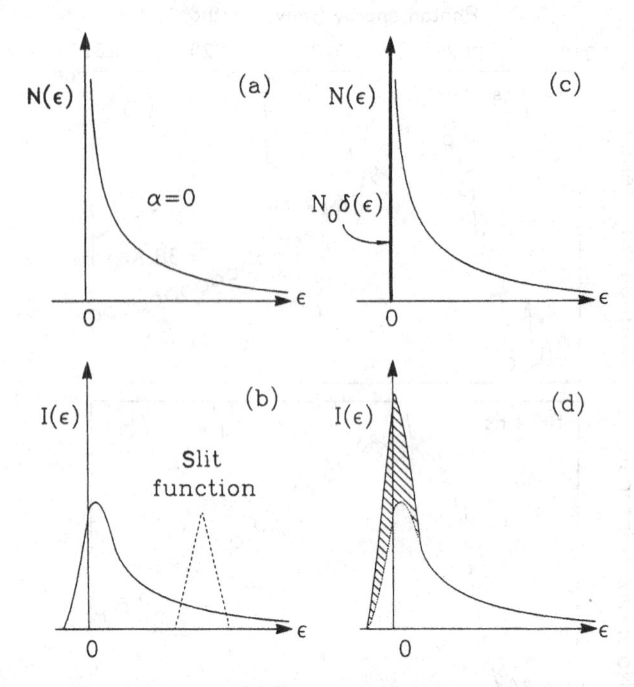

Fig. 10. (a) Excited-states Bose–Einstein distribution for a gas at constant potential with α=0. (b) The distribution of (a) convolved with a triangular spectral-resolution function. (c) The distribution of (a), plus a delta-function peak at $E = 0$. (d) The distribution in (c), convolved with the same spectral resolution function.

state spectrum is integrable, even at $\mu = 0$, and that there is always some experimental limit to the spectral resolution. This limitation can be intrinsic to the gas– e.g., from collisional broadening – or simply from the finite resolution of the spectrometer. These broadening mechanisms will effectively integrate over the (hypothetical) infinity at $E = 0$, yielding a finite excited-state spectrum even when $\mu = 0$. To illustrate this important point, Fig. 10(a) shows $N(E)$ for $\alpha = 0$, and Fig. 10(b) plots the same spectrum convolved with a triangular "slit function", which is characteristic of a grating spectrometer with equal input and output slits. If we add to the excited-state spectrum a perfectly sharp delta-function spike at $E = 0$ with, say, half of integrated intensity as the excited-state spectrum, then the slit-convolved spectrum shown in Fig. 10(d) is obtained. The contribution from the condensate is shaded. This spectrum is quite distinguishable from the $\mu = 0$ excited-state spectrum, and the fraction of condensate (here $1/3$) can be easily extracted.

In what follows, we will examine the behavior of the excitonic gas at high densities and low lattice temperature ($T_{\text{latt}} = 2$ K). The systematic studies described below were motivated by early results of Hulin *et al.* [41], who observed orthoexciton spectra – integrated over time and space – that could be described by quantum-statistical distributions. As we shall see, subsequent time- and space-resolved spectra provide overwhelming evidence for quantum statistics of excitons in Cu_2O and uncover an unusual "quantum saturation" effect, still not totally understood. At the highest excitation levels, a rapid expansion of the gas is observed together with interesting anomalies in the paraexciton spectra, suggesting superfluidity of the excitonic gas. Most recently, compelling spectroscopic evidence for BEC of paraexcitons has been obtained for uniaxially stressed crystals.

7 Quantum Statistics of Excitons at High Gas Densities

Figure 11 shows phonon-assisted spectra of orthoexcitons following a high-power, 100-ps Ar^+ laser pulse. The crystal is immersed in a superfluid helium bath at 2 K. The luminescence is detected by photon counting with 90 ps time resolution. The temperature of the gas at each sampling time can be ascertained by fitting the high-energy tail of the spectrum. The decay of the gas temperature determined in this way is plotted as the open circles in Fig. 12. An interesting discovery is that *the decay rate of the temperature depends upon the initial density of the gas,* which is controlled by changing the spot size of the laser beam. In the experiment of Fig. 11, the spot size is about 20 μm, corresponding to 10^8 W/cm^2 for the 20 nJ pulse. If the spot size is increased to 2 mm, the power density is $(100)^2$ smaller and the resulting temperature evolution is plotted as the solid dots. The cooling rate of about 100 ps is consistent with the acoustic-phonon emission process discussed in Sections 3 and 5. At high density, the gas retains a high temperature over a much longer period. Possible sources of the inhibited cooling at high density will be discussed below.

The spectra of Fig. 11 show clear evidence of quantum statistics. The dashed curves are fits of Eq. (20) to these high-power spectra. Initially the spectrum is nearly Maxwellian (compare to Fig. 5), but at later times it takes on the more sharply peaked characteristic of a Bose gas in the quantum regime. Indeed, at all times the gas is well described by Bose–Einstein statistics, and over a broad time interval the chemical potential assumes a highly quantum value of $\mu \approx -0.2k_BT$. A value of $\alpha \approx 0.2$

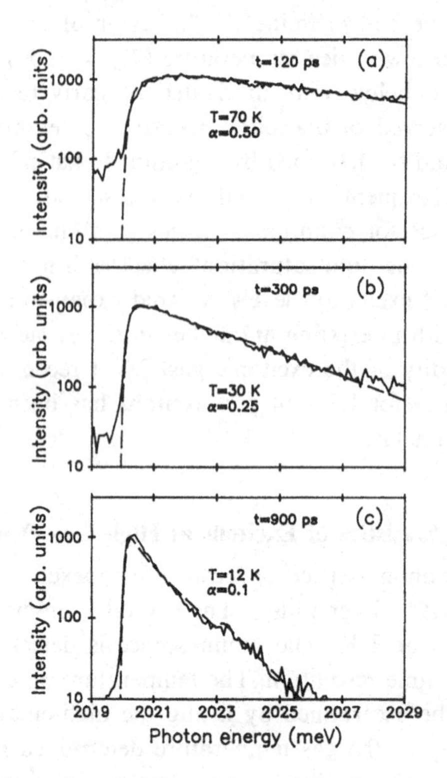

Fig. 11. The orthoexciton luminescence at three densities plotted semilogarithmically, fit to the ideal Bose–Einstein distribution (20), with the values of α and T shown. The fit distribution in each case is convolved with triangle function to take into account the finite spectral resolution. (From Ref. [38].)

corresponds to a gas density which is a factor of about three lower than the critical density for condensation.

Given μ and T, the density of the gas is determined by integrating $N(E)/V$ using (21). In Fig. 13 the spectroscopically determined μ and n are plotted as a function of time after the pulse. Part (a) shows that the rise and fall of $n(t)$ closely follows that of the integrated intensity $I_{\text{tot}}(t)$, implying that the effective volume of the gas is roughly constant over the time span of the measurement. The rapid, non-exponential decay of $I_{\text{tot}}(t)$ at subnanosecond times implies that some new relaxation or recombination mechanism occurs at high density.

Most fascinating, during the cooling cycle, the degree of degeneracy measured by α remains constant for a long period of time (Fig. 13(b)), despite the large changes in density and temperature. This implies that

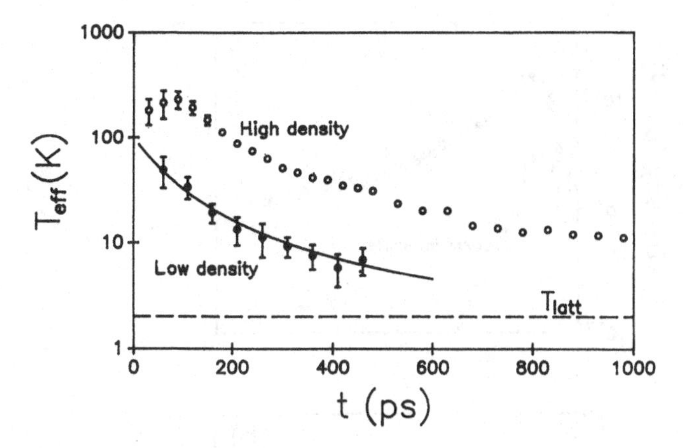

Fig. 12. The effective temperature as a function of time for the case of high density ($n > 10^{19}$ cm^{-3}) and low density ($n < 10^{17}$ cm^{-3}). The solid line in the low density case is the prediction of the deformation-potential theory for phonon emission discussed in the text. (From Ref. [38].)

the gas is evolving along a line parallel to the ($\mu = 0$) phase boundary for BEC, as shown by the open circles in Fig. 14. The effect of the gas to remain close to the phase boundary over a wide range of densities and temperatures has been termed "quantum saturation" by Snoke *et al.* [42]. This remains one of the major puzzles of this remarkable quantum gas.

Quantum saturation is even more pronounced in a "long-pulse" (10 ns) experiment. In this case, the orthoexciton gas is observed with about 1 ns resolution *during and after* the excitation pulse. Again, the gas temperature and chemical potential are found to be well defined throughout the experiment. As the excitation pulse evolves, the temperature and density of the gas *increase* along a $\mu \approx -0.1k_BT$ line to maximum values (at about the peak of the pulse) and then decay along the same line, as shown by the solid dots in Fig. 14. In these experiments, the signal levels are high enough to observe the phonon-assisted spectra of the paraexcitons simultaneously. At medium pulse intensities, the paraexcitons exhibit a higher degree of degeneracy, as predicted in Fig. 6. At high pulse intensities, interesting anomalies appear in the paraexciton spectrum which were attributed to the formation of a condensate.

Clues to the origin of the quantum saturation effect may be extracted from the observations of an increased orthoexciton decay rate and a reduced cooling rate (Figs. 12 and 13). Figure 15 explicitly plots the

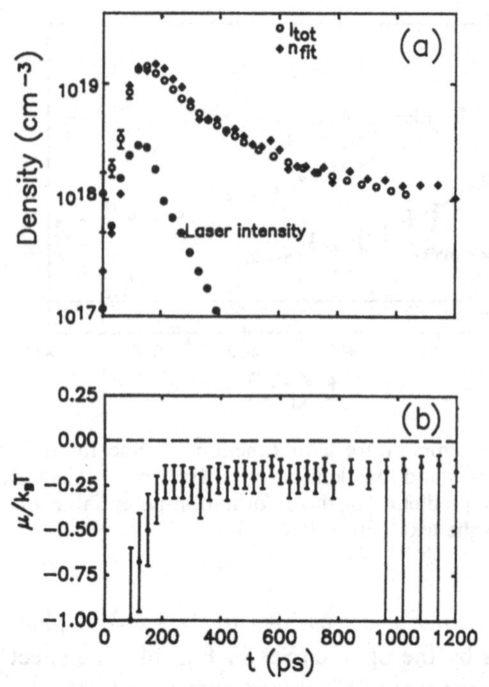

Fig. 13. The thermodynamic parameters of the orthoexciton gas in the case shown in Fig. 11. At late times the uncertainty in the chemical potential μ grows as the temperature lowers (Fig. 12) and the line narrows. The smallest possible α is plotted as a solid dot, but, within the spectral resolution, the corresponding spectra at late times are almost compatable with a classical distribution, as indicated by the large error bars. (From Ref. [38].)

orthoexciton decay rate as a function of density. The linear dependence indicates a two-body process for the destruction of orthoexcitons. Two possible mechanisms are as follows. At high densities the probability increases that the collision of two excitons can induce a non-radiative decay of one of the excitons. In this "Auger recombination process", the recombination energy (essentially $E_{\text{gap}} - E_x$) is given to the remaining electron–hole pair as kinetic energy. The hot particles can always transfer some of their kinetic energy to the remaining particles in the gas, effectively heating the gas [43]. Another density-dependent heating mechanism which might occur is collision-induced conversion of two orthoexcitons into paraexcitons, which would raise the kinetic energy of the gas by 24 meV (twice the exchange energy) per event. (See also the discussion by Nozières in this volume.) While this spin-flip process

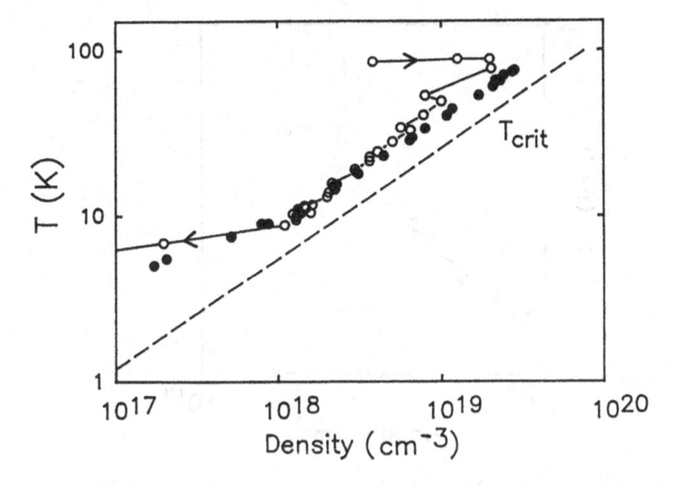

Fig. 14. The quantum saturation of the orthoexciton gas, corresponding to a temperature rise with increasing density, $T \propto n^{2/3}$. Solid circles are deduced from fits to spectra during and after long (10-ns) laser pulses. (From Ref. [42].) Open circles are deduced from fits to spectra following short (100-ps) laser pulses. (From Ref. [38].) The dashed line gives the critical densities for BEC of the $g = 3$ orthoexcitons.

and/or Auger heating may be important in explaining the temperature rise of the gas at high densities, it is difficult to see how they could completely explain why the gas follows the quantum-statistical boundary so closely.

Can the rise in temperature simply be due to heating of the lattice? This explanation seems improbable due to the following argument: A density of $2 \cdot 10^{19}$ cm^{-3} implies that the $3 \cdot 10^{10}$ excitons (the number of photons in a 0.01 μJ pulse) occupy a volume of $1.5 \cdot 10^{-9}$ cm^3. If one assumes a heat capacity of Cu_2O given by $5.9 \cdot 10^{-3}$ J/kg-K $(T/K)^3$ [44] and that all of the excess photon energy (0.375 eV) imparted to the excitons is converted to phonons, then the local temperature rise of the crystal would be about 20 K – not enough to explain the 90 K rise in the gas temperature. Also, there is no apparent band shift of the exciton line (cf. Fig. 2), indicating that T_{latt} remains less than 15 K. A similar argument applies to the long-pulse case, where both the number of particles and gas volume are about an order of magnitude larger.

An interesting issue raised in earlier work [34] takes note of the fact that at constant average energy, the temperature of an ideal boson gas

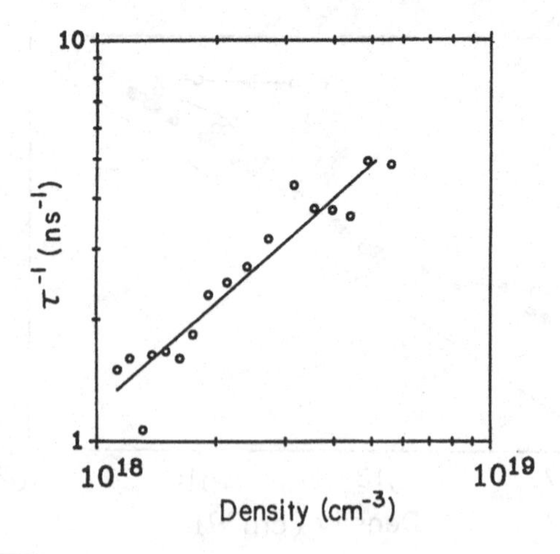

Fig. 15. The decay rate of orthoexcitons as a function of density, from the same data as Fig. 12. (From Ref. [38].)

increases by a factor of two as the gas increases in density from the classical to the quantum regime, simply due to the shape of the distribution (20). Therefore if the rate of energy loss is constant per particle, then increasing the gas density automatically increases the temperature. Later work [45] showed, however, that increased degeneracy enhances the rate of energy loss by acoustic phonon emission via the $(1 + f_{\mathbf{k}})$ final-states factor, so that quantum statistics alone will lead to at most a 20% increase in the gas temperature and will not give the "saturation" effect.

The question of the time scale for BEC, discussed in other articles in this book, does not apply to the heating effect seen here. As shown in earlier work based on solution of the Boltzmann equation for two-body scattering [34, 6], a boson gas in the *normal* state can evolve via elastic particle–particle or inelastic particle–phonon scattering to a highly degenerate distribution within a few scattering times. Once the gas has a large population in states near zero kinetic energy, the Boltzmann equation breaks down and the question remains of how long the particles need to organize into a true delta function. As long as the gas can be considered weakly interacting, however, a slow process of long-range organization would not yield a substantial increase in the average energy of the gas.

8 Effects of Inhomogeneities in the Exciton Gas

In all of the high-density experiments, the excitation light is absorbed within about $1\mu m$ of the crystal surface. The resulting density and temperature of the exciton gas are therefore inherently non-uniform. The spectral fitting, on the other hand, has assumed a single temperature and chemical potential. The fact that the spectra are so well characterized by these two parameters implies that at a given time the energy distribution of the gas is well described by an average temperature and chemical potential.

To check this conclusion, Ref. [13] analyzed spectra resulting from several spatial distributions of n and T. Since a full hydrodynamic model of the density and temperature profile of the exciton gas in steady state has not appeared [46], these calculations simply assumed "realistic" distributions of the gas. In one model, the density was chosen to be a Gaussian, $n(z) = n_{\max} \exp(-(z/z_o)^2)$, decreasing with distance, z, from the surface, and T was assumed to be a constant. For a given $n_{\max}$ and T, Bose–Einstein spectra corresponding to $n(z_i)$ at a series of z_i were generated and summed to predict the overall energy distribution of the inhomogeneous gas. The composite spectrum yielded a good fit to the "saturated" experimental spectra when $n_{\max} = 0.8n_c$ (or $\alpha_{\min} = 0.025$). In a second model, a Gaussian form for the density was assumed and the local temperature was assumed to vary as the 2/3 power of the density, as observed in the quantum saturation effect. Again a good fit was obtained yielding $\alpha_{\min} = 0.1$ for a typical spectrum.

These calculations may have a bearing on why the orthoexciton chemical potential appears to "saturate" at about -0.1 to -0.2 $k_B T$, rather than $\mu = 0$. The value $\alpha = 0.1$ corresponds to an average density of the gas which is about two times smaller than n_c, which is consistent with a spatial distribution varying from $n_{\max} \approx n_c$ to zero. It may be that the quantum saturation occurs when the orthoexciton gas is actually condensed in the region of highest density. If so, why would the condensate fraction remain small, and what causes the gas to warm exactly along the phase boundary? An interesting suggestion [47], is that when a condensate is formed, it undergoes superfluid flow to a local minimum in potential (perhaps associated with defects or the crystal surface). The resulting increase in density produces an enhanced Auger recombination or ortho-para conversion, destroying the condensate and heating the gas; i.e., keeping the gas precisely along the phase boundary! Any process which converts condensate particles into heat has this interesting property.

It is interesting to note that the constant α implies that the gas follows a line of constant entropy per particle, an "isentrope". A gas in free expansion follows an isentrope, and the exciton gas created at the surface approximates a free expansion when it moves away from the surface. We note, however, that (1) the orthoexciton saturation is seen during times when the effective volume of the gas is constant and density is rising or falling due to particle generation (e.g. the data of Ref. [42], shown in Fig. 14) or recombination (e.g. Fig. 13(a), following 100 ps pulses) and (2) the excitons can lose energy at all times to the phonon field, so that entropy is not strictly conserved.

Finally, we address the low-energy tail observed in the orthoexciton spectra at high densities, e.g., Fig. 11. A possible source of this luminescence at $E < 0$ is spectral broadening due to collisions. A convolution of (20) with a Lorentzian of full width at half maximum of ~ 0.1 meV can account for this spectral feature. This width corresponds to a collision time of 10 ps, which is consistent in the classical approximation (26) with a scattering length on the order of the Bohr radius of the order of 7 Å. Alternatively, a low-energy tail has been interpreted by Haug and Kranz [31] as arising from interactions with the Bose condensate. In that work, the experimental data of Hulin *et al.* [41] were fit to a spectrum which included a low-energy tail arising from a 5% condensate fraction, as well as substantial symmetric Lorentzian broadening which was attributed to instrumental resolution. The fact that the broadening seen in the low-energy tail is *not* proportional to density, as expected for standard collision broadening, but instead increases disproportionately at late times when the gas appears more degenerate [13], would seem to support this interpretation. Although the instrumental spectral broadening in these recent experiments is much less than that assumed by Haug and Kranz, one could argue that a small fraction of condensate leads to the low-energy tail even though a sharp peak due to the condensate is obscured by spatial inhomogeneities (see also Ref. [48]).

9 Enhanced Degeneracy of Orthoexcitons in Stressed Cu_2O

How can one coerce the "saturated" orthoexciton gas into crossing the phase boundary and condensing? One approach is to shift the phase boundary to lower densities! Eq. (25) and Fig. 6 show that the critical density for BEC is directly proportional to the ground-state multiplicity g.

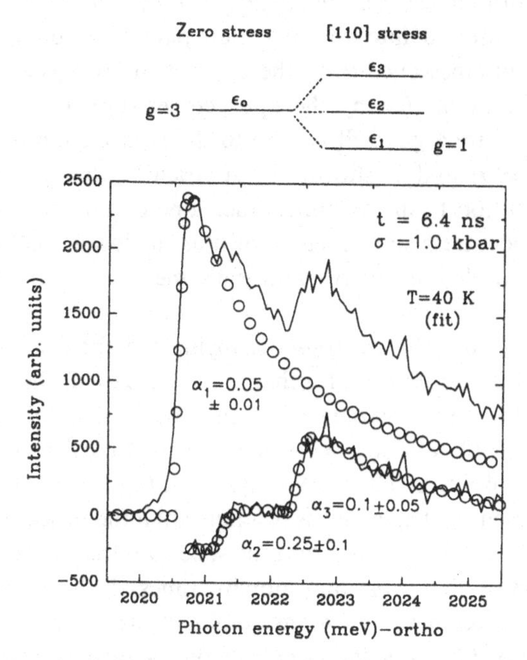

Fig. 16. Orthoexciton luminescence from Cu_2O under [110] stress.

If the orthoexciton triplet can be split by applied stress or magnetic field such that most of the particles relax to the lowest level, then the ground-state multiplicity is effectively reduced from 3 to 1. Such a reduction in the ground-state multiplicity displaces the orthoexciton phase boundary a factor of three towards lower density, making it coincident with the paraexciton phase boundary. If the experimental $n(T)$ curve were to remain the same as in Fig. 14, then the orthoexciton gas would indeed cross the new phase boundary.

By uniformly stressing the crystal, Lin and Wolfe [49] have indeed been able to effectively reduce g from 3 to 1. Uniaxial stress applied along the crystalline [110] direction reduces the crystal symmetry from cubic to orthorhombic. The energy splitting of the triplet has been measured by Waters *et al.* [50] and fit theoretically by Trebin *et al.* [51]; for stress along the [110] axis the orthoexciton levels shift by approximately 0, ± 1.5 meV/kbar, and the paraexciton level does not shift, to first order [52]. The orthoexciton triplet is split into three levels, as shown in Fig. 16 for a stress of $\sigma = 1.0$ kBar $= 10^8$ Pa. Intense 10-ns Ar^+ laser pulses are used. The upper solid curve in this figure is the orthoexciton spectrum obtained without selecting polarization of the luminescence. If a linear polarizer is inserted into the detection path, it is possible to nearly eliminate the luminescence from the upper two levels, obtaining a spectrum close to that plotted by the open circles (upper trace). In fact, these open circles are a good fit of (20) to the polarization-selected spectrum, yielding the α_1 and T shown. By subtracting this theoretical Bose–Einstein distribution from the upper trace, one obtains the contributions from the upper stress-split levels, shown as the lower solid trace. These excited states are then fit to (20) with the same T to obtain the α_2 and α_3 shown.

We see immediately that the excitons occupying the three levels are not in chemical equilibrium with each other. That is, they do not share the same chemical potential. In particular, the chemical potential for the uppermost level is only about $\mu = -0.1 k_B T = -0.4$ meV, relative to its zero-kinetic-energy point at 2.0223 eV. Physically, this means that the conversion time between the triplet levels is longer than the measurement time (6.4 ns). Nevertheless, it appears from the spectrum that most of the excitons are in the lowest level. For later sampling times, the luminescence from the lowest level becomes even more dominant, as the excitons in the split-off levels relax. Let us now concentrate on the distribution from this ground level.

A quite remarkable spectrum appears when the gas density is the highest, near the peak of the excitation pulse. A polarizer is used to eliminate residual luminescence from the two excited states. The raw spectrum is plotted both linearly and logarithmically as the solid curves in Fig. 17 (a) and (b), respectively. The fit to the Bose–Einstein distribution (20), convolved with the spectrometer slit resolution, is exceptionally good, yielding $\alpha = 0.015$ and $T = 62$ K. For comparison to classical statistics, a Maxwell–Boltzmann distribution with $T = 62$ K is plotted as a dashed line in both graphs.

Still, there is no evidence of a condensate in the orthoexciton spectrum. The experimental $n(T)$ has apparently been "pushed" to lower density (or higher temperature) by the new phase boundary. The solid dots in Fig. 18 are the results of analyzing the time-resolved spectra as a function of sampling time. We see that the orthoexciton gas in the stressed crystal displays a high degree of quantum saturation, closely paralleling the $g = 1$ phase boundary plotted as the solid line in this figure. For comparison, the data for the unstressed case is plotted as the open circles, and the dashed line is the $g = 3$ phase boundary. In view of these data, there is little doubt that the saturation effect is indeed

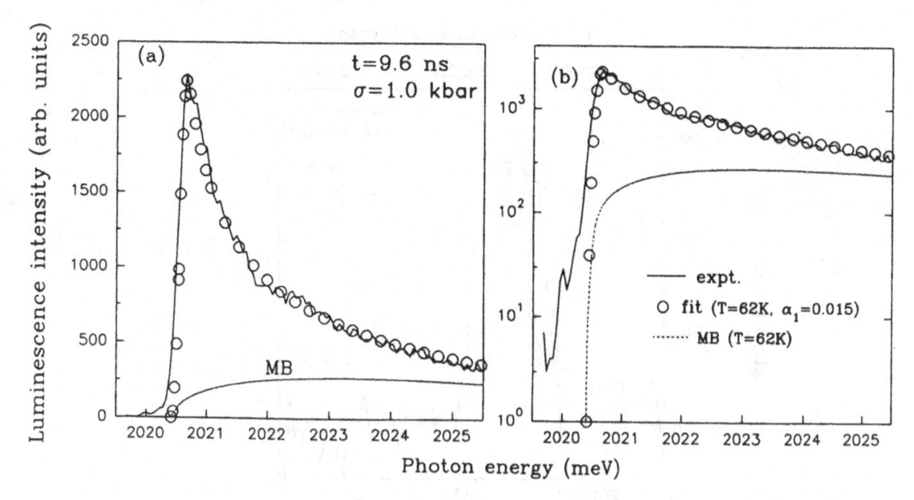

Fig. 17. Solid line: orthoexciton energy distribution in the lowest spin state in stressed Cu_2O following intense laser excitation. (Luminescence from higher spin states has been subtracted via a polarizer.) Open circles: fit to the ideal-gas Bose–Einstein distribution. (a) Linear plot. (b) Semilog plot.

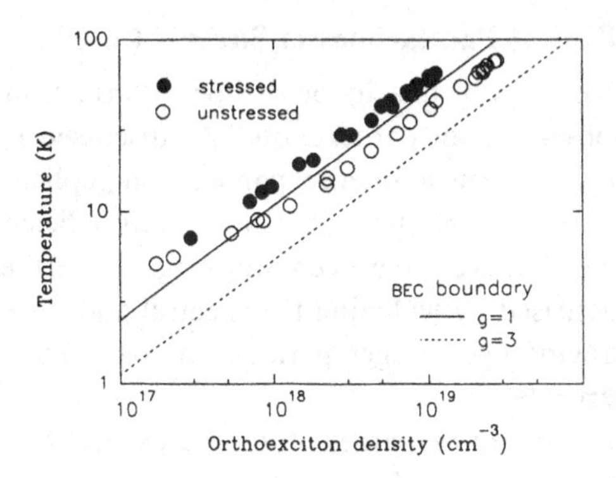

Fig. 18. Comparison of the orthoexciton "saturation" in the case of $g = 1$ (stressed Cu_2O) and $g = 3$ (unstressed Cu_2O).

associated with the degenerate quantum nature of the gas and is not solely due to an "accidental" coincidence of classical heating processes (e.g., Auger or ortho–para conversion) with the BEC phase boundary.

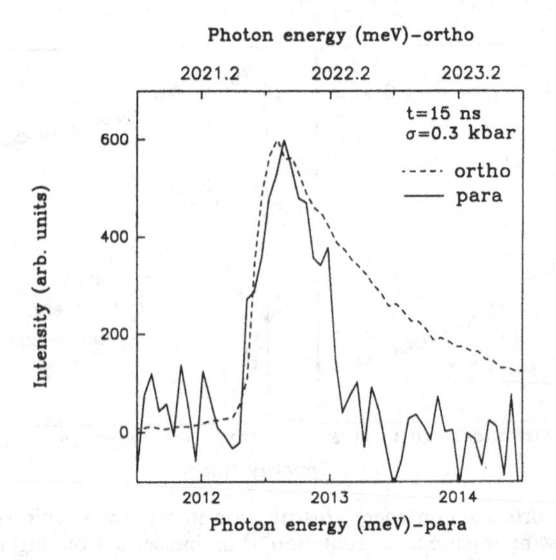

Fig. 19. The orthoexciton (dashed line) and paraexciton (solid line) energy distributions during laser-pulse excitation at photon energy of 2.4 eV in a [110] stressed crystal. The two spectra are normalized to the same height and shifted in energy for comparison.

10 BEC of Paraexcitons in Stressed Cu_2O

Our attention has been fixed primarily on orthoexcitons because they are so much easier to observe than paraexcitons. As discussed in Section 2 and listed in Table 2, the most intense paraexciton replica is assisted by the Γ_{25}^- phonon and is 500 times less radiatively efficient than the Γ_{12}^- orthoexciton line that we have been studying. Nevertheless, when the signal level is increased by widening the spectral and time resolution somewhat and averaging over longer periods, the paraexciton luminescence becomes observable.†

Despite the poor visibility of paraexcitons, we expect that their density will be comparable to that of the orthoexcitons at early times, $t <$ 10 ns. As the orthoexcitons convert down to paraexcitons, the paraexciton population should greatly exceed that of the orthoexcitons at low temperatures. Also, the paraexciton lifetime exceeds that of the orthoex-

† The paraexciton luminescence is readily observable in a μsec-timescale experiment (e.g. Section 3 and Ref. [23]) because one integrates the recombination photon flux over times comparable to the lifetime of the paraexciton. In the pulsed experiments, the gas density is only comparable to the quantum density for 20 ns or so following the peak of the laser pulse (at $t = 10$ ns). The slow radiative rate of the paraexcitons (Table 3) does not produce many photons in a few nanoseconds.

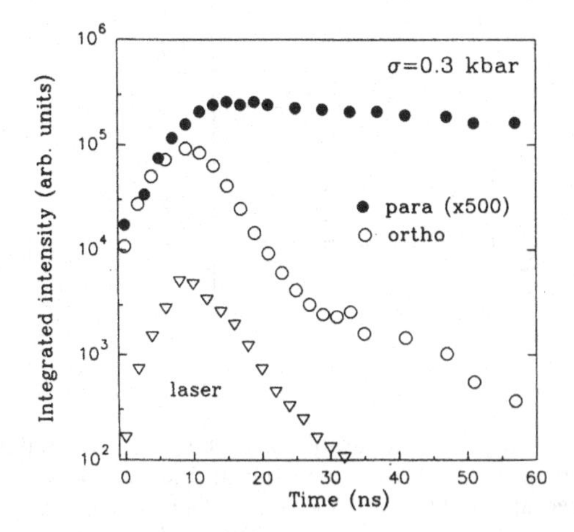

Fig. 20. The relative intensity of the ortho and para luminescence. Triangles give the intensity of the laser vs. time, multiplied by an arbitrary constant.

citon by two to three orders of magnitude. Of course, the excitonic gas will be expanding in the crystal as time progresses, which will increase the volume occupied by the gas. As usual, we must resort to experiment to find out if the paraexciton density can exceed the critical density for BEC before diffusing away.

The simultaneous ortho and paraexciton spectra at a sampling time of 15 ns and an applied stress of $\sigma = 0.3$ kBar are plotted in Fig. 19. The two spectra are superimposed such that their $E = 0$ points coincide.† The paraexciton spectrum is significantly narrower than the orthoexciton spectrum and displays little evidence of the high-energy tail associated with the excited-state spectrum. The orthoexciton spectrum is well fit to the excited-state spectrum (20) with the parameters $\alpha = 0.1$ and $T = 19$ K.

The relative number of ortho and paraexcitons is determined from their spectrally integrated intensities, as plotted in Fig. 20 for an applied stress of $\sigma = 0.3$ kbar. The paraexciton intensity has been multiplied by a factor of 500 to compensate for the different radiative efficiencies. The intensity of the laser pulse is also shown. As expected, the ortho and paraexciton populations are roughly comparable at early times, but near the peak

† For the orthoexciton, the photon energy for $E = 0$ is accurately determined in fitting the data to the excited-state spectrum (20). For the paraexciton, we find $E = 0$ by fitting a Maxwell–Boltzmann spectrum (13) to the late-time spectra, which can be averaged over a long time window to enhance the signal-to-noise ratio.

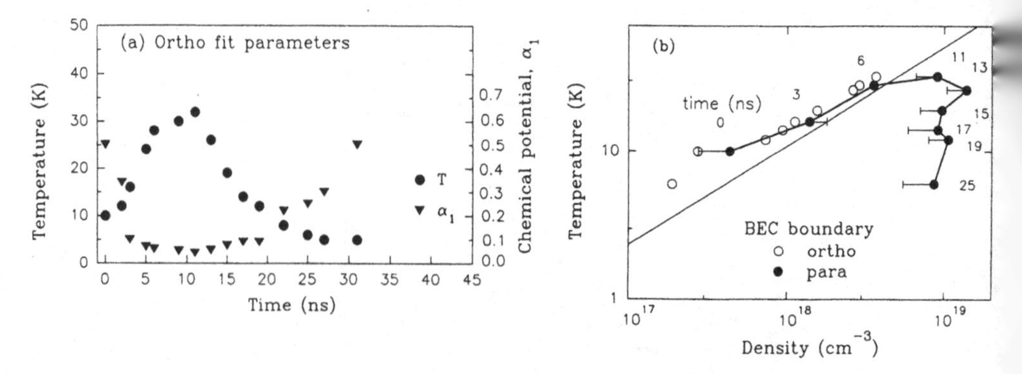

Fig. 21. (a) Time evolution of the temperature and chemical potential determined from the orthoexciton spectrum during and after an intense 10-ns laser pulse of a Cu_2O crystal stressed along [110] at 0.3 kbar. (b) Density and temperature for the ortho and para gases during the same experiment. Labels for the para data points indicate the elapsed nanoseconds.

of the pulse the number of paraexcitons begins to depart significantly from that of the orthoexcitons. As the orthoexcitons continue to decay and down-convert to paraexcitons, the paraexciton population grows and then remains stable.

In Fig. 21(a) we plot the time evolution of the temperature and chemical potential determined from the orthoexciton spectrum. If we make the reasonable assumption that, over the period of these measurements, the orthoexciton and paraexciton components share the same temperature and volume,† then we can obtain a rather accurate estimate of the paraexciton density. As usual, the orthoexciton densities are determined by integrating (20) using the parameters in Fig. 21(a), and the results are plotted as the open circles in Fig. 21(b). The concurrent paraexciton density is then determined by multiplying the orthoexciton density by the ratio of intensities plotted in Fig. 20. The solid dots in Fig. 21(b) show that the paraexciton density determined in this way crosses the BEC phase boundary at a time between 6 and 11 ns, which coincides with the emergence of the narrow paraexciton spectrum.

In Fig. 22 we show an analysis of the paraexciton spectrum in terms

† That the orthoexciton and paraexciton gas occupy nearly the same volume during the (10 ns) excitation pulse is quite reasonable considering that there are frequent interparticle collisions to keep them together at high densities, and that the orthoexcitons convert to form a good fraction of the paraexcitons. An imaging experiment with high spatial resolution in an unstressed crystal has shown that the ortho and paraexciton gases coexist spatially at least up to $t = 25$ ns (see Ref. [13].)

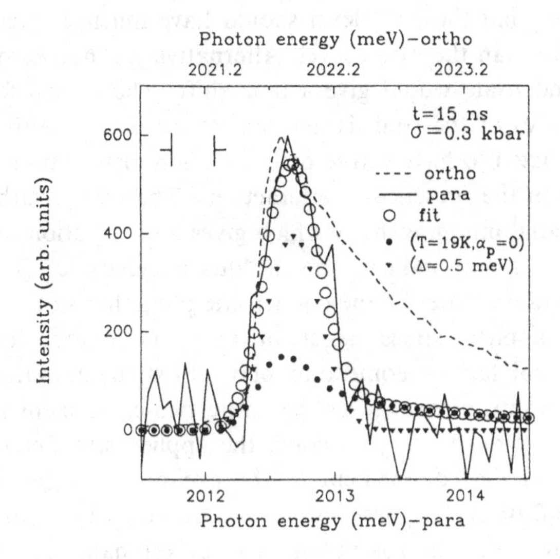

Fig. 22. The data of Fig. 19, fit to the BEC distribution discussed in the text.

of a Bose–Einstein condensate. The simple procedure of assuming a delta-function at $E = 0$, convolved with the slit function as illustrated in Fig. 10, does not yield a satisfactory fit. The extra luminescence at low energy, however, is well fit by a Gaussian line of with 0.5 meV width, blue shifted by 0.2 meV, convolved with the instrumental slit function of 0.3 meV. This is plotted as the triangles in Fig. 22. The distribution of thermally excited particles is blue shifted by the same amount in this fit, although a fit with no blue shift of the excited particles is also possible. The sum of the low-energy component and the $\alpha = 0$ excited-state component, with intensity determined by the ratio of ortho and para intensities in Fig. 20(a), is plotted as the open circles. According to Fig. 21(b), the condensate has a density which is about a factor of 3 above the phase boundary at 25–30 K and the density remains at about 10^{19} cm^{-3} over the measurement time, coinciding with a relatively constant blue shift of the low-energy component.

We are compelled to conclude that this new component is the long-sought excitonic condensate. The simplest explanation of the blue shift of the condensate is that it represents the repulsive interactions of the paraexcitons [48]. Indeed, the orthoexciton spectrum does not exhibit this blue shift, and this is consistent with the predictions of Moskalenko *et al.* [53] based on many-body theory that takes into account exciton–

exciton interactions, that the para level should have much greater blue shift at high density than the ortho level. Alternatively, a non-zero drift velocity of the condensate would give a blue shift. The reason for the finite width (0.5 meV) of the peak is unclear. A condensate with long-range order is predicted to have a true delta-function distribution in the ground state, even in the presence of interactions. The finite width may be the result of spatial inhomogeneity which gives a distribution of blue shifts of the gas, or it may indicate that a "quasicondensate" [47] has appeared but long-range order of the condensate phase has not.

Why does the applied stress assist in the condensation of the *para*excitons? Several factors come into play. First, by lowering the orthoexciton multiplicity, stress causes the entire two-component gas to have a higher level of degeneracy. Second, the applied stress can have a beneficial effect on the confinement of the excitons. As mentioned in Section 3, the diffusivity of paraexcitons is reduced as the symmetry of the crystal is lowered. The gas volume can be estimated by simply dividing the integrated intensity of the orthoexcitons (the open circles in Fig. 20) by the spectroscopically determined density at each time. The volume of the gas obtained this way increases slowly over the entire measurement interval. In contrast, the volume obtained by Snoke *et al.* [13] from spatial imaging in an *unstressed* crystal shows a considerable expansion during this same interval. Third, an increased exciton–phonon coupling under crystal stress allows more efficient cooling of the gas. The relative importance of these factors – or some not yet identified – may be determined by using magnetic field, rather than stress, to reduce the orthoexciton multiplicity.

11 Confinement of Excitons in a Parabolic Well

In all of the above, we have discussed BEC assuming a uniform potential, that is, "particles in a box". An interesting alternative situation is an ideal gas of bosons trapped in a parabolic potential of the form,

$$V(r) = Ar^2, \tag{27}$$

where r is the distance from the center of the well and A is the well parameter. As discussed previously [54] and by others in this volume, this kind of potential allows the interesting possibility of a *spatial* condensation as a telltale of BEC of both atomic and excitonic gases. In other words, BEC of particles trapped in a three-dimensional parabolic well is predicted to appear as a very narrow spatial distribution of particles at the exact energy minimum of the well, not only as a zero-kinetic-energy peak in the energy distribution.

In the low density, classical case, the chemical potential is the sum of the "internal" chemical potential (see the discussion of Eq. (22)),

$$\mu_{\rm int} = -k_B T \ln(n_Q/n), \tag{28}$$

and an "external" chemical potential given by $\mu_{\rm ext} = Ar^2$. In equilibrium, $\mu_{\rm int} + \mu_{\rm ext}$ = constant over the well, leading to the spatial distribution of the density,

$$n(r) = n(0)e^{-Ar^2/k_B T}, \tag{29}$$

a Gaussian profile. This expression is based on Maxwell–Boltzmann statistics. It turns out that the spatial distribution of non-condensed particles exhibiting Bose–Einstein statistics is not much different from this form [54].

The single-particle states of the harmonic-oscillator potential given above comprise an energy ladder, $E_s = (s + 3/2)\hbar\omega$, with $\omega = (2A/m)^{1/2}$ and $s = 1,2,\ldots$. Each level has a degeneracy of $(s+1)(s+2)/2$, implying that the density of states for $s >> 1$ is proportional to E^2. The full expression in this limit is

$$D(E) = \frac{gE^2}{2(\hbar\omega)^3}, \tag{30}$$

where g is the spin multiplicity of the particle. The chemical potential is determined by the condition $N(E) = \int D(E)f(E)dE$, where $f(E)$ is the distribution function (18). Setting $\mu = 0$ in this equation yields the critical condition for BEC,

$$N_c = 1.202g(k_B T/\hbar\omega)^3. \tag{31}$$

For a given temperature, when the total number of particles in the well exceeds N_c, the excess particles will theoretically occupy the ground state. Herein lies the qualitative departure from the uniform-potential case: since the ground state of the harmonic oscillator is a narrow Gaussian wavefunction centered at the bottom of the well, BEC corresponds to a spatial condensation rather than a k-space condensation.

The spatial extent of the ground state for an ideal gas, calculated simply as the wavefunction of a single particle in the ground state of the well, is so small that the condition (6) for the excitons to remain bosons, $na^3 \ll 1$, would not hold if a significant fraction of the gas were condensed in this state. In this case, one cannot ignore the effect of

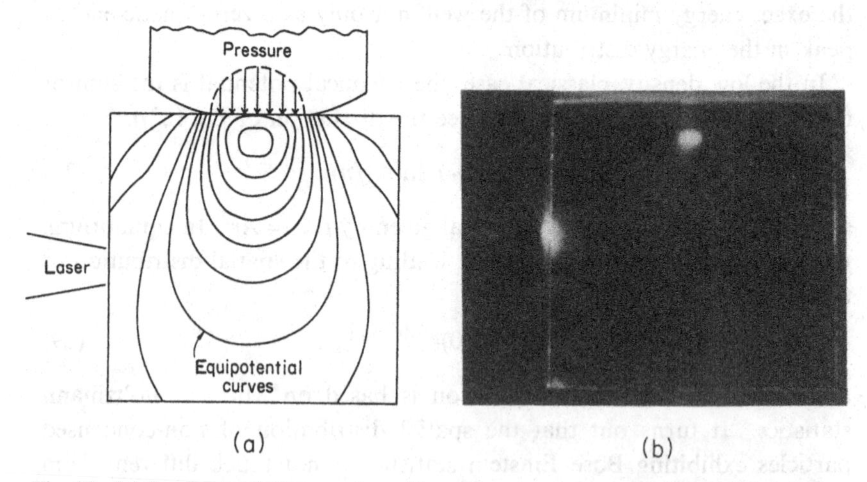

Fig. 23. (a) Equipotential curves for a Hertzian contact stress. (b) Drift of excitons into a three-dimensional parabolic well created in a Cu_2O crystal with Hertzian contact stress. This photoluminescence image of paraexcitons is obtained for continuous excitation at the left surface with an Ar^+ laser. (From Ref. [54].)

interactions among the particles in the condensate. As shown by Hijmans *et al.* [55], in the limit of weak interaction, assuming a condensate density much greater than non-condensate density near the center of the well, the spatial extent of the condensate is fixed by the relation

$$n_0(r)U + V(r) = \mu, \tag{32}$$

where U is the interaction vertex (proportional to the scattering length), $V(r)$ is the external potential and μ is the chemical potential determined by $n'(r_c)U + V(r_c) = \mu$, where n' is the excited-states density. The radius r_c of the region in which the condensate density is non-zero is determined by the condition that the integral over the condensate density profile over this region is equal to the number of condensate particles. The volume of the condensate therefore depends on the condensate fraction. Although the volume of the condensate taking into account interactions is much larger than the single-particle (non-interacting) ground state, so that the condition $na^3 \ll 1$ holds, it should still be much smaller than the excited-states spatial distribution and therefore should give a clear telltale for the presence of a condensate, if its lifetime remains sufficiently long.

A parabolic potential well can be created for excitons by non-uniformly stressing the crystal. Figure 23(a) shows a schematic of the equipotential

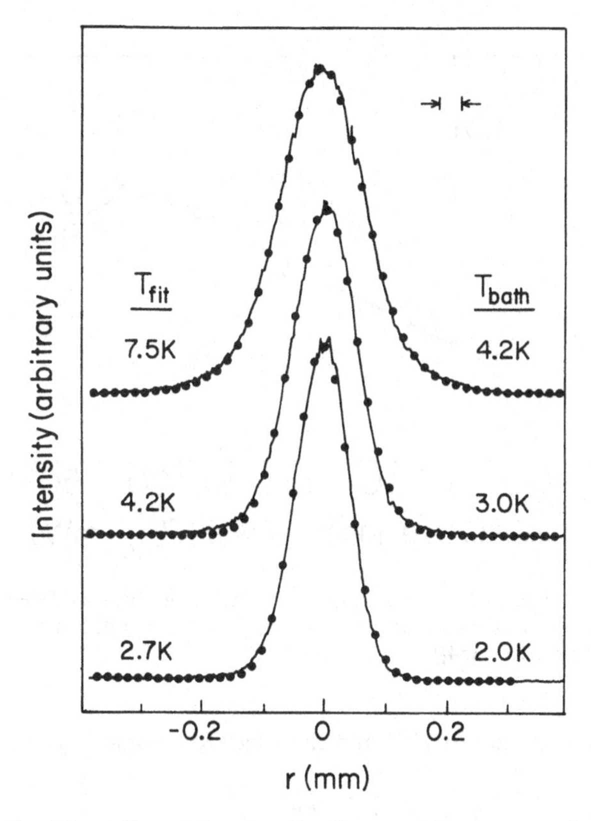

Fig. 24. Spatial profiles of the photoluminescence from paraexcitons in the potential well, during cw excitation with a dye laser. Here r is the distance from the center of the well. (From Ref. [54].)

curves caused by a Hertzian-contact geometry. A shear-stress maximum – or potential-energy minimum – is formed just below the contact area of the rounded stress rod. Part (b) of this figure shows the drift of excitons in Cu_2O towards the stress maximum from the surface where they are created. The bright spot is a large collection of excitons in the parabolic well. This "strain-confinement" technique has previously been exploited for studying excitonic phases in Ge and Si [56].

Figure 24 shows spatial profiles of the photoluminescence from (long-lived) paraexcitons in the potential well. The dots are fits to (29), adjusted for stress dependence of the paraexciton radiative efficiency. The fit temperatures are somewhat higher than the helium bath temperature, indicating some heating in the well. According to (29), the widths of these

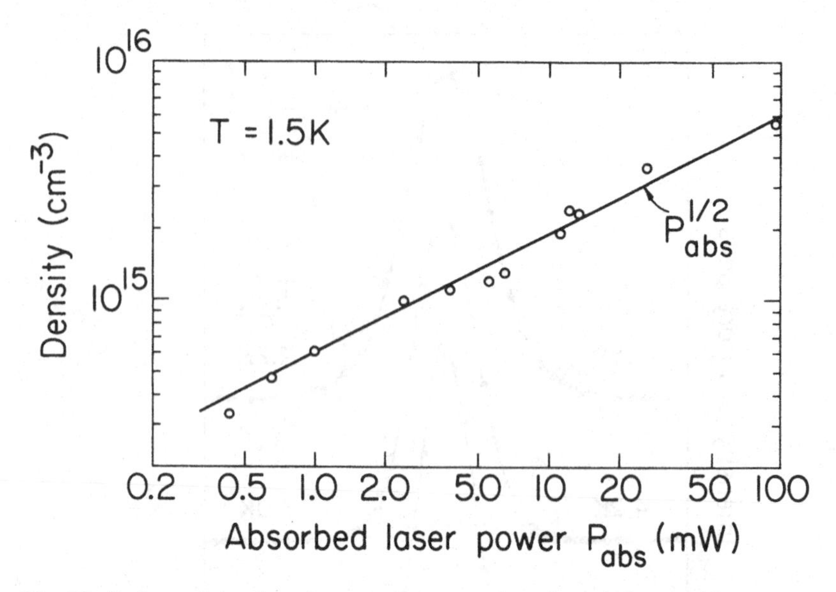

Fig. 25. Estimated density of paraexcitons as a function of absorbed laser power, when exciting the well directly with a dye laser. Here r is the distance from the centre of well. (From Ref. [54])

profiles are proportional to $T^{1/2}$ and the effective volume is given by

$$V_{\text{eff}} = \frac{\pi}{6}\left(\frac{2.77k_B T}{A}\right)^{3/2} . \tag{33}$$

Clearly, these excitons at relatively low density are behaving as a nearly ideal classical gas. The effective volume of the gas is considerably larger than in the "inertial-confinement" case discussed previously, so it is more difficult to reach the condition for BEC.† However, in the strain well the paraexcitons exhibit a much cooler temperature than in the inertial-confinement case.

When the well is excited directly with a dye laser, by tuning the laser wavelength exactly to the energy-gap minimum in the well, the gas density in the strain well can be obtained simply by measuring the number of absorbed photons. This estimate also requires the measured lifetime, τ, and effective volume, V_{eff}. The dependence of the gas density, n, on cw absorbed-power is plotted in Fig. 25. This $P_{\text{abs}}^{1/2}$ dependence is expected

† The condensation condition for the two cases can be compared by noting that $n_c \sim N_c/V_{\text{eff}}$.

for Auger-limited lifetimes, where

$$\frac{dn}{dt} = -Bn^2 + (\text{const})P_{\text{abs}} = 0 \tag{34}$$

under steady-state excitation. For the gas temperature in this experiment, the calculated critical density, $n_c \approx N_c/V_{\text{eff}}$, equals about 10^{17} cm^{-3}, which is one to two orders of magnitude above the highest experimental density. Trauernicht *et al.* [37] discuss the difficulties in reaching the phase boundary due to Auger recombination and heating of the gas. One of the indicated solutions is to lower the bath temperature in order to reduce the critical density. While quantum degeneracy has not yet been reached, the strain-well approach still looks promising for the creation and study of a quantum gas in a confining potential.

12 Conclusions

Despite their short lifetimes, free excitons in a high-purity crystal of Cu_2O form a nearly ideal gas of particles. The excitonic gas obeys the kinetic theory of gases, it moves via drift or diffusion through the crystal long after the generating laser is off, it can be trapped in a potential minimum, and on timescales longer than a nanosecond it exhibits a Maxwell–Boltzmann or Bose–Einstein energy distribution characteristic of an internal thermal equilibrium. Because the exciton lifetime is much longer than the particle scattering time, the laws of equilibrium thermodynamics apply. Indeed, the finite lifetime of the excitons has a distinct benefit: due to the radiative decay of excitons, it is possible to observe their energy distribution directly.

The search for BEC of excitons has progressed stepwise. First, it has been proven that excitons behave statistically as bosons– namely, the energy distribution narrows with increasing density, in contrast to the broadening of the fermion distribution with increasing density. Second, the BEC phase boundary has been mapped over more than an order of magnitude of density variation, via the "quantum saturation" effect. Third, it has been shown that the density of excitons can exceed the critical density for condensation at a given temperature, under certain conditions. Finally, an extremely narrow energy distribution ($\Delta E \ll k_B T$) of paraexcitons in the ground state has been observed under conditions of $n > n_c$.

A full understanding of these results awaits a complete theoretical treatment of the weakly interacting exciton gas, including the hydrodynamics of the expanding boson gas and the energy distribution of a finite-lifetime, two-component gas with phonon and Auger interactions. Perhaps the most significant lack in present theory is an exact calculation or an accurate measurement of the exciton–exciton interaction potential: the proper form can not be simply extrapolated from that of hydrogen, due to the important effects of the light mass of the hole and the electron–hole exchange. This interaction potential goes into the calculation of the biexciton binding energy, the ortho–para splitting and interconversion, and exciton–exciton scattering.

Nevertheless one of the beauties of this system is the fact that the laws for an ideal Bose gas work so well over such a large range of density and temperature, attested by the fact that the spatial diffusion data and the spectra are well fit using simple Maxwell–Boltzmann or Bose–Einstein distributions. All of this analysis supports the conclusion at this point that paraexcitons do indeed undergo Bose–Einstein condensation.

Acknowledgments. We acknowledge the essential collaboration of A. Mysysrowicz. The Illinois work is supported by National Foundation Grant DMR 92-07458. One of us (D.W.S.) thanks the Aerospace Sponsored Research program for support for this work.

References

[1] S. Nikitine, J.B. Grun, and M. Sieskind, J. Phys. Chem. Solids **17**, 292 (1961).
[2] M. O'Keefe, J. Chem. Phys. **39**, 1789 (1963).
[3] A. Gotzene and C. Schwab, Solid State Comm. **18**, 1565 (1976).
[4] J.W. Hodby *et al.*, J. Phys. C **9**, 1429 (1976).
[5] P.Y. Yu and Y.R. Shen, Phys. Rev. Lett. **32**, 939 (1974); Phys. Rev. B **12**, 1377 (1975).
[6] D.W. Snoke, D. Braun, and M. Cardona, Phys. Rev. B **44**, 2991 (1991).
[7] V.T. Agekyan, Phys. Stat. Sol. (a) **43**, 11 (1977).
[8] P.D. Bloch and C. Schwab, Phys. Rev. Lett. **41**, 514 (1978).
[9] E.g. A.L. Fetter and J.D. Walecka, *Quantum Theory of Many-Particle Systems* (McGraw-Hill, New York, 1971), p. 259.
[10] P.L. Gourley and J.P. Wolfe, Phys. Rev. B **25**, 6338 (1982).
[11] M.M. Beg and S.M. Shapiro, Phys. Rev. B **13**, 1728 (1976).
[12] Y. Petroff, P.Y. Yu, Y.R. Shen, Phys. Rev. Lett. **29**, 1558 (1972); Phys. Rev. B **12**, 2488 (1975).

[13] D.W. Snoke, J.P. Wolfe and A. Mysyrowicz, Phys. Rev. B **41** 11171 (1990).
[14] K. Reimann and K. Syassen, Phys. Rev. B **39**, 11113 (1989).
[15] The measured relative absorption ratio of 50 in Ref. [8]is reduced by a factor of three to take into account the orthoexciton spin multiplicity.
[16] J.L. Birman, Sol. State Comm. **13**, 1189 (1978).
[17] F. Bassani and M. Rovere, Solid State Comm. **19**, 887 (1976).
[18] D.W. Snoke, A.J. Shields, and M. Cardona, Phys. Rev. B **45**, 11693 (1992).
[19] D.W. Snoke, J.P. Wolfe, and D.P. Trauernicht, Phys. Rev. B **41**, 5266 (1990).
[20] J.S. Weiner *et al.*, Solid State Comm. **46**, 105 (1983).
[21] A. Mysyrowicz, D. Hulin, and A. Antonetti, Phys. Rev. Lett. **43**, 1123 (1979).
[22] K. Reimann, J. Lum. **40/41**, 475 (1988); A.V. Akimov and A.A. Kaplyanskii, *Proc. 5th Int. Conf. on Phonon Scattering*, A.C. Anderson and J.P. Wolfe, eds. (Springer, Berlin, 1986), p. 449.
[23] D.P. Trauernicht, J.P. Wolfe, and A. Mysyrowicz, Phys. Rev. Lett. **52**, 855 (1984).
[24] D.P. Trauernicht and J.P. Wolfe, Phys. Rev. B **33**, 8506 (1986).
[25] E.g. D.I. Blokhintsev, *Quantum Mechanics* (Dordrecht-Holland, Amsterdam, 1964), p. 493ff.
[26] E.g. F.K. Richtmyer, E.H. Kennard, and John N. Cooper, *Introduction to Modern Physics*, 6th edition (McGraw-Hill, New York, 1969) p. 571.
[27] For a review see J.C. Hensel, T.G. Phillips, and G.A. Thomas, Solid State Physics **32**, H. Ehrenreich, F. Seitz, and D. Turnbull, eds. (Academic, New York, 1977).
[28] T.J. Greytak and D. Kleppner, in *New Trends in Atomic Physics*, G. Grynberg and R. Stora, eds., (North-Holland, Amsterdam, 1984).
[29] R.C. Casella, J. Phys. Chem. Solids **24**, 19 (1963).
[30] A. Mysyrowicz, D. Hulin, and E. Hanamura, in *Ultrafast Phenomena VII*, (Springer, Berlin, 1990).
[31] H. Haug and H.H. Kranz, Z. Phys. B **53**, 153 (1983).
[32] A. Quattropani and J.J. Forney, Il Nuovo Cimento **39**, 569 (1977).
[33] A.I. Bobrysheva and S.A. Mosklenko, Phys. Stat. Sol. (b) **119**, 141 (1983); S.A. Moskalenko *et al.*, J. Phys. C **18**, 989 (1985); Phys. Stat. Sol. (b) **129**, 657 (1985); A.I. Bobtysheva *et al.*, Phys. Stat. Sol. (b), **167**, 625 (1991).

[34] D.W. Snoke and J.P. Wolfe, Phys. Rev. B **39**, 4030 (1988).

[35] D.W. Snoke, W.W. Rühle, Y.-C. Lu, and E. Bauser, Phys. Rev. B **45**, 10979 (1992).

[36] *Semiconductors Probed by Ultrafast Laser Spectroscopy* (Academic Press, New York, 1984).

[37] D.P. Trauernicht, A. Mysyrowicz, and J.P. Wolfe, Phys. Rev. B **28**, 3590 (1983); D.P. Trauernicht, J.P. Wolfe, and A. Mysyrowocz, Phys. Rev. B **34**, 2561 (1986).

[38] D.W. Snoke and J.P. Wolfe, Phys. Rev. B **42**, 7876 (1990).

[39] D.W. Snoke, J.-L. Lin, and J.P. Wolfe, Phys. Rev. B **43**, 1226 (1991).

[40] Bose narrowing was shown for excitons in stressed Ge by V.B. Timofeev, V.D. Kulakovskii, and I.V. Kukushkin, Physica B+C **117/118**, 327 (1983).

[41] D. Hulin, A. Mysyrowicz, and C. Benoit a la Guillaume, Phys. Rev. Lett. **45**, 1970 (1980).

[42] D.W. Snoke, J.P. Wolfe, and A. Mysyrowicz, Phys. Rev. Lett. **59**, 827 (1987).

[43] A. Mysyrowicz, D. Hulin, and C. Benoit a la Guillaume, J. Lum. **24/25**, 629 (1981).

[44] L.V. Gregor, J. Phys. Chem. **66**, 1645 (1962).

[45] D.W. Snoke, University of Illinois at Urbana-Champaign Ph.D. thesis, 1990.

[46] For a hydrodynamic model of excitons in Cu_2O without Auger recombination and phonon emission, see B. Link and G. Baym, Phys. Rev. Lett. **69**, 2959 (1992).

[47] See the contribution of Yu. Kagan in this volume.

[48] H. Shi, G. Verechaka and A. Griffin, Phys. Rev. B **50**, 1119 (1994) have derived an expression for the decay luminescence spectrum for an excitonic BEC in the weakly interacting gas approximation, incorporating the low-energy tail and the blue shift of the spectrum.

[49] J.-L. Lin and J.P. Wolfe, Phys. Rev. Lett. **71**, 1223 (1993).

[50] R.G. Waters *et al.*, Phys. Rev. B **21**, 1665 (1980).

[51] H.-R. Trebin, H.Z. Cummins, and J.L. Birman, Phys. Rev. B **23**, 597 (1981).

[52] A. Mysyrowicz, D.P. Trauernicht, J.P. Wolfe, and H.-R. Trebin, Phys. Rev. B **27**, 2562 (1983).

[53] S.A. Moskalenko *et al.*, Phys. Stat. Sol. (b) **129**, 657 (1985).

[54] See D.P. Trauernicht, J.P. Wolfe and A. Mysyrowicz, Phys. Rev. B **34**, 2561 (1986).

[55] T.W. Hijmans, Yu. Kagan, G.V. Shlyapnikov, and J.T.M. Walraven, to be published; see also D.A. Huse and E.D. Siggia, J. Low Temp. Phys. **46**, 137 (1982) and V.V. Goldman, I. Silvera, and A.J. Leggett, Phys Rev. B **24**, 2870 (1981).

[56] J.P. Wolfe and C.D. Jeffries, in *Electron–Hole Droplets in Semiconductors*, C.D. Jeffries and L.V. Keldysh, eds. (North-Holland, Amsterdam, 1987).

14

Bose–Einstein Condensation of Excitonic Particles in Semiconductors

A. Mysyrowicz
LOA-ENSTA
Ecole Polytechnique
F-91120 Palaiseau
France

Abstract

The case of excitons as candidates for Bose–Einstein condensation is discussed, and experimental results in CuCl and Cu_2O are presented. In CuCl, spectral analysis of the luminescence from biexcitons as a function of their density reveals a gradual evolution from classical statistics towards a quantum degenerate regime. The appearance of a sharp emission line below a critical temperature and above a critical density is attributed to the presence of a laser-induced Bose–Einstein condensate of excitonic molecules. This interpretation is supported by pump-probe experiments which show that additional particles injected in the presence of a biexciton condensate are drawn into it.

In Cu_2O, free exciton luminescence spectral analysis of ortho- and paraexcitons reveals a gradual evolution from a classical to a Bose quantum degenerate regime with increasing particle densities. Orthoexciton densities close to the critical density for condensation are obtained at high incoherent excitation. Under similar pumping, paraexciton densities exceeding the critical value are inferred from luminescence intensity ratios. Anomalous transport properties of paraexcitons, such as ballistic propagation over macroscopic distances and formation of soliton-like excitonic packets are discussed as evidence for excitonic superfluidity.

1 Introduction

1.1 Excitons

The lowest electronically excited state of a non-metallic crystal corresponds to the promotion of one electron from the top of the highest fully occupied valence band to the bottom of the next empty conduction band. A correct evaluation of the required energy must include the Coulomb correlation between the promoted electron and all other electrons left behind in the valence band. A very fruitful conceptual approach (see the article by Wolfe *et al.* in this volume) consists in treating this problem by the equivalent point of view of the Coulomb interaction between a negatively charged particle, the electron in the conduction band and a positively charged particle, the electron vacancy or positively charged hole left in the valence band. The problem reduces to the same two-particle Schrödinger equation as for the hydrogen or positronium atom,

$$\left[-\frac{\hbar^2}{2m_c}\nabla_e^2 - \frac{\hbar^2}{2m_h}\nabla_h^2 - \frac{e^2}{\varepsilon_0|r_e - r_h|}\right]\Phi = E\Phi . \tag{1}$$

It is well-known that discrete solutions exist corresponding to bound electron–hole pair states. These bound states are called excitons and obey the hydrogenic relation

$$E_{n_p} = E_G - E_B\frac{1}{n_p^2} + \frac{\hbar^2 K^2}{2M_x}, \tag{2}$$

where $E_B = m_r e^4/2\hbar^2\varepsilon_0^2 = e^2/2a_x\varepsilon_0 = \hbar^2/2m_r a_x^2$ is the exciton binding energy, $a_B^2 = \varepsilon_0\hbar^2/m_r e^2$ is the Bohr radius and $m_r = m_e m_h/(m_e + m_h)$ is the reduced effective mass, ε_0 is the static dielectric constant of the medium, n_p is the principal quantum number, m_e and m_h are the effective electron and hold masses, and $M_x = m_e + m_h$ is the translational mass of the exciton [1]. Thus, in this point of view, supply of one quantum of excitation energy to the crystal corresponds to the creation of one exciton, a quasi-particle with the same internal energy structure as the hydrogen atom or the positronium atom. For optically active excitons, this energy supply can occur through absorption of a photon which occurs typically over a penetration depth less than 1μm.

As a function of photon energy, the absorption leads to a hydrogenic series of absorption lines below the energy band gap converging towards an ionization continuum of free electron–hole pairs at the band gap energy. It is worth noting the difference between this absorption mechanism and that occurring in atomic spectroscopy. In crystals the absorption of light corresponds to the creation of excitons and not to electronic transitions between terms of a series (this last process can occur with excitons but the corresponding photon energy lies in the infrared region, in view of the small exciton binding energy $\approx$ meV) [1]. It is also important to keep in mind that optically inactive excitons exist in all semiconductors. The number of possible excitonic states which can be formed with a given set of valence and conduction bands is obtained from group-theory arguments; there are at least two: a paraexciton with opposite electron and hole spins which is optically inactive and an orthoexciton with parallel electron and hole spins which can be optically active. Even if they do not contribute to the optical response of the material near the band gap, paraexcitons are nevertheless formed when the crystal is excited in the ionization continuum. Indeed, after creation of unbound electron–hole pairs, fast relaxation of the excess energy of the free carriers occurs on a subpicosecond time-scale through emission of phonons followed by association of the free electrons and holes into the various varieties of excitons. Of prime concern in the context of BEC is the fact that optically active excitons are generally not the species lowest in energy and therefore are not the best candidates for Bose condensation.

Besides an internal energy structure similar to that of hydrogen or positronium atoms, excitons possess another attribute of real particles, namely the ability to move freely inside the crystal. In this motion, the center-of-mass of the electron–hole pair propagates through the crystal carrying an electronic energy of the order of the band gap. Under normal circumstances (low particle densities), the exciton transport is diffusive with a diffusive coefficient being determined in high purity samples by scattering with lattice phonons.

Local exciton densities reaching 10^{19} cm^{-3} can be achieved easily by optical pumping of the crystal with modest pulsed lasers, since the excited volume of the crystal can be as small as 10^{-10} cm^3. For an exciton of lifetime 10^{-9}s, an optical pulse of the same duration with an energy content of a few nanojoules should be sufficient in principle.

1.2 Biexcitons

It is possible to pursue the analogy between excitons and positronium or hydrogen atoms one step further and consider the chemistry occurring at higher densities when excitons start to interact. Like hydrogen atoms, excitons can form molecules called biexcitons. The formation of a biexciton due to the coupling of a pair of excitons and its analogy to a hydrogen molecule has been experimentally well established. An excitonic molecule in its ground state is similar to a hydrogen paramolecule with two holes of opposite spins sharing two electrons, also with opposite spins. It can also propagate through the sample, carrying a total energy:

$$E_{xx} = 2E_x^0 - D_{xx} + \frac{\hbar^2 K_{xx}^2}{2M_{xx}} = E_{xx}^0 + \frac{\hbar^2 K_{xx}^2}{4M_x}, \tag{3}$$

where $2E_x^0 = 2(E_g - B_x)$ is the internal energy of two noninteracting excitons at zero wave vectors, E_{xx}^0 is the biexciton energy at zero wave vector, D_{xx} is the biexciton binding energy (the energy required to dissociate it into two free excitons), and K_{xx} and M_{xx} are the center-of-mass wave vector and mass of the biexciton, respectively.

The binding energy of the biexciton depends on the ratio of effective masses, $\sigma = m_e/m_h$ [2]. The limiting case $\sigma = 0$ corresponds to the hydrogen molecule, with a well-established experimental binding energy. No experimental verification of the other limit, $\sigma = 1$ (molecule of positronium) exists yet. However, variational calculations predict a biexciton binding energy smaller by one order of magnitude.

Due to the small mass of excitons, typically of the order of the free electron or less, quantum fluctuations are very important in determining the internal structure of biexcitons. The biexciton zero-point energy is comparable to the molecular binding energy itself. As a consequence, vibrational levels are not stable and at most one excited level exists in crystals with simple band structures (the orthobiexciton in the first rotational level). This is in sharp contrast to the rich manifold of rotational and vibrational levels of hydrogen molecules. This simplicity of the molecular excitation spectrum is favorable for Bose condensation since it prevents formation of a molecular liquid phase, as will be discussed shortly.

Biexcitons cannot be created by optical means directly into their ground state through a one-photon absorption process, since the required photon energy is almost twice the band gap energy and therefore corresponds to a region of strong absorption in the exciton ionization continuum. It is possible to create biexcitons directly from the crystal ground state by two-photon absorption. As shown by Hanamura [3], the two-photon absorption cross section corresponding to the direct creation of a parabiexciton has giant values, of the order $\beta \approx 1$ cm· MW^{-1}, or 5–6 orders of magnitude more than for typical two-photon interband transitions in crystals. Such huge values reflect the fact that a two-photon process is well adapted to the creation of a biexciton, because two electrons separated by a mean distance equal to the biexciton radius must be simultaneously promoted from the valence to the conduction band. Consequently, densities of biexcitons reaching 10^{19} cm^{-3} can be obtained with a laser of intensity less than 1 MW/cm^2 [4]. In the context of Bose condensation, it is important to note that the biexciton population is nearly spatially uniform, since the two-photon attenuation of a laser decreases as a function of crystal thickness z as

$$I(z) = \frac{I_0}{1 + \beta z I_0}, \tag{4}$$

instead of the usual exponential attenuation from Beer–Lambert law in a one-photon process. If two independent laser beams are used to generate biexcitons directly, the wavevector of the created particles can be controlled between $2K_0$ and 0 by changing the angle between the incoming beam, where K_0 is the magnitude of the incident photon wavevector.

Biexcitons can also be created indirectly through free carrier (or exciton) generation followed by biexciton formation [5]. In a steady state regime, the ratio of biexciton to exciton populations N_{xx}/N_x is dictated by the values of their respective chemical potentials. Biexciton densities reaching 10^{20} cm^{-3} have been obtained through indirect band-to-band pumping [6]. Note however that under indirect creation process, the biexciton density is spatially nonuniform since it is proportional to the local exciton density, itself spatially nonuniform. Also, the excess pump photon energy leads to local heating of the exciton and biexciton gas.

2 Bose–Einstein Condensation of Excitonic Particles

Blatt *et al.* [7] and, independently, Moskalenko [8] were the first to point out that excitons consisting of an even number of fermions have an integer spin value and as such are candidates for Bose condensation. Theories of excitonic condensation (and superfluidity) have been given by Keldysh and Kozlov [9], Haug and Hanamura [10] and Comte and Nozières [11]. In the theories of Keldysh and Nozières and coworkers, the starting point consists of a system of degenerate electrons and holes of arbitrary density, the Bose model of ideal particles being just the low density limit. At high carrier densities, the treatment is formally similar to the BCS theory of superconductivity except that collective pairing of electrons and holes occurs via Coulomb attraction (instead of the phonon-mediated Cooper pairing of electrons in metals), making the system unstable against the formation of a dielectric superfluid at sufficiently low temperatures. Thus, it appears that excitonic superfluidity is not restricted by the condition $n_x a_x^3 \ll 1$ but should be possible over a wide range of electron–hole densities at low temperature. It must be stressed, however, that occurrence of excitonic superfluidity in the high density limit necessitates a direct band gap, because collective pairing requires approximately equal but opposite electron and hole momenta.

Haug and Hanamura [10] have treated the problem in the spirit of the Bogoliubov model, in the limit $n_x a_x^3 \ll 1$, in which excitons are viewed as approximate point bosons. Several assumptions are implicit in this model. A first assumption is a predominantly repulsive exciton–exciton potential. Otherwise, excitons aggregate at increasing densities into more complex entities, the same problem which prevents Bose condensation of a normal gas of hydrogen molecules. In crystals with indirect band gap structure (with the extreme of valence and conduction band at different points of the Brillouin zone), it is energetically favorable for the system to separate into a low density exciton phase coexisting with regions of a high density phase of a two-component electron–hole liquid [12]. An increase of injected excitons in this case merely increases the size of the electron–hole liquid drops, without changing the free exciton density. Therefore, unless some artifice is used to prevent exciton liquefaction, Bose condensation in the dilute limit is not possible in indirect gap materials. In direct gap materials with simple band structure, where exciton liquefaction into drops is not expected [12], the direct and exchange Coulomb interactions leads to an overall attraction and in principle to biexciton formation. However, a more detailed examination is required when the ratio of electron to hole mass $\sigma \approx 1$. In crystals with large electron–hole exchange energy, the biexciton can become unstable against dissociation in two paraexcitons [13].

On the other hand, in the other limit $\sigma \ll 1$, the formation of excitonic molecules is unavoidable and is experimentally well established. Biexcitons, with four fermi constituents, also fall into the category of bosons and therefore can undergo Bose condensation themselves. It is therefore necessary to examine the biexciton–biexciton potential in order to determine whether the van der Waals attraction between molecules can lead to a first order phase transition into a dielectric liquid before BEC occurs in the gas phase. According to the quantum theory of corresponding states [14], this should not be the case if the ratio of zero-point kinetic energy to potential energy, expressed by the reduced quantum parameter $\eta = \hbar^2/M\varepsilon f^2$, is larger than 3.5, assuming a Lennard–Jones interaction potential $U(r) = 4\varepsilon[(f/r)^{12} - (f/r)^6]$. Here ε is the depth of the potential well and f the collision diameter. In actual excitonic systems η is much larger than 3.5 [12], because of the small particle mass (leading to high zero kinetic energy) and therefore the system should remain a gas even at $T = 0$. In that case, the results obtained for a weakly interacting exciton gas can be applied to the gas of excitonic molecules with little modification and Bose–Einstein condensation is expected for the biexcitons.

Another assumption in the theory is to treat excitons as particles of infinite lifetime. This condition is approximately fulfilled in most semiconductors to the extent that the exciton decay time $\tau \gg \tau_c$, where τ_c is the mean collision time between particles. Typically τ_c is of the order 10^{-11}s or less for densities above 10^{16} cm^{-3}. On the other hand, the time τ_{ac} required to establish true equilibrium with respect to the lattice temperature is in general found to be very long, $\tau \ll \tau_{ac}$, although a quasiequilibrium is brought about in a time comparable with the exciton-acoustic phonon relaxation time $\tau_{ac} \approx 10^{-9} - 10^{-10}$ s, even in the absence of interparticle collisions. An additional problem is the time required for an initially nonequilibrium system to undergo Bose condensation. The understanding of the dynamics of Bose condensation is the subject of current theoretical research (see the articles by Kagan and Stoof in this book).

Finally, a last assumption underlying the nearly ideal Bose gas treatment of excitons is a negligible interaction between excitons and the radiation field. If the exciton–photon coupling is strong, then the true elementary electronic excitations of the crystals are better described in terms of polaritons [15]. By inspection of Fig. 1, it is immediately apparent that no accumulation of polaritons near $K = 0$ is possible, and therefore Bose condensation of dipole active excitons is impossible [16]. Excitonic molecules, on the other hand, do not suffer from this complication even if they are formed from dipole-active excitons, because they have a quadrupole moment and consequently do not interact in first order with the photon field.

3 Experimental Results in CuCl

This compound is a typical direct-band-gap semiconductor with dipole-allowed interband transitions [17]. The crystal structure is zincblende (cubic with no inversion symmetry) and belongs to the point group T_d. This type of band structure with no degeneracy of the valence and conduction bands near $K = 0$ is the simplest possible. The value of the forbidden gap is 3.43 eV at 4.2 K. The $n_p = 1$ exciton states formed with the upper valence band of symmetry Γ_7 and lower Γ_6

conduction band are split by electron-hole exchange interaction into a dipole-allowed orthoexciton of symmetry Γ_5 at 3.208 eV responsible for a strong absorption line and a lower lying optically inactive paraexciton Γ_2 state at 3.202 eV. The binding energy of the exciton in CuCl, $E_B =$ 190 meV, is very large when compared to most semiconductors, and the exciton radius is correspondingly small, $a_B = 0.7$ nm. The electron-to-hole mass ratio σ is small, $\sigma = 0.2$, favoring biexciton formation. The internal energy of the biexciton is 6.372 eV, so that the energy required to dissociate it into two free paraexcitons is $D_{xx} = 29$ meV. The biexciton effective mass is known with high precision, $M_{xx} = 5.29m_0$ [18]. The critical density for BEC estimated from the ideal gas model is 10^{17} cm^{-3} at 2 K. Because of its large binding energy, the biexciton is stable up to at least $T \approx 100$ K. Also the fluid of biexcitons has been observed to be stable up to densities $n \approx 10^{20}$ cm^{-3}, at which point a sharp Mott-like transition into a conducting plasma state is observed [6].

3.1 Biexciton Luminescence

An elementary step in the radiative decay process of a thermal biexciton is as follows. A biexciton disintegrates into a free exciton and a photon. The energy and momentum conservation conditions governing this decay process are

$$E_{xx}(K_{xx}) = \hbar\omega + E_x(K_x), \tag{5}$$

where

$$E_{xx}^0 + \frac{\hbar^2 K_{xx}^2}{4M_x} = \hbar\omega + \frac{\hbar^2 K_x^2}{2M_x}, \tag{6}$$

and

$$\mathbf{K}_{xx} = \mathbf{K}_x + \mathbf{q}, \tag{7}$$

where $\mathbf{q}$ is the emitted photon wave vector.

The biexciton luminescence lineshape can be expressed as

$$I(\hbar\omega) = C\int d^3K_{xx}\int d^3K_x|W_{ij}(\mathbf{K}_{xx},\mathbf{K}_x)|^2 f(\mathbf{K}_{xx}) \cdot\delta(\mathbf{K}_{xx}-\mathbf{K}_x-\mathbf{q})\delta[\hbar\omega-(E_{xx}-E_x)], \tag{8}$$

where W_{ij} is the matrix element for the transition between the biexciton and exciton branch, $f(K_{xx})$ is the biexciton statistical distribution, C is a constant which includes the collection efficiency to the detector, and the delta functions express the energy and momentum conservation. The dependence of W_{ij} on the direction of polarization vectors of the photons and excitons and the angle between emitted photon and restituted exciton can be found in Ref. [19].

In a crystal with dipole-allowed excitons, such as CuCl, it is necessary to take into account the coupling of the final exciton state with the radiation field. This affects the luminescence spectrum in two ways [15]. First, it leads to a splitting between longitudinal and transverse excitons which are the final states in the transition. The splitting is given by the relation

$$\frac{E_L^2(0)}{E_T^2(0)} = 1 + 2\pi\beta, \tag{9}$$

where the parameter β characterizes the oscillator strength of the excitonic transition. The longitudinal and uncoupled transverse exciton energies are

$$\begin{aligned} E_L(K) &= E_L(0) + \frac{\hbar^2 K^2}{2M_L}, \\ E_T(K) &= E_T(0) + \frac{\hbar^2 K^2}{2M_T}. \end{aligned} \tag{10}$$

Therefore, one expects two luminescence bands, terminating in the transverse and longitudinal branches, called M_T and M_L as shown in Fig. 1. Secondly, due to its coupling to the radiating field, the transverse exciton branch acquires a strong dispersion around K_0, where K_0 is the wavevector of a photon of the same energy as shown in Fig. 1. In this region of strong dispersion, the transverse exciton becomes a polariton (a mixed exciton–photon mode). In a classical gas most biexcitons have wavevectors $K_{xx} \gg K_0$ so that the polariton effect can be neglected. In this case, (8) reduces to the simple form

$$I(\hbar\omega) \propto (E_{xx}^0 - E_x^0 - \hbar\omega)^{1/2} e^{-(E_{xx}^0 - E_x^0 - \hbar\omega)/k_B T}. \tag{11}$$

Thus, the emission from a classical biexciton gas takes the form of two bands with inverted Maxwell–Boltzmann shape. Photons of lower energies correspond to decaying biexcitons with larger kinetic energy, and vice versa (see Fig. 1.) An example of a biexciton luminescence spectrum obtained in a thin crystalline CuCl film of high quality under weak pumping conditions such that the density of biexcitons $n_{xx} \ll n_{cr}$, together with a best fit assuming a classical distribution of biexcitons, is shown in the upper curve of Fig. 2 [19].

In a quantum degenerate regime, the polariton effect becomes important since most biexcitons have small wavevectors and the final state in

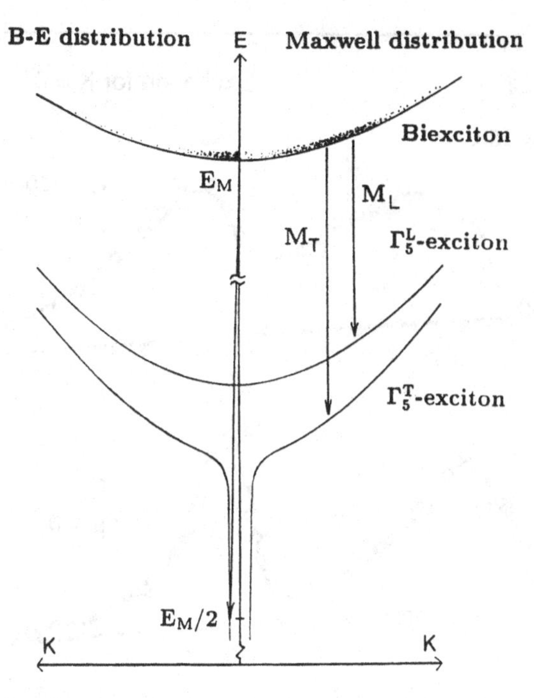

Fig. 1. Excitation diagram of a crystal with dipole-active excitons. The upper curve represents the dispersion of the biexciton. The lower two curves give the dispersion of the longitudinal exciton and the transverse polariton.

the transition correspond therefore to the region of strong dispersion of the polariton. The polariton dispersion is given by

$$\frac{\hbar^2 c^2 K^2}{\varepsilon_\infty E^2} = 1 + \frac{4\pi\beta}{1 - E^2/E_T^2(K)}. \tag{12}$$

An example of a high density biexciton luminescence spectrum recorded in a high quality 4.5 μm thin crystalline film of CuCl held at $T = 25$ K is shown in the lower curve of Fig. 2. In order to reduce heating of the film, the excess energy imparted to the biexciton gas is minimized by using two-photon excitation tuned to the biexciton resonance. This excitation scheme has the additional advantage of creating a spatially uniform population of biexcitons in the sample, thereby avoiding difficulties associated with particle migration. A large sample surface of radius $\approx 100 - 500$ μm is excited and only the emission from the central part is collected through a small pinhole of radius 10 μm put in direct contact

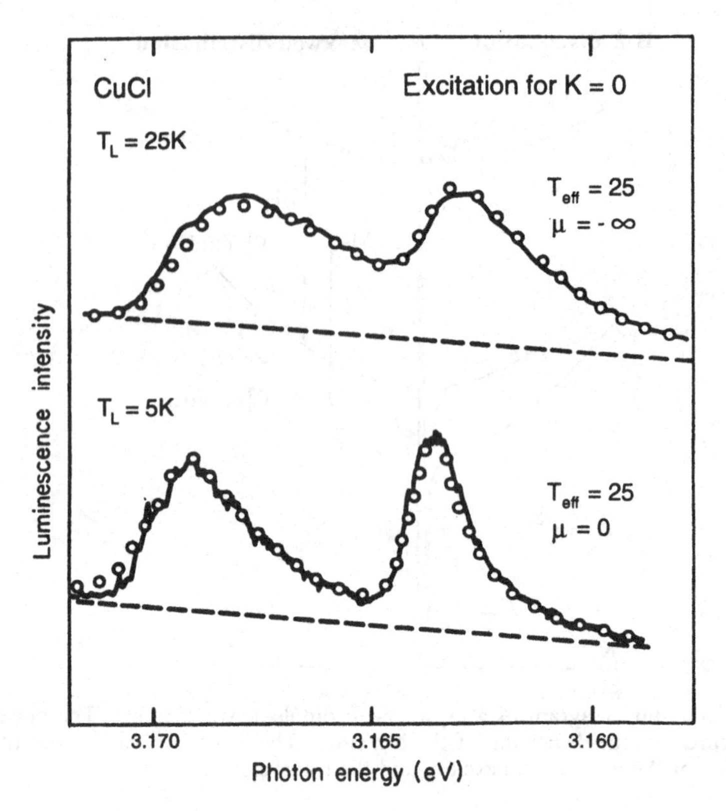

Fig. 2. Luminescence spectrum of CuCl under weak (upper) and strong two-photon excitation. The open circles are calculated lineshapes assuming a classical and a Bose quantum distribution with $\mu = 0$.

with the sample surface, so that lateral spatial nonuniformities become negligible [19]. A best fit to the measured spectrum is obtained by using a value $\mu = 0$ for the chemical potential in the Bose–Einstein function

$$f(E) = \frac{1}{e^{(E-\mu)/k_B T} - 1} \tag{13}$$

for the particle distribution, indicating that the critical density for BEC is reached. Additional measurements for intermediate pump intensities show a gradual evolution between these two limiting cases. Also, measurements as a function of sample temperature show a disappearance of quantum degeneracy in the biexciton population above $T \approx 40$ K, irrespective of the pump intensity, in good agreement with the predictions from the ideal Bose gas [19].

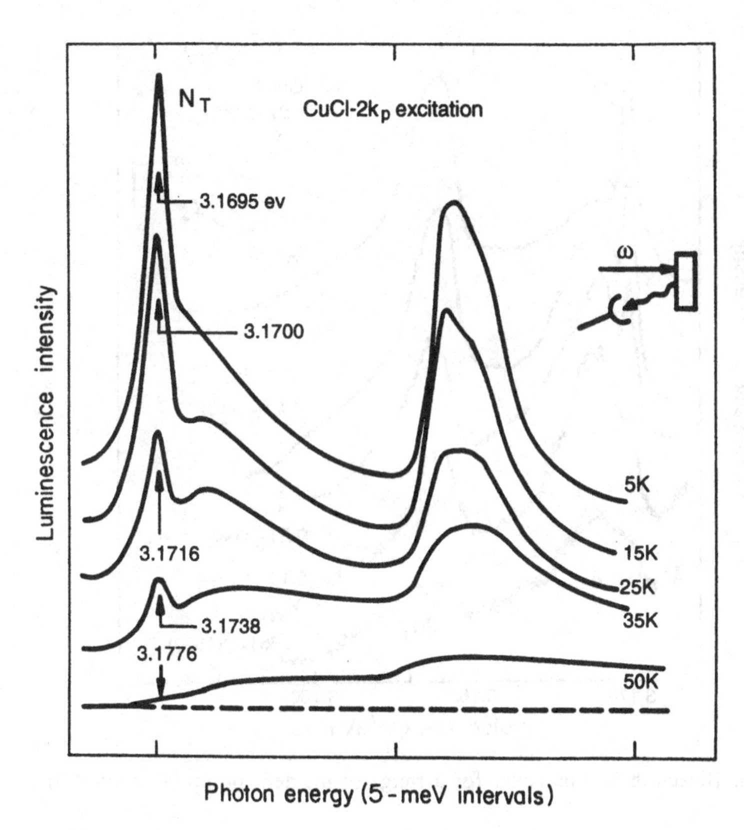

Fig. 3. Biexciton luminescence with single beam excitation for several temperatures of the sample. The spectra are shifted so that the sharp line N_T lines coincide. The shift is due to the change of band gap energy with T.

Above a critical input intensity a sharp line, denoted N_T, grows on the high energy side of the M_T band if a single-beam, resonant two-photon excitation is used. This line shows remarkable properties, such as a large spatial anisotropy, even though the crystal has cubic symmetry. It is intense in the backward detection geometry (with respect to the incident beam), but very weak in the forward geometry [19]. Further, the strength of the line shows a marked dependence upon sample temperature (see Fig. 3), with a threshold temperature for appearance varying with input excitation in the manner shown in Fig. 4. The appearance of N_T has been interpreted as evidence for the occurrence of Bose–Einstein condensation in the biexciton gas [19], with the condensation taking place at the wavevector imposed by the incident excitation. Since the radiative decay

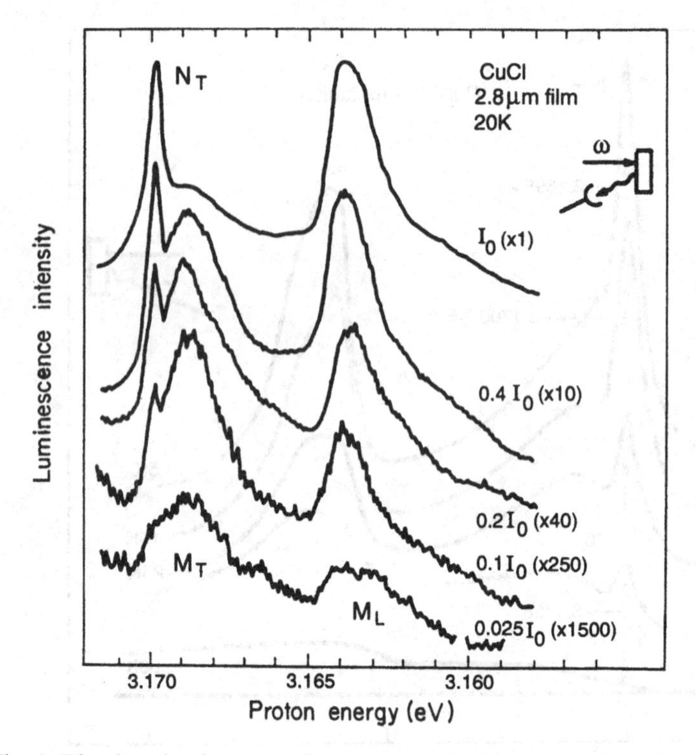

Fig. 4. Biexciton luminescence for a range of incident pump laser intensities.

of the condensate at $K_{xx} = 2K_0$ leaves a final exciton with a wavevector between K_0 and $3K_0$, depending on the direction of the emitted photon, the polariton aspect of the final exciton is now crucial. Inspection of the polariton dispersion curve of CuCl shows that in the backward detection geometry, the remaining transverse polariton at $3K_0$ is essentially exciton-like, with the complementary emitted photon lying on the high energy side of the M_T band. In the opposite limiting case (forward detection geometry), the final polariton is photon-like; in fact, the whole decay process is better described in this latter case as a two-photon decay, with both emitted photon-like polaritons having the same energy $\hbar\omega = E_{xx}/2$. This explains the spatial anisotropy of the condensate emission, directly reflecting the macroscopic occupation of a biexciton state with finite wavevector along a preferential direction.

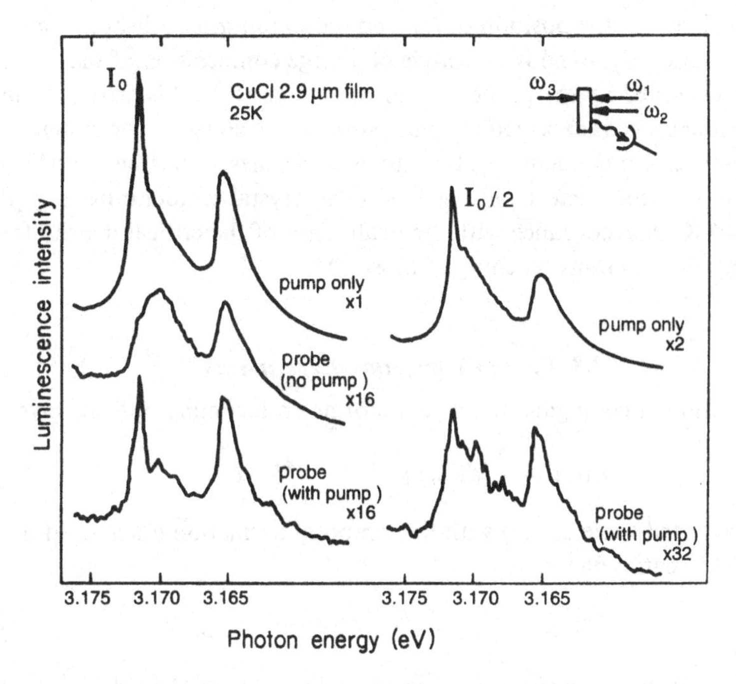

Fig. 5. Biexciton luminescence for excitation and detection conditions described in the text. Traces labeled probe (no pump) and probe (with pump) are obtained with synchronous differential detection.

3.2 Pump-Probe Experiments

The presence of a Bose condensate of biexcitons can be verified by pump and probe experiments. Two weak, counterpropagating probe beams, each of which is detuned by a small amount $E = 2$ meV to either side of the biexciton resonance, are used to inject a small probe density of biexcitons through the simultaneous absorption of one photon from each of the probe beams. The resulting molecular luminescence of Fig. 5 (left-hand side, middle curve) shows the expected inverted Maxwell–Boltzmann shape indicative of a low density gas. In a second step, a strong pump beam induces a condensate at $2K_0$. The corresponding luminescence spectrum displays the appearance of the narrow, spatially anisotropic line due to the luminescence from the condensate (Fig. 5, left-hand side, upper curve). If the probe biexcitons are now injected in the presence of the intense pump beam, the recombination radiation from probe particles only (as detected with differential techniques) is substantially modified (Fig. 5, left-hand side, lower curve). The observation of a sharp emission at the position of N_T and the simultaneous disappearance of the broader M_T band is indicative of a large compression of the phase volume occupied by the probe-excited biexcitons [19]. This attraction in momentum space is expected if probe particles are added to the system in the presence of a condensate of excitonic molecules. It has been verified that this attraction effect only occurs if the crystal temperature is kept below 40 K, in accordance with the evaluation of the critical density for a density of biexcitons reaching at most 10^{20} cm^{-3}.

3.3 Phase Conjugation Experiments

By definition, a conjugate mirror transforms an incoming optical wave

$$E(r,t) = \frac{1}{2}[e_0(x,y)e^{i(\omega t - K_0 z + \varphi_0)} + c.c.] \tag{14}$$

into a reflected wave $E_r(r,t)$ with the temporal oscillation unchanged but a reversed spatial phase

$$E_r(r,t) = \frac{1}{2}[e_0^*(x,y)e^{i(\omega t + K_0 z - \varphi_0)} + c.c.]. \tag{15}$$

This amounts to an operation of time inversion performed on the incident wave, leaving the spatial part unchanged, as can be seen immediately by writing the complete expression for the waves explicitly.

$$\begin{aligned} E_r(r,t) &= \frac{1}{2}[e_0^*(x,y)e^{i(\omega t + K_0 z - \varphi_0)} + e_0(x,y)e^{-i(\omega t + K_0 z - \varphi_0)}] \\ &= E(r,-t) \end{aligned} \tag{16}$$

Therefore, the reflected wave $E_r(r,t)$ behaves as if it retraces the path of the incoming wave back in time, as in a movie played backwards. This intriguing property leads to automatic compensation, after reflection, of distortions of the phase front experienced by an incident wave on its way to the conjugate mirror. It is possible to take advantage of this property to perform spectroscopy in the space domain instead of the frequency domain. With a small aperture one can discriminate a coherent wave of well-defined K-vector reflected by conjugate mirror from other radiations of the same frequency but different K-vectors originating from the same mirror (see Fig. 6). A true phase conjugate mirror is obtained by inducing a coherent, spatially uniform oscillation at frequency ω. The medium acts as a two-photon amplifier, amplifying an incident beam (ω, K_i). In nonlinear optics, the macroscopic oscillation at 2ω is induced by two counterpropagating beams of the

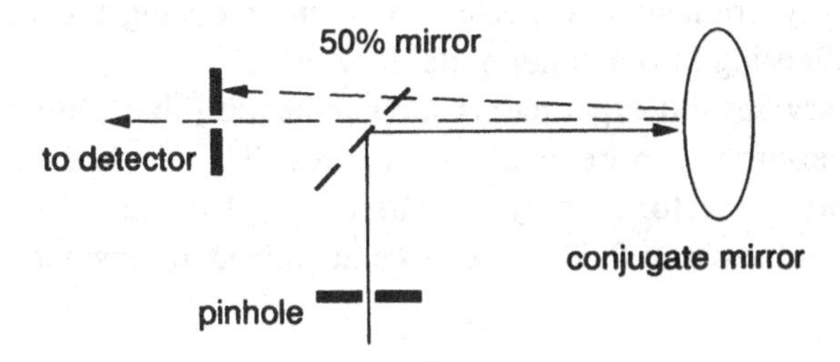

Fig. 6. Schematic diagram of a phase conjugate mirror allowing spatial discrimination of the emission from the mirror.

same intensity and frequency ω. However, a Bose condensate of biexcitons at $K = 0$ can also provide a coherent oscillation at 2ω and consequently fulfil the same function. Therefore by detecting a phase conjugate signal $(\omega, -K_i)$ upon irradiation of a crystal with an incident probe beam (ω, K_i), one obtains evidence for the presence of a biexciton condensate.

This method has been recently implemented in CuCl [20]. A new aspect of the experiment is the pumping scheme used to obtain a Bose–Einstein condensation of biexcitons. In the experiment, described in more detail by Hasuo *et al.* in this volume, an incoherent pump source is used to create the biexciton population indirectly, via exciton generation followed by molecular pairing. An increase of phase conjugate signal is observed, indicative of an increased coherent oscillation at 2ω. This is at first sight surprising, since injection of random particles should normally lead to an effective collision-induced dephasing of a prepared coherent state. This opposite trend is attributed to condensation of part of the biexciton population. However, analysis of the conjugate reflectance indicates a small condensate fraction, of the order of a few percent at best, despite the fact that the estimate local density of biexcitons should reach values comparable to those obtained with resonant pumping in the luminescence experiments described above. In the opinion of the author, the origin of the observed saturation of the condensed fraction at highest pump intensities results from two detrimental factors. First, the indirect pumping scheme delivers excess energy per absorbed photon, leading to local heating of the excitonic fluid mixture. Second, in contrast to a resonant two-photon excitation scheme with homogeneous volume excitation, the density of biexcitons is spatially nonuniform, being initially confined to the proximity of the sample surface. Fluid expansion driven by the density gradient is expected to occur, reducing the number of particles collapsing into a condensate at $K = 0$.

In summary, several experiments show indication that a Bose condensation of biexcitons can be achieved in CuCl. The condensate is stable even at finite K vectors, implying that a superfluid motion of a spatially uniform biexciton fluid takes place at sufficiently low temperatures $T < 40$ K.

4 Experimental Results in Cu_2O

This compound is well known for the variety of its excitonic features which provide a model case for the theory of Wannier excitons [21]. The crystal structure is cubic, with inversion symmetry. Consequently, parity conservation rules apply in the optical transitions. The band structure of Cu_2O consists of a non-degenerate upper valence band (symmetry Γ_7^+) and a lower conduction band (Γ_6^+) with extrema located at the center of the Brillouin zone at $K = 0$ and separated by an energy gap of 2.175 eV at 2 K. The $n = 1$ exciton is split, because of electron–hole exchange interaction, into a lower lying singly degenerate Γ_2^+ paraexciton ($X_p = 2.022$ eV at 2 K) and a triply degenerate Γ_5^+ orthoexciton ($X_0 = 2.034$ eV at 2 K) [22]. As in CuCl, the exciton binding is large, $E_B \simeq 150$ meV, and the corresponding exciton Bohr radius is small, $a_x = 0.7$ nm. The large electron–hole exchange energy $E_{xc} = 12$ meV and nearly equal electron-to-hole effective mass ratio $\sigma = m_e/m_h = 0.73$ conspire to prevent biexciton formation [13], so that free paraexcitons remain the particles with lowest energy per electron–hole pair even at high exciton densities. Because of the positive parity of both valence and conduction bands, ortho- and paraexcitons do not interact, in the dipole approximation, with the radiation field. As a consequence, the complication introduced by the polariton effect is avoided and both species acquire a long radiative lifetime.

The orthoexciton decay time τ_0 is limited at low temperatures $T <$ 10 K by ortho-paraexciton down conversion to a value $\tau_0 \approx \gamma^{-1}$, where $\gamma = 10^8 s$ is the ortho-para conversion rate [23]. A much longer decay time is measured at higher temperatures, $\tau_0 = 1.7 \mu s$ at 70 K [23]. At this temperature the populations of orthoexcitons and paraexcitons are in thermal equilibrium, so that the effective decay time for both species is governed by the relation $1/\tau = 1/\tau_0 + 1/\tau_p$. On the other hand, it is known from absorption experiments that the ratio of transition probabilities of ortho- and paraexcitons from the crystal ground state differs by a factor of 500 [24] so that the effective decay rate becomes $\tau \approx \tau_0$ and consequently $\tau_p \simeq 500\tau$. Therefore, one deduces a paraexciton radiative lifetime τ_p of 0.85 ms at 70 K. This value, unusually large for excitons in a direct gap semiconductor, is due to the combined effect of parity and spin flip selection rules. The paraexciton lifetime measured at 2 K was found to be 13 μs [23]. Note that in these experiments τ_p was underestimated, since the decay was registered from paraexcitons created near the sample front surface and therefore did not take into account migration of particles out of the observation region. Recently, a long exciton decay time could be observed even at room temperature [25], confirming the stability of excitons in Cu_2O.

The mass of the orthoexciton is 2.7 m_0 [26]; the paraexciton mass is assumed to be the same. From the ideal Bose model, one estimates a critical density for Bose condensation $n_c < 10^{17}$ cm^{-3} at $T = 2$ K. In the absence of migration, this density should be attained locally with a focussed cw optical pump source of mW power tuned to a region of strong absorption in the crystal.

4.1 Exciton Luminescence

Guided by these considerations, initial attempts to observe quantum statistics focussed on the spectral analysis of the paraexciton luminescence using cw optical pumping of the crystal [23]. The only possible radiative channel for paraexcitons is through simultaneous emission of a photon and a momentum- and parity-conserving phonon Γ_{25}^- of energy 10.8 meV, hereafter denoted $X_p - \Gamma_{25}^-$. Since the participating optical phonon has no dispersion of its energy versus wavevector at zone center[26], and since the transition matrix element for the decay process is also independent of K, the spectral shape of this luminescence line directly yields the statistical distribution of excitons present in the crystal. Indications of a deviation from classical statistics could be observed in Ref. [23]. However, the paraexciton density extracted from lineshape analysis was lower by one order of magnitude than expected from an evaluation of the absorbed photon density. This difference was attributed to exciton migration out of the excited volume, confirmed by later transport measurements.

Study of exciton statistics through spectral analysis of the orthoexciton luminescence $X_0 - \Gamma_{12}^-$, where the participating phonon Γ_{12}^- has an energy of 14 meV, is much easier to perform since the radiative probability is larger by a factor of 500. Gradual deviation from a Maxwell–Boltzmann towards a Bose lineshape become apparent in the orthoexciton emission

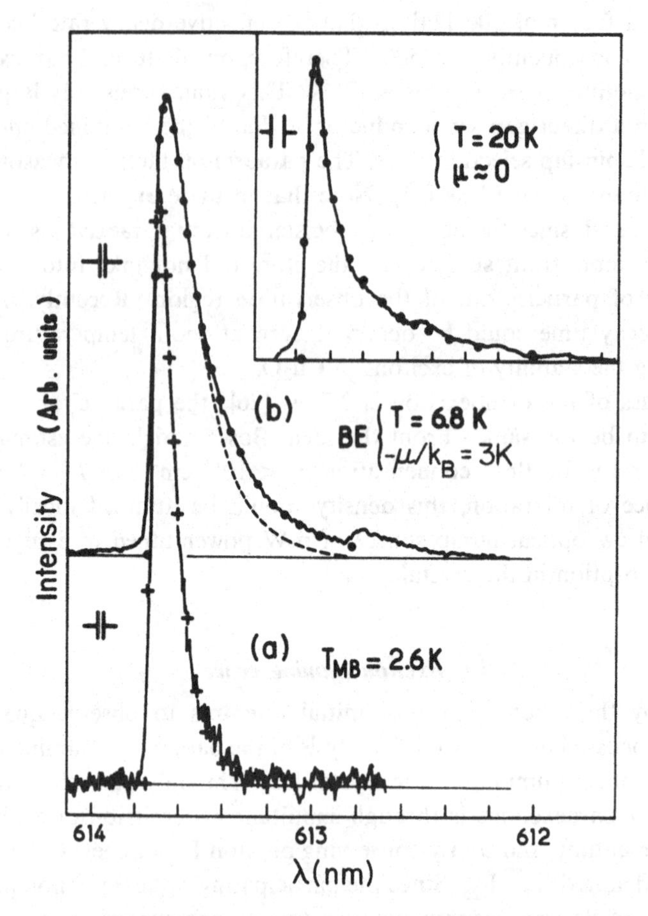

Fig. 7. Luminescence spectrum of the orthoexciton in Cu_2O under cw excitation showing gradual evolution from classical to Bose distribution. The inset shows a spectrum obtained with pulsed excitation (after Ref. [27]).

if the cw excitation rate is increased keeping all other experimental conditions identical [27]. (See Fig. 7.) The increase of quantum degeneracy is accompanied by an increase of effective temperature associated with the gas of particles. Under very intense excitation using a pulsed dye laser (peak intensity $I_0 \simeq 10^7$ W/cm^2) it is possible to reach situations where the chemical potential of the gas is very close to zero (see inset of Fig. 7) [27]. Time-resolved spectra [28] are discussed in this volume by Wolfe, Lin and Snoke. Excellent fits to a Bose–Einstein distribution are observed with values of the chemical potential close to zero, indicating that the critical density for orthoexcitons is reached. The rise of exciton gas temperature with increasing particle density prevents the gas from Bose condensing, although the critical density is reached asymptotically at $T_{eff} = 10$ K.

Time-resolved spectra of the paraexcitons recorded under similar conditions show that the degree of degeneracy is higher than for orthoexcitons, implying that the paraexciton density is higher. The ratio of ortho- to paraexciton population can be deduced from the ratio of the respective luminescence intensities. It could be shown in this way that the paraexciton density in the sample is higher than n_c [28].

4.2 Exciton Transport in Cu_2O

Perhaps the most interesting aspect of an excitonic Bose condensate is the possibility of observing superfluidity, a dragfree motion of part of the excitonic fluid through the sample. As pointed out by Gergel *et al.* [30], this should translate into a change of the penetration distance inside the crystal of an exciton fluid initially located close to the surface. Instead of a diffusive motion, with a penetration depth $l_d = \sqrt{D\tau}$ where D is the diffusion coefficient and τ the exciton lifetime, a ballistic propagation over a distance reaching $l_s \approx v_{cr}\tau$, where v_{cr} is the critical Landau velocity, should be observed.

Information on excitonic transport properties at the early stages following exciton formation can be obtained from time- and space-resolved luminescence studies [29]. A small part of the sample near the front surface is imaged on the front slit of a monochromator. The band pass of the monocromator is adjusted so as to select the emission line from the species to be examined. Using pulsed excitation and time-resolved detection of the projected sample image as a function of distance from the excitation spot provides the sought information. Side view detection is convenient to detect the expansion of the excitonic cloud at the early stages towards the interior of the sample, following front surface excitation.

Results shown in Fig. 8 show a ballistic propagation at early times with a speed dependent on initial density. At the highest densities supersonic speeds are detected. Link and Baym [31] have performed a theoretical analysis of the expansion rate and have concluded that these results indicate a dragfree hydrodynamic flow, consistent with the idea that excitonic superfluidity is involved.

Time- and space-resolved detection of excitons at much higher pen-

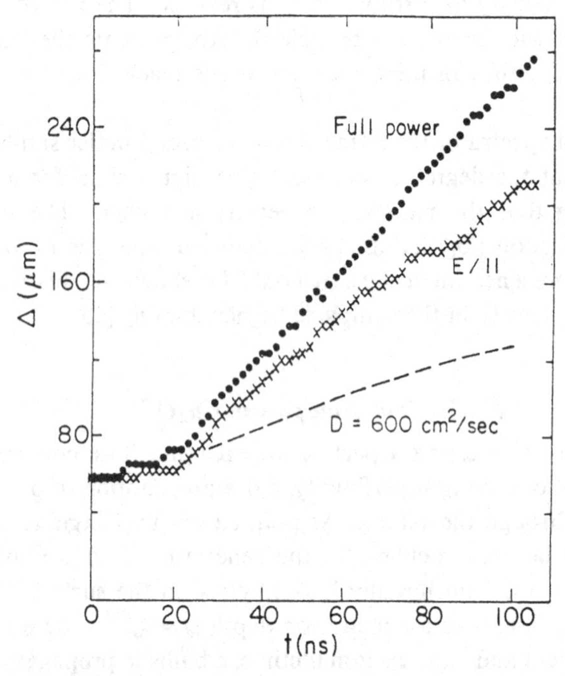

Fig. 8. Expansion of the orthoexciton cloud towards the interior of the crystal, as measured by time- and space-resolved luminescence, for three different initial exciton densities, where Δ is the distance of half maximum of the luminescence intensity from the surface of the crystal. The dashed line corresponds to the expected behaviour of a diffusive motion with a diffusion constant as shown.

etration depths inside the crystal are obtained by the exciton-mediated photovoltaic effect [32], in which an electric current is generated in the absence of an external applied voltage at a semiconductor–metal contact when excitons reach the interface. The external current results from the dissociation of excitons into free electrons and holes at the interface because of the built-in potential difference due to the presence of a Schottky barrier. This method provides a local detector with excellent spatial resolution since the active dissociation layer is less than 1 μm thick. It also has good time resolution if the electrodes at the contact are impedance-adapted for fast detection.

Application of this detection method has revealed a very unusual excitonic transport regime in Cu_2O [33]. At sufficiently high particle densities and low temperatures, optically inactive paraexcitons initially created with random *k*-vectors near the front surface of a single crystal

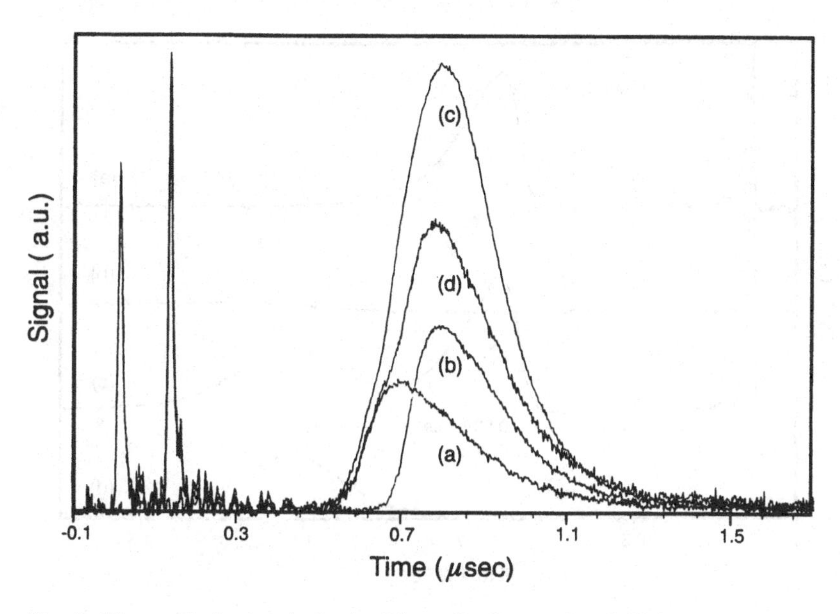

Fig. 9. Photovoltaic signal obtained in a Cu_2O sample of thickness $d = 2.5$ mm held at 1.95 K. Delay between optical pump pulses is $\delta t = 125$ ns. Both pump pulse intensities are ≈ 100 KW/cm^2. The delayed traces correspond to (a) the signal obtained at the sample back surface with the pump pulse alone, (b) the probe pulse alone, (c) by sequential illumination of the sample. Trace (d) is the algebraic sum of pump and probe induced signals [(a) + (b)]. (From Ref. [33].)

were observed to propagate ballistically through the sample in the form of a packet of limited size over unusually large distances, approaching 1 cm. This anomalous propagation at constant speed is highly temperature sensitive and takes place under excitation conditions corresponding to initial supercritical densities near the front surface. This anomalous transport has been attributed to excitonic superfluidity [33], and is further discussed by Fortin *et al.* in this volume.

Evidence for interaction between two ballistic packets, has been obtained very recently [34]. As shown in Fig. 9, the exciton-mediated photovoltaic signals obtained in a crystal excited with two successive short optical pulses of equal intensities (curve c) is larger than the algebraic sum of the signals from the individual pulses (curve d). Results obtained [34] with two optical pulses of unequal intensities and larger time separation are shown in Fig. 10. When both pulses illuminate the sample sequentially, a second excitonic wavepacket appears instead of the weak diffusive response observed in the presence of the second pulse alone.

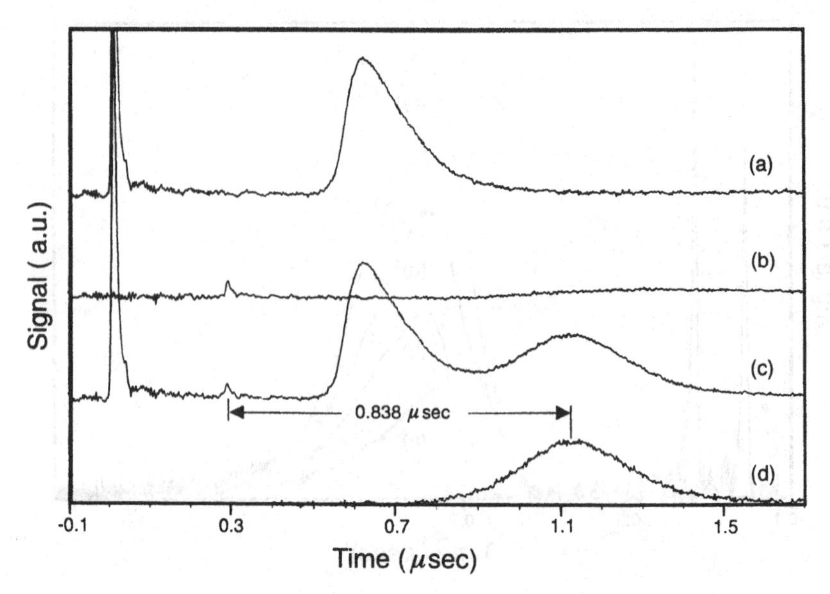

Fig. 10. Photovoltaic signal of a Cu_2O sample of thickness $d = 2.5$ mm held at 2K. Trace (a) is obtained with pump pulse alone. Trace (b) probe pulse alone. Trace (c) pump and probe pulse in sequence with $\delta t = 278$ ns. Trace (d) algebraic difference between traces (c) and (a). The pump intensity is of the order 100 kW/cm^2. The prompt signals at $t = 0$ and $t = 0.278\mu s$ are fiducials for the pump and probe optical pulses.

Formation of this second packet is attributed to quantum attraction of thermal excitons left in the trail of the first packet towards the second propagating packet. However, this interpretation is only suggestive at the present time.

A theoretical model starting from first principles describing the motion of a spatially nonuniform Bose condensate amid a gas of thermal particles has been very recently developed by Hanamura [35]. The equation of motion of the condensate is derived to second order in perturbation theory by taking into account the interaction of condensed excitons with normal excitons as well as lattice phonons. A stationary solution for the envelope of the superfluid macroscopic wavefunction is given by the symmetric solitary wave of intensity

$$\phi^2 = |\phi_0|^2 \text{sech}\left(\frac{|R - v_s t|}{R_0}\right).$$

Comparison of the experimental results with the prediction of the model reveals excellent agreement [34]. One interesting aspect of the propagation of such a soliton pulse is its velocity, $v_s = 2.98 \times 10^5$ cm/s, which is significantly smaller than the velocity of sound in the crystal, $v_l = 4.5 \times 10^5$ cm/s. Further investigations are needed to understand more fully the formation and dynamics of the spatially non-uniform condensate on which our interpretation rests.

5 Conclusions

There is growing evidence that excitonic particles in semiconductors can exhibit features expected in a weakly interacting Bose gas. Luminescence and pump/test experiments show features which are explained in a natural way by the ideal, or nearly ideal, Bose gas model. Anomalous transport of excitons initially confined close to the crystal surface is observed. This ballistic transport over remarkably long distances suggests that a superfluid phase of excitons is involved.

References

[1] Typical values of a_B and E_B range from a few meV in Si or Ge to a fraction of one eV in alkali or copper halides with corresponding Bohr radius ranging from several tens of nm to less than one nm.

[2] O. Akimoto and E. Hanamura, Phys. Soc. Japan **33**, 1537 (1972).

[3] E. Hanamura, Solid State Comm. **12**, 951 (1973); A.L. Ivanov and H. Haug Phys. Rev. B **48**, 1490 (1993).

[4] G.M. Gale and A. Mysyrowicz, Phys. Rev. Lett. **32**, 727 (1974).

[5] N. Nagasawa, T. Mita and M. Ueta, J. Phys. Soc. Japan **41**, 929 (1976).

[6] D. Hulin, A. Mysyrowicz and A. Antonetti, J. Luminescence **30**, 290 (1985).

[7] I.M. Blatt, K.W. Boer and W. Brandt, Phys. Rev. **126** 1691 (1962).

[8] S.A. Moskalenko, Soviet Physics Solid State **4**, 199, (1962).

[9] L.V. Keldysh and A.N. Kozlov, JETP **27** 521, (1968). L.V. Keldysh, Y.V. Kopaev, Soviet Physics Solid State **6**, 2219 (1965).

[10] E. Hanamura and H. Haug, Physics Reports C**33**, 209 (1979).

[11] C. Comte and P. Nozières, J. Physique (Paris) **43**, 1069 (1982).

[12] L.V. Keldysh in *Proc. 9th Int. Conf. on Physics of Semiconductors* (Moscow 1968) p. 1303; see also T.M. Rice, J. Haensel, T. Phillips, G.A. Thomas, in Solid State Physics **32**, 1 (1977).

[13] F. Bassani and M. Rovere, Solid State Comm. **19**, 887 (1976).
[14] L.H. Nosanow, J. de Physique **41** C7-1 (1980).
[15] J.J. Hopfield, Phys. Rev. **112**, 1555 (1958).
[16] R.S. Knox, *Theory of Excitons* (Academic Press, New York, 1963).
[17] For a review see A. Goldmann, Phys. Stat. Sol. B **81** 9 (1977).
[18] For a review of the literature on biexcitons in CuCl, see, for instance, C. Klingshirn and H. Haug, Physics Reports **70** 315 (1981).
[19] L.L. Chase, N. Peyghambarian, G. Grynberg and A. Mysyrowicz, Phys. Rev. Lett. **42**, 1231 (1979); N. Peyghambarian, L.L. Chase and A. Mysyrowicz, Phys. Rev. B **27** 2325 (1983).
[20] M. Hasuo, N. Nagasawa, T. Itoh, and A. Mysyrowicz, Phys. Rev. Lett. **70**, 1303 (1993).
[21] For a review of excitonic properties of Cu_2O see, for instance, S. Nikitine in *Optical Properties of Solids*, S. Nudelman and S.S. Mitra, eds. (Plenum, New York, 1969).
[22] I.I. Kreingold and V.L. Makarov, Soviet Physics Solid State **15**, 890 (1973).
[23] A. Mysyrowicz, D. Hulin and A. Antonetti, Phys. Rev. Lett. **43**, 1123, 1275(E) (1979).
[24] P.D. Bloch and C. Schwab, Phys. Rev. Lett. **41**, 514 (1978).
[25] D.W. Snoke, A.J. Shields and M. Cardona, Phys. Rev. B **45**, 11693 (1992).
[26] P. Yu and Y.R. Shen, Phys. Rev. Lett. **32**, 939 (1974).
[27] D. Hulin, A. Mysyrowicz and C. Benoit a la Guillaume, Phys. Rev. Lett. **45**, 1970 (1980).
[28] D. Snoke, J.P. Wolfe and A. Mysyrowicz, Phys. Rev. Lett. **59**, 827 (1987).
[29] D. Snoke, J.P. Wolfe and A. Mysyrowicz, Phys. Rev. B **41**, 11171 (1990). See also article by Wolfe *et al.* in this volume.
[30] V.A. Gergel, R.F. Kazarinov and R.A. Suris, JETP **27**, 159 (1968).
[31] B. Link and G. Baym, Phys. Rev. Lett. **69**, 2959 (1992).
[32] E. Tselepis, E. Fortin and A. Mysyrowicz, Phys. Rev. Lett. **59**, 2107 (1987).
[33] E. Fortin, S. Fafard and A. Mysyrowicz, Phys. Rev. Lett. **70**, 3951 (1993).
[34] A. Mysyrowicz, E. Fortin, E. Benson, S. Fafard and E. Hanamura, to be published; M. Inoue, E. Hanamura, J. Phys. Soc. Japan **41**, 771 (1976).
[35] E. Hanamura, Solid State Comm. **91**, 889 (1994).

15

Crossover from BCS Theory to Bose–Einstein Condensation

Mohit Randeria
Argonne National Laboratory
MSD 223, 9700 S. Cass Avenue
Argonne IL 60439
USA

Abstract

This article gives a detailed review of our current theoretical understanding of the crossover from cooperative Cooper pairing to independent bound-state formation and Bose–Einstein condensation in Fermi systems with increasing attractive interactions.

1 Introduction

There are two well-known paradigms for understanding the phenomena of superconductivity and superfluidity:

(I) BCS theory [1], in which the normal state is a degenerate Fermi liquid that undergoes a pairing instability at a temperature $T_c \ll \epsilon_F$, the degeneracy scale. The formation of Cooper pairs and their condensation (macroscopic occupation of a single quantum state) both occur simultaneously at the transition temperature T_c.

(II) Bose–Einstein condensation (BEC) of bosons at a T_c of the order of their degeneracy temperature. At a fundamental level these bosons (for example, ^{4}He or excitons) are invariably composite objects made up of an even number of fermions. The composite particles form at some very high temperature scale of the order of their dissociation temperature T_{dissoc}, and these "pre-formed" bosons then condense at the BEC $T_c \ll T_{\text{dissoc}}$.

In most cases of experimental interest the system under consideration clearly falls into one category or the other. For instance, ^{3}He is a Fermi superfluid described by (I) whereas ^{4}He is a Bose superfluid (II). Essentially all cases of superconductivity in metals that we understand reasonably well are much closer to (I) than to (II), as evidenced by the existence of fermionic quasiparticles above T_c. Nevertheless, as the rest of this review will show, it is of great interest to consider models which interpolate between these two extremes: the weak attraction, high density limit is described by BCS theory while the strong coupling, low density limit consists of tightly bound bosons. One outcome of such a study would be a deeper understanding of the phenomena of superconductivity, even for systems which are clearly closer to one of the limits. In addition, there are experimental systems which may be in a regime intermediate between these two extreme limiting cases, and then one is forced to study the crossover from collective Cooper pairing to the formation and condensation of independent bosons.

In the remainder of this section we will first give a very quick review of the historical development of ideas in this subject, then discuss the reasons for the recent resurgence of interest in this problem, and conclude with an outline of the rest of the paper. Readers who would like a preview of the main results of this article may wish to look at the summary in the concluding section.

1.1 History

The problem of BCS versus Bose–Einstein condensation is an old one. Some of the earliest attempts in the pre-BCS era to theoretically understand superconductivity in metals were in terms of Bose–Einstein condensation, an approach which was perhaps taken farthest by Schafroth, Blatt and Butler [2]. Then came the striking success of the Bardeen–Cooper–Schreiffer (BCS) theory [1] and superconductivity in metals was finally understood as a pairing instability of a Fermi liquid state. In the immediate aftermath of BCS the differences with BEC were emphasized: that the Cooper pairs, which were highly overlapping in real space, should not be thought of as composite bosons, and that the correlations in the BCS condensate were best described in terms of "momentum space pairing" rather than "real space pairing" in a highly degenerate Fermi system.

There are, of course, many similarities in the behavior of Fermi and Bose superfluids insofar as their macroscopic, coherent properties are concerned. Off-diagonal long range order (ODLRO) [3] in an appropriate density matrix is a unifying concept in the study of these two types of superfluids: the macroscopic occupation of a single quantum state forms the basis for understanding *all* bulk superfluid systems. Despite these similarities between the BCS condensate and the Bose condensate, there are equally obvious differences at the microscopic level. In particular, the normal states above T_c are completely different.

Perhaps the first discussion of the possibility of a crossover from a BCS state to BEC as a function of some parameter, in this case decreasing carrier density, was by Eagles [4] in the context of a theory of superconductivity in low carrier concentration systems such as $SrTiO_3$ doped with Zr. This remarkable paper has gone virtually unrecognized and many results obtained in subsequent papers were, in fact, anticipated by Eagles.

In a seminal paper, Leggett [5] studied a dilute gas of fermions at $T = 0$ with an attractive interaction and showed within a variational approach that there was a smooth crossover from a BCS ground state with Cooper pairs overlapping in space to a condensate of tightly bound diatomic molecules. One of his main motivations was to ask to what extent one might be able to describe Cooper pairs in superfluid ^{3}He as giant diatomic molecules. (See also [6–8].)

Some years later this question was taken up by Nozières and Schmitt-Rink [9] motivated by the problem of exciton condensation [10] where it might, in fact, be possible experimentally to go from one limit to the other by varying the density of carriers. These authors extended the previous analysis to lattice models and, more importantly, to finite temperatures. Using a diagrammatic formulation they showed that within their approximation, the transition temperature T_c evolves smoothly as a function of the attractive coupling from the BCS to the Bose limit.

The papers of Leggett and of Nozières and Schmitt-Rink have had a major influence on all the subsequent work in this field, and we shall return to a more detailed discussion of these later in this review.

1.2 Current Interest

There has been a resurgence of interest in superconductivity after the discovery of the copper-oxide based materials with high transition temperatures [11]. It is now clear that in the copper-oxide systems, the size of the pairs, as estimated from the Ginzburg–Landau coherence length (since these systems are believed to be in the clean limit), is only a few times the lattice spacing [12]. This is in marked contrast to materials well described by conventional BCS theory where the pair size greatly exceeds the lattice spacing, or the average distance between carriers. In this sense, the new superconductors are different from BCS systems, and the pairing in these systems is likely to be in a regime intermediate between Cooper pairs and composite bosons [13].

The nature of the normal state just above T_c in the BCS–Bose crossover raises a very interesting question [14]: how does one crossover from a Fermi liquid at weak coupling to a normal Bose liquid of tightly bound pairs in the strong coupling limit? Monte Carlo simulations of attractive fermions in two dimensions have recently shown [14] that at moderate couplings where one still has a degenerate Fermi system the spin correlations are very different from that expected of a Fermi liquid, showing characteristic "spin-gap" anomalies. These results, which will be described at length below, are particularly interesting because many of the underdoped cuprates show precisely these anomalies.

It must be emphasized that the theoretical problem of high T_c superconductivity in its full complexity goes beyond the simple ideas described in this paper. The present approach does not address issues such as the mechanism that gives rise to such high transition temperatures, the proximity to a metal-insulator transition, the role of antiferromagnetic spin fluctuations, and the symmetry of the order parameter. However, because the ideas presented here are simple and general, the author believes that they are relevant to the high T_c systems independent of microscopic details. For a complete description, one must have a microscopic theory that gives the correct symmetry of the order parameter and includes the physics of short coherence length superconductors discussed here.

We must also mention several other topics of current interest where questions closely related to the ones discussed in this paper arise. (1) The physics of tightly-bound real space pairs has been explored in detail in the context of bipolaron theories [15, 16]. These studies (see also the review by Ranninger in this volume) are directly relevant to the Bose limit of the crossover problem. (2) The problem of Bose condensation of excitons [10] has already been alluded to. The possibility of biexciton formation and phase separation due to formation of electron–hole droplets presumably complicates a simple crossover scenario. (3) The crossover from itinerant to local-moment magnetism has been studied intensively in the context of metallic magnetism [17], and more recently for the insulating state of the half-filled repulsive Hubbard model [18]. (4) A problem of great current interest is the superconductor–insulator transition in disordered systems [19]. The present considerations might have some bearing upon such questions as: To what extent do purely bosonic models describe this transition and the two phases on either side of it? What is the role of the fermionic degrees of freedom? (5) Issues related to the BCS–Bose crossover arise in the study of the phases of hydrogen under high pressure [20]. (6) Related questions also arise in nuclear physics, for example in the study of the equation of state of extended nuclear matter [21].

1.3 Outline

The remainder of this article is organized as follows. In Section 2 we describe two models for studying the BCS–Bose crossover.

In Section 3 we first introduce the functional integral formalism which will be used in the first part of the paper. Using this we address the question of pair formation versus condensation. The problems associated with a finite temperature analysis in two dimensions are mentioned briefly.

We discuss time-dependent Ginzburg–Landau theory in Section 4, and describe how the dynamics of the pairs just above T_c evolves from the BCS to the Bose regime.

We turn to the broken symmetry state in Section 5 focusing mainly on the $T = 0$ case. The crossover in the ground state and single-particle excitation spectrum is discussed using a variational formulation.

In Section 6 we describe the evolution of the collective modes using a generalized RPA analysis.

In Section 7 we discuss numerical results obtained by the quantum Monte Carlo method, focusing on recent results on the normal state correlation functions in the crossover regime in two dimensions.

A summary and conclusions are presented in Section 8.

The main focus of this review will be the work done by the author and his collaborators (in rough chronological order): J.M. Duan, L.Y. Shieh, J.R. Engelbrecht, L. Belkhir, N. Trivedi, C. Sá de Melo, R. Scalettar and A. Moreo. An attempt has been made to reference much of the recent literature relevant to the models of Section 2 but the author apologizes in advance to those whose work has been left out.

2 Preliminaries

In this section we introduce two models, one in the continuum and the other on a lattice, of fermions with attractive two-body interactions. Our aim is to see how, as a function of increasing attraction, these systems evolve between two apparently very different regimes (see Fig. 1): BCS theory in weak coupling with a Fermi liquid normal state and tightly

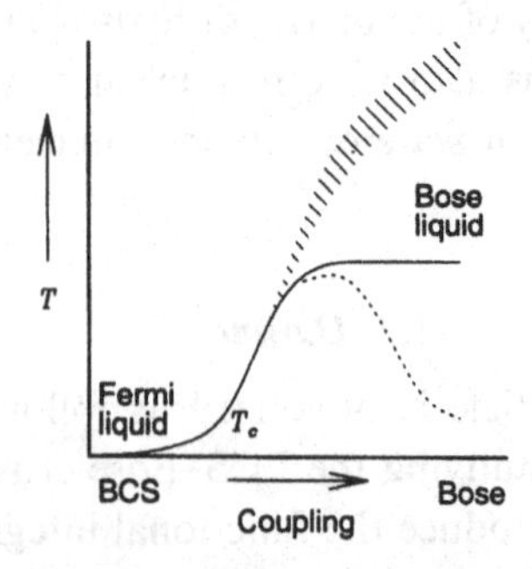

Fig. 1. Schematic phase diagram of the attractive fermion problem. The full curve is the transition temperature T_c for a continuum model (see text), and the dashed curve is T_c for the lattice (attractive Hubbard) model. The shaded region represents a crossover below which pair correlations are important. In the Bose limit, the crossover scale is the pair dissociation temperature.

bound pairs which Bose condense from a normal Bose liquid in strong coupling.

The continuum model has the Hamiltonian density

$$\mathscr{H} = \bar{\psi}_\sigma(x)\left[-\frac{\nabla^2}{2m} - \mu\right]\psi_\sigma(x) - g\bar{\psi}_\uparrow(x)\bar{\psi}_\downarrow(x)\psi_\downarrow(x)\psi_\uparrow(x). \quad (1)$$

The chemical potential μ fixes the average density n. We choose units in which $\hbar = k_B = 1$ and the system is in a unit volume.

Since we want to describe the Bose regime in terms of its constituent fermions, we have to allow the magnitude of the attraction g to be arbitrary. In addition, we cannot use a simple BCS-like cutoff $\omega_D \ll \epsilon_F$ since we must allow for the possibility of *all* the fermions to be affected by the interactions, not just a small fraction within a shell around the Fermi energy. We will thus study a dilute Fermi gas in which the range of the attractive interactions is much smaller than the average interparticle spacing, and the attraction can be characterized by a shape-independent parameter, the scattering length a_s in three dimensions. We will describe exactly how the "renormalized" coupling a_s replaces the "bare" g in Section 3.

While the continuum model described above is very useful for analytical approaches, the attractive Hubbard model

$$H = -t\sum_{ij\sigma} c^+_{i\sigma}c_{j\sigma} - U\sum_i n_{i\uparrow}n_{i\downarrow} - \mu\sum_i (n_{i\uparrow} + n_{i\downarrow}) \quad (2)$$

is in many ways simpler. There is no need for the coupling constant renormalization: the on-site attraction U is cutoff in momentum space by the bandwidth $W = 2zt$, where z is the coordination number. Another appealing feature is that even in the large-U limit there is no phase separation [9]; one only has pairs and not quartets or larger clusters. For large U two fermions form an on-site singlet and other fermions are prevented by the Pauli exclusion principle from coming on that site. (In the continuum model it is difficult to prove that phase separation does not occur. If it does, one can certainly imagine modifying the interactions between fermions which prevent this from happening while not affecting the BCS–Bose crossover physics that we are interested in).

In the weak coupling limit, the lattice and continuum models describe essentially the same physics, since the Cooper pairs corresponding to (2) turn out to be much larger than the lattice spacing a. In the strong coupling limit significant differences arise, since once the pair size becomes comparable to a, lattice effects naturally become very important. For $U/t \gg 1$ the effective low energy description [16] of (2), in the subspace of empty and doubly occupied sites, is in terms of a hard-core Bose lattice gas with hopping $t_{\text{eff}} \sim t^2/U$. Since the bosons become heavier with increasing coupling the superconducting T_c decreases as t^2/U in strong coupling (see Fig. 1). This is in contrast to the continuum case where the boson mass is $2m$ in the strong coupling limit and, as we shall show, T_c saturates at the BEC value.

While in the continuum model there is a single dimensionless parameter $1/k_F a_s$, the attractive Hubbard Hamiltonian is characterized by two: the coupling U/t and the filling factor f, controlled by μ. Note that $f = \langle n \rangle/2$, i.e., one-half the average site occupancy. We will only be concerned here with systems away from half-filling, $f \neq 1/2$, where one has a superconducting ground state; for $f = 1/2$ the ground state has both superconducting and charge-density wave order [16].

Most, but not all, of the analytical results described below will be for the continuum model (1), while all of the numerical progress on the crossover question has been in the context of the lattice Hamiltonian (2).

3 T_c and Above: Pair Formation vs. Condensation

We will begin with a description of a functional integral formulation of the continuum model (1) which permits us to make useful approximations at all couplings and temperatures both above and below T_c. Within a single formalism this allows us to recover all the previously known results – which had been obtained using different techniques in different regimes

– and to go beyond them. All of the results from the functional integral approach described below were obtained in collaboration with Sá de Melo and Engelbrecht, for detailed references, see [22, 23].

The partition function Z, at a temperature β^{-1}, is written as an imaginary time functional integral [24] with action

$$S = \int_0^\beta d\tau \int d\mathbf{x}(\bar{\psi}_\sigma(x)\partial_\tau \psi_\sigma(x) + \mathscr{H}), \tag{3}$$

where $x = (\mathbf{x}, \tau)$. Decoupling the fermion interaction with the Hubbard–Stratonovich field $\Delta(\mathbf{x}, \tau)$, which couples to $\bar{\psi}\bar{\psi}$, and integrating out the fermions we obtain $Z = \int D\Delta D\bar{\Delta} \exp - S_{\text{eff}}[\Delta, \bar{\Delta}]$. The effective action

$$S_{\text{eff}}[\Delta(x)] = \frac{1}{g}\int_0^\beta d\tau \int d\mathbf{x}|\Delta(x)|^2 - \operatorname{Tr}\ln \mathbf{G}^{-1}[\Delta(x)]. \tag{4}$$

is written in terms of the inverse Nambu propagator

$$\mathbf{G}^{-1}(x, x') = \begin{pmatrix} -\partial_\tau + \nabla^2/2m + \mu & \Delta(x) \\ \Delta^*(x) & -\partial_\tau - \nabla^2/2m - \mu \end{pmatrix} \delta(x - x'). \tag{5}$$

We now analyze this functional integral in terms of static, uniform saddle-points and fluctuations about these. The trivial saddle-point $\Delta \equiv 0$, which is stable at sufficiently high T for all couplings, will become unstable below a certain temperature which we denote, for now, by T_0 (rather than T_c for reasons which will become clear). The transition temperature is then defined by the solution of $\delta S_{\text{eff}}/\delta\Delta\,[\Delta = 0] = 0$, which can be written as $1/g = \sum_{\mathbf{k}} \tanh(\xi_{\mathbf{k}}/2T_0)/2\xi_{\mathbf{k}}$.

Before proceeding further, we need to describe how we regulate the ultraviolet divergence in the above equation. The idea is to replace the bare g by the low energy limit of the the two-body T-matrix (in the absence of a medium). In three dimensions (3D) we use [25]

$$m/4\pi a_s = -1/g + \sum_{|\mathbf{k}|<\Lambda} (2\varepsilon_{\mathbf{k}})^{-1}, \tag{6}$$

which defines the s-wave scattering length a_s. Recall that as a function of the bare interaction, $1/a_s$ increases monotonically from $-\infty$ for a very weak attraction to $+\infty$ for strongly attractive interaction. Beyond the two-body bound state threshold in vacuum ($1/a_s = 0$), a_s is the "size" of this bound state with binding energy $E_b = 1/ma_s^2$. The dimensionless coupling constant in the dilute gas model is then $1/k_F a_s$, which ranges from $-\infty$ in the weak coupling BCS limit to $+\infty$ in the strong coupling Bose limit.

Using (6) we obtain the equation for the transition temperature in terms of the renormalized coupling a_s:

$$-\frac{m}{4\pi a_s} = \sum_{\mathbf{k}} \left[\frac{\tanh\left(\xi_{\mathbf{k}}/2T_0\right)}{2\xi_{\mathbf{k}}} - \frac{1}{2\varepsilon_{\mathbf{k}}} \right], \tag{7}$$

with $\xi_{\mathbf{k}} = \varepsilon_{\mathbf{k}} - \mu$ and $\varepsilon_{\mathbf{k}} = |\mathbf{k}|^2/2m$. There are two unknowns in this equation, T_0 and μ, and thus we need another equation, $N = -\partial\Omega/\partial\mu$, which fixes the chemical potential for a given density. Note that unlike the BCS analysis in which one may set $\mu = \epsilon_F$ (the non-interacting Fermi energy) at the outset, in the crossover problem μ will turn out to be a very strong function of the coupling as one goes into the Bose regime where all the particles are affected by the attractive interactions.

3.1 Saddle Point Analysis: Pair Formation

The saddle-point estimate T_0 is obtained by using the approximation $\Omega_0 = S_{\text{eff}}\left[\Delta = 0\right]/\beta$ for the thermodynamic potential. The number equation is then given by

$$n = n_0(\mu, T) \equiv \sum_{\mathbf{k}} \left(1 - \tanh\left(\frac{\xi_{\mathbf{k}}}{2T}\right)\right). \tag{8}$$

In the weak coupling limit, $1/k_F a_s \to -\infty$ and we find the BCS results $\mu = \epsilon_F$ and $T_0 = 8e^{-2}\gamma\pi^{-1}\epsilon_F \exp\left(-\pi/2k_F|a_s|\right)$, where $\gamma \simeq 1.781$.

The equations can also be solved analytically in the strong coupling limit, where we find that the roles of the gap and number equations are reversed: the gap equation (7) determines μ, while the number equation (8) determines T_0. In this limit $1/k_F a_s \to +\infty$ and one finds tightly bound pairs with binding energy $E_b = 1/ma_s^2$. The non-degenerate Fermi system has $\mu \simeq -E_b/2$ and one finds $T_0 \simeq E_b/2\ln\left(E_b/\epsilon_F\right)^{3/2}$.

This unbounded growth of the "transition temperature" is an artifact of the approximation and there is, in fact, no sharp phase transition at T_0 (outside of weak coupling where $T_0 = T_c$ ignoring the small effects of thermal fluctuations). The point is that the saddle-point approximation becomes progressively worse with increasing coupling: the trivial saddle-point $\Delta = 0$ can only describe a normal state consisting of essentially non-interacting fermions. While this is adequate for weak coupling, in the strong coupling limit unbound fermions are obtained in the normal state only at very high temperatures. In this limit, where the system is completely non-degenerate, a simple "chemical equilibrium" analysis (boson $\rightleftharpoons$ 2 fermions) yields a dissociation temperature

$T_{\text{dissoc}} = E_b / \ln\left(E_b/\epsilon_F\right)^{3/2}$. The logarithm in the denominator represents an entropic contribution favoring broken pairs and leads to $T_{\text{dissoc}} < E_b$, the binding energy. We thus see that in strong coupling T_0 is related to the pair dissociation scale rather than the T_c ($\ll T_0$) at which coherence is established.

A numerical solution of (7) and (8) yields T_0 smoothly interpolating between the two limiting cases analyzed above. This temperature represents a "pairing scale" below which pair correlations become important; for sufficiently large attraction these correlations describe the formation of real bound states. For temperatures above the pairing scale T_0, the system may be described as essentially consisting of unbound fermions.

3.2 Beyond Saddle Point: Pair Condensation

Since the saddle-point estimate for T_c fails with increasing coupling, we must include the effects of fluctuations about $\Delta(\mathbf{x}, \tau) = 0$. We show that treating the fluctuations at the Gaussian level produces a sensible strong coupling answer for T_c *provided* their frequency dependence is retained. This is, in fact, simply another way to state Nozières and Schmitt-Rink's (NSR) results since the Gaussian theory for T_c is shown to be identical to their diagrammatic calculation [9]. One of the nice things about the functional integral approach is that it provides a systematic way to check the adequacy of the Gaussian approximation by looking at the fourth order corrections (see below) and, in principle, to systematically go beyond it. (In addition, the functional integral approach also works below T_c.)

Before plunging into the details of the calculation, it may be worth commenting on static spatial ("classical") fluctuations versus dynamic ("quantum") fluctuations. The role of spatial fluctuations is well-known from the study of critical phenomena. Dynamic fluctuations, however, become crucial in determining T_c with increasing coupling because the bosonic degrees of freedom present in the strong coupling normal state can only be adequately described by retaining the full frequency dependence of the fluctuations about the trivial saddle-point. In some crude sense, the formation of tightly bound pairs is a quantum fluctuation which must be included in the functional integral.

To second order in Δ the action is

$$S_{\text{eff}} \simeq S_{\text{eff}}[0] + \sum_{\mathbf{q}, iq_l} |\Delta(\mathbf{q}, iq_l)|^2 / \Gamma(\mathbf{q}, iq_l),$$

where

$$\Gamma^{-1}(\mathbf{q}, iq_l) = \sum_{\mathbf{k}} \left[\frac{1 - n_{\mathbf{k}} - n_{\mathbf{k+q}}}{iq_l - \xi_{\mathbf{k}} - \xi_{\mathbf{k+q}}} + \frac{1}{2\varepsilon_{\mathbf{k}}} \right] - \frac{m}{4\pi a_s}. \tag{9}$$

Here $n_{\mathbf{k}}$ is a Fermi function and $iq_l = i2l\pi/\beta$ are the Matsubara Bose frequencies. The resulting expression for the thermodynamic potential $\Omega = \Omega_0 - \beta^{-1} \sum_{\mathbf{q}, iq_l} \ln \Gamma(\mathbf{q}, iq_l)$, is identical to that obtained by NSR [9] by summing particle–particle ladders for the free energy. Following these authors, it is convenient to rewrite Ω in terms of a phase shift defined by $\Gamma(\mathbf{q}, \omega \pm i0) = |\Gamma(\mathbf{q}, \omega)| \cdot \exp(\pm i\delta(\mathbf{q}, \omega))$. The new number equation $N = -\partial\Omega/\partial\mu$ is then given by

$$n = n_0(\mu, T) + \sum_{\mathbf{q}} \int_{-\infty}^{+\infty} \frac{d\omega}{\pi} \, n_B(\omega) \frac{\partial \delta}{\partial \mu}(\mathbf{q}, \omega), \tag{10}$$

where the "free" n_0 is defined in (8) and $n_B(\omega) = 1/(\exp(\beta\omega) - 1)$ is the Bose function. The second term in (10) describes the effects of pair correlations, and, for sufficiently large attraction, the formation of two-body bound states.

We note in passing that in the non-degenerate limit, say at very high temperatures, the Gaussian formalism reduces to the classic analysis of Beth and Uhlenbeck on the second virial coefficien [26]. To see this, note that, when $n_{\mathbf{k}} \ll 1$, Γ coincides with the familiar T-matrix describing two-particle scattering in the absence of a medium. The resulting expression for Ω then relates the thermodynamics to the two-body scattering phase shift, as first described by Beth–Uhlenbeck. In this sense the Gaussian theory is a generalization of their result to two-body scattering in a medium [9].

The temperature T_c at which long range order is established is defined by the solution of (7) together with (10). In weak coupling, the results are essentially unaffected by the inclusion of Gaussian fluctuations in the number equation and T_c is the same as the saddle-point T_0 obtained earlier. A numerical solution of the equations shows that with increasing coupling, T_c deviates from T_0 and goes to a finite strong coupling limit which coincides with the BEC transition temperature (see Fig. 2). The chemical potential $\mu(T_c)$ for the fermions also smoothly interpolates between the weak and strong coupling limits as shown in Fig. 3. In fact, the equations may be solved analytically in the Bose limit. From (7) we find $\mu(T_c) = -E_b/2$, which is one-half the energy required to break a pair. Further, (10) may be simplified (using the properties of Γ described

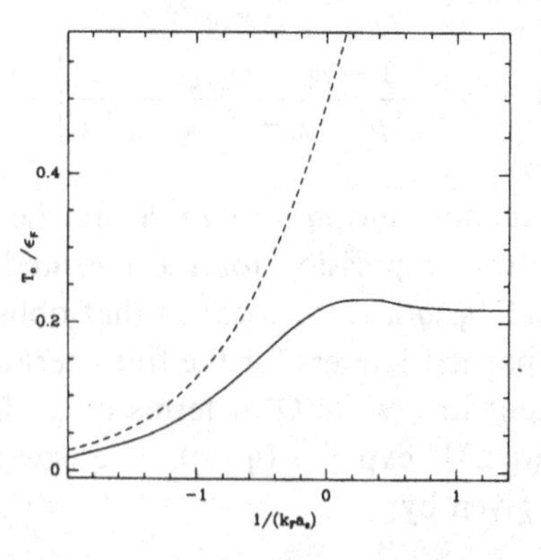

Fig. 2. The transition temperature T_c (full line) and the crossover scale T_0 (dashed line) plotted as a function of the dimensionless coupling $1/k_F a_s$ for an attractive Fermi gas in three-dimensions. The BCS limit corresponds to $1/k_F a_s \to -\infty$ and the Bose limit to $1/k_F a_s \to \infty$. All temperatures are in units of the noninteracting Fermi energy ϵ_F. T_0 is obtained from a solution of the saddle-point equations, while T_c is obtained from the Gaussian approximation described in the text. (From Ref. [22].)

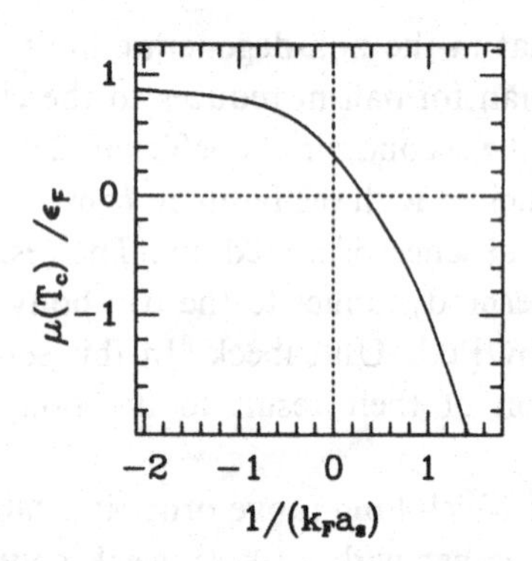

Fig. 3. The chemical potential μ/ϵ_F, obtained from the Gaussian approximation, plotted as a function of the dimensionless coupling $1/k_F a_s$ for an attractive Fermi gas in three-dimensions. (From Ref. [22].)

below; see also Ref. [9]) to obtain $T_c = [n/2\zeta(3/2)]^{2/3}\pi/m = 0.218\epsilon_F$, which is the BEC result for bosons of mass $2m$ and density $n/2$.

To understand how this strong coupling result emerges from our formalism, note that in this limit $\Gamma(\mathbf{q}, z)$ has an isolated pole on the real axis, for each q, representing a two-body bound state with center-of-mass momentum $\mathbf{q}$. Since the energy required to break a pair is very large, this pole is widely separated from the branch cut representing the continuum of two-particle fermionic excitations. The low energy physics for $T \ll T_{\text{dissoc}}$ is thus dominated by this pole, and one can approximate $\Gamma(\mathbf{q}, iq_m) \simeq R(\mathbf{q})/\left[iq_m - \omega_b(\mathbf{q}) + 2\mu\right]$, where $\omega_b(\mathbf{q}) \simeq -E_b + |\mathbf{q}|^2/4m$ is the dispersion relation for a pair of mass $2m$. The partition function, after rescaling the Δ field, may be rewritten as

$$Z = Z_0 \int d\bar{\phi} d\phi \exp\left[\sum_{\mathbf{q}, iq_l} \bar{\phi}_q \left(iq_l - \omega_b(\mathbf{q}) + 2\mu\right) \phi_q\right],$$

which is nothing but the partition function of a free Bose gas. Thus we see how the BEC result is obtained in the strong coupling limit. In the next section we will go beyond the Gaussian approximation and see that the fourth order term describes a repulsive two-body interaction between the bosons. As is well-known, this repulsion stablizes the low temperature phase, but does not substantially affect the condensation temperature.

It may be worth commenting on the fact that the numerical result for T_c (see Fig. 2) appears to be a non-monotonic function of $1/k_F a_s$ with a maximum value at intermediate coupling which is slightly larger than the BEC value. We note that this is where the Gaussian approximation becomes the most questionable (see next section), and even within this approximation the numerical accuracy is the least. Nevertheless, this qualitative feature was also seen in Nozières and Schmitt-Rink's results for a somewhat different model with a separable potential. The very large maximum obtained in some other calculations [27] is certainly an artifact of the approximate form used for the number equation.

Finally, we must mention two other approaches to the finite temperature crossover problem. Haussmann [28] has generalized the Nozières–Schmitt-Rink diagrammatic approach using a self-consistent Green's function formulation which appears to be promising. Gyorffy and coworkers [29] have used the coherent potential approximation (CPA) to study the attractive Hubbard model. This method correctly emphasizes the phase coherence aspect of the transition. However, while this theory gives reasonable results in the strong coupling regime, it appears not to work for sufficiently small attraction, at least in its present form.

3.3 Two Dimensions (2D)

The case of two dimensions has recently attracted much attention partly because of its possible relevance to the layered high temperature superconductors. We will discuss the $T = 0$ formulation of the crossover in the two-dimensional (2D) ground state and collective modes in later sections. Here we discuss the question of the T_c crossover. As is well-known, phase fluctuations destroy off-diagonal *long* range order (ODLRO) in a strictly 2D system. Thus T_c must be interpreted as the Kosterlitz–Thouless (KT) temperature below which there is algebraic order and a finite superfluid density.

Most of the analyses [30] of the T_c crossover in 2D have been based on a Nozières–Schmitt-Rink approach. As shown above, this is simply a Gaussian approximation to the functional integral which perhaps explains the difficulties faced in these calculations. While incorporating the all-important temporal fluctuations, one is also including the spatial fluctuations at the Gaussian level. On the one hand, Gaussian fluctuations destroy ODLRO in 2D and if one searches for a "T_c" at which LRO sets in one should get zero, and, on the other, they are completely inadequate to describe the KT transition. It is not an easy matter to develop a theory of the KT transition starting with fermions with attractive interactions even in the weak coupling limit. Therefore, it is perhaps not surprising that an analytical treatment of the finite T crossover problem in 2D is still an open problem. We will describe numerical results from quantum Monte Carlo simulations in Section 7. For a completely different approach starting from the broken symmetry state and using the KT jump in the superfluid density ρ_s to determine T_c (see Ref. [31]).

4 Time-Dependent Ginzburg–Landau Theory near T_c

In this section, we will discuss the effective low energy, long-wavelength action describing the bosonic degree of freedom $\Delta(\mathbf{x}, t)$, and the resulting equation of motion: the time-dependent Ginzburg–Landau (TDGL) equation. These results are based on Refs. [22, 23]; the results for the static Ginzburg–Landau theory were first given by Dreschler and Zwerger [27]. There are two distinct aims of such a study. (1) By looking at the Δ^4 terms in the action one can assess the validity of the Gaussian theory described earlier. (2) It is of great interest to understand the evolution of the low frequency dynamics of the bosonic degrees of freedom above T_c, since the behavior in the two limits is very different: diffusive in the BCS regime and propagating in the Bose limit.

We begin by expanding the action (4) to fourth order in $\Delta(q)$, where

$q = (\mathbf{q}, iq_m)$, to obtain

$$S_{\text{eff}} = \sum_q \Gamma^{-1}(q)|\Delta(q)|^2 \tag{11}$$
$$+\frac{1}{2}\sum_{q_1,q_2,q_3} b(q_1,q_2,q_3)\Delta(q_1)\Delta^*(q_2)\Delta(q_3)\Delta^*(q_1-q_2+q_3)+\ldots.$$

The Gaussian piece was already studied above (see (9)), and the coefficient of the nonlinear term is given below in terms of fermion propagators. It suffices to set the external frequencies and momenta – the arguments of b – to zero, so that

$$b \equiv b(0,0,0) = \sum_{\mathbf{k}}\sum_{ik_n} \frac{1}{(ik_n - \xi_{\mathbf{k}})^2}\frac{1}{(ik_n + \xi_{\mathbf{k}})^2}, \tag{12}$$

where $ik_n = i(2n+1)\pi/\beta$ is a Matsubara Fermi frequency.

To study the evolution of the TDGL, we consider $\delta S_{\text{eff}}/\delta\Delta^*(q) = 0$ near T_c for Δ slowly varying in space and time. For the *static* part of the problem, one needs to make the expansion $\Gamma^{-1}(\mathbf{q}, 0) = a + c|\mathbf{q}|^2/2m + \ldots$. It is straightforward to find general expressions for the coefficients a, b and c. These are in the form of momentum integrals which can be evaluated numerically once T_c and $\mu(T_c)$ are determined from the equations of the previous section. We will not quote these general results here (which can be found in Refs. [22, 23, 27]) and will only give the BCS and Bose limiting values below, where the integrals are analytically tractable.

We next turn to the much more interesting question of the time-dependence of the linearized TDGL equation. Given the questions about the validity of various microscopic derivations [32] of TDGL in the weak coupling BCS regime, it is important to make some clarifying remarks. Our main aim is to study the dynamics of the pairs above T_c, which is in some sense a simple regime; the complications arise below T_c if one really wants to study the nonequilibrium dynamics of the order parameter. For our purposes it is valid to use the Matsubara technique. Further, by studying fluctuations about the broken symmetry state we can argue [23] that our results are also valid just below T_c provided

$$|\Delta_0| \ll \omega \ll \min\{T_c, |\mu_c|\}, \tag{13}$$

so that Δ_0, the non-trivial solution of of the saddle-point equation (see next section), is the smallest energy scale in the problem.

We need to make a low frequency expansion of $Q(iq_l) \equiv \Gamma^{-1}(\mathbf{q} = 0, iq_l) - \Gamma^{-1}(0,0)$ after analytic continuation. (Note that in Ref. [27] an expansion in Matsubara frequencies was made before analytic continuation, thus yielding unphysical results.) We find that $Q(\omega+i0^+) = Q'(\omega) - iQ''(\omega)$ with

$$Q'(\omega) = \kappa\omega\,\mathscr{P}\int_0^\infty d\epsilon \frac{\sqrt{\epsilon}\tanh\left((\epsilon-\mu)/2T_c\right)}{2(\epsilon-\mu)(\omega-2\epsilon+2\mu)} \quad ; \tag{14}$$

and

$$Q''(\omega) = \kappa 2^{-3/2}\pi\,(2\mu+\omega)^{1/2}\tanh\left(\omega/4T_c\right)\Theta\,(2\mu+\omega), \tag{15}$$

where $\kappa = N(\epsilon_F)/\sqrt{\epsilon_F}$. A low frequency ($\omega \ll T_c$) expansion of Q is possible both in the BCS and Bose limits, where the condition $\omega \ll |\mu|$ is automatically satisfied. Note, however, that for that intermediate coupling for which $\mu(T_c) = 0$, $\Gamma(\mathbf{q} = 0, z)$ has a branch point at the origin, the point about which we would have liked to expand. The vanishing energy scale $\mu(T_c)$ does not permit a low frequency expansion and a linearized TDGL description fails. Strictly speaking, condition (13) implies that a linearized TDGL analysis is not valid inside a small region $|\mu| < \Delta_0$ about $\mu(T_c) = 0$.

We can use the expansion $Q(\omega + i0^+) = -d\omega + \ldots$, provided (13) is satisfied, where

$$d = \sum_{\mathbf{k}} \frac{\tanh(\beta\xi_{\mathbf{k}}/2)}{4\xi_{\mathbf{k}}^2} + i\frac{\pi}{8}\kappa\beta\sqrt{\mu}\Theta(\mu). \tag{16}$$

Collecting the results of the small $|\mathbf{q}|$ and small ω expansions, and transforming to $(\mathbf{x}, t)$, we obtain the result

$$\left(a + b|\Delta\,(\mathbf{x},t)\,|^2 - \frac{c}{2m}\nabla^2 - id\frac{\partial}{\partial t}\right)\Delta(\mathbf{x},t) = 0. \tag{17}$$

We now discuss the two limiting cases, and the intervening singular point, in more detail.

In weak coupling, we obtain the well-known results $a = N(\epsilon_F)\ln(T/T_c)$, $b = 7\zeta(3)N(\epsilon_F)/8\pi^2T_c^2$, $c = 7\zeta(3)N(\epsilon_F)\epsilon_F/12\pi^2T_c^2$ and $d = (i\pi N(\epsilon_F)/8T_c)\cdot[1 - i(2T_c/\pi\epsilon_F)]$. By rescaling the order parameter $\Psi = \sqrt{2c}\Delta$, we obtain the conventional TDGL equation:

$$\left(\varepsilon + |\Psi|^2 - \xi_{\rm GL}^2\nabla^2 + \tau_{\rm GL}\partial_t\right)\Psi(\mathbf{x},t) = 0, \tag{18}$$

where $\varepsilon = |T - T_c|/T_c \ll 1$. The characteristic length scale is $\xi(T) = \xi_{\rm GL}\,\varepsilon^{-1/2}$ with $\xi_{\rm GL} \sim v_F/T_c$, which is the size of the Cooper pair $\xi_{\rm pair}$ (defined in the following section). The width of the Ginzburg region, defined in the usual way as that range of reduced temperatures ε for which the effects of the fourth order term become important, is very small, being of order $(T_c/\epsilon_F)^4$.

The dynamics of Ψ in the BCS limit is overdamped with characteristic time scale $\tau(T) = \tau_{\mathrm{GL}}\,\varepsilon^{-1}$, where $\tau_{\mathrm{GL}} = \pi/8T_c$. This damping arises from the continuum of fermionic excitations into which a pair can decay. There is in addition an $\mathcal{O}\left(T_c/\epsilon_F\right)$ propagating part coming from the real part of d, since our model does not impose particle–hole symmetry [33].

As the coupling increases, we see from (16) that the coefficient of the propagating piece grows while that of the damped part diminishes. The singular point $\mu(T_c) = 0$ separates the regime of damped pair excitations from that of propagating pairs. For stronger coupling, the fermionic excitations have a gap in the normal state, and an essentially propagating mode is obtained at this level of approximation [34]. The full significance of the singular point at intermediate coupling where $\mu(T_c) = 0$ is not clear at the present time. Nor is it known whether the thermally excited modes, ignored in this description, smooth out the effects of this singularity.

In the extreme strong coupling limit we find the TDGL coefficients to be given by $a = \kappa\pi\left(|\mu|^{1/2} - (E_b/2)^{1/2}\right)$, $b = \kappa\pi/32|\mu|^{3/2}$, $c = \kappa\pi/16|\mu|^{1/2}$, and $d = \kappa\pi/|\mu|^{1/2}$, where $\kappa = N(\epsilon_F)/\epsilon_F^{1/2}$. Note that we can set $\mu \simeq -E_b/2$ in b, c and d. Using the normalization $\Psi_0 = \sqrt{d}\Delta$ we can rewrite (17) as

$$-\tilde{\mu}\Psi_0 + U|\Psi_0|^2\Psi_0 - (2M)^{-1}\nabla^2\Psi_0 - i\partial_t\Psi_0 = 0. \tag{19}$$

This is simply the Gross–Pitaevskii equation for a dilute gas of bosons of mass $M = 2m$ with a repulsive interaction $U = 4\pi a_b/M$ characterized by a (boson) scattering length $a_b = 2a_s > 0$ with $n_b a_b^3 \ll 1$, where $n_b = n/2 \sim k_F^3$ (in terms of the k_F of the constituent fermions). This repulsive interaction between the composite bosons has also been obtained independently by Haussmann using a self-consistent Green's function method [28].

The chemical potential of the bosons controls the phase transition via the change in sign of $\tilde{\mu} = E_b - 2|\mu|$. The prefactors of the divergent length and time scales at this transition are given by $\xi_{\mathrm{GL}} \simeq k_F^{-1}/\sqrt{k_F a_s} \gg \xi_{\mathrm{pair}} \simeq a_s$ and $\tau_{\mathrm{GL}} \simeq T_c^{-1}/(k_F a_s)$ respectively. These are both much longer than the microscopic scales because of the diluteness condition $k_F a_s \ll 1$. Consequently the Ginzburg region ($\varepsilon \ll k_F a_s$) is again small in strong coupling.

We will find in the next section that the size of the pairs ξ_{pair} defined in terms of the $T = 0$ wavefunction is a monotonically decreasing function of the attractive interaction, as one might guess on physical grounds. The "Ginzburg–Landau" coherence length in the dilute gas model is, however, a non-monotonic function of the coupling. A numerical evaluation of the

TDGL coefficients shows that ξ_{GL} decreases with increasing interactions at first, reaches a minimum value of order of the interparticle spacing k_F^{-1} at intermediate coupling, and then increases again attaining the strong coupling form discussed above.

Closely related to the large Ginzburg–Landau lengths in the two extreme limits is the small Ginzburg region found in both the BCS and the Bose limits of the dilute gas model. Thus the neglect of the fourth order contributions is justified in both limits. However, in the intermediate coupling regime, where $k_F\xi_{GL} \sim 1$, one does not have a small parameter which controls the calculation: the Ginzburg region is of order $\varepsilon = (T-T_c)/T_c \sim \mathcal{O}(1)$. The only reason to trust the intermediate coupling results, at least qualitatively, is that they smoothly interpolate between two non-trivial limiting cases. At T=0 the variational principle provides further justification, as will be seen in the following section.

5 Broken Symmetry State – Mainly T=0

The functional integral approach described above works equally well [23] in the broken symmetry state for all $T < T_c$; however, except for a few brief remarks at the end, we shall mainly focus on a $T = 0$ variational approach [5]. The two approaches are in fact completely equivalent at $T = 0$.

5.1 Ground State Crossover

In the absence of a small parameter, it is very useful to have a variational formulation. This approach works for both the continuum and the lattice models of Section 2. We begin with the BCS variational wavefunction

$$|\Psi\rangle = \prod_{\mathbf{k}} \left(u_{\mathbf{k}} + v_{\mathbf{k}} c^{\dagger}_{\mathbf{k}\uparrow} c^{\dagger}_{-\mathbf{k}\downarrow} \right) |0\rangle, \tag{20}$$

where $|u_{\mathbf{k}}|^2 + |v_{\mathbf{k}}|^2 = 1$. It is guaranteed to give the standard results in the BCS limit; the reason why one might expect it to work even in the the Bose limit of tightly bound pairs is the following. We can rewrite

$$\begin{aligned} |\Psi\rangle &= \text{const.} \times \prod_{\mathbf{k}} \left(1 + \varphi_{\mathbf{k}} c^{\dagger}_{\mathbf{k}\uparrow} c^{\dagger}_{-\mathbf{k}\downarrow} \right) |0\rangle \\ &= \text{const.} \times \exp\left(\sum_{\mathbf{k}} \varphi_{\mathbf{k}} c^{\dagger}_{\mathbf{k}\uparrow} c^{\dagger}_{-\mathbf{k}\downarrow} \right) |0\rangle, \end{aligned} \tag{21}$$

where $\varphi_{\mathbf{k}} = v_{\mathbf{k}}/u_{\mathbf{k}}$. The N-particle projection of the last expression yields

$$|\Psi_N\rangle \propto \left(\sum_{\mathbf{k}} \varphi_{\mathbf{k}} c^{\dagger}_{\mathbf{k}\uparrow} c^{\dagger}_{-\mathbf{k}\downarrow}\right)^{N/2} |0\rangle, \tag{22}$$

which looks formally indentical to a condensate of composite bosons with an internal pair wavefunction φ. Thus in a limit in which the "size" of the pairs described by φ is much less than the average spacing between the fermions, $|\Psi\rangle$ in (21) describes a condensate of composite bosons.

The calculation then proceeds along the usual BCS route [1], with one difference which we note below. We define, as usual, $v_{\mathbf{k}}^2 = 1 - u_{\mathbf{k}}^2 = \frac{1}{2}(1 - \xi_{\mathbf{k}}/E_{\mathbf{k}})$ where $E_{\mathbf{k}} = \sqrt{\xi_{\mathbf{k}}^2 + \Delta_{\mathbf{k}}^2}$ with $\xi_{\mathbf{k}} = \varepsilon_{\mathbf{k}} - \mu$. The gap function $\Delta_{\mathbf{k}}$ is determined by the standard gap equation $\Delta_{\mathbf{k}} = -\sum_{\mathbf{k}'} V_{\mathbf{k},\mathbf{k}'}\Delta_{\mathbf{k}'}/2E_{\mathbf{k}'}$, where $V_{\mathbf{k},\mathbf{k}'}$ represents the attractive interaction.

An important difference with the standard BCS method is that the chemical potential μ for the fermions must be determined self-consistently with the gap function [4, 5]. In the weak coupling limit $\mu = \epsilon_F$, where ϵ_F is the Fermi energy of the non-interacting system to an excellent approximation. However with increasing attraction the size of the Cooper pairs decreases and consequently the momentum occupation probability $n_{\mathbf{k}} = 2v_{\mathbf{k}}^2$ broadens out considerably, resulting in a significant shift in μ. This is determined by the number equation $\sum_{\mathbf{k}} n_{\mathbf{k}} = N$, which can be written as

$$\sum_{\mathbf{k}} \left(1 - \frac{\xi_k}{E_{\mathbf{k}}}\right) = N. \tag{23}$$

5.2 Three Dimensions (3D)

To be concrete we first focus on the attractive Fermi gas (1) in 3D. The gap parameter can be shown to be $\mathbf{k}$ independent in the low energy limit, so that $\Delta_{\mathbf{k}} = \Delta$. (The gap and number equations derived above work equally well for the attractive Hubbard model (2); see Section 6 and Refs. [9, 16, 35].) Using the regularization procedure (6) we obtain the "renormalized" gap equation [25]

$$\sum_{\mathbf{k}} \left(\frac{1}{\varepsilon_{\mathbf{k}}} - \frac{1}{E_{\mathbf{k}}}\right) = \frac{2N(\epsilon_F)}{k_F a_s}. \tag{24}$$

The order parameter Δ and chemical potential μ at $T = 0$ are obtained by solving (23) and (24). In the weak coupling BCS limit ($1/k_F a_s \to -\infty$) we find $\mu = \epsilon_F$ and $\Delta = 8e^{-2}\epsilon_F \exp\left(-\pi/2k_F|a_s|\right)$ so that $\Delta = \pi T_c/\gamma$,

($\gamma \simeq 1.781$). In the opposite strong coupling limit ($1/k_F a_s \to +\infty$), we expect $\mu < 0$ and $|\mu| \gg \Delta_0$. Analyzing (24) and (23) under these conditions, which we verify at the end, the integrals are tractable. We obtain $\Delta = (16/3\pi)^{1/2} \epsilon_F/\sqrt{k_F a_s}$, and $\mu = -E_b/2 + 4\pi a_s n/m$, where $E_b = 1/ma_s^2$ is the pair binding energy. It is worth recalling that in strong coupling $\mu(T_c) = -E_b/2$, and the shift in $\mu(T = 0)$ is simply a result of the repulsive interaction $U = 4\pi a_s/m$ between the bosons.

A numerical solution [23] of the gap and number equations shows a smooth (but in this particular case, a fairly rapid) crossover from the BCS to the Bose limit. Note that there is no singularity at either the two-body bound state threshold ($1/k_F a_s = 0$) or at the point at which $\mu = 0$.

The size of the pairs may be estimated using

$$\xi_{\text{pair}}^2 = \langle \psi_{\mathbf{k}} | r^2 | \psi_{\mathbf{k}} \rangle / \langle \psi_{\mathbf{k}} | \psi_{\mathbf{k}} \rangle, \tag{25}$$

where the "internal pair wave function" is [36] $\psi_{\mathbf{k}} = \Delta_{\mathbf{k}}/2E_{\mathbf{k}}$, which is related to, but distinct from, the wave function $\varphi_{\mathbf{k}}$ in (22). Evaluating the matrix element (using $r^2 \to \nabla_{\mathbf{k}}^2$) we find that in the BCS limit $\xi_{\text{pair}} \sim v_F/\Delta \gg k_F^{-1}$, which is the interparticle spacing. With increasing coupling the pair size decreases monotonically, with a limiting form $\xi_{\text{pair}} \sim a_s \ll k_F^{-1}$ in the Bose regime.

There is an amusing feature in the evolution of the gap to single-particle (fermionic) excitations [4, 5]. This is given by the minimum of the Bogoliubov quasiparticle energy $E_{\mathbf{k}}$ within the band, namely,

$$E_{\text{gap}} = \min_{\varepsilon_{\mathbf{k}} \geq 0} \left[(\varepsilon_{\mathbf{k}} - \mu)^2 + \Delta^2 \right]^{1/2}. \tag{26}$$

From this it immediately follows that

$$E_{\text{gap}} = \begin{cases} \Delta & \text{for } \mu > 0 \\ \left(\mu^2 + \Delta^2\right)^{1/2} & \text{for } \mu < 0 \end{cases} \tag{27}$$

(for a simple way to see this, see Fig. 6 of Ref. [13]). As pointed out by Leggett [5], this has an important consequence for the case of non-s-wave pairs. Even if one has nodes in the gap function in the BCS limit, there is a non-vanishing gap in the composite boson limit which simply corresponds to the ionization energy of the diatomic molecule. The reader can find a detailed discussion of the crossover in non-s-wave superconductors in Ref. [7] (for the 3D case) and Ref. [13] (in two dimensions).

5.3 Two Dimensions (2D)

It was shown by Miyake [8] and by Duan, Shieh and the author (RDS) [13] that, in a 2D dilute gas model, the existence of a two-body bound state in vacuum is a *necessary* (and sufficient) condition for a Cooper instability. For purely attractive two-body potentials $V(r)$ this is not surprising since an infinitesimal attraction leads to a bound pair in vacuum. However, as emphasized by RDS, when $\int d^2r V(r) > 0$ (e.g., hard-core plus attraction, where this integral is infinite), one has to cross a finite threshold in the attraction before a bound state forms in vacuum (absence of a medium), and the non-trivial result is that one has to cross the same threshold for a pairing instability in a degenerate Fermi system. In addition RDS showed that this result was special to the s-wave channel: an l-wave bound state in vacuum is *not* a necessary condition for a pairing instability with $l > 0$ in 2D.

The above result is intimately related to the jump discontinuity in the 2D density of states (DOS) at the bottom of the band, or the related logarithmic singularity in the 2D propagator and T-matrix. The low energy s-wave T-matrix for two-particle scattering in vacuum (i.e., in the absence of a medium), which we shall need below, has the form

$$T_0(\omega) = (4\pi/m)\left[-\ln(\omega/E_a) + i\pi\right]^{-1}, \tag{28}$$

where E_a is the binding energy of the two-body bound state. The continuum model in 2D has another feature – a constant DOS – which greatly simplifies the crossover calculation [4, 8, 13]. For the dilute gas model in 2D this allows an exact solution of the $T = 0$ gap and number equations for all couplings leading to very simple and transparent results [13].

The variational gap equation $1/g = -\sum_{\mathbf{k}} 1/2E_{\mathbf{k}}$, with $\Delta_{\mathbf{k}} = \Delta$, has an ultraviolet divergence which is regulated by using the T-matrix. Unlike the 3D case (6), however, we cannot take the zero energy limit at the outset since the two-dimensional T-matrix (28) vanishes logarithmically. We find the "renormalized" gap equation

$$\frac{4\pi}{mT_0(2\omega)} = \int_0^\infty d\varepsilon_{\mathbf{k}} \left[\frac{1}{\varepsilon_{\mathbf{k}} - \omega - i0^+} - \frac{1}{E_{\mathbf{k}}}\right], \tag{29}$$

which is independent of ω. The only way in which the attractive interaction enters is via the energy of the bound state in vacuum, which must exist as described above.

The dimensionless coupling constant in the 2D case – the analog of $1/k_F a_s$ in 3D – is given by E_a/ϵ_F, which varies from 0 (in the BCS limit) to ∞ (in the Bose limit). It is then easy to solve the above gap equation together with (23) for arbitrary E_a/ϵ_F to obtain

$$\Delta = \sqrt{2\epsilon_F E_a} \quad \text{and} \quad \mu = \epsilon_F - E_a/2. \tag{30}$$

This simple analytical result clearly shows a smooth crossover from the BCS to the Bose limit. Note that the essential singularity of the BCS limit is buried inside the dependence of the binding energy E_a on the attractive interaction. For a detailed discussion of the 2D crossover problem, including an extension to higher angular momentum pairing, the reader is referred to Ref. [13].

5.4 Functional Integral Derivation of T=0 Results

We briefly describe, following Ref. [23], how one arrives at (24) and (23) from the functional integral of Section 3. Below T_c one has a stable uniform, static solution $\Delta(\mathbf{x}, \tau) = \Delta \neq 0$ of the saddle-point equation $\delta S_{\text{eff}}[\Delta]/\delta\Delta = 0$. Using (4) and after performing the Matsubara sum, this can be rewritten as the familiar gap equation

$$\frac{1}{g} = \sum_{\mathbf{k}} \frac{\tanh(\beta E_{\mathbf{k}}/2)}{2E_{\mathbf{k}}}. \tag{31}$$

The number equation is obtained, as before, by using the saddle-point approximation for the thermodynamic potential $\Omega_0 = S_{\text{eff}}[\Delta_0]/\beta$ in $N = -\partial\Omega/\partial\mu$. This leads to

$$n = n_0 \equiv \sum_{\mathbf{k}} \left\{1 - \frac{\xi_{\mathbf{k}}}{E_{\mathbf{k}}} \tanh\left(\frac{\beta E_{\mathbf{k}}}{2}\right)\right\}. \tag{32}$$

In the $T = 0$ limit, the above equations clearly reduce to the gap and number equations derived from the variational approach above.

It is important to emphasize that the saddle-point approximation is much more reliable for $T \ll T_c$ than it was for $T \geq T_c$. Outside of the weak coupling regime, the important physics of pairing correlations, and eventually bound state formation, above T_c is in fluctuations beyond the trivial saddle-point. In the broken symmetry state, however, the non-trivial saddle-point of S_{eff} includes the non-perturbative effects of bound state formation and condensation. At least for $T \ll T_c$, the fluctuations about the non-trivial saddle-point make only a small correction even for strong coupling.

6 Collective Excitations

Gaussian fluctuations about the broken symmetry saddle-point represent the collective excitations of the superconductor. The functional integral approach is a very elegant way to study the crossover problem for collective modes, see Ref. [23]. We present here a simpler derivation using the old-fashioned RPA [37] based on the work of Belkhir and the author [35]. We should also mention that related results on the evolution of the collective modes have been obtained by many authors for the BCS–Bose crossover [38, 39], for the crossover in the exciton condensation problem [40], and for the crossover from spin-density wave (SDW) to local moment antiferromagnetism [18, 41].

The RPA analysis of collective modes has also been extended to charged systems. While long range Coulomb interactions have been largely ignored up to this point in our discussion, we discuss them here since they have a profound impact on the collective modes [37, 42]. At the end of the section we will briefly mention some results on charged systems based on Ref. [35]. For a comprehensive treatment of plasmons in layered systems, especially in the context of the BCS–Bose crossover, the reader is referred to the recent work of Côté and Griffin [39].

Our study of collective modes is motivated by the following reasons. The important excitations in the BCS limit involve broken pairs. With increasing attraction these are pushed to very high energies, and in the Bose limit it is the collective modes which are the dominant low-energy excitations. Thus one might expect that a proper description of the intermediate coupling regime must include both broken pairs and collective modes.

Our starting point is the attractive Hubbard Hamiltonian (2) written in momentum space

$$H = \sum_{k,\sigma}(\varepsilon_k - \mu)c^+_{k\sigma}c_{k\sigma} - \frac{U}{M}\sum_{kk'q} c^+_{k+q\uparrow}c^+_{k'-q\downarrow}c_{k'\downarrow}c_{k\uparrow}, \tag{33}$$

where $\varepsilon_k = -2t\sum_{i=1}^{d}\cos(k_i a)$ and M is the total number of lattice sites. The "reduced" Hamiltonian $H_{\rm bcs}$ includes only the $k' = -k$ part of the interaction term in (33), and thus only describes the interaction between pairs with zero center-of-mass momentum.

The variational solution described above can be recast in the form of a linearization of the equations of motion for $c_{k\sigma}$ and $c^+_{k\sigma}$ with the dynamics governed by $H_{\rm bcs}$. This is the well-known Hartree–Fock–Bogoliubov scheme and one arrives at the familiar number and gap equations. Δ and the chemical potential $\tilde{\mu}$ are given by the solution of $1/U = (2M)^{-1}\sum_{\mathbf{k}} E_{\mathbf{k}}^{-1}$ and $f = (2M)^{-1}\sum_{\mathbf{k}}\left(1 - \xi_{\mathbf{k}}/E_{\mathbf{k}}\right)$. Here $\xi_{\mathbf{k}} = \varepsilon_k - \tilde{\mu}$, $E_{\mathbf{k}} = (\xi_{\mathbf{k}}^2 + \Delta^2)^{1/2}$, and $f = N/2M$ is the filling factor. The only real difference with the continuum model treated in earlier sections is that one must take into account the Hartree shift in the chemical potential: $\tilde{\mu} = \mu + fU$, which can be significant for large U.

To determine the collective mode spectrum we study the time evolution of density fluctuations $c_{k+q\sigma}^{+}c_{k\sigma}$. Its equation of motion is coupled to that of pairs with a finite center of mass momentum $c_{k+q\uparrow}^{+}c_{-k\downarrow}^{+}$ and $c_{-k-q\downarrow}c_{k\uparrow}$, resulting from the particle–hole mixing due to the condensate: $\langle c_{k\uparrow}^{+}c_{-k\downarrow}^{+}\rangle \neq 0$. The overall strategy of the RPA is simple: first, to linearize the equations of motion [37] with respect to the mean field ground state, and, second, to diagonalize the resulting Anderson–Rickayzen equations by finding the appropriate eigenoperators for the collective coordinates. At the mean field level one has diagonalized only the H_{bcs} part of the full Hamiltonian $H = H_{\mathrm{bcs}} + H_{\mathrm{int}}$. At the RPA level, we treat the small fluctuations introduced by H_{int} which describes the interaction between the Bogoliubov quasiparticles.

The actual implementation of this idea is algebraically messy. We omit all details, referring the reader to Ref. [35], which generalizes the weak coupling calculation of Bardasis and Schrieffer [43] to the crossover problem. The BCS limit is considerably simplified by an approximate particle–hole symmetry arising from the fact that only fermions within a thin shell symmetric about the Fermi energy are affected by the pairing. For arbitrary U, one does not have such a p–h symmetry and the calculations are more complicated.

To determine the collective excitation spectrum $\omega(q)$ we take the matrix elements of the equations of motion between the ground state (which is the BCS state renormalized by the zero-point motion of the collective modes), and a state containing exactly one quantum of excitation. The resulting secular equation is

$$\begin{vmatrix} [1 + U I_{E,n,n}(q)] & U I_{\omega,n,l}(q) & 2U I_{E,n,m}(q) \\ U I_{E,n,m}(q) & [1 + U I_{E,l,l}(q)] & 2U I_{\omega,l,m}(q) \\ (U/2) I_{E,m,n}(q) & (U/2) I_{\omega,m,l}(q) & [1 + U I_{E,m,m}(q)] \end{vmatrix} = 0, \tag{34}$$

with

$$I_{a,b,c}(q) = \frac{1}{M}\sum_{k} \frac{a(k,q)b(k,q)c(k,q)}{\omega(q)^2 - E_{k,q}^2}. \tag{35}$$

Here a, b, c denote any one of the following quantities: the excitation energy $\omega(q)$, or the quasiparticle energy $E_{k,q} = E_{k+q}+E_k$, or the coherence factors $l(k,q) = u_k u_{k+q} + v_k v_{k+q}$, $m(k,q) = u_k v_{k+q} + v_k u_{k+q}$, and $n(k,q) = u_k u_{k+q} - v_k v_{k+q}$. We now solve (34) for $\omega(q)$, restricting attention to the long wavelength $q \to 0$ regime.

In weak coupling, the integrals $I_{\omega,n,l}$ and $I_{E,n,m}$ vanish as a result of the p–h symmetry discussed above, and we are left with a 2×2 determinant to solve. The small q and ω expansion of the remaining terms in (34) is conveniently written in terms of four quantities: $x = \sum_k E_k^{-3}$, $y = \sum_k (\nabla_k \xi)^2/E_k^3$, $w = \sum_k \xi\, \nabla_k^2 \xi / E_k^3$, and $z = \Delta^2 \sum_k (\nabla_k \xi)^2/E_k^5$. The momentum sums can be readily evaluated to obtain the $\mathbf{q} \to 0$ dispersion relation in d dimensions

$$\omega(q) = \left[\langle v_F^2\rangle/d\right]^{1/2} (1 - UN(0))^{1/2}\, q. \tag{36}$$

Here $\langle v_F^2\rangle = N_v(0)/N(0)$ is the mean squared Fermi velocity, where the density of states $N(\xi) = (2\pi)^{-d} \int d^d k \delta(\xi_k - \xi)$, and $N_v(\xi) = (2\pi)^{-d} \int d^d k$ $(\nabla_k \xi_k)^2 \delta(\xi_k - \xi)$. Note that $\langle v_F^2\rangle^{1/2}$ is ta times a dimensionless function of the filling f (for plots of this in 2D and 3D see Ref. [35]). The lattice result (36) is essentially the same as Anderson's continuum result [37] (where $\sqrt{\langle v_F^2\rangle} = p_F/m$), since the size of the bound pairs is much larger than the lattice spacing a in weak coupling.

In the strong coupling limit the gap and the number equations can be solved, to leading order in $\alpha = 2t/U \ll 1$, to obtain $\Delta = U[f(1-f)]^{1/2}\left[1 - d\alpha^2\right]$ and $\tilde{\mu} = U(2f-1)\left[1 + 2d\alpha^2\right]/2$. To make contact with the Bogoliubov theory for the dilute Bose gas we focus on the low density limit $f \ll 1$. The elements of the determinant in (34) can be expanded as follows: $1 + U I_{E,n,n} = 4f\left[1 + 4d\alpha^2\right]$, $1 + U I_{E,m,m} = 1 - 16 f d\alpha^2$, $U I_{E,n,m} = -2f^{1/2}\left[1 + d\alpha^2\right]$, $U I_{\omega,n,l} = -\omega\left[1 + 2d\alpha^2\right]/U$, and $U I_{\omega,m,l} = 2\omega f^{1/2}\left[-1 + 7d\alpha^2\right]/U$. We also find, independent of the filling, $1 + U I_{E,l,l} = \alpha^2 q^2 a^2/2 - \omega^2/U^2$. We thus obtain, for $f \ll 1$ and $U/t \gg 1$, the excitation dispersion relation

$$\omega(q) = \sqrt{4df}\,(4t^2 a/U)\, q. \tag{37}$$

To make contact with Bogoliubov theory, we note that in the strong coupling limit the fermions bind into on-site singlet pairs, and the attractive Hubbard model can be mapped onto a system of hard core bosons described by $H_{\text{bose}} = \left(2t^2/U\right)\sum_{i,j}(-b_i^+ b_j + n_i n_j)$. The pairs move only via virtual dissociation which leads to a t^2/U hopping amplitude, and the hard core constraint comes from the Pauli principle for the constituent fermions. In the long wavelength limit, the collective excitation of a dilute ($n_b a_b^3 \ll 1$) 3D Bose gas is the sound mode with dispersion

$\omega = \left(4\pi n_b a_b / m_b^2\right)^{1/2} q$. Here the density of bosons $n_b = f/a^3$, the effective mass $m_b \simeq 1/t_b a^2 = U/2t^2a^2$, and the boson scattering length a_b is of the order of the lattice spacing a. This leads to a speed of sound which is essentially the same as that in (37), up to numerical factors of order unity, which are not unexpected given the crude estimate of a_b. Quite remarkably, starting with interacting fermions and using RPA we were able to reach the regime of a low density Bose gas in the strong coupling limit.

For intermediate couplings we have solved the RPA equation (34) numerically as a function of the coupling U/t and for various fillings f; see Ref. [35] for details. The numerical results show that the spectrum smoothly interpolates between the Anderson mode in weak coupling and the Bogoliubov sound mode in strong coupling.

The evolution of the eigenvector corresponding the collective mode is also of some interest. For $q = 0$ the eigenvector is a pure phase mode, for all couplings, as required by Goldstone's theorem. However, as shown in Ref. [23], for small q there is a small admixture of the amplitude mode. The magnitude of this mixing is controlled by, in addition to q, the deviation from p–h symmetry, and therefore grows with the coupling.

6.1 Charged Systems

We briefly summarize the results of an analysis [35] which includes the effects of the long ranged Coulomb interaction

$$V_c(q) = \frac{2(d-1)\pi\, e^2}{a^d q^{d-1}}, \tag{38}$$

where a^d is the unit cell volume in d-dimensions. The effective interaction in the particle–particle channel is taken to be the attraction $-U$ as before; however the bare V_c is used in the particle–hole channel. This can be shown to modify the equations of motion as follows: in the third row of (34) each factor of U is replaced by $V_c(q)$, while the first two rows remain unchanged.

For charged systems we work only in the high density regime, since at sufficiently low density one would expect an instability to a Wigner crystal of pairs (which is clearly outside the scope of this analysis). For the lattice model under consideration we work exactly at half-filling $f = 1/2$ since this simplifies the algebra considerably. We only quote the final results for $q \to 0$. In the weak coupling limit in 3D one finds the well-known result [37] that the Anderson mode of the neutral system is pushed up to the plasma frequency: $\omega^2 = 8\pi e^2 \langle v_F^2 \rangle N(0)/3$. For $U/t \gg 1$ in 3D we find $\omega^2 = (4\pi e^2/a^3)(4t^2a^2/U)$. Using the boson mass $m_{\mathrm{B}} = U/2t^2a^2$, charge $e_{\mathrm{B}} = 2e$, and density $n_{\mathrm{B}} = 1/2a^3$, this result may be rewritten as $\omega^2 = 4\pi n_{\mathrm{B}} e_{\mathrm{B}}^2/m_{\mathrm{B}}$ which is exactly the plasma frequency of a dense charged bose gas [44].

In 2D the plasmon does not have a gap and we find a $\sqrt{q}$ dispersion. The weak coupling result in 2D is $\omega^2 = 2\pi e^2 \langle v_F^2 \rangle N(0) q$, while in the Bose limit we find $\omega^2 = \left(8\pi e^2 t^2/U\right) q$ which may be rewritten as $\omega = \left(2\pi e_{\mathrm{B}}^2 n_{\mathrm{B}}/m_{\mathrm{B}}\right)^{1/2} \sqrt{q}$. The smooth evolution of the plasmon from the BCS to the Bose limit is demonstrated by a numerical solution of the RPA equations in Ref. [35].

7 Numerical Studies: Normal State Crossover

The absence of a small parameter in the intermediate coupling regime, the importance of dynamic fluctuations, and the special problems of two dimensions discussed in Section 3, all suggest that numerical simulations will play a very important role in elucidating the BCS–Bose crossover problem. Quantum Monte Carlo (QMC) methods for strongly correlated systems have been intensively investigated in recent years. While much of the work on lattice fermions has focussed on the repulsive Hubbard model [45], there have been some studies of the 2D *attractive* model relevant to the present subject [46, 47, 14].

Very briefly, the finite temperature QMC scheme is as follows. Using an imaginary-time coherent state path integral formulation, and a Hubbard–Stratanovich decoupling, the fermion problem is written as a functional integral over the bosonic decoupling fields with a complicated non-local action. The bosonic path integral is then simulated using Monte Carlo techniques. The determinants obtained from integrating out the fermion fields can usually take on both positive and negative values, thus leading to large statistical errors – the "sign problem" that plagues fermion Monte Carlo calculations. An important feature of the attractive Hubbard model is the absence of the sign problem [48]: for a charge decoupling, in which the auxiliary field couples to $(n_{i\uparrow} + n_{i\downarrow})$, the determinants are always positive definite.

Scalettar *et al.* [46] first studied the phase diagram of the 2D attractive Hubbard model using QMC, and presented preliminary evidence for superconductivity away from half-filling, and the coexistence of SC and CDW, together with a vanishing T_c, at half-filling $\langle n \rangle = 1$. Moreo and Scalapino [47] demonstrated the existence of off-diagonal algebraic order for $\langle n \rangle \neq 1$ and, using finite size scaling, estimated the Kosterlitz–Thouless T_c. As a function of U and $\langle n \rangle$, the maximum T_c was found to be around $0.1t$.

Anomalous Normal State Properties: Spin Gaps

The normal state, just above T_c, in the $U/t \ll 1$ limit is a Fermi liquid, while that in the opposite $U/t \gg 1$ limit is a normal Bose liquid. We now describe the results of a QMC study [14] of how the normal state evolves from a Fermi liquid-like regime to a Bose regime as a function of increasing attraction. Understanding when and how deviations from Fermi liquid behavior arise in strongly correlated Fermi systems is a major topic of current research in many-body physics. The deviations from Fermi liquid behavior discussed below are, in a sense, generic to the normal state of a short coherence length superconductor.

We show that for intermediate coupling strengths the normal state clearly deviates from a canonical Fermi liquid. Our main results are: (1) the uniform, static spin susceptibility χ_s is strongly temperature dependent, with $d\chi_s/dT > 0$, for $T_c < T < T_p$, where T_c is the superconducting transition temperature, and T_p a "pairing" scale below which strong singlet pairing correlations develop. (T_p, which will be defined below, is closely related to T_0 of Section 3.) (2) This is accompanied by a "spin gap", i.e, a reduction in the low frequency spectral weight, leading to an NMR relaxation rate $1/T_1T$ which tracks $\chi_s(T)$. This is very different from the Korringa law in a Fermi liquid, namely $1/T_1T \sim \chi_s^2$, with temperature independent Pauli susceptibility χ_s.

Our results provide a natural explanation for the anomalous spin gap behavior of the NMR Knight shifts and relaxation rates observed [49, 50, 51, 52] above T_c in many of the underdoped high T_c superconductors. We discuss in more detail below the implications of our results for experiments on the reduced oxygen YBCO systems, especially why the spin-gap anomalies are observed only in the underdoped materials and not in the optimally doped ones. We also note that normal state pseudo-gaps have also been observed in the optical conductivity [53] of underdoped cuprates.

We emphasize that our results, which are perhaps easiest to rationalize in the non-degenerate, pre-formed boson limit at large-U, persist well into the intermediate-U regime where one has a degenerate Fermi system. This is important to establish because the materials which exhibit these anomalies also have a Fermi surface as measured by angle-resolved photoemission spectroscopy (ARPES) [54]. Thus the spin gaps cannot come from "pre-formed pairs" but, rather, from strong singlet correlations.

We now give the details of our results. The numerical simulations are on systems of size up to 8×8 with some data obtained from 16×16 lattices. Most of the data is at quarter-filling $\langle n \rangle = 0.5$ for temperatures $T > 0.1t$; note that $T_c \simeq 0.05t$ for $U/t = 4$ at this filling [47]. The chemical potential μ is adjusted as a function of the interaction and temperature to obtain the desired $\langle n \rangle$; see Fig. 1 of Ref. [14]. We find $d\mu/dT < 0$, as one usually expects, for all T and U, in contrast to the result of summing ladder diagrams [30] in 2D (see end of Section 2 for a discussion of problems with finite temperature calculations in 2D). Further, at low T the results agree with the $T = 0$ mean field treatment [35] described in previous sections. Comparing the chemical potential with the temperature we can check whether or not one has a degenerate Fermi system, which is important for the reasons given above.

The uniform, static spin susceptibility $\chi_s \equiv \chi(q \to 0, \omega = 0)$ as a function of U and T, is shown in Fig. 4. For $U = 0$, χ^0 has the expected Pauli behavior for low temperatures and crosses over to a Curie-like form at higher temperatures (with a small bump at $T \simeq 0.5$ due to the log singularity in the DOS at the band center). With increasing U the overall magnitude of χ_s decreases, which may be understood (at high T) within RPA with $\chi^{\text{RPA}} = \chi^0/(1 + U\chi^0)$. However, the behavior $d\chi_s/dT > 0$ observed in the data is the result of strong singlet pair fluctuations in the normal state and is, of course, completely absent in the RPA. For large U one obtains tightly bound singlet pairs which contribute to χ_s only when they are ionized. What is remarkable is that such correlations persist down to $U = 4$ where one has a degenerate Fermi system.

The temperature at which the Monte Carlo data deviates from the RPA can be used to define a "pairing" scale T_p which is clearly larger (for all the values of U studied) than the transition temperature T_c at which coherence is established. We expect that T_p (defined from χ_s) is essentially the same as the crossover scale T_0 of Section 3. The anomalous behavior in the normal state discussed here is in the regime $T_c < T < T_p$. For small U, T_p is expected to be the same as the mean-field transition temperature T_c^0, and one can describe the regime $T_c < T < T_c^0$ in terms of superconducting fluctuations [55]. For large U, the two scales are widely separated with $T_c^0 \sim t^2/U$, while $T_p \sim U$ gives an estimate of the pair binding energy. The anomalous temperature dependences in

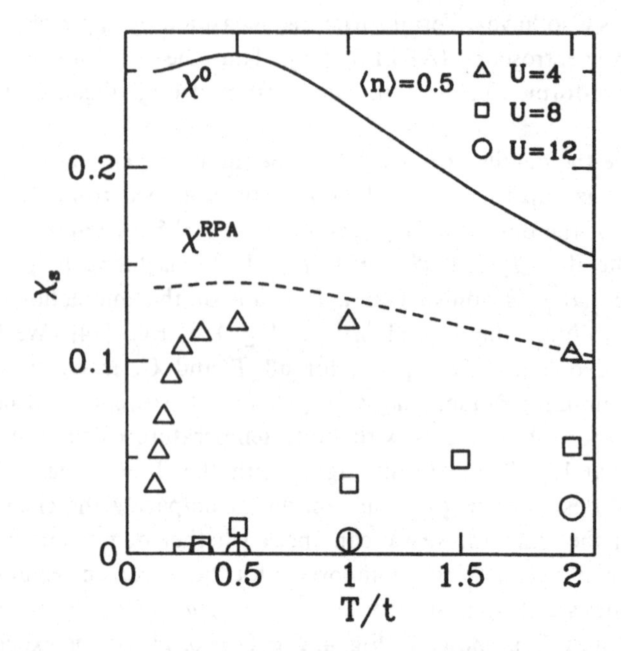

Fig. 4. The normal state spin susceptibility for the 2D attractive Hubbard model at quarter-filling $\langle n \rangle = 0.5$ plotted as a function of temperature, for various U values. The full curve is for the non-interacting system. The $U/t = 4$ and 8 data are from Monte Carlo simulations on an 8×8 lattice, and the $U/t = 12$ data are on a 4×4 lattice. The error bars are of the size of the symbols. The RPA curve is fit to the $U/t = 4$ data (not shown in the diagram) in the high temperature regime $2t < T < 4t$ using a "renormalized" $\tilde{U} = 3.25$. (From Ref. [14].)

the normal state then clearly arise from the formation of singlet pairs without any coherence, and *not* from superconducting fluctuations.

A static susceptibility with $d\chi_s/dT > 0$, though suggestive, is by itself not sufficient to establish a "spin-gap", namely a reduction in the low frequency spectral weight $\mathrm{Im}\chi(\mathbf{q}, \omega)$, which is probed by the NMR relaxation rate $1/T_1T = \lim_{\omega \to 0} \sum_{\mathbf{q}} \mathrm{Im}\chi(\mathbf{q}, \omega)/\omega$. In general, it is very difficult to obtain $\mathrm{Im}\chi(\mathbf{q}, \omega)$ from imaginary-time Monte Carlo data, since it requires analytic continuation from Matsubara to real frequencies. However, we use a trick that allows us to extract $1/T_1T$ directly from the imaginary time spin–spin correlation function measured at $\tau = 1/2T$, provided the temperature is much smaller than the characteristic frequency of the spectral weight $\mathrm{Im}\chi(\mathbf{q}, \omega)$ (see Ref. [14] for details).

The relaxation rate $1/T_1T$ determined from this analysis is plotted in Fig. 5, it is found to decrease as T is lowered. In Fig. 6 we show

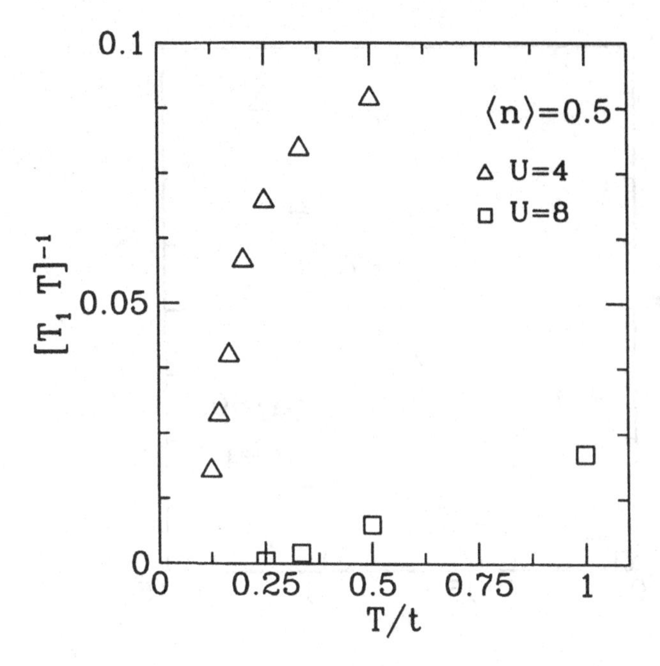

Fig. 5. The normal state NMR relaxation rate $1/T_1T$ (in arbitrary units) for the 2D attractive Hubbard model, at quarter-filling $\langle n \rangle = 0.5$, for $U/t = 4$ and 8 on 8×8 lattices. (From Ref. [14].)

that $1/T_1T$ indeed tracks the static susceptibility χ_s, thus establishing spin-gap behavior in the normal state. It is worth noting that, for the two values of U studied, the data for $1/T_1T$ vs. χ_s appear to lie on the same curve.

Is the opening up of a "spin gap" accompanied by a charge gap, or a pseudo-gap? In this model the answer appears to be yes, although more detailed work is necessary. The single-particle density of states obtained via analytic continuation in Ref. [57] gives some evidence for a gap-like structure developing above T_c. We have also studied the momentum distribution function $\langle n_k \rangle$. While $\langle n_k \rangle$ is clearly broader than what would be expected for a Fermi gas ($Z = 1$) with thermal smearing, it is difficult to conclude (from fitting the Monte carlo data to various functional forms) whether the observed behavior necessarily implies a gap. However the measured $\langle n_k \rangle$ again clearly demonstrates that the system is in a degenerate Fermi regime.

Finally, we discuss the applicability of our results to normal state

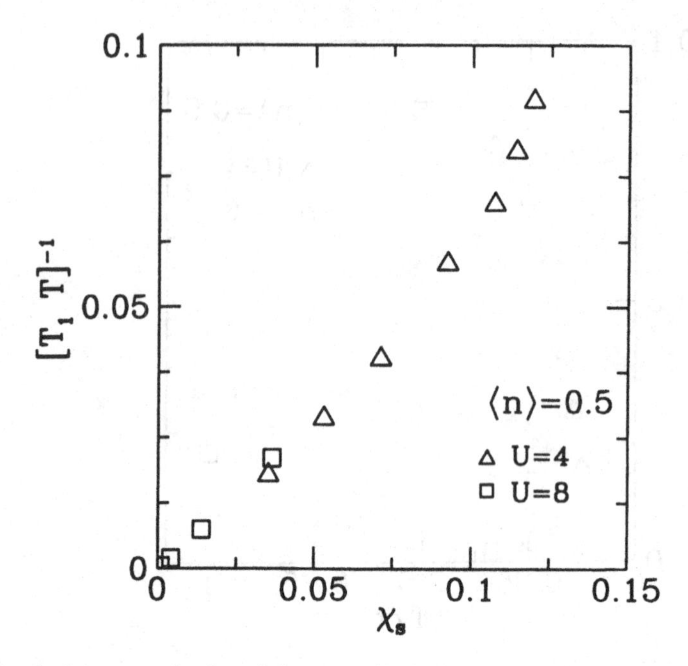

Fig. 6. A parametric plot of the normal state NMR relaxation rate $1/T_1T$ (from Fig. 5) versus the spin susceptibility (from Fig. 4) for the 2D attractive Hubbard model. For a Fermi liquid all of the data, for a given U, should collapse to a single point; instead the two quantities track each other: $1/T_1T \sim \chi_s$. (From Ref. [14].)

NMR experiments [49, 50, 51] on the $YBa_2Cu_3O_{6+x}$ systems; while these are the best studied systems, spin-gap behavior is not [52] restricted to them. Two issues (at least) need to be discussed: (1) the role of antiferromagnetism (AFM), and (2) why spin gaps are seen in some materials, and not others.

In the $T_c = 60$K system ($x \simeq 0.65$) the Knight shift, which probes χ_s, increases by a factor of four as T increases from T_c to 300K, which we would identify as T_p. For the O and Y sites (where the form-factors [56] filter out the AFM contribution) $1/T_1T$ indeed tracks χ_s below 300K. Our model, which has no AFM, is able to explain these spin gap features naturally. For the Cu site, $1/T_1$ is larger in magnitude and shows spin gap behavior only below 150K. This would appear to require a combination of AFM fluctuations and a spin gap opening up due to enhanced pairing correlations. Clearly our simple model does not have any non-trivial **q**-dependence of the spin gap.

In the $T_c = 90\text{K}$ ($x \simeq 1.0$) material, spin gap behavior is not seen (or is seen over a very narrow regime). To discuss the range $T_c < T < T_p$ over which these anomalies will be seen, one needs to know how T_c and T_p change as a function of doping (x). Since the attractive Hubbard model is *not* a *microscopic* model for YBCO (e.g., half-filling in this model has nothing to do with the magnetic insulator at $x = 0$), we can only make qualitative remarks. Let us assume that once YBCO is metallic, increasing x is analogous to increasing the carrier concentration n in the model. For fixed U, T_c increases with n, till $n \simeq 0.85$, and then drops to zero at $n = 1.0$ in 2D, as shown in Ref. [47]. On the other hand, T_p is expected to be relatively independent of n for small n, to the extent that one has independent pair formation, and could even drop with increasing n due to the Pauli principle blocking the formation of independent pairs. This crude argument suggests why the system with the highest T_c, and highest carrier concentration, might have the smallest window between the pairing temperature and T_c.

8 Summary and Conclusions

We have reviewed the subject of the crossover from cooperative Cooper pairing to the Bose condensation of tightly bound pairs as the strength of the attractive interaction increases. Most of the discussion centered around a continuum model of an attractive Fermi gas or the attractive Hubbard model on a lattice. A variety of techniques – variational, diagrammatic, equations of motion, functional integrals and quantum Monte Carlo – have been used to study this problem. We briefly summarize the main physical ideas that emerge from these studies.

First and foremost: at any given coupling, there is only one finite temperature phase transition in this problem at T_c, which is from a normal state to a superfluid state with ODLRO or, in two-dimensions, algebraic order. Even though the weak and strong coupling limits appear to be very different at first sight, there are no phase transitions as a function of coupling (see Fig. 1).

In the broken symmetry state, below T_c, as a function of increasing interactions, the system evolves from a BCS state, which is a condensate of large, overlapping Cooper pairs, to a Bose condensate of tightly bound composite bosons. Variationally, all that happens is that the pair wave function shrinks in the relative coordinate; the large kinetic energy cost required to do this being more than compensated by the even larger gain from the attractive interaction. The important low energy excitations in the BCS limit are broken pairs with an exponentially small gap. With increasing coupling these get pushed to higher energies. The Anderson–Bogoliubov collective mode for the neutral BCS system evolves smoothly into the Bogoliubov–Beliaev sound mode for the Bose gas. Inclusion of long range Coulomb interactions modifies these modes into plasmons, which also evolve smoothly from the BCS to the Bose limit.

The nature of the transition in the two limiting cases is rather different. In the BCS case, the formation of the Cooper pairs and their condensation occurs at the same temperature, which is T_c. The first deviations from weak coupling may be understood in terms of superconducting fluctuations. With increasing attraction the crossover scale T_0 at which pair correlations become manifest and the transition temperature T_c get increasingly separated. In the strong coupling regime, T_0 is roughly the temperature below which real bound states begin to appear. For $T \gg T_0$ the bound states are thermally dissociated. In the Bose limit, the formation of the bound pairs and their condensation are two independent processes with widely separated energy scales; the only phase transition is at the condensation T_c.

The normal state just above T_c, in the BCS limit, is a Fermi liquid, while that in the opposite strong coupling limit is a normal Bose liquid of tightly bound pairs. How does the system evolve from a Fermi liquid to a Bose liquid as a function of the interaction strength? In two dimensions, numerical simulations are beginning to provide an answer. For moderate values of the attraction, the normal state, even though it is still in a degenerate Fermi regime, already shows anomalous magnetic correlations due to the opening up of pseudo-gap in the normal state. These effects are strikingly similar to anomalies seen in normal state NMR and optical conductivity experiments on underdoped high T_c systems.

In conclusion, the problem of the crossover from BCS theory to Bose–Einstein condensation is a very rich one, and there are possibly new surprises in the intermediate coupling regime that we have yet to uncover.

Acknowledgements. I would like to thank Jan Engelbrecht, Carlos Sá de Melo, Nandini Trivedi and Lotfi Belkhir for invaluable input; much of the work presented here was done in collaboration with them. I am grateful to Tony Leggett for first introducing me to this problem and for many conversations over the years. Finally I would like to thank Alexei Abrikosov and Allan Griffin for their interest and encouragement.

This work was supported by the US Department of Energy, Office of Basic Energy Sciences under contract W-31-109-ENG-38.

References

[1] J. R. Schreiffer *Theory of Superconductivity* (Benjamin/Cummings, Reading, 1964).

[2] For a review, see J. M. Blatt *Theory of Superconductivity* (Academic, New York, 1964) and references therein.

[3] C. N. Yang, Rev. Mod. Phys. **34**, 694 (1962).

[4] D. M. Eagles, Phys. Rev. **186**, 456 (1969).

[5] A. J. Leggett in *Modern Trends in the Theory of Condensed Matter*, A. Pekalski and R. Przystawa, eds. (Springer-Verlag, Berlin, 1980); and J. Phys. (Paris) Colloq. **41**, C7–19 (1980).

[6] N. D. Mermin and P. Muzikar, Phys. Rev. **21**, 980 (1980).

[7] M. G. McClure, D. Phil. thesis, University of Sussex (1981).

[8] K. Miyake, Prog. Theor. Phys. **69**, 1794 (1983).

[9] P. Nozières and S. Schmitt-Rink, J. Low Temp. Phys. **59**, 195 (1985).

[10] See C. Comte and P. Nozières, J. Phys. (Paris) **43**, 1069, 1083 (1982); H. Haug and S. Schmitt-Rink, Prog. Quant. Elect. **9**, 3 (1984); and references therein.

[11] J. Bednorz and K.A. Muller, Z. Phys. B **64**, 189 (1986); M.K. Wu *et al.*, Phys. Rev. Lett. **58**, 908 (1987).

[12] See *High Temperature Superconductivity, Proceedings of the Los Alamos Conference*, K. Bedell *et al.*, eds. (Addison-Wesley, Redwood City, Calif., 1990).

[13] M. Randeria, J. Duan, and L. Shieh, Phys. Rev. Lett. **62**, 981 (1989); Phys. Rev. **41**, 327 (1990).

[14] M. Randeria, N. Trivedi, A. Moreo, and R. T. Scalettar, Phys. Rev. Lett. **69**, 2001 (1992) and in preparation.

[15] A. S. Alexandrov and J. Ranninger, Phys. Rev. **23**, 1796 (1981); S. Robaszkiewicz, R. Micnas, and K. A. Chao, Phys. Rev. **23**, 1447 (1981).

[16] R. Micnas, J. Ranninger, and S. Robaszkiewicz, Rev. Mod. Phys. **62**, 113 (1990) and references therein.

[17] See the reviews by H. Capellmann, and by T. Moriya in *Metallic Magnetism*, H. Capellmann, ed. (Springer, Berlin, 1986); G. G. Lonzarich and L. Taillefer, J. Phys. C **43**, 4339, 1083 (1985).

[18] J. R. Schrieffer, X. G. Wen, and S. C. Zhang, Phys. Rev. B **39**, 11663 (1988).

[19] M. P. A. Fisher, G. Grinstein, and S. M. Girvin Phys. Rev. Lett. **64**, 587 (1990); A. F. Hebard and M. Paalanen, Phys. Rev. Lett. **69**, 1604 (1992).

[20] K. Moulopoulos and N. W. Ashcroft, Phys. Rev. Lett. **66**, 2915 (1991).

[21] M. Schmidt, G. Röpke, and H. Schulz, Ann. Phys. **202**, 57 (1990); T. Alm, B. L. Friman, G. Röpke, and H. Schulz, Nucl. Phys. A **551**, 45 (1993); see also G. Röpke, this volume.

[22] C. A. R. Sá de Melo, M. Randeria, and J. R. Engelbrecht, Phys. Rev. Lett. **71**, 3202 (1993).

[23] J. R. Engelbrecht, C. A. R. Sá de Melo, and M. Randeria (preprint, 1993).

[24] V. N. Popov, *Functional Integrals and Collective Excitations* (Cambridge University Press, 1987); for a nice introduction and references, see V. Ambegaokar in *Percolation, Localization and Superconductivity*, A. M. Goldman and S. A. Wolf, eds. (Plenum, New York, 1984).

[25] For a more careful derivation, see M. Randeria, J. Duan and L. Shieh, Phys. Rev. **41**, 327 (1990).

[26] See, e.g., E. M. Lifshitz and L. P. Pitaevskii, *Statistical Physics*, Part I (Pergamon, Oxford, 1980), Section 77.

[27] M. Drechsler and W. Zwerger, Ann. Physik **1**, 15 (1992); W. Zwerger, (preprint 1993).

[28] R. Haussmann, Z. Phys. B **91**, 291 (1993), and private communication.

[29] B. L. Gyorffy, J. B. Staunton, and G. M. Stocks Phys. Rev. B **44**, 5190 (1991).

[30] S. Schmitt-Rink, C.M. Varma and A. Ruckenstein, Phys. Rev. Lett. **63**, 445 (1989); J. Serene, Phys. Rev. B **40**, 10873 (1989); A. Tokumitu, K. Miyake, and K. Yamada, Prog. Theor. Phys. Suppl. **106**, 63 (1992); Phys. Rev. B **47**, 11988 (1993).

[31] P.J. H. Denteneer, G. An, and J. M. J. van Leeuwen, Europhys. Lett. **16**, 5 (1991).

[32] E. Abrahams and T. Tsuneto, Phys. Rev. **152**, 416 (1966); see also M. Cyrot, Rep. Prog. Phys. **36**, 103 (1973) and references therein.

[33] H. Ebisawa and H. Fukuyama, Prog. Theor. Phys. **46**, 1042 (1971).

[34] Even in a purely bosonic system the dynamics of the order parameter will be damped sufficiently close to T_c (see P. Hohenberg and B. I. Halperin, Rev. Mod. Phys. **49**, 435 (1977), section VI E). Our

analysis, however, is restricted to be valid only outside the critical region.

[35] L. Belkhir and M. Randeria, Phys. Rev. B **45**, 5087 (1992); Phys. Rev. B **49**, 6829 (1994).

[36] See, e.g., A. J. Leggett in *Quantum Liquids*, J. Ruvalds and T. Regge, eds. (North Holland, Amsterdam, 1978), p. 167.

[37] P. W. Anderson, Phys. Rev. **112**, 1900 (1958).

[38] J. Sofo, C. A. Balseiro, and H. E. Castillo, Phys. Rev. B **45**, 9860 (1992); T. Kostryko and R. Micnas, Phys. Rev. B **46**, 11025 (1993); A. S. Alexandrov and S. G. Rubin, Phys. Rev. B **47**, 5141 (1993).

[39] R. Côté and A. Griffin, Phys. Rev. B **48**, 10404 (1993).

[40] S. Schmitt-Rink, D. S. Chemla, and H. Haug, Phys. Rev. B **37**, 941, (1988); R. Côté and A. Griffin, Phys. Rev. B **37**, 4539 (1988).

[41] H. Monien and K. Bedell, Phys. Rev. B **45**, 3164 (1992); P.J. H. Denteneer and J. M. J. van Leeuwen, Europhys. Lett. **22**, 413 (1993).

[42] For a review and references, see P. C. Martin in *Superconductivity*, R. D. Parks, ed. (M. Dekker, New York, 1969), Part 1, p. 371.

[43] A. Bardasis and J. R. Schrieffer, Phys. Rev. **121**, 1050 (1961).

[44] L. Foldy, Phys. Rev. **125**, 2208 (1962).

[45] See the review by D. J. Scalapino in *High Temperature Superconductivity, Proceedings of the Los Alamos Conference*, K. Bedell *et al.*, eds. (Addison-Wesley, Redwood City, Calif., 1990).

[46] R. T. Scalettar *et al.*, Phys. Rev. Lett. **62**, 1407 (1989).

[47] A. Moreo and D. J. Scalapino, Phys. Rev. Lett. **66**, 946 (1991).

[48] J. E. Hirsch, Phys. Rev. B **28**, 4059 (1983).

[49] W. W. Warren *et al.*, Phys. Rev. Lett. **62**, 1193 (1989); R. E. Walstedt *et al.*, Phys. Rev. B **41**, 9574 (1990).

[50] M. Takigawa *et al.*, Phys. Rev. B **43**, 247 (1991).

[51] H. Alloul, T. Ohno, and P. Mendels, Phys. Rev. Lett. **63**, 1700 (1989).

[52] H. Zimmerman *et al.*, *Physica* C **159**, 681 (1989); R. E. Walstedt, R. F. Bell and D. B. Mitzi, Phys. Rev. B **44**, 7760 (1991).

[53] J. Orenstein *et al.*, Phys. Rev. B **42**, 6342 (1990); Z. Schlesinger *et al.*, Phys. Rev. Lett. **67**, 2741 (1991); C. C. Homes *et al.*, Phys. Rev. Lett. **71**, 1645 (1993).

[54] R. Liu *et al.*, Phys. Rev. B **46**, 11056 (1992).

[55] In 2D where the Kosterlitz–Thouless T_c is much lower than T_c^0, it is interesting to ask if deviations from Fermi liquid behavior persist even as $U \to 0$. Finite size errors, which are more pronounced at small U, make it difficult to address this question via Monte Carlo calculations.

[56] B. S. Shastry, Phys. Rev. Lett. **63**, 1288 (1989); A. J. Millis, H. Monien, and D. Pines, Phys. Rev. B **42**, 167 (1990); N. Bulut. D. Hone, D. J. Scalapino and N. E. Bickers, Phys. Rev. B **41**, 1797 (1990).

[57] A. Moreo, D. J. Scalapino, and S. R. White, Phys. Rev. B **45**, 7544 (1992); A. Moreo, N. Trivedi, and M. Randeria (unpublished).

16

Bose–Einstein Condensation of Bipolarons in High-T_c Superconductors

J. Ranninger

Centre de Recherches sur les Très Basses Températures,
CNRS, BP 166
38042 Grenoble-Cédex 9
France

Abstract

The short coherence lengths and the low carrier concentrations in high temperature superconductors (HT_cSC) quite naturally favor a scenario where preformed bound electron pairs undergo a Bose–Einstein condensation (BEC). There are numerous experimental indications in the normal state properties (anomalous diamagnetic susceptibility, NMR spin-lattice relaxation, Knight shift and crystal electric field transitions, as well as entropy non-linear in T and strong local lattice deformations seen by EXAFS), all of which indicate a characteristic temperature T_{PB} associated with the breaking of such preformed electron pairs and the existence of a pseudogap above T_c. As the number of charge carriers is increased by chemical doping, T_{PB} decreases and the critical temperature T_c, where superconductivity sets in, in this so-called "underdoped regime", increases. The maximum value for T_c is reached when $T_c \lesssim T_{PB}$. Upon further doping, going into the so-called "overdoped regime", T_c decreases. We take this as an indication that the superconducting state depends on the existence of preformed electron pairs. The normal state properties of the "underdoped regime" seem to show significant differences from Fermi liquid behavior, while in the "overdoped regime" they seem to be typical of ordinary metals.

In order to capture this empirically established scenario in HT_cSC, we discuss a phenomenological model based on a mixture of localized bosons (bound electron pairs such as bipolarons) and itinerant fermions (electrons), with a local exchange between bosons and fermion pairs. Superconductivity in such a system materializes when, due to the boson–fermion exchange mechanism, a BCS-like gap opens up in the electron spectrum. At the same time, the initially localized bosons acquire itinerancy and condense inside the energy gap of the fermions. The dependence of T_c on the total *number of carriers is reminiscent of the BEC temperature of lattice bosons, with an effective mass determined by the boson–fermion pair exchange term. Above T_c, the bosons behave as resonant states inside the Fermi sea, which is manifest by the appearance of a pseudogap and a blocking of the chemical potential close to the bosonic level.*

1 Introduction

Since the experimental discovery of the first superconducting material (Hg) by Kamerlingh Onnes in 1911 [1], progress in producing superconducting materials with critical temperatures higher than about 20 K stagnated until the discovery of the high T_c superconductors by Bednorz and Müller and Chu *et al.* [2].

As far as the classical (pre-high T_c superconductors) are concerned, they could be described by the BCS theory [3] based on the idea of pairing of electrons in k-space – called Cooper pairs. Within the weak coupling limit of this theory, the critical temperature T_c is given by:

$$T_c \simeq 1.13\omega_0 e^{-1/\lambda_{\text{eff}}}, \tag{1.1}$$

where $\lambda_{\text{eff}} = \mu^* - \lambda$, μ^* denotes the effective Coulomb repulsion and λ is the phonon-mediated electron–electron attraction. The prefactor in the expression for T_c being proportional to ω_0 – the characteristic phonon frequency of the material – perfectly describes the isotope effect $T_c \propto M^{-1/2}$ (M being the ionic mass), thus confirming the phonon-induced origin of electron pairing.

It was quickly understood [4] that, on the basis of the weak coupling theory of superconductivity, T_c could not be expected to exceed 30–40 K. The reason for this lies in the interdependence of the parameters entering this theory. The effective attraction given by λ_{eff} comes about via a subtle decrease in efficiency of the Coulomb repulsion $\mu \rightarrow \mu^*$ due to retardation effects. In fact, the quantity

$$\mu - \lambda = 4\pi e^2 N(\epsilon_F)/\langle q^2 \epsilon(0, \mathbf{q})\rangle \tag{1.2}$$

is generally positive; $N(\epsilon_F)$ denotes the density-of-states at the Fermi level and $\epsilon(0, \mathbf{q})$ is the dielectric constant for zero frequency and momentum $\mathbf{q}$. The average $\langle \ldots \rangle$ over $\mathbf{q}$ is taken over the Fermi surface of the material. Due to the retardation of the phonon-mediated electron–electron interaction, μ is reduced to $\mu^* = \mu/(1 + \mu ln(\epsilon_F/\omega_0))$, which provides the possibility of $\mu^* - \lambda < 0$ and hence makes electron–electron pairing possible.

At first sight T_c, given by expression (1.1), could be increased by increasing the characteristic phonon frequency ω_0. Since, however, $\lambda = |v|^2 N(\epsilon_F)$, where the electron–phonon coupling $|v|^2 \sim \omega_0^{-2}$, this would effectively lead to a decrease rather than an increase of T_c. Likewise an increase of the density of states $N(\epsilon_F)$ – at first sight favorable for increasing T_c – renders the system unstable via global lattice transformations before T_c can be substantially increased. Increasing the effective electron–phonon coupling $|v|^2$ may indeed be the only possibility of increasing T_c – if we remain within the electron–phonon-mechanism of superconductivity.

These inherent limitations of achieving values of T_c much higher than 30–40 K within the picture of weak coupling electron–phonon mediated superconductivity has led a large number of theorists to look for pairing mechanisms outside the scope of electron–phonon interaction in order to describe the superconductivity in HT_cSC. There may be no need for this, either on theoretical grounds or on the basis of experimental facts which we have up to now.

In principle, strong electron–phonon coupling can give values of T_c much higher than 30–40 K. This reasoning is based on extensions of the weak-coupling BCS theory to its strong-coupling formulation in terms of the Eliashberg theory [5]. This theory is a self-consistent theory, in principle valid for arbitrarily high values of electron–phonon coupling but assuming the validity of the adiabatic approximation (or the Migdal hypothesis), implying $k_B T_c/\epsilon_F \ll 1$ and $\omega_0/\epsilon_F \ll 1$, where ϵ_F denotes the Fermi energy. As we first showed [6], this hypothesis is violated for $\lambda \gtrsim 1$ when the lattice becomes locally (rather than globally) unstable and electrons become heavily dressed particles – so-called small polarons. Such small polarons tend to attract each other and form locally bound pairs, called bipolarons, which are real-space pairs, in contrast to Cooper pairs. Superconductivity then results from a BEC of such bipolarons (which are effectively bosons) in the sense of Schafroth superconductivity [7].

The short coherence length ξ (a few tens of Angstroms) and the low concentration of charge carriers in the HT_cSC, which makes ξ only slightly larger than the interparticle distance, favors such a picture of local pair superconductivity in the form of a BEC. The experimental evidence that such pairs really exist in the semiconducting parent compounds of HT_cSC is strong support – although not a proof – for a BEC of bipolarons in such materials.

In Section 2, we shall review the experimental evidence for local electron pairing in the semiconducting parent compounds of HT_cSC . These experiments strongly suggest that in the metallic phase of these compounds, a charge transfer mechanism between bipolarons and pairs of highly correlated electrons is at play. This leads us naturally to a generic model of an interacting boson–fermion mixture for HT_cSC. The qualitative feature of such a model will be discussed in Section 3, where we shall show under what circumstances a BEC of bipolarons could possibly be realized in HT_cSC. The compatibility of such a picture with a number of experiments in these materials is discussed in Section 4.

2 Experimental Evidence for Bipolarons in HT_cSC

Tightly bound electron pairs on small molecular clusters making up crystalline lattices have been experimentally verified and studied in great detail over the last 20 years. The most striking examples are $Ti_{4-x}V_xO_7$ [8], $Na_xV_2O_5$ [9] and WO_{3-x} [10]. In $Ti_{4-x}V_xO_7$, diatomic molecular units occur in the form of Ti^{3+}–Ti^{3+} and Ti^{4+}–Ti^{4+}, each Ti being surrounded by an oxygen octahedral ligand environment. The extra two electrons sitting on Ti^{3+}–Ti^{3+} units are tightly bound, which is manifest from the much shortened ($\sim$ 10%) intramolecular distance for those units as compared to the Ti^{4+}–Ti^{4+} molecular units. The electron pairs located on the Ti^{3+}–Ti^{3+} units have been called bipolarons because of their lattice-induced pairing mechanism. In Ti_4O_7, at low temperature, Ti^{3+}–Ti^{3+} pairs alternate with Ti^{4+}–Ti^{4+} pairs, thus forming a semiconducting charge ordered state of bipolarons. Upon increasing the temperature (to about 140 K), these bipolarons become dynamically disordered, showing a thermally activated resistivity and a diamagnetic susceptibility. At still higher temperatures (at about 150 K), a first-order phase transition leads to a metallic state of single (unbound) electrons characterized by a Pauli susceptibility. Upon doping Ti_4O_7 with V ($Ti_{4-x}V_xO_7$), the temperature region of dynamically disordered bipolarons can be sizeably enlarged to the detriment of the charge ordered state. In $Na_xV_2O_5$ and WO_{3-x}, the situation is very similar, involving V^{4+}–V^{4+}, V^{5+}–V^{5+} and W^{5+}–W^{5+}, W^{6+}–W^{6+} molecular units respectively.

These experimental results led us early on [6] to investigate the possibility of bipolaronic superconductivity as a concrete example for a manifestation of a Schafroth superconductivity which could potentially be realized in systems with strong electron–lattice coupling. In principle it is conceivable that the above-mentioned materials show bipolaronic superconductivity at sufficiently low temperatures provided that the bipolaronic motion occurs via tunneling rather than thermal activation. One of the biggest hindering factors for BEC to occur in these materials is related to doping the system away from its charge ordered state. This has been done so far by ion substitution (Ti $\rightarrow$ V) on sites where bipolaron formation normally occurs. This evidently introduces strong impurity effects in the bipolaronic system, hindering the onset of BEC of the bipolarons.

Materials where bipolaronic superconductivity could occur should be sought in systems where the concentration of bipolarons can be varied over a wide regime without disturbing the subsystem in which the bipolarons move, i.e. by doping outside the bipolaronic subsystem. Such a situation might be approached in the cuprate HT_cSC and possibly $Ba_{1-x}K_xBiO_3$ and the fullerenes.

In the following, we shall focus on hole-doped cuprate HT_cSC, which consist of CuO_2 planes and adjacent layers containing so-called apical oxygens. The electrons in the CuO_2 layers are highly correlated due to strong on-site Coulomb repulsion on Cu atoms and, moreover, are strongly hybridized between the Cu $3d_{x^2-y^2}$ and O $2p_{x,y}$ orbitals. For the low doped semiconducting parent compounds of HT_cSC such as $YBa_2Cu_3O_{6+x}$, $Bi_{2-x}Sr_2Ca_2Cu_3O_{10+y}$, $La_{2-x}Sr_xCuO_4$ ..., this leads to a charge transfer gap. Doping HT_cSC by ion implantation occurs exclusively outside the correlated electron system and sensitively affects the apical oxygen network, on which it leads to the formation of bipolarons. Experimental evidence for this comes from resonant Raman scattering, photo-luminescence and photo-absorption, as well as photo-modulation studies.

For undoped parent compounds, resonant Raman scattering is observed involving vibrations of the apex oxygens polarized orthogonally to the CuO_2 planes. This Raman active mode (475 cm^{-1} in $YBa_2Cu_3O_6$) changes significantly upon low doping of the parent compound [11]; it hardens (507 cm^{-1} in $YBa_2Cu_3O_{6.15}$). The resonant Raman scattering becomes more pronounced and a large number of clearly detectable overtones appear when the incident frequency of the laser is in the resonant region, *i.e.* around 1.16 eV. This energy being less than the charge transfer gap (1.7 eV in $YBa_2Cu_3O_6$) indicates that one is probing states inside this charge transfer gap. Site selective ^{18}O isotope doping [11] reveals that the doping-induced 507 cm^{-1} modes are related to vibrations of two O(4)–Cu(1)–O(4) units on either side of a dopant oxygen O(1) which leads to a change of $Cu^{+}(1)$ to $Cu^{++}(1)$ and ultimately causes a change in bond length of the O(4)–Cu(1)–O(4) unit, as well as a significant change in its vibrational frequency. The distribution of spectral weights of the overtones is characteristically different from those observed in strongly anharmonic crystals, an unmistakable signature of polarons [11,12].

Photo-luminescence spectra carried out on the same samples [11] indicate the presence of some narrow electronic band inside the charge transfer gap. The observed photo-luminescence can be understood in terms of electrons which have been excited across the charge transfer gap into the conduction band and then recombine radiatively into in-gap states of the O(4)–Cu(1)–O(4) units. This necessarily involves a charge transfer between the CuO_2 planes and the apex oxygen network. Evidence that a direct link exists between dramatic doping-induced changes in the vibrational frequency (involving the apex oxygens) and the associated charge excitations comes from photo-induced infrared absorption and photo-modulation studies [12]. There, additional charge carriers are created in the semiconducting parent compounds by laser illumination. Such photo-induced charge carriers lead to the suppression of modes which characterize the semiconducting phase and to the appearance of modes which normally exist only in the metallic compounds [13]. This is a strong indication of the existence of polaronic charge carriers in the metallic compounds. Moreover a broad spectrum of electronic origin appears in the frequency regime of the charge transfer gap in the insulating compounds. The intensities of the infrared absorption of the newly created phonon mode and of the near infrared absorption of the electronic excitation scale in the same way with the strength of the laser illumination [12]. Since this scaling varies as the square root of the laser power, this implies recombination processes involving two charge carriers and hence provides evidence for bipolarons as stable entities. These optical experiments are a clear proof of strong coupling of charge carriers to the displacement of the apex oxygens polarized orthogonally to the CuO_2 planes and bipolaron formation involving adjacent apex oxygen units.

Although all these optical experiments were carried out exclusively on the semiconducting parent compounds, they are nevertheless strongly indicative of the physics which we can expect for the metallic compounds. Upon increasing the chemical doping, the semiconducting state breaks down, resulting in a metallic state triggered by intrinsic charge fluctuations between the CuO_2 planes and the apex oxygen network. The type of optical experiments performed on the semiconducting parent compounds are such that they test these intrinsic charge fluctuations in an out-of-equilibrium situation. This is manifest in:

(i) the strong *resonant* enhanced Raman scattering of the apex oxygen mode orthogonal to the CuO_2 plane,
(ii) the effect of photo-induced carriers leading to the creation of phonon modes in the insulating parent compounds which are characteristic of the metallic compounds,
(iii) the strong correlation between these photo-induced modes and the appearance of electronic excitations in the charge transfer gap,
(iv) the possibility of inducing a metallic (superconducting) state upon intense photo-induced doping of the semiconducting parent compounds [14].

The broad spectrum of electronic origin appearing in the frequency regime of the charge transfer gap upon photo-doping finds its counterpart in the metallic compounds in the strong non-Drude behavior of the optical conductivity showing a similarly broad mid-and near-infrared absorption band [15].

Charge fluctuations between the CuO_2 layers and the apex oxygen network in the metallic compounds lead to dynamical fluctuations of the interatomic distance between the Cu ions in the CuO_2 planes and the adjacent layers containing the apex oxygen atoms. Such effects have been seen by EXAFS in a great variety of high T_c materials in the metallic compounds [16,17]. It should be noted that in metallic $YBa_2Cu_3O_{6+x}$, these distance fluctuations are linked to bond fluctuations of the O(4)–Cu(1)–O(4) units varying between 1.80 and 1.94 Å for the Cu(1)–O(4) distance [16]. These two distances correspond to those of undoped $YBa_2Cu_3O_6$ consisting of O(4)–Cu^+(1)–O(4) units and those of low doped insulating $YBa_2Cu_3O_{6+x}$, with oxygens intercalated between two such units and with Cu^+(1) being replaced by Cu^{++}(1). In the optical experiments discussed above, we have pointed out the dramatic change in vibrational frequency associated with such bond length changes.

From the detailed analysis of the experiments discussed above, we arrive at the following picture. For low doped materials, pairs of localized charge carriers are created which remain tied to small molecular units containing the apex oxygens as the essential ingredients. It is an experimental fact that upon exceeding a certain rate of chemical doping, the sites on which bipolarons form rearrange themselves and order along chains of very long length [18]. Such a rearrangement of a fixed number of bipolaronic sites (or doping atoms) forces the system to activate a charge transfer from the CuO_2 planes towards the dopant atoms in order to bind those dopant atoms in their new positions [19]. This is directly

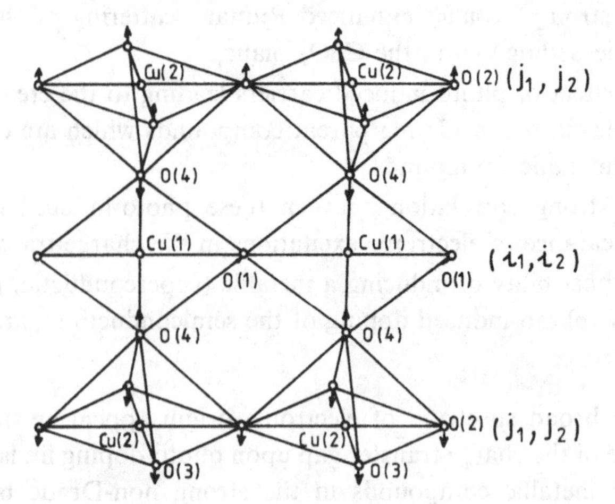

Fig. 1. Schematic plot of a molecular cluster on which bipolaron↔electron pair charge transfer takes place in metallic $YBa_2Cu_3O_{6+x}$. The bipolarons are formed on the complex involving the two O(4)–Cu(1)–O(4) (i_1, i_2) units on either side of a dopant O(1) atom. The electrons reside in the Cu(2)–2O(3)–2O(4) units (j_1, j_2).

evident from the photo-emission data [20] which measure abrupt changes of the number of $Cu^{++}(1)$ atoms as a function of chemical doping and also confirm equally abrupt increases of the overlap of electronic wave functions between the Cu(2) ions in the CuO_2 planes and the O(4) apex oxygens [21]. The reason for the ordering of the chemically doped atoms leading to an homogeneous distribution of bipolaronic sites and the onset of a metallic state must lie in the energy gained by the bipolarons due to delocalization in the normal state and the possibility of achieveing a macroscopic occupation of the $q = 0$ state in the superconducting state due to a BEC of bipolarons. Judging from the EXAFS measurements, this charge transfer is of a dynamical nature [16,17], leading to a situation where a bipolaron on any arbitrary site of the atomic clusters involving the apex oxygens can be exchanged by a pair of electrons in the adjacent CuO_2 planes. This charge transfer can be described by:

$$\left(\tilde{c}^{+}_{i_1\uparrow}\tilde{c}^{+}_{i_2\downarrow}c_{j_1\downarrow}c_{j_2\uparrow} + h.c.\right), \tag{2.1}$$

where $i_{1,2}$ in (2.1) represent two adjacent molecular units on the apex oxygen network and $j_{1,2}$ two adjacent Cu(2)–2O(2)–2O(3) clusters in the CuO_2 planes (see Fig. 1, for a likely realization of this situation in metallic $YBa_2Cu_3O_{6+x}$). The bipolarons are expected to exist in the form of intersite bipolarons $b^+_{i_1 i_2} = \tilde{c}^+_{i_1\uparrow}\tilde{c}^+_{i_2\downarrow}$ consisting of two polaronic charge carriers:

$$\tilde{c}^+_{i\sigma} = c^+_{i\sigma}|\phi(x - x_0)\rangle_i, \tag{2.2}$$

made of the bare charge carriers surrounded by strong local deformations of the molecular units on which they are located (the O(4)–Cu(1)–O(4) unit in $YBa_2Cu_3O_{6+x}$). The state $|\phi(x - x_0)\rangle_i$ denotes the displaced oscillator function characterizing the deformed molecular unit, where x_0 measures the degree of deformation. Provided this deformation is very local and large – as it usually is for polaronic charge carriers – $\tilde{c}_{i\sigma}$ and $c_{i\sigma}$ can be considered as commuting operators.

3 BEC in a Hybridized Boson–Fermion Mixture

On the basis of the above analysis of the experimental situation of HT_cSC, we now construct a generic model which we expect contains the essential physics in those materials [22]. The number of charge carriers being generally small (10^{21}–10^{22} cm^{-3}), we can safely neglect the spatial overlap of different bipolarons. To a first approximation, this amounts to replacing the intersite bipolarons by simple bosons defined on effective cluster sites ℓ, such as the one depicted in Fig. 1. We can then write the effective hybridization term in the form: $(b^+_\ell c_{\ell\downarrow}c_{\ell\uparrow} + h.c.)$, where b^+_ℓ corresponds to $\tilde{c}^+_{\ell\uparrow}\tilde{c}^+_{\ell\downarrow}$, and define a generic model to describe HT_cSC in form of the following interacting boson–fermion mixture hamiltonian:

$$H = \sum_{\langle \ell \neq \ell' \rangle \sigma} t_{\ell\ell'} c^+_{\ell\sigma} c_{\ell'\sigma} + \Delta \sum_\ell b^+_\ell b_\ell + \frac{v}{\sqrt{\Omega}} \sum_\ell (b^+_\ell c_{\ell\downarrow} c_{\ell\uparrow} + h.c.)$$

$$-\mu \left(2\sum_\ell b^+_\ell b_\ell + \sum_{\ell,\sigma} c^+_{\ell\sigma} c_{\ell\sigma} \right). \tag{3.1}$$

The charge transfer between bipolarons and electron pairs has strength v and the common chemical potential ensures conservation of charge carriers; Ω denotes the normalization volume. The energy of the bipolaronic level is given by Δ, which should roughly correspond to T_{PB}.

We note that, already, the simplest case of a non-interacting boson–fermion mixture ($v \equiv 0$) leads to deviations from the BEC of a pure Bose gas. To show this, we consider free bosons with a finite dispersion $E_k = \xi^B_k + \Delta - 2\mu$ and free fermions with a dispersion $\epsilon_k = \xi^F_k - \mu$ (see Fig. 2) with $\xi^F_k = \hbar^2 k^2 / 2m_F$ and $\xi^B_k = \hbar^2 k^2 / 2m_B$, where $m_{F(B)}$ denotes the fermion

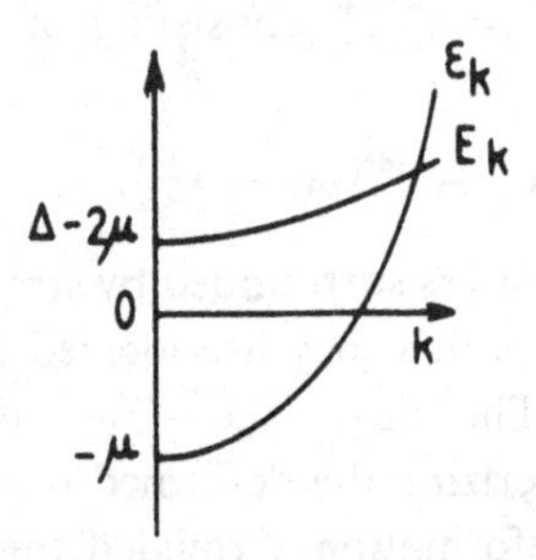

Fig. 2. Schematic plot of the fermion (ϵ_k) and boson (E_k) dispersion for the non-interacting boson–fermion model.

(boson) mass. Using the standard three-dimensional density of states for free bosons and fermions given by $g^0_{F(B)} = \sqrt{2}m^{3/2}_{F(B)}\epsilon^{1/2}\Omega/2\pi^2\hbar^3 \equiv g\epsilon^{1/2}$, we have the following condition for the conservation of total particle number:

$$n = 2g\left[\int_0^\infty d\epsilon\epsilon^{1/2}n_F(\epsilon-\mu) + \left(\frac{m_B}{m_F}\right)^{\frac{3}{2}}\int_0^\infty d\epsilon\epsilon^{\frac{1}{2}}n_B(\epsilon+\Delta-2\mu)\right] + 2n_0(T), \tag{3.2}$$

where n_0 denotes the number of condensed bosons. For low T (i.e. $T \ll \mu$) one finds:

$$n = 2g\left[\frac{2}{3}\mu^{\frac{3}{2}} + \frac{\pi^2}{12}(k_BT)^2\mu^{-\frac{1}{2}} + (k_BT)^{\frac{3}{2}}\left(\frac{m_B}{m_F}\right)^{\frac{3}{2}}I_{\frac{1}{2}}(\varphi)\right] + 2n_0(T), \tag{3.3}$$

with $\varphi = \exp(\beta(\Delta-2\mu))$, $I_{1/2}(\varphi)$ is a Riemann ζ function and $\mu = \Delta/2$ for $T \leq T_{BEC}$. From (3.3), we note that there is a critical value of the density of charge carriers ($n_c = 4g/3 \times (\Delta/2)^{3/2}$) below which BEC is not possible. The BEC temperature T_{BEC} is then determined by:

$$n = n_c + 2g\left[\frac{\pi^2}{12}(k_BT_{BEC})^2\left(\frac{\Delta}{2}\right)^{-\frac{1}{2}} + (k_BT_{BEC})^{\frac{3}{2}}\left(\frac{m_F}{m_B}\right)^{\frac{3}{2}}I_{\frac{1}{2}}(1)\right]. \tag{3.4}$$

In Fig. 3(a,b), we plot T_{BEC} as a function of n and $n_0(T)$ as a function of T for $n \geq n_c$. Thus we see that in the non-interacting boson–fermion mixture, the fraction of condensed bosons is always bigger than that for a simple free Bose gas due to the extra thermally excited bosons near T_{BEC}.

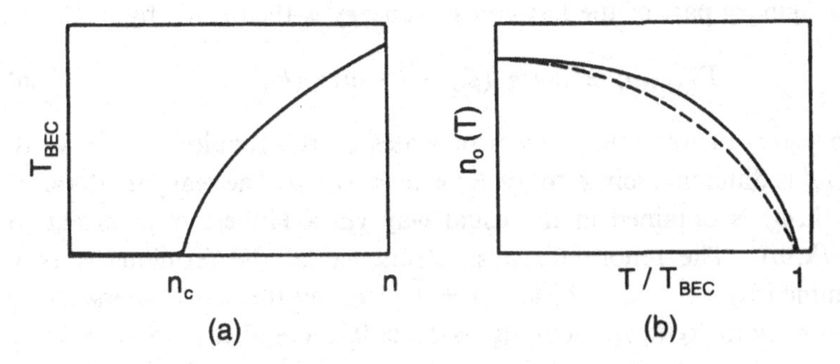

Fig. 3. (a) T_{BEC} as a function of the total carrier concentration n in the non-interacting boson–fermion model. (b) The condensed fraction $n_0(T)$ of bosons as a function of temperature in the non-interacting boson–fermion model (continuous line) and the pure Bose gas (dashed line).

We now return to the case of an interacting boson–fermion mixture based on the Hamiltonian (3.1), but assuming from the outset itinerant rather than localized bosons. We do this in anticipation of a fully self-consistent treatment [23] of the Hamiltonian (3.1), where, as a result of the boson–fermion exchange term, the initially totally localized bosons are renormalized into propagating – although heavily damped – quasiparticles.

In order to describe the normal state properties of such a system, as well as its transition towards a superconducting state, we must consider the diagrams for the self-energy $\Sigma_F(\mathbf{k}, \omega)$ for the fermions (electrons in the CuO_2 planes) and $\Sigma_B(\mathbf{q}, \omega)$ for the bosons (bipolarons in the apex oxygen network) given in Fig. 4 and solve the coupled equations for the fermion and boson Green's functions in a self-consistent way [23].

For an interacting Bose system, the chemical potential lies below the unrenormalized energy spectrum of the bosons, i.e. $E_0 \equiv \Delta - 2\mu > 0$. Considering the boson and fermion self-energies in their simplest RPA approximation, i.e. in terms of free fermion and boson Green's functions we obtain:

$$\Sigma_F(\mathbf{k}, \omega) = \frac{v^2}{\Omega} \sum_q \frac{n_F(\epsilon_{\mathbf{k}-\mathbf{q}}) + n_B(E_\mathbf{q})}{\omega - E_\mathbf{q} + \epsilon_{\mathbf{k}-\mathbf{q}} + i\eta},$$

$$\Sigma_B(\mathbf{q}, \omega) = \frac{v^2}{\Omega} \sum_k \frac{1 - n_F(\epsilon_\mathbf{k}) - n_F(\epsilon_{\mathbf{k}-\mathbf{q}})}{\omega - \epsilon_{\mathbf{k}-\mathbf{q}} - \epsilon_\mathbf{k} + i\eta}. \quad (3.5)$$

For $E_0 > 0$ and $T \to 0$, $n_B(E_q)$ is negligible in this expression for $\Sigma_F(\mathbf{k}, \omega)$.

The imaginary part of the fermion self-energy is then given by

$$\Gamma_F(\mathbf{k},\omega) \simeq 2\pi v^2 g_F(E_0+\mu-\omega)n_F(E_0-\omega), \tag{3.6}$$

where $g_F(\omega)$ denotes the density of states of the fermions. The width $\Gamma_F(\mathbf{k},\omega)$ is different from zero for $E_0 < \omega < E_0+\mu$. The real part $R_F(\mathbf{k},\omega)$ of $\Sigma_F(\mathbf{k},\omega)$ is obtained in the usual way via a Hilbert transformation of $\Gamma_F(\mathbf{k},\omega)$. The renormalized spectrum $\omega_\mathbf{k}$ of the fermions is then determined by $\omega_\mathbf{k} - \epsilon_\mathbf{k} - R_F(\mathbf{k},\omega_\mathbf{k}) = 0$, given by the set of intersections of $\omega - \epsilon_\mathbf{k}$ with $R_F(\mathbf{k},\omega)$. As long as $E_0 > 0$, there always exists a wave vector $\mathbf{k}_F$ such that for $\mathbf{k} < \mathbf{k}_F$, we have $\omega_\mathbf{k} \leq 0$ and for $\mathbf{k} > \mathbf{k}_F$, we have $\omega_\mathbf{k} \geq 0$. This determines the renormalized Fermi vector $\mathbf{k}_F$. As E_0 approaches zero, $\mathbf{k}_F$ increases and a situation is finally reached for which no $\mathbf{k}_F$ can be found within the Brillouin zone for which $\omega_{\mathbf{k}\geq\mathbf{k}_F} > 0$. This signals the breakdown of the Fermi liquid behavior, manifest in the absence of a discontinuity of $n_F(\omega_\mathbf{k})$.

The limit $E_0 \to 0$ is subtle in the sense that $n_B(E_q)$ in the expression for $\Sigma_F(\mathbf{k},\omega)$ diverges for $q \to 0$. Rewriting $n_B(E_\mathbf{q})$ as:

$$n_B(E_\mathbf{q}) = n_B(E_\mathbf{q}) + \delta_{q0}(n_0 - n_B(E_\mathbf{q})), \tag{3.7}$$

the dominant contribution to $\Sigma_F(\mathbf{k},\omega)$ in the limit $T \to 0$ is determined by:

$$\Sigma_F(\mathbf{k},\omega) \sim \frac{n_0 v^2}{\omega + \epsilon_\mathbf{k} + i\eta}. \tag{3.8}$$

This yields the poles

$$\omega_\mathbf{k} = \pm\gamma_\mathbf{k} = \pm\sqrt{\epsilon_\mathbf{k}^2 + n_0 v^2} \tag{3.9}$$

in the fermion Green's function and a corresponding Fermi distribution function of the form:

$$n_\mathbf{k} = \frac{1}{2}\left(1 - \frac{\epsilon_\mathbf{k}}{\gamma_\mathbf{k}}\right). \tag{3.10}$$

The expressions (3.9) and (3.10) have exactly the BCS form for the superconducting ground state but with a gap proportional to the number of condensed bosons n_0.

Let us now turn to the properties of boson Green's function in the same limiting case, i.e. $T \to 0$ and $E_0 \to 0$. From the discussion of the instability of the fermionic system, we know that it is triggered by the macroscopic occupation of the $q = 0$ mode. The imaginary part of the

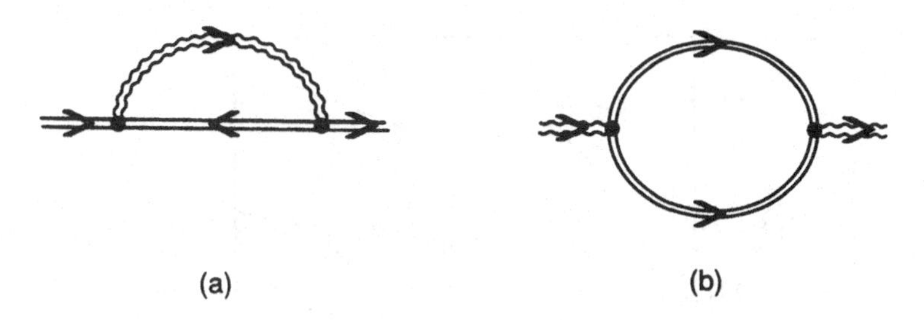

Fig. 4. Fermionic (a) and bosonic (b) self-energy diagrams in terms of fully dressed self-consistent fermion and boson Green's functions.

boson self-energy for $q = 0$ is given by:

$$\Gamma_B(q = 0, \omega) = 2\pi \frac{v^2}{\Omega} \sum_{\mathbf{k}} \tanh \frac{1}{2}\beta\epsilon_{\mathbf{k}}\delta(\omega - 2\epsilon_{\mathbf{k}}), \tag{3.11}$$

which for $\omega = 0$ is identically zero, signaling the onset of undamped bosonic excitations. This imposes the following constraint on the renormalized boson spectrum:

$$E_0 = R_B(q = 0, \omega = 0) = \frac{v^2}{2\Omega} P \sum_{\mathbf{k}} \frac{1}{\epsilon_{\mathbf{k}}} \tanh \frac{\beta\epsilon_{\mathbf{k}}}{2}, \tag{3.12}$$

where $R_B(\mathbf{q}, \omega)$ denotes the real part of the boson self-energy $\Sigma_B(\mathbf{q}, \omega)$. Rewriting the sum over $\mathbf{k}$ in (3.12) as an integral over the density of states of the fermions, we obtain:

$$E_0 = \frac{v^2}{2} \int_{-\infty}^{\infty} d\epsilon g_F(\epsilon) \frac{1}{\epsilon - \mu} \tanh \frac{\beta}{2}(\epsilon - \mu). \tag{3.13}$$

In the limit $T \to 0$, the integrand on the rhs of (3.13) diverges for $\epsilon = \mu$ unless $g_F(\epsilon = \mu) \equiv 0$. This implies the opening up of a gap in the one-particle electron spectrum which is consistent with the conclusions drawn earlier from the properties of the fermion Green's function, (3.6)–(3.9).

The arguments developed above clearly demonstrate the mutually induced superconducting state in the boson–fermion mixture being due to a BEC of the bosons which occurs with the simultaneous opening of a gap in the excitation spectrum of the fermions. A self-consistent mean field theory of this problem has previously been given [24,25].

A fully self-consistent treatment of this problem in the present scheme – taking into account, in addition to the self-energy diagrams for the normal phase (Fig. 4), all those including anomalous fermion and boson Green's functions – will be discussed elsewhere [23]. The critical temperature

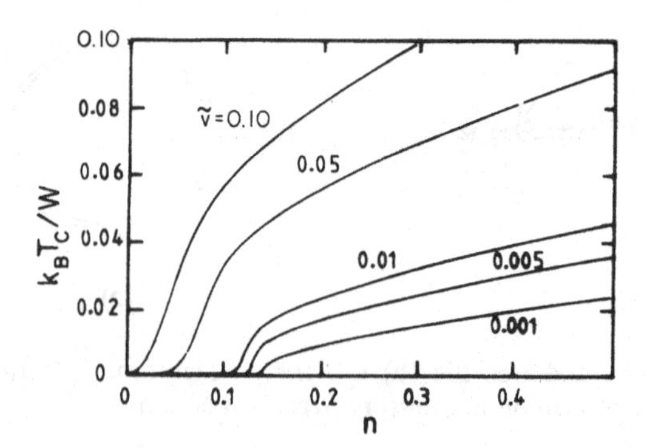

Fig. 5. Transition temperature T_c as a function of the total number of charge carriers and for several values of the boson–fermion pair coupling $\tilde{v} = v/W$.

T_c, signaling the onset of superconductivity, is obtained as the limiting value for μ for a given particle concentration. Up to this density, there exist undamped solutions of the self-consistently treated problem in the normal state (involving only the self-energies illustrated in Fig. 4).

In Fig. 5 we plot T_c as a function of the total number of particles $n = 2n_F + 2n_B$ for various values of the dimensionless boson–fermion pair coupling constant $\tilde{v} = v/W$; W denoting the fermion band width. We notice that the density dependence of $T_c(n)$ can be divided into two limiting regimes. For low carrier concentrations $n < n_{cr}(\tilde{v})$, one obtains BCS-like behavior where the electron–electron attraction comes about by an exchange of virtual bosons – just as in BCS for virtual phonons. For $n > n_{cr}(\tilde{v})$, $T_c(n)$ is similar to that obtained for the non-interacting boson–fermion mixture (see Fig. 3(a)) and hence is essentially determined by the BEC of the bosons in the system. The changeover from one to the other regime occurs rather abruptly, leading to a sizeable increase of T_c upon increasing n only very little. This fact clearly demonstrates the interest in achieving a BEC in conection with efforts to obtain large values of T_c. In the remainder of this section, we shall sketch the main results of this treatment.

The fully self-consistent treatment of this problem shows the development of a pseudogap in the normal state as the superconducting state is approached upon lowering the temperature. At the same time, the initially flat (q-independent) distribution function for bosons at high temperatures gets redistributed, shifting its intensity more and more to smaller q values as T approaches T_c (see Fig. 6(a)). This is a precursor

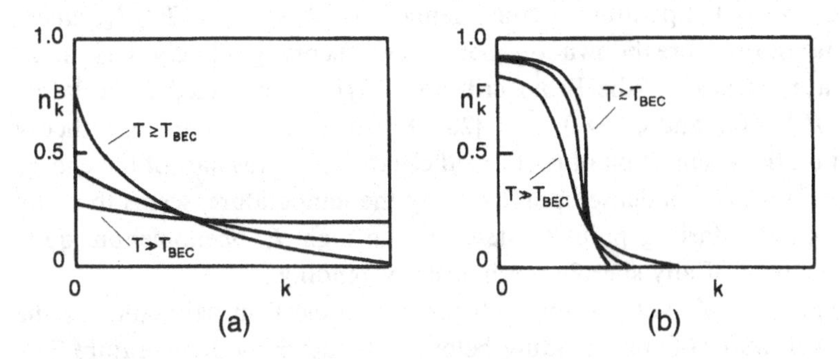

Fig. 6. Schematic plot of (a) the boson occupation number $n_{\mathbf{k}}^{B}(T)$ and (b) the fermion occupation number $n_{\mathbf{k}}^{F}(T)$, for the fully self-consistently treated interacting boson–fermion model.

effect of a BEC, similar to that seen in liquid ^{4}He upon approaching T_λ from above.

The appearance of a pseudogap in the fermionic density of states above T_c leads to a renormalization of the Fermi distribution function. This shows non-Fermi liquid properties with a noticeable redistribution of spectral weight involving the entire Brillouin zone, as illustrated schematically in Fig. 6(b). The wavevector $\mathbf{k}_F$ for the Fermi subsystem, and hence the number of fermions, varies only a little as a function of the total number of carriers. This is due to the pinning of the chemical potential in the vicinity of E_0. On the contrary, the number of bosons varies practically linearly with an increase in the total number of particles.

As concerns the superconducting state, the fully self-consistent calculation involving normal as well as anomalous fermion and boson Green's functions yields an acoustic dispersion relation (in the absence of a long-range Coulomb interaction) for the bosonic modes near $T \sim 0$, thus signaling a superfluid phase.

4 Experimental Suggestions for Verification of a BEC in HT$_c$SC

Because of the short coherence lengths (a few tens of Å) and the low carrier concentrations (10^{20}–10^{21} cm^{-3}), fluctuations are expected to play a major role and it is therefore highly unlikely that a mean field theory could be adequate to describe the superconducting transition in HT$_c$SC. Both the critical regime defined by $\zeta^G = |T_G - T_c|/T_c$, where mean field exponents cease to be valid (the Ginzburg criterion [26]), as well as the wider temperature regime, defined by $\zeta^B = |T_B - T_c|/T_c$, where a quantitative breakdown of mean field theory gradually sets in as one approaches T_c (the Brout criterion [27]), are expected to be large: $\zeta \sim 0.01 - 0.1$ and $\zeta^B \sim 0.1 - 1$ [28]. For these reasons alone, a theory built on the assumption of preformed electron pairs (as part of the charge carriers) which condense upon lowering the temperature, seems to be an appropriate starting point to describe the high T_c phenomenon, quite independent of any specific mechanism of binding.

Experimental evidence for such preformed electron pairs and for the physical origin of their binding below a characteristic temperature T_{PB} comes from:

(i) The comparative study of bipolaronic properties in the insulating and metallic state, as discussed in Section 2.

(ii) Measurements of the magnetic susceptibility [29] (using SQUID and torque measurements) showing strong diamagnetic contributions coming from the charge carriers in the CuO_2 planes. These diamagnetic contributions start to set in at temperatures T_{dia} which are several times the critical temperatures and, upon approaching T_c, outweigh the van Vleck contribution close to T_c. This temperature-dependent diamagnetism can be expressed in terms of a universal Curie–Weiss law $\chi_{\rm dia} = -C_{\rm dia}/(T - \theta_{\rm dia})$ with $T_c/\theta_{\rm dia} \sim 1$, for a wide variety of HT_cSC. It can be excluded that this diamagnetism is related to superconducting fluctuations since in certain samples they can be seen up to 30 times T_c. By increasing the number of carriers, $T_{\rm dia}$ diminishes, thus reducing more and more the temperature region of observable diamagnetism.

(iii) The dependence of the normal state entropy [30] as a function of temperature shows strong non-linear behavior with an inflexion point at a characteristic temperature T_{PB} which strongly depends on the total number of charge carriers. As the number of charge carriers is increased, entering the so-called "overdoped regime", T_{PB} tends to T_c. For sufficiently high temperatures, the entropy can be reconciled with that of itinerant fermions having small band width of the order of 1000 K. It would be highly desirable to see to what extent T_{dia} and T_{PB} are correlated as a function of doping. The present status of the experiments unfortunately does not permit such a quantitative comparison.

(iv) Indirect evidence for the existence of preformed electron pairs comes from the experimental verification of a pseudogap in the Fermi sea. As discussed in the previous section, such a pseudogap has to be expected whenever long-lived electron pair states are present. Experimental evidences for such a pseudogap include the temperature dependence of the NMR Knight shift K_s, which appears to be dominated by pseudogap transitions above as well as below T_c [31]. Similar findings are obtained by inelastic neutron scattering techniques, measuring the temperature dependence of the linewidth of the crystal electric field transitions [32].

These experimental results strongly support the picture of a hybridized boson–fermion mixture with bosons (bipolarons) breaking up into individual fermions at T_{PB} ($\sim T_{\text{dia}}$?). It is universal behavior of HT$_c$SC that T_c increases linearly with the number of charge carriers [33] and varies inversely with their mass [34] up to some critical concentration, beyond which one enters the "overdoped regime" and T_c starts to decrease, sometimes rapidly going to zero.

The "underdoped regime" shows a number of non-Fermi liquid properties such as: a violation of the Korringa law for NMR, a scaling of $1/T_1T$ with χ rather than χ^2 [36] (where T_1 denotes the NMR relaxation rate and χ the magnetic susceptibility), a non-Drude behavior of the optical conductivity and the absence of a T^2 law for the resistivity at low temperatures [35].

In contrast to the above, the "overdoped regime" seems to have normal Fermi liquid properties and an entropy which varies linearly with temperature for $T \geq T_c$, compatible with ordinary itinerant fermions [30].

The fact that T_{PB} (and seemingly also T_{dia}) tends to T_c as one approaches the "overdoped" metallic normal state could be an indication that the maximum of T_c which can be obtained in HT$_c$SC is limited by the existence of preformed local pairs below T_{PB} ($\sim T_{dia}$).

The study of correlation functions within the Bose–Einstein model is presently not sufficiently advanced to verify the above-mentioned characteristics. We have, however, information from a somewhat related model– the negative U Hubbard model [25] – in which the generic transition from weak coupling $|U|/W \ll 1$ BCS-like behavior to strong coupling $|U|/W \gg 1$ local pair superconducting behavior [37] has been fairly well studied. Here U represents an on-site attraction between electrons with opposite spin (in other words, the pair breaking energy $\sim T_{PB}$) and W denotes their band width. The maximum of T_c in this model is close to $T_{PB} \sim U$ for a certain $|U|/W$ of the order of unity. Larger values for $|U|/W$ describe the "underdoped regime" with a normal state between T_c and T_{PB} which shows many of the anomalous normal state properties [38], listed above, and which in many ways are reminiscent of a purely bosonic system [39].

Concerning the physical foundations of the boson–fermion model, we propose the following decisive experiments which would either verify or disprove the relevance of this model:

(1) A verification of the disappearance of polaronic effects upon entering the overdoped regime for $T > T_c$ using resonant Raman spectroscopy, photo-luminescence, photo-induced Raman measurements, the appearance of polaron-induced local lattice modes and EXAFS, measuring polaron-tunneling induced local fluctuations of intramolecular distances.
(2) A verification of the disappearance of the diamagnetic contribution to the magnetic susceptibility (not due to superconducting fluctuations) upon entering the "overdoped" regime. Tendencies suggesting this disappearance have been observed [29].
(3) A detailed analysis of the doping dependence of the chemical potential μ by photo-emission spectroscopy. According to the boson–fermion model, μ is expected to be pinned by the presence of the bosonic resonant state inside the Fermi sea. Our preliminary calculations on this issue suggest a rather weak doping dependence of μ, provided the pairing energy Δ does not vary with doping. Experimentally, this question is unsettled. There are experiments in which μ is practically unchanged with doping [40] and those in which μ behaves more like doped semiconductors [41], showing rather significant changes with doping. High resolution photo-emission experiments testing the dependence of μ on the carrier concentration and on the temperature, as well as the existence of a pseudogap, would be strong tests for the existence of bosonic states (pre-existing electron pairs) above T_c.

Let us now turn to the properties of the superconducting state. Experimentally there are a number of features [35] which clearly deviate from ordinary BCS-like behavior, such as:

(1) Unusually strong effects of the superconducting state on the lattice properties. Apart from the classical shift of phonon frequencies with energies above and below the gap energy [42], one observes strongly correlated motion of the molecular vibrations (seen by ion channeling experiments [43]) and a freezing of the vibrational energy of individual

atoms (seen by resonant neutron scattering [44]) which are triggered by the onset of the superconducting state [45].

(2) An unusually large gap ($2\Delta/k_B T_c \sim 5-8$) which in practice does not vary with temperature and whose signal fades away as one approaches T_c from below. This is seen from tunneling spectroscopy [46], EELS [47] and reflectivity measurements [48]. These lattice effects may be of importance in limiting the fluctuations of the purely electronic condensation.

(3) An essentially frequency-independent optical conductivity (for frequencies less than the superconducting gap) which shows the same temperature behavior as $1/T_1 T$ [49].

(4) Specific heat of the purely electronic contribution which scales surprisingly well with that of ^{4}He over a wide temperature regime $0 < |T - T_c|/T_c < 0.2$ [50], as shown in Fig. 7. The scaling of the specific heat of HT_cSC to that of liquid ^{4}He permits one to determine the number of "effective bosons" n_b in such systems. The number derived in this way compares well with the number of charge carriers determined independently by Hall measurements [51]. Using the formula for the Bose condensation temperature for quasi two-dimensional systems,

$$k_B T_{BEC}^{2D+\epsilon} = \frac{\hbar^2 2\pi n d}{m_{ab} \ln(2 k_B T_{BEC}^{2D+\epsilon} m_c d^2/\hbar^2)} \tag{4.1}$$

permits one to derive the effective mass of the "effective bosons" m_{ab} in the CuO_2 planes. Here m_c denotes the "effective boson" mass and d the size of the molar volume along the c direction. Upon inserting into (4.1) the experimentally determined transition temperature for $T_{BEC}^{2D+\epsilon}$ and for n_B the "effective boson" density derived from the comparison of the specific heat of several HT_cSC with that of liquid ^{4}He, one obtains $m_{ab} \sim$ 3-5 free electron masses if we assume $m_c/m_{ab} \sim 100$. This value agrees reasonably well with the effective masses deduced from optical experiments [54]. Finally, using the expression for the penetration depth of charged bosons

$$\lambda_{H,c} = \sqrt{m_{ab} c^2/16\pi n_B e^2} \tag{4.2}$$

for the magnetic field orthogonal to the CuO_2 planes, and inserting the values for m_{ab} and n_B derived above, one obtains values such as $\lambda_{H,c} \sim$ 2500–5000 Å which are quite compatible with those derived by independent measurements [51,52].

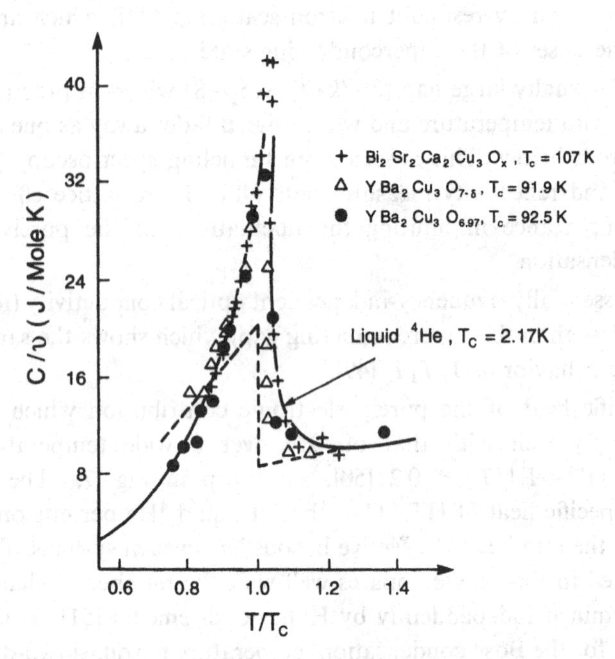

Fig. 7. Comparison of the experimentally determined specific heats of liquid ^{4}He and of $YBa_2Cu_3O_{6+x}$ and $Bi_2Sr_2Ca_2Cu_3O_3O_{10+x}$. The experimental points denoted by $+$, Δ, $\circ$ and $\bullet$ correspond to Refs. [51], [52] and [53] and [30] respectively.

It is remarkable that reasonable estimates of the concentration and mass of the charge carriers in HT_cSC can be determined by simply comparing the specific heat to that of liquid ^{4}He and using the BEC temperature as the critical temperature for superconductivity. At present we do not have any understanding of why the specific heat of HT_cSC should fit that of liquid ^{4}He over such a wide regime in temperature, well beyond the point where universality and scaling can be expected. Although this comparison itself is not a proof for a BEC in HT_cSC, it nevertheless is one more indication for the existence of preformed electron pairs in these materials.

Recent angle-resolved photo-emission studies [55] measuring the superconducting gap and low T measurements of the penetration depth which varies linearly with T [56] both appear to be indicative of d-wave symmetry of the superconducting order parameter. The fact that d-wave pairing rather than any other pairing should occur in HT_cSC is a new feature which should help to clarify the origin of the large values of T_c. It is clear that for systems with a short coherence length, the superconducting order parameter must adapt to the local underlying crystal structure. On the basis of an appropriate phenomenological model such as the extended Hubbard model with on-site repulsion and nearest neighbor attraction [57], one indeed obtains a whole series of superconducting ground states with d, p and extended s-wave pairing depending on the concentration of charge carriers. The boson–fermion model presented above would also lead to symmetries of the gap which are not necessarily s-wave pairing. We simply have to take into account in this model the precise position (i_1, i_2 in (2.1)) of bipolarons, stretched over adjacent sites, and the induced pairing fluctuations of the electrons involving square configurations of four CuO_2 units (j_1, j_2 in (2.1)) in the basal planes of the cuprate HT_cSC.

5 Conclusions

The origin of high T_c superconductivity is presently still a matter of great controversy. Certainly, in these materials we are faced with strong correlations between the electrons, which most theorists consider as crucial to explain high T_c superconductivity in terms of a non-lattice-driven mechanism. As we have stated, there is extensive experimental evidence for strong coupling of electrons to certain local lattice modes and consequent polaron formation. On the basis of these experiments, we are led to a picture in which localized bipolarons form one of the two subsystems of HT_cSC, whereas the strongly correlated electrons are confined to the other subsystem. A charge exchange, involving bipolarons and electron pairs, is triggered when one exceeds a certain doping concentration. At that point, an insulator–metal transition sets in, monitored by the binding of extra dopant atoms in those materials. The metallic regime consists of two apparently physically different, regions:

(1) an "underdoped regime", in which T_c increases monotonically as a function of the charge carrier concentration, and
(2) an "overdoped regime", in which T_c decreases with increased carrier concentration.

Specific heat, susceptibility, reflectivity, NMR Knight shift and crystal electric field excitations indicate that, in the "low doped regime", there exists a pseudogap, which we interpret as being due to a bosonic resonant state of bipolarons inside the Fermi sea. With increased doping, the binding energy T_{PB} of such a bipolaronic state decreases and eventually coincides with the superconducting transition temperature T_c. At that point, the maximum value of T_c is reached. Upon further doping, T_c starts to decrease. This whole scenario has been described in terms of a model of a boson–fermion mixture with an exchange of boson $\leftrightarrow$ fermion pairs. Preliminary results of our theoretical investigation of this model have been presented here [23].

Most of our theoretical arguments for developing the boson–fermion scenario for HT_cSC were based on experimental evidence coming from the hole-doped cuprates. There is recent experimental evidence that polaronic charge carriers also exist in electron-doped cuprates [58] as well as in fullerenes [59]. This strengthens our conviction that the whole high T_c superconductivity phenomenon has a unique mechanism, such as the one described here.

Of course, for a satisfactory theory of high T_c superconductivity, the problem of strong electron–phonon coupling has to be solved in conjunction with the strong electron correlation problem. One of the key questions to be addressed in this connection would be a careful study of the correlation between the pseudogap exhibited in the electron spectrum and the pseudo spin gap seen in the magnetic susceptibility of the insulating and superconducting cuprate HT_cSC. The latter has been extensively studied by inelastic neutron scattering [60] and NMR [61]. The microscopic origin of this spin gap and its evolution in the highly doped superconducting samples is presently not understood. In principle, any formation of singlet pairs would produce such a pseudo spin gap, whatever its origin might be: magnetic, purely correlation driven, negative U-centers or bipolarons, etc. A prime question to answer would be to check if a pseudogap in the electron spectrum exists in the non-magnetic HT_cSC. If so, then it would be conceivable that it is the existence of such a pseudogap which is at the origin of the pseudo spin gap in systems with magnetic correlations. Such a verification would be decisive for our understanding of the origin of pairing in HT_cSC.

Acknowledgements. I would like to thank C. Berthier, C. di Castro, D. Mihailovic, M. Miljak, Ph. Nozières, J.M. Robin and N. Schopohl for many valuable discussions, and T. Schneider for acquainting me with the experimental results on the crystal field excitations and Knight shift data in HT_cSC.

References

[1] H. Kamerlingh Onnes, Leiden Commun. **1206**, 1226 (1911).

[2] J.G. Bednorz and K.A. Müller, Z. Phys. B **64**, 189 (1986); C.W. Chu *et al.*, Phys. Rev. Lett. **58**, 405 (1987).

[3] J. Bardeen, L. Cooper and R. Schrieffer, Phys. Rev. **108**, 1175 (1957).

[4] For a review on this subject, see V.L. Ginzburg, Physica C **209**, 1 (1993).

[5] G.M. Eliashberg, Sov. Phys. JETP **11**, 696 (1960); D. Rainer, Prog. Low Temp. Phys. **10**, 371 (1986).

[6] A.S. Alexandrov and J. Ranninger, Phys. Rev. B **23**, 1796 (1981); *ibid.* **24**, 1164 (1981); A.S. Alexandrov, J. Ranninger and S. Robaszkiewicz, Phys. Rev. B **33**, 4526 (1986).

[7] M.R. Schafroth, Phys. Rev. **100**, 463 (1955).

[8] B.K. Chakraverty and C. Schlenker, J. Physique **37**, C4–353 (1976); S. Lakkis *et al.*, Phys. Rev. B **14**, 1429 (1976).

[9] B.K. Chakraverty, M.J. Sienko and J. Bonnerot, Phys. Rev. B **17**, 3781 (1978).

[10] O.F. Schirmer and E. Salje, J. Phys. C **13** L1067 (1980); R. Gehlig and E. Salje, Phil. Mag. B **47**, 229 (1983).

[11] V.N. Denisov *et al.*, Phys. Rev. B **48**, 16714 (1993).

[12] C. Taliani *et al.*, in *Electronic Properties of High T_c Superconductors and Related Compounds*, Springer Series of Solid State Physics **99**, H. Kuzmany, M. Mehring and J. Fink, eds. (Springer, Berlin,1990), p. 280.

[13] D. Mihailovic *et al.*, Phys. Rev. B **42**, 7989 (1990).

[14] V. Kudinov *et al.*, Phys. Lett. A **151**, 358 (1990); V.V. Eremenko *et al.*, Physica C **185**, 961 (1991); G. Yu *et al.*, Phys. Rev. B **45**, 4964 (1992).

[15] J. Orenstein *et al.*, in *Electronic Properties of High T_c Superconductors and Related Compounds*, Springer Series of Solid State Physics **99**, H. Kuzmany, M. Mehring and J. Fink, eds. (Springer, Berlin,1990), p. 254 ; K. Kamarao *et al.*, ibid. p. 260.

[16] J. Mustre de Leon *et al.*, in *Lattice Effects in High T_c Superconductors*, Y. Bar Yam, T. Egami, J. Mustre de Leon and A.R. Bishop, eds. (World Scientific, Singapore, 1992), p. 93.

[17] A. Bianconi *et al.*, *Lattice Effects in High T_c Superconductors*, Y. Bar Yam, T. Egami, J. Mustre de Leon and A.R. Bishop, eds. (World Scientific, Singapore, 1992), p. 65.

[18] D. De Fontain *et al.* , Nature **343**, 544 (1991); J.D. Jorgensen, ibid. **349**, 565 (1991).

[19] J. Ranninger, Physica Scripta **T 39**, 61 (1991).

[20] H. Tolentino *et al.*, Physica C **192**, 115 (1992); G. Uimin and J. Rossat-Mignod, Physica C **199**, 251 (1992).

[21] A. Bianconi, *Proc. Int. Conf. on Superconductivity* (Bangalore), S.K. Joshi, C.N.R. Rao and S.V. Subramanian, eds. (World Scientific, Singapore, 1990), p. 488.

[22] We do not discuss alternative mechanisms which have been proposed for HT_cSC.

[23] J.M. Robin, Ph.D. Thesis, University of Grenoble; J. Ranninger and J.M. Robin, to be published.

[24] J. Ranninger and S. Robaszkiewicz, Physica B **135**, 468 (1985); S. Robaszkiewicz, R. Micnas and J. Ranninger, Phys. Rev. B **37**, 180 (1987).

[25] R. Micnas, J. Ranninger and S. Robaszkiewicz, Rev. Mod. Phys. **62**, 113 (1990).

[26] V.L. Ginzburg, Sov. Phys. Solid State **2**, 18244 (1960).

[27] R. Brout, Phys. Rev. **118**, 1009 (1960).

[28] A. Kapitulnik *et al.*, Phys. Rev. B **37**, 537 (1988).

[29] M. Miljak *et al.*, Solid State Commun. **85**, 519 (1993) and references therein.

[30] J.M. Loram *et al.*, Phys. Rev. Lett. **71**, 1740 (1993).

[31] N. Winzek *et al.*, Physica C **205**, 45 (1993).

[32] R. Osborn and E.A. Goremychkin, Physica C **185**, 1179 (1991).

[33] Y.J. Uemura *et al.*, Phys. Rev. Lett. **62**, 2317 (1989).

[34] R. Tournier *et al.*, J. Mag. Mag. Mat. **76**, 552 (1988).

[35] P. Allen, Comments on Condensed Matter Physics **15**, 327 (1992).

[36] R.E. Walstedt *et al.*, Phys. Rev. B **41**, 9574 (1990); ibid **44** 7760 (1991); M. Horvatic *et al.*, Physica C **185/9**, 1139 (1991).

[37] Ph. Nozières and S. Schmitt-Rink, J. Low Temp. Phys. **59**, 195 (1985).

[38] M. Randeria *et al.*, Phys. Rev. Lett. **69**, 2001 (1992).

[39] A.S. Alexandrov and N. Mott, Supercond. Sci. Technol. **6**, 215 (1993).

[40] J.W. Allen *et al.*, Phys. Rev. Lett. **64**, 595 (1990); Rong Liu *et al.*, Phys. Rev. B **46**, 11056 (1992).

[41] M.A. van Veenendaal *et al.*, Phys. Rev. B **47**, 446 (1993).

[42] C. Thomsen *et al.*, Sol. State Comm. **75**, 219 (1990).

[43] L.E. Rehn *et al.*, in *Lattice Effects in High T_c Superconductors*, Y. Bar Yam, T. Egami, J. Mustre de Leon and A.R. Bishop, eds. (World Scientific, Singapore, 1992), p. 27 ; K. Yamaya *et al.*, ibid p. 33.
[44] H.A. Mook *et al.*, Phys. Rev. Lett. **65**, 2712 (1990).
[45] For a review on anomalous lattice properties in HT$_c$SC, see J. Ranninger, Z. Phys. B **84**, 167 (1991).
[46] J.R. Kirtley, Int. J. Mod. Phys. **4**, 201 (1990).
[47] B.N. Persson and J.E. Demuth, Phys. Rev. B **42**, 8057 (1990).
[48] L.D. Rotter *et al.*, Phys. Rev. Lett. **67**, 2741 (1991); C.C. Homes *et al.*, Phys. Rev. Lett. **71**, 1645 (1993).
[49] R.T. Collins *et al.*, Phys. Rev. B **43**, 8701 (1991).
[50] A.S. Alexandrov and J. Ranninger, Sol. State Comm. **81**, 403 (1992).
[51] W. Schnelle *et al.*, Physica C **161**, 123 (1989).
[52] W. Schnelle *et al.*, Physica C **168**, 465 (1990).
[53] M.B. Salamon *et al.*, Physica C **168**, 283 (1990).
[54] D. Mihailovic *et al.*, Phys. Rev. B **42**, 7989 (1990); Xiang-Xin Bi and P. Ecklund, Phys. Rev. Lett. **70**, 2625 (1993).
[55] Z.X. Shen *et al.*, Phys. Rev. Lett. **70**, 1553 (1993).
[56] W.N. Hardy *et al.*, Phys. Rev. Lett. **70**, 3999 (1993).
[57] R. Micnas *et al.*, Phys. Rev. B **37**, 9410 (1988).
[58] P. Calvani *et al.*, to be published.
[59] M. Knupfer *et al.*, Phys. Rev. B **47**, 13944 (1993); V.N. Denisov *et al.*, Sov. Phys. JETP **75**, 158 (1992); V.N. Denisov *et al.*, Synthetic Metals **55–57**, 3050 (1993).
[60] J. Rossat-Mignod *et al.*, Physica C **169**, 158 (1991).
[61] C. Berthier *et al.*, Physica C **185/9**, 1141 (1991).

17

The Bosonization Method in Nuclear Physics

F. Iachello

Center for Theoretical Physics, Sloane Physics Laboratory
Yale University
New Haven, CT 06511-8167

Abstract

The bosonization method developed in nuclear physics in the last 20 years is briefly reviewed.

1 Introduction

In the last 20 years, a considerable amount of work has gone into the development of bosonization methods for highly correlated finite fermion systems. This work has been stimulated by the phenomenological successes of the interacting boson model introduced in 1974 in which atomic nuclei composed of nucleons (fermions) are treated in terms of an interacting system of bosons (fermion pairs). In this article the logic scheme of the bosonization method in nuclear physics will be briefly reviewed. The purpose here is to provide the basic references upon which the review is built and sources where further references can be found. Bosonization methods were introduced much earlier than 1974 in connection with infinite Fermi systems. The method developed in nuclear physics has some similarities with those developed in other areas of physics, but it also has major differences. In nuclear physics one maps not only operators but also states and seeks a description not only of the ground state but also of the entire excitation spectrum. Furthermore, number projection plays a very important role contrary to the case of infinite systems where number projection is not relevant. These differences, and others, will be briefly remarked upon in this article.

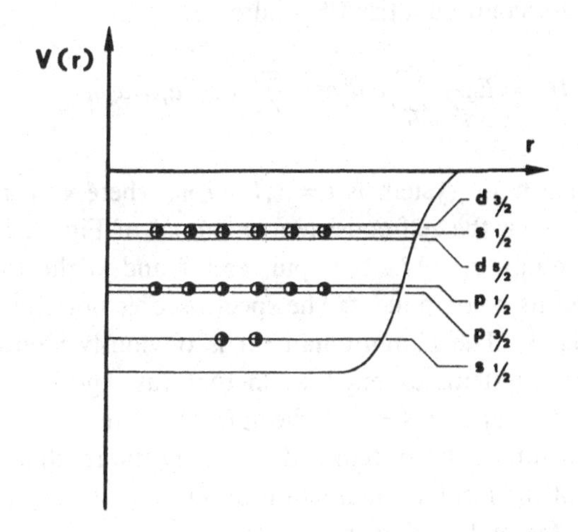

Fig. 1. The self-consistent potential $V(r)$ in nuclei. The single-particle levels are labelled by the spectroscopic notation ℓ_j.

2 Strongly Interacting Finite Fermion Systems

2.1 Fermions and the Shell Model

Quite often in physics one has to deal with a system of interacting fermions. In some cases, this system is finite, i.e. with a number of particles of the order of $n \sim 100$. The best examples of finite Fermi systems are atomic nuclei, where $n \sim 2-250$. In first approximation, one can imagine that the individual fermions (nucleons) move in the average field generated by all others. The average (or mean) field can be obtained by a Hartree–Fock (or similar) calculation. The mean field in nuclei has the shape depicted in Fig. 1. Although the mean field provides a description of some properties of the ground state of the system, it does not help much in the study of the excitation spectrum, which is often dominated by the correlations (residual interactions). This means that one has to deal with a finite Fermi system whose Hamiltonian is

$$H_F = E_0 + \sum_{p=1}^{n} \epsilon_p + \sum_{p<p'}^{n} v_{pp'}, \tag{1}$$

where ϵ_p is the single particle energy and $v_{pp'}$ the residual interaction between particles, p and p'. The Hamiltonian (1) describes a "shell-model". Introducing a label i to characterize the single-particle states, one can write a second quantized Hamiltonian

$$H_F = E_0 + \sum_i \epsilon_i a_i^\dagger a_i + \sum_{ii'kk'} v_{ii'kk'} a_i^\dagger a_{i'}^\dagger a_k a_{k'} \quad . \tag{2}$$

The label i for a finite system is $i = v, l, s, j, m$, where v is the principal quantum number in the self-consistent potential of Fig. 1, l the orbital angular momentum, $s = 1/2$ the spin, and j and m the total angular momentum and its z-component. The spectroscopic notation vl_j is often used, as in Fig. 1. The Hamiltonian (2) is obviously identical to that of infinite Fermi systems except that in that case the label $i \equiv \mathbf{k}, s, \sigma$, where $\mathbf{k}$ is the momentum, $s = 1/2$ the spin and σ its z-component. The excitation spectrum of the system is dictated by the residual interaction. The behavior of the residual interaction in nuclei is shown schematically in Fig. 2. Here the matrix elements

$$W_J = \langle (1f_{7/2})^2 J \mid v \mid (1f_{7/2})^2 J \rangle \tag{3}$$

for two identical particles occupying the single-particle level $1f_{7/2}$ is shown as a function of the total angular momentum $J = 0, 2, 4, 6$. (In contrast to atoms where the appropriate coupling scheme is $L - S$, in nuclei the dominant coupling scheme is $j - j$.) As one can see from Fig. 2, the interaction is large and attractive when $J = 0$, is still attractive (but not so large) when $J = 2$, and it is small or repulsive when $J \geq 4$.

2.2 *Fermion Pairs*

The properties of the residual interaction shown in Fig. 2 suggest an approximation scheme to deal with the excitation spectra of nuclei. This approximation scheme consists in introducing *correlated* fermion pairs. The second quantized forms of the pair operators are

$$\begin{aligned}
S^\dagger &= \sum_j \alpha_j S_j^\dagger , & S_j^\dagger &= \left(a_j^\dagger \times a_j^\dagger\right)_0^{(0)} , \\
D_\mu^\dagger &= \sum_{jj'} \beta_{jj'} D_{jj',\mu}^\dagger , & D_{jj',\mu}^\dagger &= \left(a_j^\dagger \times a_{j'}^\dagger\right)_\mu^{(2)} , \\
G_\mu^\dagger &= \sum_{jj'} \gamma_{jj'} G_{jj',\mu}^\dagger , & G_{jj',\mu}^\dagger &= \left(a_j^\dagger \times a_{j'}^\dagger\right)_\mu^{(4)} , \\
&\ldots
\end{aligned} \tag{4}$$

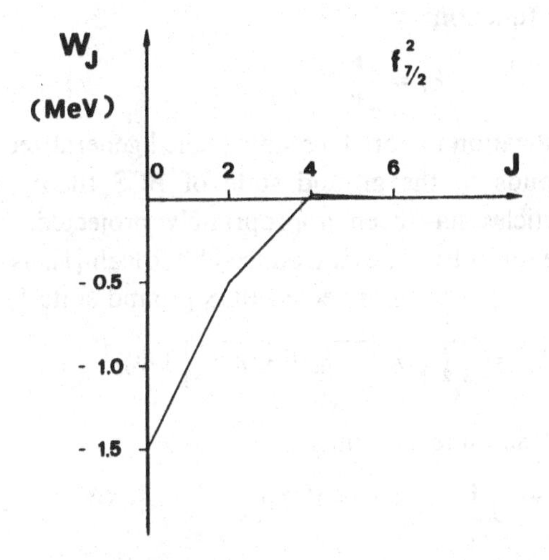

Fig. 2. Matrix elements W_J of the residual nucleon-nucleon interaction for two identical particles occupying the single particle level $1f_{7/2}$, as a function of the total angular momentum, J, of the two particles.

Here the crosses denote angular momentum couplings,

$$\begin{aligned} S^{\dagger}{}_{j} &= \sum_{m,m'} \langle j,m,j,m' \mid 0,0 \rangle a^{\dagger}{}_{j,m} a^{\dagger}{}_{j,m'} \\ &= \sum_{m} \frac{(-1)^{j-m}}{\sqrt{2j+1}} a^{\dagger}{}_{j,m} a^{\dagger}{}_{j,-m}\,, \\ D^{\dagger}{}_{jj',\mu} &= \sum_{m,m'} \langle j,m,j',m' \mid 2,\mu \rangle a^{\dagger}{}_{j,m} a^{\dagger}{}_{j',m'}\,, \\ G^{\dagger}{}_{jj',\mu} &= \sum_{m,m'} \langle j,m,j',m' \mid 4,\mu \rangle a^{\dagger}{}_{j,m} a^{\dagger}{}_{j',m'}\,, \\ &\dots \end{aligned} \tag{5}$$

One may note that the angular momentum zero pairs, called S pairs, are similar to the Cooper pairs of superconductivity

$$b^{\dagger}_{\mathbf{k}} = c^{\dagger}_{\mathbf{k},\uparrow} c^{\dagger}_{-\mathbf{k},\downarrow} \tag{6}$$

where $c^{\dagger}$ is the creation operator for electrons. The others (D pairs, G pairs,...) represent a generalization of the pairs of superconductivity to larger angular momenta. The number of pairs is obviously

$$N = \frac{n}{2}\,, \tag{7}$$

where, for the moment, n=even.

States of the interacting fermion system are constructed by acting with pair operators on the vacuum. If only S-pairs are retained, there is only one state with wave function

$$| \Psi \rangle = \frac{1}{\mathcal{N}} (S^\dagger)^N \, | \, 0 \rangle \, , \tag{8}$$

where $\mathcal{N}$ is a normalization factor. This state (called generalized seniority zero state) corresponds to the ground state of BCS theory in which the number of particles has been appropriately projected. It is the generalization of the seniority states introduced by Racah [1], as discussed by Talmi, and others [2]. The unprojected BCS ground state [3]

$$| \Psi \rangle = \prod_{\mathbf{k}} \left[\sqrt{1 - h_{\mathbf{k}}} + \sqrt{h_{\mathbf{k}}} b_{\mathbf{k}}^\dagger \right] | \, \Phi_0 \rangle \tag{9}$$

is, when written for the finite system,

$$| \Psi \rangle = \prod_{j,m>0} \left[u_j + v_j (-)^{j-m} a^\dagger{}_{j,m} a^\dagger{}_{j,-m} \right] | \, 0 \rangle \, . \tag{10}$$

Unfortunately, as shown by many authors, number projection plays a crucial role in finite systems, since the fluctuations are of order $\sqrt{n}$ and $n \sim 100$. BCS is thus not appropriate for finite systems.

The more and more pair states are retained, the more and more the full Hilbert space of the original fermion system is recovered. If S and D pairs are retained, the wave functions are of the type

$$\frac{1}{\mathcal{N}} \left(S^\dagger \right)^{N-n_d} \left(D^\dagger \right)^{n_d} | \, 0 \rangle \, , \; n_d = 0, \ldots, N \, , \tag{11}$$

which include (8) but are more general. If S, D and G pairs are retained, one has

$$\frac{1}{\mathcal{N}} \left(S^\dagger \right)^{N-n_d-n_g} \left(D^\dagger \right)^{n_d} \left(G^\dagger \right)^{n_g} | \, 0 \rangle \, . \tag{12}$$

It should be remarked that introduction of at least D pairs is crucial in nuclear physics, in view of the attractive nature of the $J = 2$ pairing (quadrupole pairing).

By introducing correlated pairs, one truncates the original fermionic space F into a smaller space of finite dimensions, called B in Fig. 3. The complete fermionic space contains, in addition to the correlated pair states, states in which the correlated pairs are broken, for example, when S and D pairs are retained, the complete space is

$$\begin{array}{l} \left(S^\dagger \right)^{N-n_d} \left(D^\dagger \right)^{n_d} | \, 0 \rangle \\ \left(S^\dagger \right)^{N-n_d-1} \left(D^\dagger \right)^{n_d} a^\dagger{}_j a^\dagger{}_{j'} | \, 0 \rangle \\ \left(S^\dagger \right)^{N-n_d-2} \left(D^\dagger \right)^{n_d} a^\dagger{}_j a^\dagger{}_{j'} a^\dagger{}_{j''} a^\dagger{}_{j'''} | \, 0 \rangle \\ \ldots \, . \end{array} \tag{13}$$

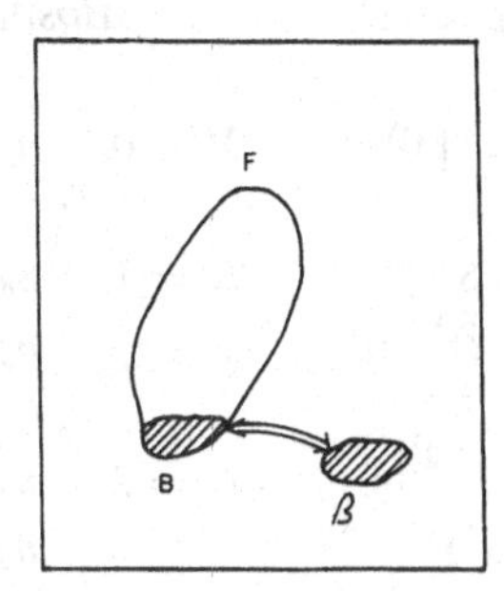

Fig. 3. The truncation-mapping procedure. F is the fermionic space, B the pair space and $\mathscr{B}$ the bosonic space.

The broken pair states are those containing explicitly the fermion operators ${a^\dagger}_j$. Of course, in constructing the space (13) one has to be careful about overcompleteness and orthogonality (called the decoupling problem) but this problem is important only for very small systems.

2.3 Mapping Fermions into Bosons

Once fermion pairs have been introduced, the next step is to map fermion pair states into boson states [4]. Mapping is a one to one correspondence in which to each fermion pair state there corresponds a boson state. When S and D pairs are retained, the mapping is done through the introduction of s and d boson operators, s, d_μ ($\mu = 0, \pm1, \pm2$), satisfying Bose commutation relations

$$\left[s, s^\dagger\right] = 1 \; ; \; \left[d_\mu, d^\dagger_{\mu'}\right] = \delta_{\mu,\mu'} \; ; \; \left[s, d^\dagger_\mu\right] = 0 \, . \tag{14}$$

The fermion pair states (11) are mapped into boson states

$$\frac{1}{\mathscr{N}} \left(s^\dagger\right)^{N-n_d} \left(d^\dagger\right)^{n_d} \mid 0\rangle. \tag{15}$$

Explicitly, the first few fermion pair states and their mappings are:

Fermion pair space B			*Boson space B*		
$n=0,$	$v=0$	$\mid 0\rangle$	$N=0,$	$n_d=0$	$\mid 0\rangle$
$n=2,$	$v=0$	$S^\dagger \mid 0\rangle$	$N=1,$	$n_d=0$	$s^\dagger \mid 0\rangle$
	$v=2$	$D^\dagger \mid 0\rangle$		$n_d=1$	$d^\dagger \mid 0\rangle$
$n=4,$	$v=0$	$S^{\dagger 2} \mid 0\rangle$	$N=2,$	$n_d=0$	$s^{\dagger 2} \mid 0\rangle$
	$v=2$	$S^\dagger D^\dagger \mid 0\rangle$		$n_d=1$	$s^\dagger d^\dagger \mid 0\rangle$
	$v=4$	$\mathscr{P}_v D^{\dagger 2} \mid 0\rangle$		$n_d=2$	$d^{\dagger 2} \mid 0\rangle$
	...			...	

(16)

The mapping is such that the number of bosons N is half the number of fermions

$$N = \frac{n}{2} \tag{17}$$

and the number of d-bosons is half the fermion generalized seniority

$$n_d = \frac{v}{2}. \tag{18}$$

As mentioned above, some subtle problems arise in the construction of the pair space, requiring the introduction of projection operators to take into account the correct orthogonality properties of the states. These projection operators are denoted by $\mathscr{P}_v$ in Eq. (16). The truncation and subsequent mapping procedure is shown schematically in Fig. 3.

2.4 Hamiltonian (and Other) Operators

Mapping of states is not sufficient to determine the properties of the system. The next step is to map operators. For example, in order to compute the excitation spectrum, one needs to construct, starting from the fermion Hamiltonian (2), the boson Hamiltonian

$$H_{\mathscr{B}} = \mathscr{E}_0 + \sum_{\tau} \varepsilon_\tau b_\tau^\dagger b_\tau + \sum_{\tau\tau'\eta\eta'} u_{\tau\tau'\eta\eta'} b_\tau^\dagger b_{\tau'}^\dagger b_\eta b_{\eta'}, \tag{19}$$

where $b_\tau^\dagger$ denotes generically a boson operator. Although in general the mapping from fermions to bosons can be done for the entire system, in practice only the valence shells are usually considered. The valence shells of nuclei between particle numbers 50 and 82 are shown in Fig. 4.

In order to obtain the boson Hamiltonian (19) one performs two steps:

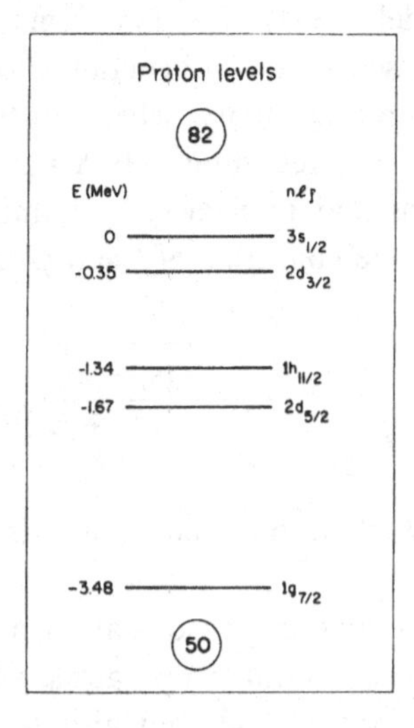

Fig. 4. The valence shells of nuclei between nucleon numbers 50 and 82.

(i) Determine the pair structure constants α_j, $\beta_{jj'}$, which appear in the correlated pairs, Eq. (4). Several methods have been suggested, the simplest one being that of diagonalizing H_F in the two-particle valence space. This method gives the structure constants α_j, $\beta_{jj'}$, and the single-particle boson energies, ε_τ.

(ii) Determine the boson-boson interaction $u_{\tau\tau'\eta\eta'}$. Again, several methods have been suggested, the simplest one being that of equating matrix elements of the fermion Hamiltonian, H_F, in the fermion space of the left-hand side of Eq. (16), with matrix elements of the boson Hamiltonian, H_B, in the boson space of the right-hand side of Eq. (16), schematically written

$$\langle H_F \rangle = \langle H_B \rangle \tag{20}$$

for four particles (two bosons). This method is referred to as OAI mapping [4]. The boson–boson interaction appears here as the contribution of two terms, the direct and the exchange terms. The exchange term, which is very important for finite systems, arises from the composite nature of the bosons (Pauli principle).

At the end of this procedure one has replaced a strongly interacting finite *fermion* system by a strongly interacting *boson* system. The bosonization of a strongly interacting finite fermion system is also called the Interacting Boson Approximation (IBA) and it leads to a model of nuclei called the

Interacting Boson Model (IBM) [5]. Two remarks are in order here. First, the method discussed here is a generalization of methods used in superconductivity. Superconductivity, when written in the language of Sections 2–5, corresponds to retaining only S pairs, to considering only J=0 pairing interactions and to neglecting number projection, hereby assuming that the structure constants of the S pairs are given by

$$\alpha_j = \sqrt{\frac{1}{2}\left[1 - \frac{(\varepsilon_j - \lambda)}{[(\varepsilon_j - \lambda)^2 + \Delta^2]^{1/2}}\right]}. \tag{21}$$

In Eq. (21), ε_j are the single fermion energies, λ is the Fermi energy and Δ the pairing gap.

The second remark is that the bosonization method can be applied not only to particle–particle fermion operators, where the product $a^\dagger a^\dagger$ is replaced by a boson operator $b^\dagger$, but also to particle–hole fermion operators [6], where the product $a^\dagger a$ is replaced by a boson operator, $b^\dagger$. The equivalent of the bosonization method then becomes, in the infinite system, the theory of excitons.

3 Strongly Interacting Finite Boson Systems

3.1 Bosons and the Shell Model

The procedure of Section 2 replaces a strongly interacting fermion system by a strongly interacting boson system with Hamiltonian

$$H_{\mathscr{B}} = \mathscr{E}_0 + \sum_{\sigma}^{N} \varepsilon_\sigma + \sum_{\sigma<\sigma'}^{N} v_{\sigma\sigma'}. \tag{22}$$

Here ε_σ are the single-particle boson energies and the $v_{\sigma\sigma'}$ are the boson–boson interactions. A second quantized version of (22) is given in (19) where τ labels the single particle boson states. It is convenient here to use a spectroscopic notation $\tau = s, d, g, \ldots$. The Hamiltonian (22) or (19) is that of a "shell-model" for bosons. As mentioned in Section 2, in nuclei, in first approximation, only s and d bosons are important. For this reason, considerable effort has gone into the study and solution of a system composed of s and d bosons over the last 20 years.

3.2 The s–d Boson System

The s–d boson system, usually called interacting boson model [5], is characterized by the Hamiltonian

$$\begin{aligned}
H_B &= E_o + \varepsilon_s(s^\dagger \cdot \tilde{s}) + \varepsilon_d(d^\dagger \cdot \tilde{d}) \\
&+ \sum_{L=0,2,4} c'_L \left[\left[d^\dagger \times d^\dagger \right]^{(L)} \times \left[\tilde{d} \times \tilde{d} \right]^{(L)} \right]_0^{(0)} \\
&+ v'_2 \left[\left[d^\dagger \times d^\dagger \right]^{(2)} \times \left[\tilde{d} \times \tilde{s} \right]^{(2)} + \left[d^\dagger \times s^\dagger \right]^{(2)} \times \left[\tilde{d} \times \tilde{d} \right]^{(2)} \right]_0^{(0)} \\
&+ v'_0 \left[\left[d^\dagger \times d^\dagger \right]^{(0)} \times \left[\tilde{s} \times \tilde{s} \right]^{(0)} + \left[s^\dagger \times s^\dagger \right]^{(0)} \times \left[\tilde{d} \times \tilde{d} \right]^{(0)} \right]_0^{(0)} \\
&+ u_2 \left[\left[d^\dagger \times s^\dagger \right]^{(2)} \times \left[\tilde{d} \times \tilde{s} \right]^{(2)} \right]_0^{(0)} \\
&+ u'_0 \left[\left[s^\dagger \times s^\dagger \right]^{(0)} \times \left[\tilde{s} \times \tilde{s} \right]^{(0)} \right]_0^{(0)} . \qquad (23)
\end{aligned}$$

This Hamiltonian is the most general Hamiltonian invariant under rotations up to two-body boson–boson interactions. Although in principle, the bosonization procedure, starting from a fermion two-body Hamiltonian may produce a boson–boson Hamiltonian with three-body, four-body, ... terms, in practice it has been found that the higher order terms are small and therefore it is sufficient to consider the Hamiltonian (23). One is then asked to find the eigenvalues of H_B for a system of N bosons, where N is the number of active pairs. Several methods have been used to find the spectrum of (23):

(*i*) *Matrix diagonalization*

This is the direct numerical diagonalization of (23). At the present time computer programs exist that diagonalize (23) up to a number of bosons $N \sim 30$. For larger values, the matrices to diagonalize become very large and a direct diagonalization is not possible. None-the-less, since the number of active pairs in nuclei is usually less than 30, this method has been used extensively and, in combination with the bosonization method, has allowed one to compute spectra of nuclei, starting from the effective nucleon–nucleon interaction, with accuracies better than 10%. An example is shown in Fig. 5.

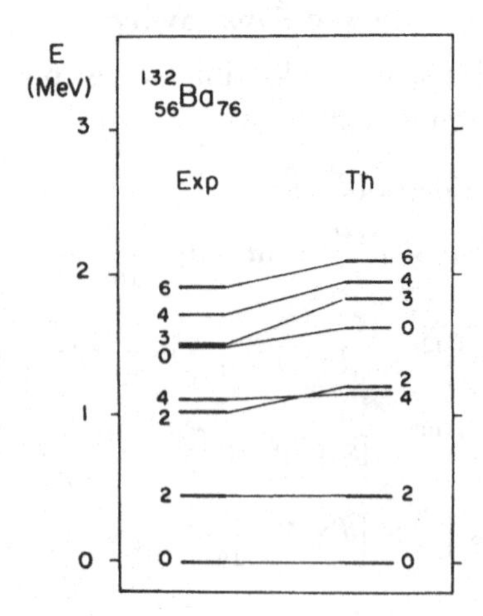

Fig. 5. *Ab initio* calculation of the excitation spectrum of ^{132}Ba using the bosonization method. On the left the experimental spectrum, on the right the calculated spectrum.

(ii) *Lie algebraic methods*

Bilinear products of boson creation and annihilation operators

$$G_{\alpha\beta} = b_{\alpha}^{\dagger} b_{\beta} \tag{24}$$

generate a Lie algebra $\mathcal{G}$. If the index α goes from 1 to n, the Lie algebra is $u(n)$. For the s–d boson system in which the index α goes from 1 to 6, the Lie algebra is $\mathcal{G} = u(6)$. The six components are the single component of the s-boson and the five components of the d-boson, d_{μ} ($\mu = 0, \pm 1, \pm 2$). The Hamiltonian H_B can be written in terms of the operators $G_{\alpha\beta}$, as

$$H_B = E_o + \sum_{\alpha\beta} \varepsilon_{\alpha\beta} G_{\alpha\beta} + \sum_{\alpha\beta\gamma\delta} u_{\alpha\beta\gamma\delta} G_{\alpha\beta} G_{\gamma\delta}. \tag{25}$$

All the powerful techniques of group theory can then be used to find the eigenvalues of H_B. This is the celebrated "Algebraic Theory" which has been used extensively in recent years to solve many-body problems. Algebraic theory is presented in some detail in Ref. [5] and will not be discussed further here.

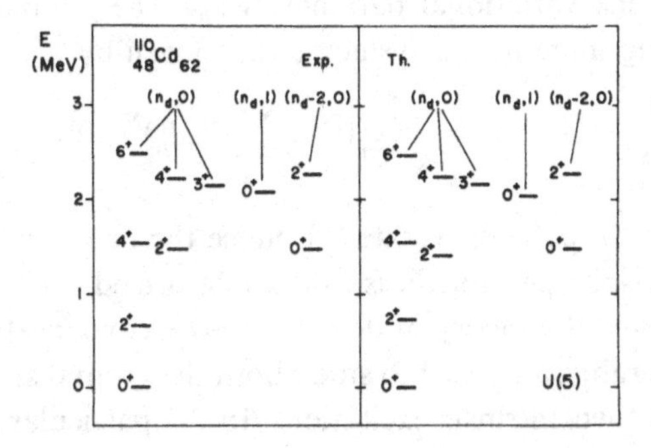

Fig. 6. An example of a spherical nucleus: ^{110}Cd. On the left the experimental spectrum, on the right the calculated spectrum. Both the experimental and calculated spectra are slightly anharmonic. These anharmonicities can be accounted for by going beyond the simple approximation (26), as discussed in Ref. [5].

(iii) *The method of Bose condensates*

This method is not used much in practice in nuclei. I will present it here in view of its relation to the subject of Bose condensation discussed in this volume, and in view of its interest for systems with N large. I will begin with the trivial case in which there is no boson–boson interaction. In this case, the ground state is the state in which all bosons occupy the lowest level (s-level)(called spherical Bose condensate),

$$\frac{1}{\sqrt{N!}}(s^\dagger)^N \mid 0\rangle \quad . \tag{26}$$

Excitations are obtained by removing one boson from the lowest level and placing it into the d-level

$$\begin{aligned} &\frac{1}{\sqrt{\mathcal{N}!}}(s^\dagger)^{N-1}d^\dagger \mid 0\rangle \,, \\ &\frac{1}{\sqrt{\mathcal{N}'!}}(s^\dagger)^{N-2}d^{\dagger 2} \mid 0\rangle \,, \\ &\dots \end{aligned} \tag{27}$$

where $\mathcal{N}$, $\mathcal{N}'$ are renormalization constants. The corresponding spectrum is a harmonic spectrum with energies $\varepsilon, 2\varepsilon, ...$ where $\varepsilon = \varepsilon_d - \varepsilon_s$. Several nuclei have been found with a spectrum of this type, one of which is shown in Fig. 6.

A more interesting case is that in which there is a boson-boson interaction. In this case one can introduce an operator

$$b_c^\dagger = (s^\dagger + \sum_\mu \alpha_\mu d_\mu^\dagger)\,, \mu = 0, \pm 1, \pm 2, \tag{28}$$

which contains some variational parameters α_μ. The ground state of a strongly interacting finite boson system is then given by

$$\frac{1}{\sqrt{N!}}(b_c^\dagger)^N \mid 0\rangle = \frac{1}{\sqrt{N!}}(s^\dagger + \sum_\mu \alpha_\mu d_\mu^\dagger)^N \mid 0\rangle, \tag{29}$$

called an intrinsic (or coherent) state [7]. Since the operator $b_c^\dagger$ contains a mixture of all boson operators, it is called a deformed Bose condensate. In the last few years, the theory of deformed Bose condensates has been developed considerably, and used to study both the ground states of finite Bose systems and their intrinsic excitations. In the particular case of s–d systems, a frame of reference can be chosen such that the expectation values of all operators do not depend on all five variables, α_μ, but only on two of them, called intrinsic (or Bohr) variables, β, γ. Starting from the six spherical boson operators $s^\dagger, d_{+2}^\dagger, d_{+1}^\dagger, d_0^\dagger, d_{-1}^\dagger, d_{-2}^\dagger$, one can then introduce six other boson operators that depend continuously on β and γ as [8]

$$\begin{aligned}
b_c^\dagger &= (1+\beta^2)^{-1/2}\left[\beta\cos\gamma d_0^\dagger + \frac{1}{\sqrt{2}}\beta\sin\gamma(d_{+2}^\dagger + d_{-2}^\dagger) + s^\dagger\right], \\
b_\beta^\dagger &= (1+\beta^2)^{-1/2}\left[\cos\gamma d_0^\dagger + \frac{1}{\sqrt{2}}\sin\gamma(d_{+2}^\dagger + d_{-2}^\dagger) - \beta s^\dagger\right], \\
b_\gamma^\dagger &= \frac{1}{\sqrt{2}}\cos\gamma(d_{+2}^\dagger + d_{-2}^\dagger) + \sin\gamma d_0^\dagger, \\
b_x^\dagger &= \frac{1}{\sqrt{2}}\left(d_{+1}^\dagger + d_{-1}^\dagger\right), \\
b_y^\dagger &= \frac{1}{\sqrt{2}}\left(d_{+1}^\dagger - d_{-1}^\dagger\right), \\
b_z^\dagger &= \frac{1}{\sqrt{2}}\left(d_{+2}^\dagger - d_{-2}^\dagger\right).
\end{aligned} \tag{30}$$

The original basis can then be transformed into a deformed boson basis, obtained by acting with the operators of (30) on a vacuum state

$$\mathscr{B}^D : \frac{1}{\mathscr{N}} b_i^\dagger(\beta,\gamma) b_{i'}^\dagger(\beta,\gamma)\cdots \mid 0\rangle. \tag{31}$$

The number operator, when written in terms of the new operators becomes

$$\hat{N} = b_c^\dagger b_c + b_\beta^\dagger b_\beta + b_\gamma^\dagger b_\gamma + b_x^\dagger b_x + b_y^\dagger b_y + b_z^\dagger b_z. \tag{32}$$

The ground state is the state of Eq. (29),

$$(N!)^{-1/2}\left[b_c^\dagger(\beta,\gamma)\right]^N \mid 0\rangle \equiv \mid N;\beta,\gamma\rangle. \tag{33}$$

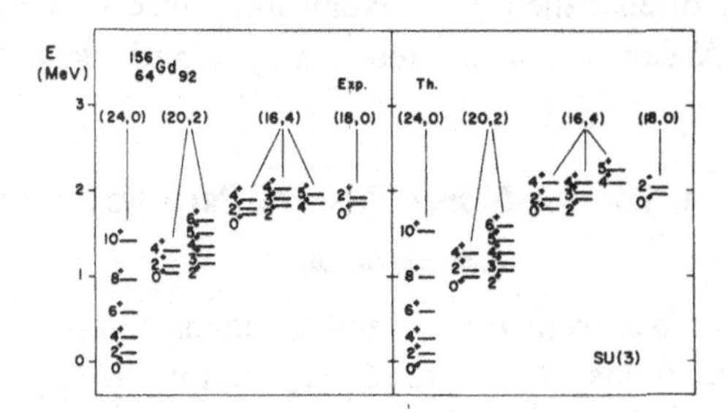

Fig. 7. An example of a deformed nucleus: ^{156}Gd.

Excitations are obtained by removing one boson from the condensate

$$\begin{aligned} &\frac{1}{\mathcal{N}}(b_c^\dagger)^{N-1} b_\beta^\dagger \mid 0\rangle\,, \\ &\frac{1}{\mathcal{N}}(b_c^\dagger)^{N-1} b_\gamma^\dagger \mid 0\rangle\,. \end{aligned} \tag{34}$$

These excitations are called intrinsic excitations. The operators $b_x^\dagger, b_y^\dagger, b_z^\dagger$ generate spurious modes instead, corresponding to rotations of the condensate. The variational parameters β, γ are determined by minimizing the energy with respect to β and γ,

$$E(N;\beta,\gamma) = \frac{\langle N;\beta,\gamma \mid H \mid N;\beta,\gamma\rangle}{\langle N;\beta,\gamma \mid N;\beta,\gamma\rangle}\,. \tag{35}$$

Several nuclei have been found whose excitation spectrum can be well described by this method, one of which is shown in Fig. 7.

Some remarks are in order here. The method of Bose condensates is different from the theory of superfluidity [9], although it shares some similarities with it. When the depletion of the condensate is small, one has

$$N_s \simeq N\,, \; N_d \simeq 0. \tag{36}$$

The Bogoliubov Hamiltonian for the s–d boson system is

$$H_B = E_0' - \varepsilon_s N + u_0' N(N-1) + \varepsilon_d (d^\dagger \cdot \tilde{d}) + v_0' N \left[d^\dagger \times d^\dagger + \tilde{d} \times \tilde{d}\right]_0^{(0)}\,. \tag{37}$$

This Hamiltonian is obtained from (23) by replacing the operators $s^\dagger$ and s by c-numbers, $\sqrt{N}$. This approximation is valid only for weakly interacting Bose systems. In nuclei, the interactions are strong and the approximation (37) is not appropriate. Another comment is that the evaluation of the ground state energy (35), and of the excitation energies, is good to leading order in N, the number of bosons. The method of Bose condensates is thus particularly appropriate where N is large. In this respect it is often called $1/N$ expansion, since corrections to the leading order in N can be constructed in a systematic way.

4 Two (or More) Types of Particles

4.1 Fermions

In some cases, one has to deal with finite systems composed of two (or more) types of fermions. Again the best example of such a system is the atomic nucleus composed of protons and neutrons. To the extent that protons and neutrons are assumed to have the same properties (a good approximation from the point of view of the strong interactions) one can consider the nucleus as if it is composed of nucleons (protons and neutrons). This is the approximation used in Sections 2 and 3. However, there are some properties of nuclei that depend explicitly on the difference between protons and neutrons. For a more refined treatment of nuclei, one has therefore to deal with a system composed of two types of particles. The shell model for two types of particles is no different from that with one type of particles. There are now single-particle levels both for protons and neutrons, as shown in Fig. 8. The bosonization method can be used without major modification for these systems.

4.2 Fermion Pairs

One introduces pairs composed of two protons and two neutrons [10]:

$$\begin{aligned} S^\dagger{}_\pi &= \sum_{j\pi} \alpha_{j\pi} S^\dagger{}_{j\pi}, & S^\dagger{}_{j\pi} &= \left(a^\dagger{}_{j\pi} \times a^\dagger{}_{j\pi}\right)^{(0)}_0, \\ S^\dagger{}_\nu &= \sum_{j\nu} \alpha_{j\nu} S^\dagger{}_{j\nu}, & S^\dagger{}_{j\nu} &= \left(a^\dagger{}_{j\nu} \times a^\dagger{}_{j\nu}\right)^{(0)}_0, \\ & \ldots, \end{aligned} \tag{38}$$

where now the indices π and ν denote protons and neutrons. One can consider in principle many pair states, but in practice, in nuclei, only s and d pairs are important.

4.3 Bosonization

The pair space is then mapped onto a boson space composed of proton bosons and neutron bosons $s^\dagger_\pi, d^\dagger_{\pi,\mu}$; $s^\dagger_\nu, d^\dagger_{\nu,\mu}$. One then obtains the boson

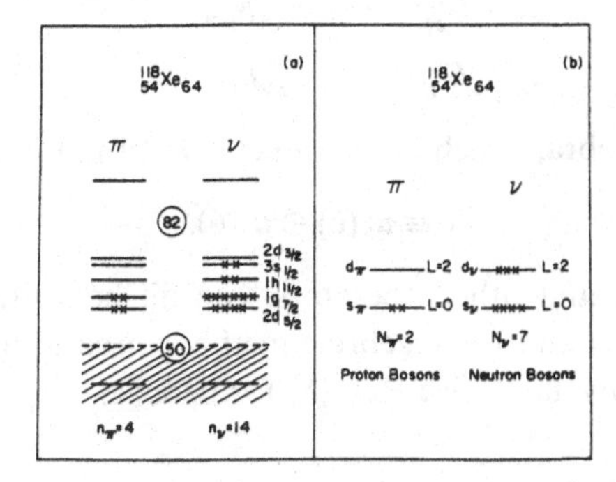

Fig. 8. The valence shells for protons and neutrons between particle numbers 50 and 82 (a), and the boson model that replaces the fermion problem after truncation and mapping (b).

Hamiltonian corresponding to a given fermion Hamiltonian. The boson Hamiltonian has the structure

$$H_B = H_\pi + H_\nu + V_{\pi\nu}, \tag{39}$$

where H_π and H_ν describe protons and neutrons, respectively, and $V_{\pi\nu}$ their interaction. The H_π and H_ν have the structure of Eq. (23) with an index π and ν attached.

4.4 Methods of Solution

The eigenvalue problem for (39) can be solved using the same techniques discussed in Section 3.

(i) Matrix diagonalization

This technique is straightforward except that the dimensions of the matrices here are even larger. Computer programs exist for the s–d boson system up to $N_\pi \sim 15$ and $N_\nu \sim 15$.

(ii) Lie algebraic methods

These techniques are particularly useful here in view of the complexity of the problem. The bilinear products

$$G^{(\pi)}_{\alpha\beta} = b^\dagger_{\alpha,\pi} b_{\beta,\pi},$$
$$G^{(\nu)}_{\alpha\beta} = b^\dagger_{\alpha,\nu} b_{\beta,\nu}, \tag{40}$$

generate a Lie algebra, which is the direct sum of $u(6)$ algebras

$$\mathcal{G} \equiv u_\pi(6) \oplus u_\nu(6). \tag{41}$$

The Hamiltonian, and other operators, can all be written in terms of the operator in (41) and Lie algebraic methods used to find their matrix elements and eigenvalues. (See Ref. [5] for details.)

(*iii*) *Bose condensates.*

Methods of Bose condensation can also be used here. The ground state of the (deformed) two-fluid system is written as

$$\mid g\rangle = (N_\pi ! N_\nu !)^{-1/2} (b^\dagger_{\pi,c})^{N_\pi} (b^\dagger_{\nu,c})^{N_\nu} \mid 0\rangle \tag{42}$$

and excitations are constructed.

5 Mixed Systems of Bosons and Fermions

Another important case that has been extensively investigated over the last 15 years is that of mixed systems of bosons and fermions [11]. In nuclei this situation appears when one wishes to treat systems with an odd number of protons and/or neutrons. In these cases, at least one particle must be unpaired. If the pairs are bosonized one ends up with a mixed system of bosons and fermions. Also to this situation belong those cases in which the correlated pairs are broken and one has explicitly two or more fermions in addition to the bosons. The bosonization method has been used extensively to deal with these situations. One starts from the purely fermionic Hamiltonian and introduces correlated pairs as in Section 2. The pair states are mapped into boson states and a Hamiltonian is constructed of the type

$$H = H_B + H_F + V_{BF}, \tag{43}$$

where H_B and H_F describe the system of bosons and fermions, respectively, and V_{BF} their interaction. Particular care must be exerted here in order to avoid overcounting. Also the boson–fermion interaction has a rather complex form in order to take into account the composite nature of the bosons. The boson–fermion interaction is illustrated by the diagrams in Fig. 9.

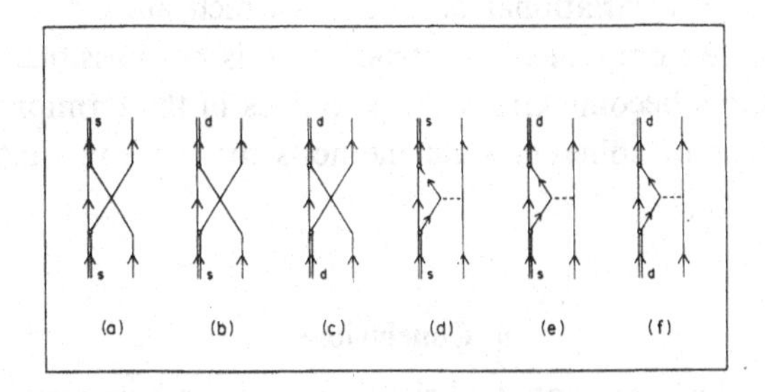

Fig. 9. The boson–fermion interaction appearing in Eq. (43). This interaction contains direct (d, e, f) and exchange (a, b, c) terms. The exchange terms arise from the Pauli principle.

The eigenvalue problem for the Hamiltonian H of Eq. (43) is solved by methods similar to those discussed previously.

(i) Matrix diagonalization

(ii) Lie algebraic methods

In this case one takes advantage of the fact that the bilinear products of creation and annihilation operators for bosons and fermions generate a graded Lie algebra composed of the set of operators of the type

$$\begin{aligned} G^{(B)}_{\alpha\beta} &= b^\dagger_\alpha b_\beta, \\ G^{(F)}_{ij} &= a^\dagger_i a_j, \\ F^\dagger_{i\alpha} &= a^\dagger_i b_\alpha, \\ F^\dagger_{\alpha i} &= b^\dagger_\alpha a_i. \end{aligned} \tag{44}$$

If n is the dimension of the bosonic space (6 for the s–d system) and m is the dimension of the fermionic space, the $(m+n)^2$ operators of Eq. (44) generate the graded algebra $u(n/m)$. The powerful machinery of group theory is then used to help find solutions of the eigenvalue problem for a mixed system of boson and fermions.

(iii) Coherent state methods

The treatment of mixed systems of bosons and fermions using the coherent state method is more complex than the corresponding case of bosons. The reason is that, while it is possible to introduce in the bosonic coherent state (29) variational parameters which are c-numbers, when one deals with fermions such an introduction is not possible. The variational parameters become Grassman variables in the fermionic case. As a result, the use of coherent state methods for fermions has not been developed much.

6 Conclusions

In this article I have presented a brief survey of the bosonization method in finite systems. The literature on this subject is very extensive and some references have been indicated in Sections 2–5. Others can be found in the reference list of Refs. [5] and [11] and in Ref. [12], from which most of the material reviewed here has been taken. The applications of the method have up to now been confined to the structure of atomic nuclei where the method has prove to be very useful. The method can be used for other finite Fermi systems. An area where, in principle, it could be very useful is that of the newly discovered metallic clusters. These, too, are finite Fermi systems, composed of valence electrons. The number of valence electrons is of the same order of magnitude as the number of nucleons in nuclei, $\sim$ 100. The residual interaction in metallic clusters may be different from that of nuclei and therefore the corresponding boson system may not be an s–d system. None the less, the bosonization method is sufficiently general that it can be applied to these situations as well. Work in this direction is in progress [13].

Acknowledgments. This work was supported in part by DOE Grant No. DE-FG02-91ER40608.

References

[1] C. Racah, Phys. Rev. **63**, 367 (1943).

[2] Y.K. Gambhir, A. Rimini and T. Weber, Phys. Rev. **188**, 1513 (1969); B. Lorazo, Nucl. Phys. A **153**, 255 (1970); I. Talmi, Nucl. Phys. A **172**, 1 (1971). See also I. Talmi, *Simple Models of Complex Nuclei*, (Harwood, Chur, 1993) and references therein.

[3] J. Bardeen, L.N. Cooper and J.R. Schrieffer, Phys. Rev. **108**, 1175 (1957).

[4] T. Otsuka, A. Arima and F. Iachello, Nucl. Phys. A **309**, 1 (1978).

[5] For a review see F. Iachello and A. Arima, *The Interacting Boson Model*, (Cambridge University Press, 1987).

[6] H. Feshbach and F. Iachello, Phys. Lett. **45B**, 7 (1980).

[7] A. Bohr and B.R. Mottelson, Physica Scripta **22**, 468 (1980); A.E.L. Dieperink, O. Scholten and F. Iachello, Phys. Rev. Lett. **44**, 1747 (1980); J. Ginocchio and M.W. Kirson, Phys. Rev. Lett. **44**, 1744 (1980).

[8] A. Leviatan, Phys. Lett. **143B**, 25 (1984).

[9] N.N. Bogoliubov, J. Phys. USSR **11**, 23 (1947).

[10] A. Arima, T. Otsuka, F. Iachello and I. Talmi, Phys. Lett. **66B**, 205 (1977); T. Otsuka, A. Arima, F. Iachello and I. Talmi, Phys. Lett. **76B**, 139 (1978).

[11] For a review, see F. Iachello and P. van Isacker, *The Interacting Boson–Fermion Model*, (Cambridge University Press, 1991).

[12] F. Iachello and I. Talmi, Rev. Mod. Phys. **59**, 339 (1987).

[13] F. Iachello, E. Lipparini and A. Ventura, *Lecture Notes in Physics* **404**, 318 (Springer, Berlin, 1992).

18

Kaon Condensation in Dense Matter

Gerald E. Brown

Department of Physics
State University of New York at Stony Brook
Stony Brook, New York 11794
USA

Abstract

The K^--meson of mass $m_K = 494$ MeV is similar to an exciton, consisting of a strange quark, as particle, and $\bar{u}$ antiquark, as hole, bound together to make the kaon. In dense nuclear matter the kaon feels an attractive mean field from the nucleons which lowers its energy appreciably. As soon as its energy is brought down to $\sim$ half of its rest mass, it becomes energetically favorable in neutron stars to replace electrons by kaons, the neutron stars becoming nuclear matter stars at higher density. The kaons form a zero momentum Bose–Einstein condensate.

The new equation of state of dense matter, with the inclusion of kaon condensation, is substantially softer than conventional ones, meaning the maximum mass of compact objects formed in the collapse of large stars is only $\sim 1.5\ M_\odot$. With this equation of state, black holes are easy to form and it is estimated that there are $\sim 10^9$ of them in our galaxy. In this sense, a large number of black holes is the "smoking gun" for kaon condensation.

1 Introduction

A Bose–Einstein condensation which is somewhat exotic, even from the nuclear and particle physics point of view, is kaon condensation, condensation of K^--mesons in a zero momentum state. Yet, aside from the fact that we must think of energies and densities orders of magnitude higher than in the case of the exciton gas, the K^- can be viewed as analogous to an exciton and a system of K^--mesons can easily be viewed as a gas of nearly free particles, because their interactions with each other are known to be weak.

The K^--meson has a mass of

$$m_K = 494 \text{ MeV} . \tag{1}$$

The kaon is a boson, which will be important for our condensation scenario. It is composed of a strange quark and a nonstrange $\bar{u}$ antiquark; i.e., $|K^-\rangle = |\bar{u}s\rangle$. Whereas we have heard, throughout this conference, about the breaking of symmetry, in order to establish a Bose–Einstein condensate, we should realize that from the viewpoint of QCD we live in a world of broken symmetry, broken chiral symmetry. One of the consequences of this is that the negative energy sea of quarks can be considered to be filled, at least up to some cutoff momentum at which asymptotic freedom takes over. The filled negative energy sea of quarks plays the role of the valence band in Cu_2O, which is much discussed in this volume (see the articles by Wolfe *et al.* and Mysyrowicz.) The $\bar{u}$ is then a hole in the valence band of up quarks, and the s is a particle in the conduction band of strange quarks. The K^- can, then, be considered to be an exciton, with the s-particle somewhat heavier than the $\bar{u}$-hole.

We shall work at temperatures very small compared with m_K, a few MeV in magnitude, and will assume for our purposes here that $T = 0$. Temperature dependent effects have not yet been calculated for kaon condensation. Our Bose–Einstein condensate is then quite trivial, a condensate at zero temperature of K^--mesons in the state of zero momentum. I will try to convince you that such a condensate is most likely formed in the collapse of a large star, and that kaon condensates are integral components of the compact objects, often referred to as neutron stars, formed in such collapses. Obviously, the question to be answered is how the interactions compensate the mass (1) of the kaon. It must be energetically favorable to introduce kaons, before they will appear.

We will develop the theoretical framework in the next section, or at least, discuss it. Before doing this, let us review what has been deduced empirically about the K^--nucleus interaction, using this knowledge to build the K^--nuclear matter interaction. The K^--nucleus interaction has been obtained recently by Friedman, Gal and Batty [1], who described the kaonic atoms; i.e., the atoms formed by a K^--meson and the nucleus, for nuclei across the periodic table. In the case of the† $\bar{K}$–nucleon interaction, there is the well-known $\Lambda(1405)$ resonance, which lies below the $\bar{K}N$ continuum, and thus would be a bound state if its decay through the (π, Λ) or (π, Σ) channels would be turned off [2, 3]. As it is, the

† The notation is that the $\bar{K}$, the antikaon, has two components, K^- and $\bar{K}^0$. Similarly $K = (K^+, K^0)$.

$\Lambda(1405)$ is a bound state in the continuum. Because of this bound state below the K^--nucleon scattering threshold, the low-energy K^--nucleon scattering amplitudes are repulsive and the optical model potential (mean field potential)

$$2m_K V_{mf} = -4\pi b\rho \,, \tag{2}$$

where b is the scattering amplitude, is repulsive. (Kinematical factors are suppressed here. They go out in the end.)

However, since the $\bar{K}$ and nucleon form a bound state, albeit in the continuum, the basic interaction between $\bar{K}$ and nucleon must be attractive.

Friedman *et al.* [1] argue that as the nuclear density is increased, the K^--nucleus interaction must turn attractive. From fitting kaonic atoms across the periodic table, they deduce that the K^- in the middle of the nickel nucleus (i.e., roughly at nuclear matter density ρ_0) experiences an attraction of

$$V_{mf}(\rho_0) \simeq -200 \text{ MeV} \,. \tag{3}$$

The precise value of this number is not so certain, because the analysis is difficult due to the fact that the kaon can disappear through reactions such as

$$K^- + N \to \Sigma^- + \pi \tag{4}$$

and others; i.e., there is a large imaginary part to the potential. Let us none the less accept that there is a large attraction of the order of that given in (3). The simplest possible generalization of (3) to higher densities is

$$V_{mf}(\rho) = -200(\rho/\rho_0) \text{ MeV}. \tag{5}$$

The energy of a K^--meson in dense nuclear matter then will be

$$\epsilon_{K^-} = 494 \text{ MeV} - 200(\rho/\rho_0) \text{ MeV} \,. \tag{6}$$

In the collapse of large stars, due to the presence of electrons a large negative chemical potential μ_- is established in the core of the collapsing star. An example of such a star was the progenitor of Supernova 1987A, which was known from observations of its luminosity and type, to be of mass

$$M_{star} = 18 \pm 2 \text{ M}_\odot \,, \tag{7}$$

measured in units of the mass of our sun $\text{M}_\odot$. In collapse calculations, the core of the star which became the compact object, often called neutron star, was estimated in many calculations to be

$$M_{core} \cong 1.5 \, M_{\odot} \, . \tag{8}$$

More important for us, the density reached in the center of the core is generally found to be several times nuclear matter density ρ_0, the precise value depending upon the equation of state employed.

During the collapse of the star and the supernova explosion, the neutrinos formed in the capture reaction

$$e^- + p \leftrightarrow n + \nu \tag{9}$$

are trapped, so that chemical equilibrium establishes the following equality of chemical potentials:

$$\mu_e + \mu_p = \mu_n + \mu_\nu \, . \tag{10}$$

However, later, as neutrinos leave, the μ_ν can be left out of this equation and we obtain

$$\mu_e = \mu_n - \mu_p \, . \tag{11}$$

The electron chemical potential, which initially is the chemical potential for negative charge $\mu_e = \mu_-$, is then determined by this chemical equilibrium. For densities of 2–4 ρ_0, it generally lies between

$$200 \text{ MeV} < \mu_e < 300 \text{ MeV} \, , \tag{12}$$

the precise value depending somewhat on the nucleon–nucleon interactions. Not only is μ_e large, but it tends to increase with density.

Since ϵ_{K^-} (see (6)) decreases with density, there will come a critical density at which

$$\epsilon_{K^-}(\rho_c) = \mu_e \, . \tag{13}$$

Above this density it will be energetically favorable for electrons to change into K^--mesons, through the reaction

$$e \to K^- + \nu \, , \tag{14}$$

the neutrinos leaving the star.† We will give results of detailed calculations later.

The energy is lowest, in a mean-field description, if the kaons so formed

† The formation of kaon condensation will happen after the era of neutrino trapping, so that neutrinos can freely leave the star.

go into a state of zero momentum. The kaon condensate can then be described by a classical wave function (see (15) below).

Calculations to date of kaon condensation have been carried out at zero temperature. However, temperatures as high as 75 MeV are reached in the collapse of stars [4]. Characteristic thermal energies are $\epsilon_{th} \sim \pi$T. Thus, ϵ_{th} is of the same order as the electron chemical potential $\mu_e \sim$ 225 MeV at condensation. Consequently, the thermal energy may hinder Bose–Einstein condensation, as it seems to hinder the condensations of spin-polarized hydrogen or excitons discussed in this volume. In the case of excitons in Cu_2O discussed by Wolfe *et al.*, modification of the crystal symmetry by applied stress seems to give the excitons a new way to eliminate thermal energy and entropy. Such a way is present in a star. Neutrinos will carry off energy quickly, reducing the entropy to that needed for Bose–Einstein condensation. But the precise time scale for all of this still needs to be worked out.

2 Calculation of Kaon Condensation

Calculations of kaon condensation have been carried out using chiral Lagrangians, based on the Gell–Mann, Levy σ-model. In fact, they have been carried out in the nonlinear σ-model, rather than the GL–linear model, but we shall ignore this (important) distinction. The salient point is that the GL–Hamiltonian plays the same role as the Ginzberg–Landau expression for the free energy; we don't even need to change the initials!

One introduces the kaon condensate as

$$\langle K^- \rangle = e^{-i\mu_- t}\, v_K, \tag{15}$$

where v_K is the amplitude of the classical condensed kaon field, taken here to be in the state of zero momentum, therefore uniform in $\mathbf{r}$. The GL equations will involve $|v_K|^2$ as a parameter, and kaon condensation will be possible when the energy can be lowered by choosing a nonzero $|v_K|^2$.

It is useful to introduce the chiral angle θ, which will serve as order parameter for the kaon condensation, by

$$\theta = \sqrt{2}|v_K|/f\ , \tag{16}$$

where f is the pion decay constant, $f = 93$ MeV. The angle θ is the chiral angle, and the significance of this angle can best be appreciated from the paper by Brown, Kubodera and Rho [5] where strangeness condensation

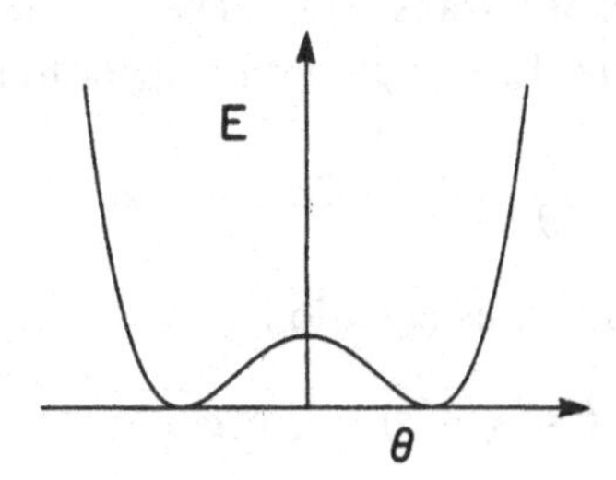

Fig. 1. Ginzberg–Landau energy (the temperature T is taken to be zero) for $\rho > \rho_c$, where ρ_c is the critical density for the phase transition.

is interpreted as rotation in θ, in the region from 0 to 90°, the effects of the bare strange quark mass being rotated out.

Let us now discuss the construction of the phase transition. The condition that there be a phase transition is [6] that the propagator

$$D(\omega) = (\omega^2 - m_K^2 - \pi(\omega))^{-1} \tag{17}$$

have a pole at $\omega = \mu_-$. Here $\pi(\omega)$ is the kaon self energy. In our model it includes only s-wave interactions. Physically, the reason that $\omega = \mu_-$ is that given in the last section; below this energy it becomes favorable for electrons to change into kaons.

We have investigated only the second order phase transition, which takes place at $\rho_c = 3.22\rho_0$, where

$$\rho_0 = 2.5 \times 10^{14}\ \mathrm{g/cm^3} \tag{18}$$

is nuclear matter density. The number density of nucleons is given by multiplying by Avogadro's constant $A_0 = 6 \times 10^{23}$, so that at $\rho_c = 3.22\rho_0$, the critical number density is

$$n_c = 4.8 \times 10^{40}\ . \tag{19}$$

Of course, such a large number is mindboggling.

The Ginzberg–Landau free energy has the form (shown in Fig. 1)

$$E = -a(\rho)\theta^2 + \frac{1}{2}\theta^4 \tag{20}$$

for $\rho > \rho_c$. Here we have rescaled θ so that the coefficient of the θ^4-term is $\frac{1}{2}$. The coefficient $a(\rho)$ has the form

$$a(\rho) = \alpha(\rho - \rho_c) \tag{21}$$

for a second order phase transition. The α comes out of our GL–Lagrangian. Minimizing the energy with respect to θ gives

$$\frac{\partial E}{\partial \theta} = 0 = -2a(\rho)\hat{\theta} + 2\hat{\theta}^3, \tag{22}$$

where $\hat{\theta}$ is the solution of this equation,

$$\hat{\theta} = \sqrt{a(\rho)} = \sqrt{\alpha(\rho - \rho_c)}\,. \tag{23}$$

The energy E is found to be

$$E = -\frac{1}{2}\alpha^2(\rho - \rho_c)^2\,. \tag{24}$$

From this one can see that the gain in energy is very rapid above $\rho = \rho_c$.

In Table 1 we show the results of calculations [7] for the scenario outlined in the previous section. For completeness, we list refs. [8, 9, 10, 11, 12] as a chronological development of the development of the input used in the GL equations.

Table 1 describes the composition of neutron stars at densities $\rho > \rho_c$. One sees that the proton fraction x rises rapidly with density towards 0.5 and that the electron fraction x_e decreases. The dense part of the core is no longer a neutron star, but nuclear matter, or a "nucleon" star. Thus, the kaon condensation scenario has replaced neutron by nucleon stars. This should have many major implications for their structure, which are presently being worked out.

The tendency to go from neutron matter towards nuclear matter is easily understood. Once the electrons can be replaced by kaons, the high energies of the former (fermions) need not be introduced in order to neutralize the charge on the protons. Other things being equal, the strong neutron–proton attraction favors equal numbers of protons and neutrons. In this way the nuclear symmetry energy can be brought lowest.

3 Lots of Black Holes are the Smoking Gun for Kaon Condensation

We would now like to discuss the effect of kaon condensation on the structure of compact objects formed following the collapse of large stars. In the last section, we saw that kaon condensation substantially lowered the energy of the dense matter. It also greatly softens the pressure, so that the maximum mass of the compact objects is lower than is usually found in the literature. Indeed, Brown [13] found $M_{\max} \cong 1.5\ M_\odot$, just about the same as (8), the estimated mass of the core formed in the

Table 1. *Results of the calculation of the energy gain with respect to neutron rich nuclear matter in beta equilibrium. Here* $u = \rho/\rho_0$, θ *is the chiral angle,* $\Delta\epsilon$ *is the gain in energy, in* MeV, μ *is the chemical potential for negative charge, in* MeV, x *is the proton fraction;* x_K, *the kaon fraction;* x_e, *the electron fraction and* x_μ, *the muon fraction.*

u	θ	$\Delta\epsilon$	μ	x	x_K	x_e	x_μ
3.22	0.0	0.0	224.0	0.163	0.000	0.096	0.067
3.72	24.1	-5.2	165.6	0.317	0.267	0.034	0.016
4.22	27.1	-15.8	127.4	0.385	0.368	0.013	0.003
4.72	27.4	-28.4	100.8	0.421	0.415	0.006	0.000
5.22	26.8	-41.9	81.3	0.444	0.441	0.003	0.000
5.72	26.0	-55.9	66.7	0.458	0.457	0.001	0.000
6.22	25.0	-70.1	55.5	0.468	0.467	0.001	0.000
6.72	24.1	-84.7	46.7	0.475	0.475	0.000	0.000
7.22	23.1	-99.4	39.7	0.480	0.480	0.000	0.000
7.72	22.3	-114.3	34.0	0.484	0.484	0.000	0.000
8.22	21.4	-129.5	29.4	0.487	0.487	0.000	0.000
8.72	20.6	-144.9	25.6	0.489	0.489	0.000	0.000
9.22	19.9	-160.5	22.4	0.491	0.491	0.000	0.000

explosion of SN 1987A. It has been suggested [14] that this core went into a black hole.

Arguments have been given [15] that the compression modulus of nuclear matter is somewhat smaller than the conventional value [16] $K_0 = 210 \pm 30$ MeV. It was found [13], however, that K_0 had to be raised to 190 MeV, the lower region of conventional values, in order to stabilize a 1.50 $M_\odot$ nucleon star. Figure 2 shows the known measured masses of neutron stars. It is quite striking that all of the accurately measured masses in binaries lie below 1.5 $M_\odot$, the largest being the 1.44 $M_\odot$ pulsar in PSR 1913+16. The mass of Vela X-1 lies somewhat higher, but the relatively large error bars reach down to $\sim$ 1.55 $M_\odot$.

In the formation of binary pulsars, after the first large star of a binary explodes in order to form a compact object, the latter and the remaining large star share a period of common envelope. During this period, accretion can be rather large, $\sim 10^{-3}$ $M_\odot$ per year [18]. Given this large possible accretion, it would seem strange if compact objects larger than 1.44 $M_\odot$ would not be seen in binaries, were they to exist. Of course, 1.44 $M_\odot$ is just the Chandrasekhar mass for an $N = Z$ star, so that the collapse of the cores of large stars begins only when they

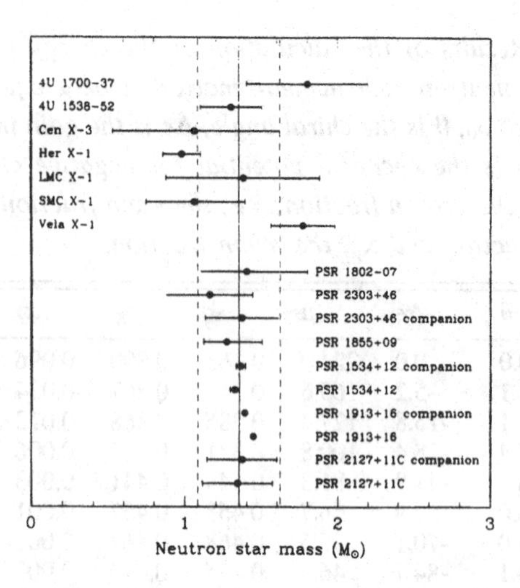

Fig. 2. Measured masses of neutron stars, from Ref. [17].

are unstable, with masses somewhat less than 1.44 $M_{\odot}$ because of some electron capture. However, with accretion following the collapse, the compact cores formed from large stars are estimated to be of the order 1.4 $M_{\odot}$. Thus, it is true that compact objects tend to be formed with such a mass. Our argument is that accretion from a companion star should produce, in some cases, a larger mass. Since these are not seen, this argues in favor of the maximum mass being not much larger than 1.44 $M_{\odot}$.

We next argue that the collapse scenario of the core of a large star is radically changed at high densities because of the kaon condensate. Let us first review the conventional collapse scenario. As can be seen from Fig. 3, the maximum masses of neutron stars at short times are less than the longtime cold maximum mass, even though the trapped neutrinos give considerable pressure. The point is that nuclear matter is substantially softer than neutron matter, and the former is turned into the latter as the trapped neutrinos leave, allowing electron capture. The conversion of protons to neutrons increases the pressure more than it is decreased by the loss of neutrinos. Thus, if the mass of the compact object exceeds 1.5 $M_{\odot}$ at early times (or any time), it will immediately collapse into a black hole. Inclusion of thermal pressures, which are presently being calculated, may raise the solid, short-time curve, slightly above the dashed, long-time one, so that there would be a small "window"

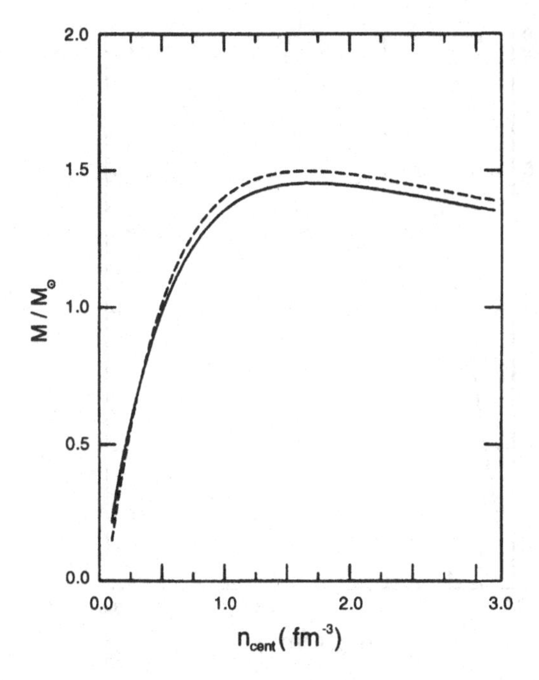

Fig. 3. Maximum masses of neutron stars in the conventional scenario of neutron matter composition [19]. The compression modulus K_0 has been chosen to be small, equal to 130 MeV, so that the late time cold maximum mass, shown by the dashed line, is equal to $M_{max} = 1.5\ M_{\odot}$. The solid line shows the maximum mass during the first few seconds during which neutrinos are trapped. A lepton fraction of $Y_\ell = 0.4$ was chosen for the early times.

of masses for a star to exist for some seconds, and then collapse into a black hole as it cooled. We shall next discuss what happens in connection with the much larger window, found in the case of the equation of state including kaon condensation.

In the case of kaon condensation, the original nuclear matter does not change into neutron matter. Indeed, at the higher densities, the fraction of protons goes to one half. The early time maximum masses, as a function of central density, are shown by the solid line in Fig. 4.

As can be seen from Fig. 4, the compact object can be stabilized, during the time that the neutrinos are trapped, and then collapse into a black hole. This means that the star can explode, returning matter to the galaxy, and then go into a black hole [20]. Such black holes will be of relatively light mass, not much larger than $1.5\ M_{\odot}$. In general, observers have looked for invisible centers of gravitational attraction, heavier than

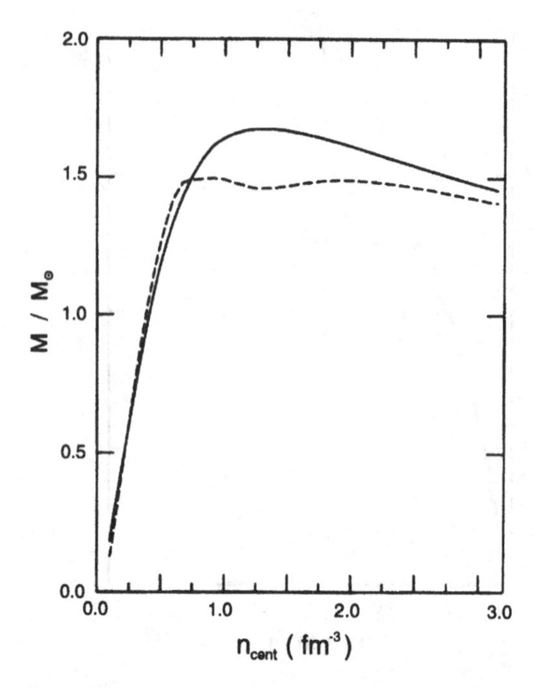

Fig. 4. As in Fig. 3 the solid line shows the short-time maximum nucleon star mass, as function of central density, with an assumed lepton fraction of Y_ℓ = 0.4, consisting of electrons and trapped neutrinos. The dashed line shows the long-time cold mass, after the neutrinos have left.

$\sim$ 3 $M_\odot$, as candidates for black holes, in order to make sure that the mass exceeded that of the maximum mass neutron star. The black holes of mass $\gtrsim$ 1.5 $M_\odot$ provide a new region of masses to look for.

Inclusion of late time accretion increases the above window for nucleosynthesis with later collapse into black holes by an estimated 0.04 $M_\odot$ [20].

The possibility of nucleosynthesis with later collapse into black holes was pointed out by Woosley *et al.* in 1986 [21]. The mechanism for stabilizing the compact object, which could later go into a black hole, over the short time necessary for nucleosynthesis was the thermal pressure which results from the conversion of gravitational energy into thermal energy as the core goes through the Kelvin–Helmholtz contraction. As noted above, this is presently being worked out with detailed equations of state. There will be a relatively small "window" for nucleosynthesis with later collapse into a black hole in the standard scenario where the compact core ends up as a neutron star. The larger "window" will be further increased in our case of the kaon condensate. We estimate that inclusion of all of the effects noted above will bring the upper limit for nucleosynthesis with later collapse into a black hole to $> 1.8\ M_\odot$, corresponding to main sequence stars of mass $\geq 30\ M_\odot$.

That main sequence stars of mass $\geq 25-30\ M_\odot$ must end up in black holes without producing nucleosynthesis, i.e., without returning matter to the galaxy, is required by the observed abundances of elements [22]. Maeder's argument is based on the measurement of $\Delta Y/\Delta Z$, the ratio of helium abundance to that of metals, in low-metallicity galaxies, especially irregular dwarf galaxies [23]. Measurement of the ratio avoids the problem of determining the normalization of the stellar mass function. Most of the total galactic stellar mass is in light stars, of mass less than $1\ M_\odot$. These contribute neither to ΔY nor to ΔZ, so they do not affect the ratio. The ratio can be measured with good accuracy [23]:

$$\frac{\Delta Y}{\Delta Z} = 4 \pm 1.3\ . \tag{25}$$

If all stable stars of mass up to $\sim 100\ M_\odot$ were to explode, returning matter to the galaxy, this ratio would lie between 1 and 2. Helium is produced chiefly by relatively light stars, metals by heavy stars, so that cutting off the production by the heavy stars going directly into black holes without nucleosynthesis increases the $\Delta Y/\Delta Z$. Maeder [22] found, using the standard initial mass function for stars, that Pagel's measured $\Delta Y/\Delta Z$ was best reproduced by a cutoff of nucleosynthesis at a main sequence stellar mass of $\sim 22.5\ M_\odot$. There is considerable uncertainty in the initial mass function, as noted by Maeder [22], so that this limit could easily be $\sim 30\ M_\odot$ or even higher.

The Brown–Bethe estimate [20] of $30\ M_\odot$ as the cutoff for stars to drop directly into black holes without nucleosynthesis and the scenario discussed above that a large range of stars below this mass can first accomplish nucleosynthesis and then collapse into (light mass) black holes, extends the estimated range of main sequence masses for which stars go into black holes down to $\sim 18\ M_\odot$ [20]. Interestingly, this possibly includes Supernova 1987A, with progenitor mass of $18 \pm 2\ M_\odot$. Indeed, arguments have been put forward [14] that SN 1987A probably went into a black hole.

A good candidate for a similar situation is the supernova remnant Cas A, where the supernova explosion — unobserved at the time — took place in the late seventeenth century. From element abundances in the ejecta it has been argued [24] that the progenitor was an $\sim 20\ M_{\odot}$ star, of roughly equal mass to the progenitor of SN 1987A. Although observers have looked with high accuracy in several different ways at the center of the supernova remnant, they have been unable to find a compact object.

The Brown–Bethe scenario suggests that about as many stars end up in black holes as nucleon stars. They estimate that there are ~ 500 million heavy mass black holes and a comparable number of light mass black holes in the Galaxy. In fact, compact remnants are seen in only ~ 20 of the roughly 150 supernova remnants in the Galaxy [25]. It may turn out that a black hole is the more likely fate than a nucleon star for the compact core of a large star.

I would like to thank M. Prakash for pointing out the large "window" for nucleosynthesis, with later collapse into a black hole, in the case of strangeness condensates, and for providing me with Figs. 3 and 4. I am grateful to Vesteinn Thorsson for providing me with Table 1 and both to him and to Mannque Rho for their collaboration on the theory of kaon condensation. I am grateful to Hans Bethe, with whom the scenario for black hole formation of ref. [20] was worked out.

This work was supported by the US Department of Energy Grant No. DE–FG02–88ER40388.

References

[1] E. Friedman, A. Gal and C. J. Batty, Phys. Lett. B **308**, 6 (1993).

[2] P. B. Siegel and W. Weise, Phys. Rev. C **38**, 221 (1988).

[3] A. Müller–Groeling, K. Holinde and J. Speth, Nucl. Phys. A **513**, 557 (1990).

[4] J. Cooperstein, in *First symposium on nuclear physics in the universe*, ORNL, TN, 24–26 Sept. (1992), M. Guidry, ed. (Adam Hilger), in press.

[5] G. E. Brown, K. Kubodera and M. Rho, Phys. Lett. B **192**, 273 (1987).

[6] G. Baym and D. K. Campbell, *Mesons in Nuclei III*, M. Rho and D. H. Wilkinson, eds. (North-Holland, Amsterdam, 1979), p. 1031.

[7] V. Thorsson, private communication.

[8] D. B. Kaplan and A. E. Nelson, Phys. Lett. B **175**, 57 (1986).

[9] H. D. Politzer and M. B. Wise, Phys. Lett. B **274**, 156 (1991).

[10] G. E. Brown, K. Kubodera, M. Rho and V. Thorsson, Phys. Lett. B **291**, 355 (1992).

[11] G. E. Brown, C.-H. Lee, M. Rho and V. Thorsson, Nucl. Phys. A **567**, 937 (1994).
[12] G. E. Brown, Nucl. Phys. A, to be published.
[13] G. E. Brown, in *First symposium on nuclear physics in the universe*, ORNL, TN, 24–26 Sept. (1992), M. Guidry, ed. (Adam Hilger), in press.
[14] G. E. Brown, S. W. Bruenn and J. C. Wheeler, Comm. Astrophys. **16**, 153 (1992).
[15] G. E. Brown, Nature **336**, 519 (1988).
[16] J. P. Blaizot, D. Gogny and B. Grammaticos, Nucl. Phys. A **265**, 315 (1976).
[17] S. E. Thorsett, Z. Arzoumanian, M. M. McKinnon and J. H. Taylor, Astrophys. Journ. Lett., **405**, L29 (1993).
[18] R. A. Chevalier, Astrophys. Journ. **346**, 847 (1989) ; R. A. Chevalier, Astrophys. Journ. **411**, L33 (1993).
[19] M. Prakash, private communication. See also V. Thorsson, M. Prakash and J. M. Lattimer, Nucl. Phys. A **572**, 693 (1994).
[20] G. E. Brown and H. A. Bethe, Astrophys. Journ. **423**, 659 (1994).
[21] S. E. Woosley and T. A. Weaver, Ann. Rev. Astron. and Astrophys. **24**, 205 (1986); J. R. Wilson, R. Mayle, S. E. Woosley and T. A. Weaver, *Proc. 12th Texas Rel. Astrophys. Symp.*, Jerusalem, Israel, 16–20 Dec., (1984).
[22] A. Maeder, Astron. Astrophys. **264**, 105 (1992).
[23] B. E. J. Pagel, E. A. Simonson, E. A. Terlevich and M. G. Edmunds, MNRAS **255**, 325 (1992).
[24] R. A. Chevalier and R. P. Kirshner, Astrophys. Journ. **233**, 154 (1979).
[25] S. Kulkarni, private communication.

19

Broken Gauge Symmetry in a Bose Condensate

A. J. Leggett
Department of Physics
University of Illinois at Urbana-Champaign
Urbana, IL 61801
USA

Abstract

This paper examines the meaning of the "phase" of a Bose condensate, and in particular the degree of validity of the often used analogy with the direction of magnetization of a ferromagnet. It focusses on two specific questions: (i) *Under what circumstances is the relative phase of two condensates well defined?* (ii) *Would it be possible in principle to set up a "standard of phase"? In the most obvious sense, the answer to* (ii) *is concluded to be no.*

As is well known, it is very fashionable nowadays to treat Bose condensation as a special case of the more general idea of spontaneously broken symmetry, which is ubiquitous in condensed-matter physics [1]. The standard account goes something like this: Just as in a magnetic material described by an isotropic Heisenberg model, the Hamiltonian is invariant under simultaneous rotation of all the spins, so in a Bose system described by the standard creation and annihilation operators $\psi^\dagger(r), \psi(r)$ it is invariant under the global $U(1)$ gauge transformation $\psi(r) \rightarrow \psi(r)e^{i\varphi}$, $\psi^\dagger(r) \rightarrow \psi^\dagger(r)e^{-i\varphi}$. Thus, at first sight, symmetry forbids either the expectation value $< \mathbf{S} >$ of the magnetization of the magnetic material, or the corresponding quantity $< \psi >$ in the Bose system, to take a finite value. Nevertheless, we know that if the material in question is ferromagnetic, then when it is below its Curie temperature and we take the thermodynamic limit in the presence of "infinitesimal" symmetry-breaking fields in the usual sense, the quantity $< \mathbf{S} >$ indeed takes a finite (in fact macroscopically large) value, and thus the direction of the magnetization is a well defined quantity. Similarly, it is argued, in the Bose case when one is below the Bose condensation temperature and takes the thermodynamic limit in a similar way, the quantity $\eta \equiv <\psi>$ ("order parameter") becomes finite (in this case proportional to $N^{1/2}$) and its phase φ thus becomes well defined. The Bose condensate is thus said to possess "spontaneously broken gauge symmetry", just as the ferromagnet below its Curie temperature possesses "spontaneously broken rotational symmetry".

The aim of the present paper is to examine in detail the concept of spontaneously broken gauge symmetry and hence, by implication, the degree of validity of the analogy between the Bose condensate and the ferromagnet. To some extent, it is a continuation of the discussion begun in Ref. [2], to which I shall sometimes refer for technical details for which there is no space here. It should be said at once that, at least as regards the vast majority of the experiments apparently feasible today, nothing of any great importance hinges on the outcome of the discussion, and indeed the points at issue might be regarded as in some sense "theological"; my reason for pursuing them nevertheless is simply that I find existing treatments of the concept of broken gauge symmetry somewhat unsatisfactory. However, as a by-product we shall find that some experiments are suggested which are not obviously beyond existing capabilities.

I wish to make three remarks before I start on the main theme. First, I shall feel free in the following to use examples, and thought-experiments, which involve not only "Bose condensation" in the strict sense but also the "pseudo-Bose condensation" of Cooper pairs in Fermi systems. In the latter system, the quantity which plays the role of the order parameter $<\psi>$ in a Bose system is related to the so-called "two-particle anomalous average" defined by

$$F_{\alpha\beta}(r,r') \equiv \langle \psi_\alpha(r)\psi_\beta(r')\rangle. \tag{1}$$

For example, in a simple BCS superconductor the natural choice for the order parameter is the quantity $F_{\uparrow\downarrow}(r,r)$, and all considerations regarding the phase, etc., then go forward just as in the Bose case.† In the case of Fermi superfluids such as ^{3}He with more complicated pairing structures, there may be more than one order parameter (as in the spin-1/2 Bose system formed by spin-polarized atomic hydrogen, see Ref. [2]) and their relative phase then becomes of interest: see Ref. [2], and below.

Secondly, for the purposes of the present paper the relevant order parameter will always be assumed to be constant in space, except when

† Except for a factor of 2 in the commutation relations, which is irrelevant to the considerations of this paper and which I will generally ignore.

there is a *discrete* discontinuity such as that across a Josephson junction. To be sure, the question of how far the concept of spontaneously broken symmetry remains valid in the presence of substantial spatial inhomogeneity is an important one, and may well be connected with some of the intriguing questions raised by Kagan [3], but to keep the discussion focused I shall not attempt to attack it here. However, one may hope that an improved understanding of the conditions for, and implications of, broken gauge symmetry in the simpler cases discussed here may eventually shed some light on these more practically important questions.

Thirdly, let us try to tighten up the Bose-ferromagnet analogy a little by focusing specifically on two variables of the ferromagnet, namely the z-component of total spin S_z and the angle $\theta \equiv \tan^{-1}(S_y/S_x)$ made by the component of spin in the xy plane with some arbitrary reference axis. We could, in fact, at this point drop the requirement that the Hamiltonian of the ferromagnet has the complete $O(3)$ symmetry and demand only that it be invariant under rotation around the z-axis. Under these conditions the formal analogy with the Bose system is complete: to make it explicit, we define the operator $S^+ \equiv S_x + iS_y$ so that we have the commutation relation

$$[S_z, S^+] = S^+, \tag{2}$$

in complete analogy with the relation for the Bose case

$$[N, \psi^+] = \psi^+, \tag{3}$$

where N is the total number operator. Thus, S_z is the analog of N and $(-)\theta$ of the "phase" φ of ψ; S_z generates "rotations" of θ just as N does of φ. We shall tentatively treat θ and φ as operators satisfying the commutation relations $[S_z, \theta] = i, [N, \varphi] = -i$. Although there are, of course, well known difficulties [4] in doing so when the number of particles involved is small, these difficulties appear to be irrelevant for systems of typical "laboratory" size (as is shown, *inter alia*, by the success of theories in which the phase is treated as a quantum-mechanical operator in predicting the rate of "macroscopic quantum tunneling" in Josephson systems, see e.g. Ref. [5]), and in any case do not appear to affect the questions that I shall consider in this paper. The meaning of the *relative* phase as an operator is clarified explicitly below. Of course, it is conceivable that in the comparison of the role of φ with that of θ, a crucial role is played by the possibility that the ferromagnet has the full $O(3)$ symmetry and not just the $U(1)$ one corresponding to rotation around the z-axis; we return briefly to the possible relevance of this circumstance at the end of the paper.

Let us first review briefly some of the considerations discussed in more detail in Ref. [2]. We first note that for an isolated system with a fixed number N of particles it is possible to define the "order parameter" $\eta(r)$ in an alternative way, namely

$$\eta(r) = \sqrt{N_0}\chi_0(r), \tag{4}$$

where $\chi_0(r)$ is the single-particle wave function into which condensation has taken place and N_0 the number of condensed particles; for the spatially homogeneous case discussed here, this reduces to the rather trivial expression

$$\eta = \sqrt{N_0}\, exp(i\varphi), \tag{5}$$

where φ is explicitly the phase of χ_0. Since the latter has no physical meaning, it is clear that neither does the phase of η, as we should of course infer, for definite N, from (3). Thus it is clear that to make sense of the concept of the "phase" of the order parameter, we must first discuss the conditions for defining *relative* phase, and then ask whether we can in some sense define an "absolute" phase by means, for example, of a universal "phase standard". Note at this point that we do not normally regard a similar problem as arising for the ferromagnet, because we are used to the idea that S_z need not be well defined (thus allowing θ to have a definite value), whereas the notion that N is not definite is more difficult to accept. Whether or not this difference of perception is really well based is a question we shall have to return to later.

Let us then consider, as in Ref. [2] , a system which has available to it two states χ_1, χ_2 into which Bose condensation may take place. An obvious example is two bulk superconductors (or superfluids) joined by a Josephson junction; in this case the states χ_1 and χ_2 would of course simply correspond to localization on one side of the junction or the other. However, it is more convenient to discuss first a case which is closer to that envisioned in section 3 of Ref. [2], namely the Cooper pairs in superfluid ^{3}He–A. If we ignore the orbital degrees of freedom, and moreover suppose that the so-called **d**-vector (for notation, see e.g. Ref. [6]) lies in the xy-plane, then the two states χ_1 and χ_2 in question correspond to pairs formed with total spin ± 1, respectively, along the z-axis: we denote these by $|\uparrow\uparrow\rangle$ and $|\downarrow\downarrow\rangle$ respectively. The most general form of wave function in which all Cooper pairs are "Bose-condensed" into a *single* wave function may thus be written schematically (a, b real)†

$$\Psi \simeq \left(ae^{i\Delta\varphi/2}\,|\uparrow\uparrow> + be^{-i\Delta\varphi/2}\,|\downarrow\downarrow>\right)^{N/2} \equiv \Psi(\Delta\varphi) \qquad (6)$$

(N being the total number of ^3He atoms). In the theory of magnetic resonance in superfluid ^{3}He [7], the ground state wave function is taken to be of the form (6) with $a = b$ and the relative phase $\Delta\varphi$ of the up- and down-spin pairs chosen so as to minimize the dipole energy.‡

What is the variable canonically conjugate to $\Delta\varphi$? It is nothing but the difference ΔN of the total number of up and down spins in the system (actually, of course, this is the total z-component of spin within a factor, but I avoid the notation S_z so as to reserve this for the ferromagnetic analogy). That ΔN and $\Delta\varphi$ are indeed canonically conjugate variables (in the limit $N \to \infty$) may be seen in a number of different ways, e.g. by going back to the microscopic definition (1) of the order parameter (see Ref. [7], section 2). For our purposes the most instructive derivation probably consists in considering the class of many-body wave functions (a subspace of the complete many-body space, of course) defined by

$$\Psi = \int c(\Delta\varphi)\Psi(\Delta\varphi)d\Delta\varphi, \qquad (7)$$

when $\Psi(\Delta\varphi)$ is defined by (6) and $c(\Delta\varphi)$ is an arbitrary complex coefficient. Note that in this expression, $\Psi(\Delta\varphi)$ plays the role of $|\,x>$ in the standard Dirac notation and $c(\Delta\varphi)$ that of the "wave function" (probability amplitude) $\psi(x)$: cf. Ref. [8], section 3.4. It is then straightforward to show by direct calculation that within the subspace of states defined by (7) the operator $-i\partial/\partial(\Delta\varphi)$ has the same matrix elements as ΔN, so that $\Delta\varphi$ and ΔN are indeed canonically conjugate variables within this subspace. It is possible to extend this result to more general states, but not worth doing in the present context.

We now return to a question raised, but not answered, in Ref. [2], namely: Why should we assume that the wave function of a Bose-condensed system which has available to it two different states (here $|\uparrow\uparrow>$ and $|\downarrow\downarrow>$) will always be of the form (6), rather than (say) of the form

$$\Psi = (|\uparrow\uparrow>)^{\frac{N+\Delta N}{2}}(|\downarrow\downarrow>)^{\frac{N-\Delta N}{2}} \equiv \Psi(\Delta N) \qquad (8)$$

† For simplicity of presentation I ignore here the notational complications introduced by antisymmetrization, etc.

‡ Note that, contrary to the case discussed in section 3 of Ref. [2], $\Delta\varphi$ does *not* represent the angle of the Cooper-pair spin in the xy-plane, but, rather, the angle of the direction $\mathbf{d}$ along which they have spin projection is 0.

(which, as mentioned in Ref. [2], is a special case of (7) with $c(\Delta\varphi) = exp(-i\Delta N.\Delta\varphi)$)? It is clear that if we put no conditions whatever on the physical situation envisaged, there is no particular reason to prefer (6) to (8) (or indeed to a myriad of other possible states). However, if we specify that the system is in or close to thermodynamic equilibrium, there is a simple reason why (6) is usually preferred. Consider for definiteness, as above, the case of superfluid ^{3}He–A. The terms in the Hamiltonian which depend on ΔN and $\Delta\varphi$ may be written

$$H = \frac{(\Delta N)^2}{2\chi_s} - g_D \cos \Delta\varphi \,. \tag{9}$$

Here χ_s is the spin susceptibility of the system (which, as it turns out, is essentially just the normal-state Pauli value), and g_D is the coupling constant due to the dipole force (which, since it does not conserve spin angular momentum alone, can have nonzero matrix elements between the states $|\uparrow\uparrow>$ and $|\downarrow\downarrow>$). The crucial observation, now, is that both χ_s and g_D are *extensive* quantities, that is, they are proportional to the total number of particles N. We can then see immediately from (9) that in the thermodynamic limit $N \to \infty$, it is always advantageous to have $\Delta\varphi$ well defined and allow ΔN to fluctuate, i.e. the wave function is indeed approximately of the form (6). A more exact consideration shows that in this limit the groundstate wave function (7) is of harmonic-oscillator form around $\Delta\varphi = 0$, with the rms fluctuation of $\Delta\varphi$ proportional to $N^{-1/2}$ and that of ΔN to $N^{1/2}$.

It is clear that a similar argument will go through for any case in which the states χ_1 and χ_2 into which Bose condensation can take place are "extensive" (in a sense obvious from the above), provided only that the term in the Hamiltonian which connects them (the generalization of g_D) is not identically zero. Thus, given this condition, the relative phase of the two condensates is indeed always well defined in the thermodynamic limit, so that the concept of broken (relative) gauge symmetry is justified.†

While this conclusion is, of course, in no way novel, the argument given above has the advantage of making it quite explicit that the occurrence of broken (relative) symmetry in the thermodynamic limit is in no sense "magic" but is simply a result of *energy* considerations, and thus that the point at which we may legitimately assume that the limit is a function of the strength of the "symmetry-breaking" interaction. If we imagine that

† It is needless to remark that similar arguments concerning the *absolute* "phase" in the presence of external "source" terms are commonplace in the literature. The advantage of the present formulation is that it uses no such unphysical assumptions.

we could somehow adjust the strength of the dipole coupling constant g_D in ^{3}He–A, then the above argument makes it clear that as g_D tends to zero, we would need to go to larger and larger volumes of the liquid to stabilize the broken relative symmetry of the up- and down-spin Cooper pairs. Indeed, it is interesting to inquire what is the order of magnitude of the "critical" volume for the actual value of the dipole coupling realized in nature. The answer is $\sim 10^{-18}\text{cm}^3$; since this is (by a numerical accident) on the order of the cube of the coherence length, it seems doubtful that one could obtain volumes of superfluid A-phase much less than this. If this should be possible (e.g. in the form of thin films), one would expect their NMR properties to be spectacularly different from those of the bulk liquid.

Now let us turn our attention to the case in which the two relevant states χ_1 and χ_2 into which Bose condensation can take place are spatially distinct; the most obvious realization of this situation is of course via the Josephson effect, and some aspects of it were already discussed in Ref. [2]. In the context of the present discussion, the salient point is that, in contrast to the "extensive" case treated above, the Josephson coupling constant E_J (the analog of g_D in the above argument) is in principle subject to adjustment by the experimenter as a function of time. We may in fact assume that when Josephson contact is made E_J is sufficiently large that the relative phase $\Delta\varphi$ of the Cooper-pair (or the atom) wave functions localized on the left or right is very well defined, while when it is broken, E_J may be effectively taken to be zero.

We first briefly revisit the question first raised by P.W. Anderson and discussed in section 5 of Ref. [2]: Do two superfluids which have never "seen" one another possess a definite relative phase? The arguments given in section 4 of Ref. [2], which show that even when the superfluids have been in contact (and have acquired a definite relative phase) in the past, this is eventually washed out, point strongly to the answer "no". However, as pointed out in that reference (section 5) any attempt to answer the question experimentally, by asking whether or not a Josephson current flows when two such "memory-free" superfluids are connected, may be frustrated by the fact that any attempt to determine this is liable, according to the precepts of the quantum theory of measurement, itself to "create" a relative phase, irrespective of whether or not there "really was" one before the measurement.† Nevertheless, there exists an alternative way of giving meaning to the question (although once it is framed in this

† A very detailed analysis of this phenomenon has recently been given by Hegstrom and Sols [9].

way, the answer is perhaps obvious): We assume that the two superfluids together constitute a closed system (i.e., cannot exchange particles with the rest of the world). The "experiment" simply consists in weighing them at separate times t_1 and t_2, which can be arbitrarily far separated, so as to determine the number difference ΔN at these times, without ever making Josephson contact between them. Evidently, if, given a sufficient time τ, the two superfluids can acquire a definite relative phase even without making Josephson contact, then for $t_2 - t_1 >> \tau$ we should find that $\Delta\varphi(t_2)$ is definite (to order $N^{-1/2}$) and hence the second measurement of ΔN should in general give a different result from the first, the fluctuation being presumably of order $N^{1/2}$. On the other hand, if the relative phase $\Delta\varphi$ is *not* definite, then there is no reason why ΔN should vary in time in the absence of contact (and indeed every reason why it should not!), and one would expect the measured values of $\Delta N(t_1)$ and $\Delta N(t_2)$ to be identified. I can see no reason whatever to doubt that it is the latter conclusion which would be found experimentally, so that in *this* (operationally defined) sense, the statement that "two superfluids which have never seen one another before nevertheless have a definite relative phase" is, I believe, false.

I now turn to what I believe is the most intriguing question in this whole business, namely the question of a "phase standard". To orient ourselves, let us note that, according to the considerations of Ref. [2], section 4, once two superfluids have been placed in Josephson contact, then even after separation they may be able to maintain a definite relative phase for a time which can in principle be quite long by laboratory standards; we may assume that by suitable engineering (e.g. placing the samples in a spaceship) this time may be made as long as we please. Then the question arises: Could we in principle "standardize" phases of (say) buckets of helium the world over by placing them successively at some central system maintained by a standards laboratory?† If this should indeed turn out to be possible, then *for all practical purposes* we could define an "absolute" phase simply as the phase relative to the standard. Let us try to make this question a bit more precise. Suppose that two buckets of helium (1 and the "standard", S) are known to have a definite relative phase $\Delta\varphi_{1S}$, e.g. because they have been placed in Josephson contact and allowed to come to equilibrium (so that in fact $\Delta\varphi_{1S} = 0$).

† Needless to say, even if the answer were to turn out to be yes, this is unlikely in the extreme to be a practical proposition, *inter alia* because of the extreme difficulty of controlling the chemical potentials acting on the different buckets sufficiently accurately to prevent an unknown drift of their relative phase.

Suppose further that the same is true of S and 2 (for the precise meaning of this statement, see below). Is it then true that 1 and 2 have a definite relative phase? Note that this question is experimentally answerable via an ensemble of experiments in which 1 and 2 are placed in contact and the "instantaneous" Josephson current between them is measured. If the relative phase is definite, the same current will flow on all occasions,† while if it is not true, the results will be randomly distributed.

It is, however, crucial to distinguish two different versions of this thought-experiment. In version (A), S is placed in contact with 1 and 2 *simultaneously* and the whole system is allowed to come to equilibrium; all contacts are then broken, and 1 is subsequently placed in contact with 2 without the mediation of S. In version (B), 1 is first placed in contact with S, with 2 absent, and sufficient time is allowed for equilibrium to be established. The 1–S contact is then broken, and subsequently contact between S and 2 is made and time allowed for the attainment of equilibrium. Finally this contact is also broken, and eventually 1 and 2 are put in contact as in version (A).

In view of space limitations, I shall give here only the results of the analysis of these two thought-experiments; the details of the calculations, which are not entirely trivial, will be presented elsewhere [10]. Case (A) is straightforward: at the end of the day, a definite phase relation is indeed established (subject, of course, to all the necessary caveats concerning time-scales, etc., see Ref. [2]) between the superfluids 1 and 2, so that on joining them a predictable Josephson current is obtained. Case (B) is a little more tricky: one might intuitively suspect that the phase relation established at the first stage between 1 and S would be "upset" in the subsequent attainment of Josephson equilibrium between S and 2, and this indeed turns out to be the case, although, surprisingly, a proper demonstration of this result requires explicit consideration of the "environment" of the system (e.g., for the superconducting case, the radiation field). Indeed, given a certain very plausible, though not rigorously proved, assumption [10] concerning the dependence on the total particle number involved of the final state of the environment, one can demonstrate that *no* definite phase relation is established at the end of the day between systems 1 and 2. Hence a "standard of superfluid phase" is possible only in sense (A), not in sense (B).

We must finally return to the ferromagnetic analogy. Let us try, as

† In order to obtain a finite Josephson current, one could apply a known chemical potential difference between 1 and 2 for a known time (so as to make $\Delta\varphi_{12}$ nonzero before placing them in contact).

a thought-experiment, repeating the arguments made in this paper with the "phase of a superfluid" replaced by the "direction of spin (in the xy-plane) of a ferromagnet". Just as we have implicitly assumed throughout that the total number of the relevant particles (electrons, He atoms, etc.) in the universe is concerned, let us for the sake of the argument make the analogous assumption of conservation of the total z-component of spin.† We would then have to conclude that, strictly speaking, the *absolute* direction of magnetization is meaningless and that we should only talk about *relative* directions of spin of different magnets. That most of us are happy in everyday life with the idea that *for all practical purposes* we can talk about "absolute" directions would then seem to imply a tacit belief in the possibility of "standardization". For the case of the superfluid phase, we have argued that such standardization is possible only in sense (A), which requires *prima facie* very special and artificial conditions: what, if any, is the salient difference in the case of magnetization?

I believe that the crucial difference between the two cases is to be sought in the very different role of what we have vaguely called the "environment". It turns out to be implicit in our arguments about phase standardization for a superfluid that the "environment" (radiation field, phonons, normal electrons or whatever) either cannot exchange particles with the "system" at all, or, even if it can, cannot exist (or is not typically found to exist) in the coherent superposition of states of different particle number (in the sense, explored above, in which a superfluid system does). That this is so is essentially just the statement that most of the world is not superfluid (let alone a superfluid state of the relevant particles)!‡ On the other hand, just about any "environment" can not only exchange angular momentum with the system, but has no difficulty in existing in a coherent superposition of states of different S_z (or rather J_z: here the exchange of spin and orbital angular momentum *does* play a vital role). More accurately, the universe can easily exist in a superposition of states which differ by the *relative* angular momentum of "system" and "environment". (That this is so is no doubt not unconnected with the question of the more general SO(3) invariance.) In other words, the typical situation as regards magnetization corresponds to "version A"

† Of course in real life spin and orbital angular momentum are not separately conserved, but taking this into account only pushes the problem one step back.

‡ To be sure, a superconductor may induce a degree of phase coherence in a normal metal adjacent to it – the so-called proximity effect (see e.g. Ref. [11], section 7.3). However, the effects die out exponentially with distance into the normal metal and thus cannot play the necessary "global" role.

of our thought-experiment rather than "version B", so the problem of standardization is automatically overcome.

A final thought: Will "version A" ever be realized in a practical sense for superfluid systems? This idea is not quite so preposterous as it sounds. If the dream of room-temperature superconductivity is attained (and perhaps even if it is not), it is not impossible that by the year 2100 there will be a network of superconducting transmission lines covering the globe, with no normal "breaks". Then, indeed, it would be possible in principle for groups, based say, in Tokyo and Washington, DC, to "mate" two pieces of superconductor which had never been in direct physical contact in such a way as to produce a predictable Josephson current on first acquaintance! Although this state of affairs would of course in no way alter any of the considerations present above, it might perhaps modify our perception of their significance somewhat.

This work was supported by the National Science Foundation under Grant No. DMR 92-14236.

References

[1] P.W. Anderson, *Basic Notions of Condensed Matter Physics* (W.A. Benjamin, Menlo Park, California, 1984).
[2] A.J. Leggett and F. Sols, Found. Phys. **21**, 353 (1991).
[3] Yu. Kagan, this volume.
[4] P. Carruthers and M. M. Nieto. Phys. Rev. Lett. **14**, 387 (1965).
[5] A. J. Leggett, J. Phys. Soc. Japan Supp. **26–3**, 1986 (1987).
[6] A. J. Leggett, Rev. Mod. Phys. **47**, 331(1975).
[7] A. J. Leggett, Ann. Phys. **85**, 11(1974).
[8] A. J. Leggett, in *Chance and Matter*, J. Souletie, J. Vannimenus and R. Stora, eds. (North-Holland, Amsterdam, 1987).
[9] R. Hegstrom and F. Sols, submitted to Phys. Rev. A.
[10] A. J. Leggett, Found. Phys., to be published.
[11] P. G. de Gennes, *Superconductivity of Metals and Alloys* (W.A. Benjamin, N.Y., 1966).

Part two

Brief Reports

20

BEC in Ultra-cold Cesium: Collisional Constraints

E. Tiesinga, A. J. Moerdijk, B. J. Verhaar, and H. T. C. Stoof
Department of Physics, Eindhoven University of Technology,
P.O. Box 513
5600 MB Eindhoven
The Netherlands

Abstract

We study necessary conditions for the observation of Bose–Einstein condensation in a magnetically trapped sample of atomic cesium gas. These constraints are due to interatomic collisions in the sample. We show that the prospects for observing Bose–Einstein condensation are favorable for a gas of ground-state Cs atoms in the highest state of the lowest hyperfine manifold. An interesting aspect of the calculations is that the scattering length for this $f = 3, m_f = -3$ *hyperfine state shows pronounced resonance structures as a function of applied magnetic field leading to variations of two orders of magnitude. Most importantly, the scattering length can change sign near the resonances. This suggests a controllable means to change the behavior of the Bose condensate because for negative values a condensate is unstable and other (quantum-)collective effects might be observed. The origin of the resonances is understood from the bound singlet and triplet rovibrational* Cs_2 *states which are perturbed due to the hyperfine and Zeeman interactions.*

It is a long standing goal to achieve quantum-collective effects in atomic Fermi- or Bose gases. The prominent reasons are that for a relatively simple system with low density, a microscopic theoretical treatment of the phase transition is still feasible, and to have an experimental testing ground for more complicated quantum-coherent effects such as superfluidity in ^{4}He and superconductivity in metals. Here, we focus on atomic species which behave as (composite) bosons. A gaseous system like this can Bose condense below a critical temperature, i.e. macroscopically occupy the one-particle ground state. For this to occur the atomic wavefunctions need to overlap, which can be expressed as $n\Lambda^3 = O(1)$, where n is the density and Λ is the temperature-dependent DeBroglie wavelength. The first system considered was a gas of ultracold hydrogen [1] stored either in a gas cell with superfluid helium-lined walls or in a wall-free magnetic trap. Although enormous progress has been made, the required density/temperature combination has not been achieved. Since the advent of laser cooling [2] other species have become available as candidates for observing Bose–Einstein condensation. In a few milliseconds the (mostly alkali-metal) atoms are cooled from room temperature to temperatures in the micro kelvin regime and after that stored magnetically. One of the first experiments to magnetically trap heavy alkali-metal atoms used ^{133}Cs [3]. In this case the attempts so far reached a density/temperature combination of 2×10^{10} cm^{-3} and 1μK. All experiments aimed at Bose–Einstein condensation are hampered by collisions between atoms. This limits the density and the lifetime of the sample to a large extent. Since storage in a magnetic trap is restricted to low-field-seeking hyperfine states of the electronic ground state, particle loss occurs either by exothermic collisions releasing sufficient kinetic energy to leave the trap or by transitions to high-field-seeking states. For sufficiently high densities recombination is an important loss mechanism as well. Although inelastic collisions limit the observable densities, the elastic collisions are essential for the formation of the condensate [4]. This is due to the fact that these collisions comprise the only process available to transfer the atoms in the one-particle ground state. Furthermore, the validity of the usual discussion of the Bose condensate in terms of a Bogoliubov transformation [5] crucially depends on a positive value of the scattering length, which is equivalent to a repulsive elastic collision at low kinetic energies. For negative values the Bogoliubov theory breaks down. This paper briefly discusses the trap-loss effects for atomic cesium and focusses mainly on the resonant behavior of the scattering length as a function of magnetic field. The latter promises to be an excellent property for the study of the Bose condensate. An extensive discussion of the density decay can be found in Ref. [6].

The starting point of the our discussion is the hyperfine level diagram (Fig. 1) of the cesium ground-state ($^2S_{1/2}$) atom. If we consider the magnetic field along the z-axis, the effective single-atom Hamiltonian comprises hyperfine and Zeeman terms

$$H^{\mathrm{hf}} = \frac{a_{\mathrm{hf}}}{\hbar^2}\mathbf{S}^e \cdot \mathbf{S}^n + (\gamma_e S_z^e - \gamma_n S_z^n)B\,, \tag{1}$$

where $\mathbf{S}^e$ and $\mathbf{S}^n$ are electron and nuclear spin, respectively. Magnetic

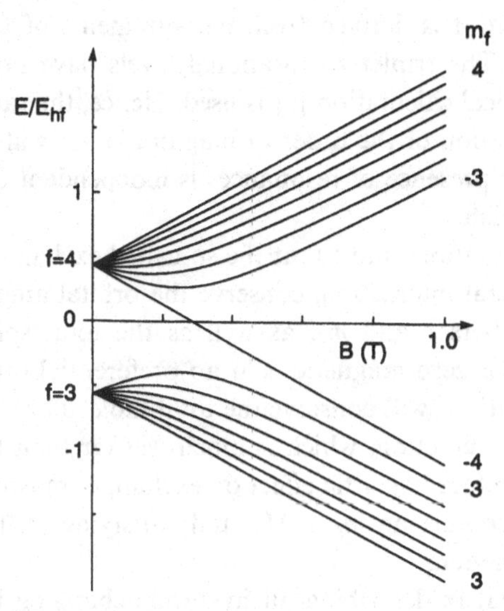

Fig. 1. Hyperfine level diagram of ^{133}Cs as a function of magnetic field. The states are labeled $|f, m_f\rangle$ and $E_{\rm hf} = 4a_{\rm hf}$.

traps generally operate with fields well below the critical value $B_c = 4a_{\rm hf}/(\gamma_e \hbar) \approx 0.33$ T, so that the total spin vector $\mathbf{f} = \mathbf{S}^e + \mathbf{S}^n$ is still conserved. Since the electron (nuclear) spin is 1/2 (7/2), the 16 hyperfine states of cesium are conveniently labeled with $|f, m_f\rangle$. The collision of two such atoms is described by the effective two-body Hamiltonian [6],[7]

$$H = \frac{\mathbf{p}^2}{2\mu} + \sum_{i=1}^{2} H_i^{\rm hf} + V^c + V^d, \qquad (2)$$

containing a kinetic energy term with μ the reduced mass and two-body interactions V^c and V^d. The central or exchange interaction V^c represents an effective description of all Coulomb interactions between the electrons and nuclei and depends only on the magnitude of the total electron spin $\mathbf{S} = \mathbf{S}_1^e + \mathbf{S}_2^e$. It can be written as a sum of singlet and triplet terms

$$V^c = V_0(r)P_0 + V_1(r)P_1, \qquad (3)$$

with P_S the projection operators on the subspaces with definite total electron spin quantum number $S = 0, 1$. The magnetic dipole–dipole interaction V^d is classically understood as the coupling between either the electronic or the nuclear magnetic moment of one atom with those of the other. The singlet Cs–Cs potential V_0 is taken from Demtröder and coworkers [8], where it is derived from measurements of Cs_2 rotation–vibration energies. The triplet rovibrational levels have not been measured and a theoretical calculation [9] is used. Hence, the rates presented below give an indication of the order of magnitude. As will be discussed below, however, the presence of resonances is independent of the precise shape of the potentials.

Symmetry considerations show that the so-called exchange transitions, induced by the central interaction, conserve the orbital angular momentum quantum numbers ℓ and m_ℓ, as well as the total spin projection $M_F = m_{f1} + m_{f2}$. For zero magnetic field no preferential direction exists and exchange transitions will conserve the total molecular spin F as well. The dipole–dipole interaction, which is much weaker than the exchange interaction and therefore has little effect on exchange transitions, induces dipole transitions conserving $m_\ell + M_F$ and satisfying $|\Delta\ell| = 0, 2$ with $\ell = 0 \rightarrow \ell = 0$ forbidden.

The scattering matrix describing all hyperfine-changing transitions is found by rigorously solving the coupled channels equations. We find that for the ultra-cold temperatures considered, a zero-temperature calculation is sufficient. This implies that the initial state of the collision has zero angular momentum. Moreover, the inelastic transitions are divided in two classes. Their rate constants differ by several orders of magnitude. Exchange transitions give a rate constant of the order of $G^c(T = 0) = 10^{-12}\,\text{cm}^{-3}/\text{s}$ and lead to a decay time of $10^2 \times (10^{10}\,\text{cm}^{-3}/n)$ seconds. Dipolar decay rate constants are much smaller and are of the order of $G^d(T = 0) = 2 \times 10^{-15}\ \text{cm}^{-3}/\text{s}$ corresponding to a decay time of $5 \times 10^4 \times (10^{10}\,\text{cm}^{-3}/n)$ seconds. The second-order spin–orbit interaction [10] also gives rise to density decay. It is very qualitatively known but most likely of the same order of magnitude as the dipole interaction. In view of the uncertainties in the central interaction already present we have omitted this decay process.

The distinction between the two classes makes those gases which decay via dipolar relaxation preferable. It turns out that a sample of magnetically trapped atomic cesium with all atoms in either the $|4,4\rangle$ or the $|3,-3\rangle$ hyperfine level solely decays via the dipole–dipole interaction, where the latter is a low-field-seeking state as long as the trapping field remains below B_c. A collision between two doubly polarized $|4,4\rangle$ hyperfine states is described by pure $S = 1$ triplet scattering, implying the absence of exchange transitions. Furthermore, a gas of $|3,-3\rangle$ atoms, with collisions described by both singlet and triplet terms, is long lived since other hyperfine states accessible via the central interaction

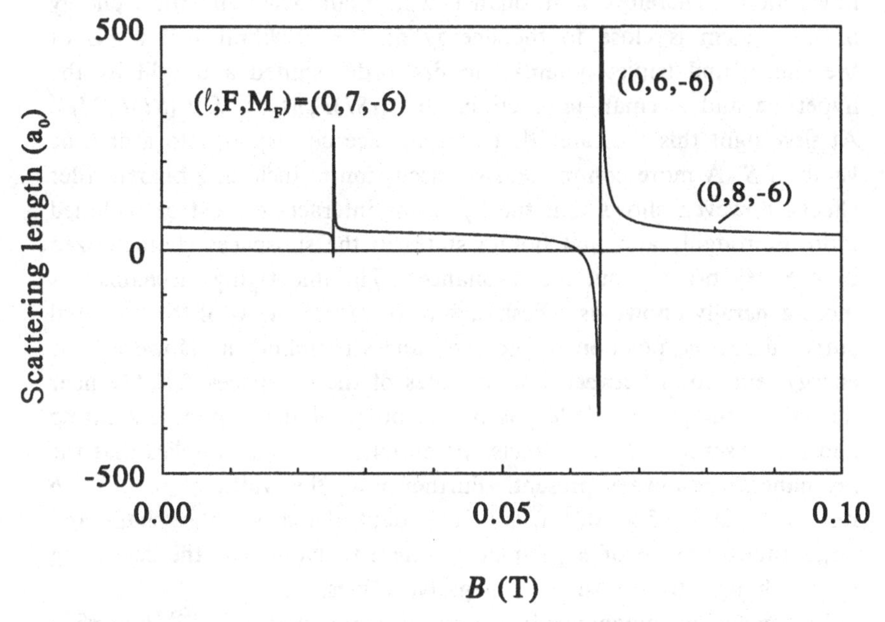

Fig. 2. Scattering length for elastic $|3,-3\rangle + |3,-3\rangle$ scattering as a function of magnetic field. Labels denote quantum numbers (ℓ, F, M_F).

are energetically forbidden as long as collision energies remain below the hyperfine splitting $4a_{\rm hf} \approx 0.4$ K. Elastic scattering, essential for thermalization of a sample as well as for the formation of the Bose condensate, is for both samples governed by the central interaction. For the doubly-polarized gas the scattering length is solely given by $\ell = 0$ scattering in the triplet potential and is therefore independent of magnetic field. The uncertainty in the singlet/triplet curve leaves the sign of the scattering length undetermined. In the case of a gas of $|3,-3\rangle$ atoms, however, resonant behavior of the scattering length (See fig. 2) is observed. This is due to a magnetic-field-dependent coupling between the singlet and triplet potential curves.

At this point it is convenient to express the collision in terms of the $|(SI)F, M_F\rangle$ basis, with total nuclear spin $\mathbf{I} = \mathbf{S}_1^n + \mathbf{S}_2^n$. Note that in this basis the central interaction is diagonal while the hyperfine- and to a lesser extent the Zeeman interaction couple the $S = 0$ and 1 potential curves. Moreover, the total energy of the initial $|3,-3\rangle + |3,-3\rangle$ state, which for low field values is approximatively given by $-9/2a_{\rm hf} + 3/8\hbar\gamma_e B+$ kinetic energy, lies below the theshold of the central interaction. Therefore a resonance will occur when the total energy in the system is close to the energy of $\ell = 0$ vibrational levels of the singlet and triplet potential to first-order shifted and split by the hyperfine and Zeeman interaction, i.e. $\langle (SI)F, M_F | \sum H^{\text{hf}} | (SI)F, M_F \rangle$. At first sight this indicates that a resonance corresponds to a specific value of S. A more comprehensive discussion[6] including higher-order effects, however, shows that the hyperfine interaction must be included more thoroughly and that bound states in the subspaces characterized by ℓ, F, M_F bring about the resonances. The underlying mechanism is more generally known as a Feshbach resonance. In view of the potential curves used, the position of the resonances is unknown. However, the energy separation between bound states of the subspaces ℓ, F, M_F near the initial energy of $-9/2a_{\text{hf}}$ is of the order of the hyperfine splitting and is not sensitive to the precise potential shape. This implies that the resonances are always present. Furthermore, the width of the $F = 6$ resonance, about 5×10^{-3} T, is independent of the potential shape and large enough to be of experimental interest. Note that the scattering length changes from positive to negative values.

In conclusion, magnetically trapped cesium in the $|3, -3\rangle$ hyperfine state is a very promising candidate for the observation of quantum-collective effects.

References

[1] I.F. Silvera and J.T.M. Walraven, Phys. Rev. Lett. **44**, 164 (1980); J.M. Doyle, J.C. Sandberg, I.A. Yu, C.L. Cesar, D. Kleppner, and T.J. Greytak, Phys. Rev. Lett. **67**, 603 (1991); O.J. Luiten, H.G.C. Werij, I.D. Setija, M.W. Reynolds, T.W. Hijmans, and J.T.M. Walraven, Phys. Rev. Lett. **70**, 544 (1993).

[2] The special issues of J. Opt. Soc. Am. B devoted to laser cooling; Vol. 2 (1985) and Vol. 6 (1989).

[3] C. Monroe, W. Swann, H. Robinson, and C. Wieman, Phys. Rev.Lett. **65**, 1571 (1990); C. Monroe, E. Cornell, C. Sackett, C. Myatt,and C. Wieman, Phys. Rev. Lett. **70**, 414 (1993).

[4] H.T.C. Stoof, Phys. Rev. Lett. **66**, 3148(1991); H.T.C. Stoof, Phys. Rev. A **45**, 8398 (1992); Yu. M. Kagan, and G.V. Shlyapnikov, Zh. Eksp. Teor. Fiz. **101**, 528 (1992); B.V.Svistunov, J. Moscow Phys. Soc. **1**, 373 (1991). Contributions of these authors on nucleation can also be found in this volume.

[5] A.L. Fetter and J.D. Walecka, *Quantum Theory of Many-Particle Systems*, (McGraw-Hill, New York, 1971).

[6] E. Tiesinga, A.J. Moerdijk, B.J. Verhaar, and H.T.C. Stoof, Phys. Rev. A **46**, R1167 (1992); E. Tiesinga, B.J. Verhaar, and H.T.C. Stoof, Phys. Rev. A, May (1993).

[7] H.T.C. Stoof, J.M.V.A. Koelman, and B.J. Verhaar, Phys. Rev. B **38**, 4688 (1988).

[8] H. Weickenmeier, U. Diemer, W. Demtröder, and M. Broyer, Chem. Phys. Lett. **124**, 470 (1992); H. Weickenmeier, U. Diemer, M. Wahl, M. Raab, W. Demtröder, and W. Müller, J. Chem. Phys. **82**, 5354 (1985).

[9] M. Krauss and W.J. Stevens, J. Chem. Phys. **93**, 4236(1990).

[10] P.S. Julienne, J. Mol. Spectrosc. **56**, 270 (1975); S.R. Langhoff, J. Chem. Phys. **61**, 1708 (1974).

21

BEC and the Relaxation Explosion in Magnetically Trapped Atomic Hydrogen

T. W. Hijmans, Yu. Kagan,† G. V. Shlyapnikov,† and J. T. M. Walraven

Van der Waals-Zeeman Laboratorium, Universiteit van Amsterdam,
Valckenierstraat 65/67
1018 XE Amsterdam
The Netherlands

Abstract

We predict and analyze non-trivial relaxational behavior of magnetically trapped gases near the Bose condensation temperature T_c. Due to strong compression of the condensate by the inhomogeneous trapping field, particularly at low densities, the relaxation rate shows a strong, almost jump wise, increase below T_c. As a consequence the maximum fraction of condensate particles is limited to a few percent. This phenomenon can be called a "relaxation explosion". We discuss its implications for the detectability of BEC in atomic hydrogen.

Magnetostatic traps offer the possibility to study gases of Bose particles in the truly dilute limit, and have proved particularly fruitful [1, 2, 3, 4, 5] in the study of atomic hydrogen (H). In these traps, proposed for H by Hess [6], the effective elimination of physical boundaries is accomplished by creating a magnetic field minimum in free space. This minimum forms a potential well for electron spin-up polarized atoms (H↑), called low-field seekers. The occurrence of Bose–Einstein condensation (BEC) in such systems introduces qualitatively different behavior from the case of a homogeneous Bose gas. This is related to the explosive increase of the dipolar relaxation rate associated with the strong compression of the condensate in an external potential. A similar phenomenon occurs in connection with three-body recombination in high density systems [7] The large condensate density resulting from this compression gives rise to an increase of the rate of inelastic (dipolar) pair collisions, in which spins of the colliding particles are flipped, producing high-field seeking

† Permanent address: Russian Research Center, Kurchatov Institute, Moscow 123182, Russia.

atoms (H↓) which are ejected from the trap. The approach to BEC is often [6, 8] described in terms of trajectories in density–temperature space. We shall see that in H, the increase in relaxation rate resulting from the appearance of the condensate, the "relaxation explosion", prevents one from penetrating deep into the BEC region and markedly alters these trajectories.

For simplicity we consider an isotropic harmonic trapping potential of the form

$$V(r) = \mu_B B_0 + \frac{m}{2}\omega^2 r^2, \tag{1}$$

where r is the distance to the trap center, and ω is the oscillation frequency. The critical BEC temperature is expressed by the relation $T_c = 3.31\hbar^2 n^{2/3}/m$, where n is the density of the gas at the trap center and m is the mass of the atom. The density profiles in external potentials were analyzed by Goldman *et al.* [9], and by Huse and Siggia [10]. The critical temperature can be expressed in terms of the total number of particles N, and the parameters of the trapping potential [11]. For the potential given by Eq. (1) we have

$$T_c = (N/g_3(1))^{1/3}\hbar\omega. \tag{2}$$

Here $g_3(1) = 1.20$, where g_ℓ is a Bose integral given by $g_\ell(\xi) = \sum_{n=1}^{\infty}(\xi^n/n^\ell)$.

Well above T_c, the number of relaxation events per unit time is given by

$$\nu_r = 2\alpha \int \mathrm{d}\mathbf{r}\, n^2(r), \tag{3}$$

where $n(r)$ is the density distribution and α is the rate constant for dipolar decay. For hydrogen $\alpha \approx 10^{-15}$ cm^3/s [12, 13]. Below T_c we should replace Eq. (3) by a more general expression [14, 15]:

$$\nu_r = 2\alpha \int \mathrm{d}\mathbf{r}\, \frac{1}{2} < \hat{\psi}^\dagger(\mathbf{r})\hat{\psi}^\dagger(\mathbf{r})\hat{\psi}(\mathbf{r})\hat{\psi}(\mathbf{r}) >, \tag{4}$$

where $\hat{\psi}$ is the field operator of the atoms and $< \hat{\psi}^\dagger\hat{\psi}^\dagger\hat{\psi}\hat{\psi} >$ is the local two-particle correlator.

A detailed calculation [16] shows that, below T_c, the relaxation rate can be expressed in terms of its value at T_c:

$$\nu_r(T) = \nu_r(T_c)(1 + 7.5(\frac{T_c}{n_c\tilde{U}})^{3/5}(\frac{\Delta T}{T_c})^{7/5}) \;\; ; T \le T_c \tag{5}$$

Here, $\Delta T = T_c - T$, and $n_c\tilde{U}$ represents the mean field interaction

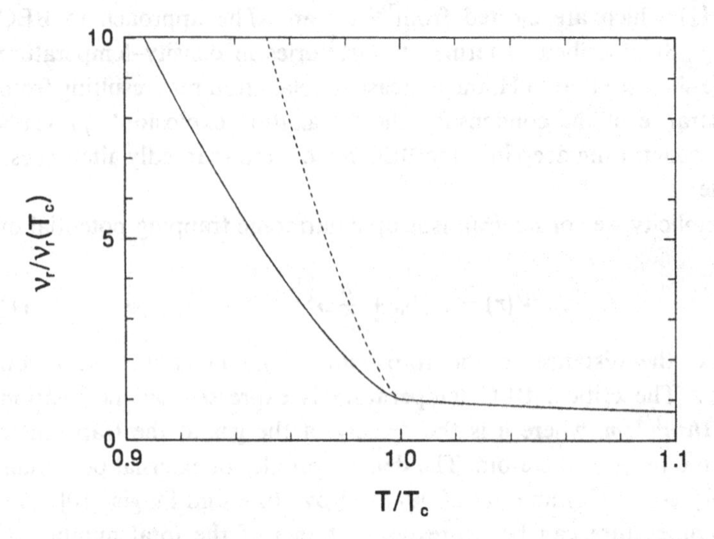

Fig. 1. Number of relaxation events per unit time for a fixed number of atoms, normalized to the value at T_c. The solid line corresponds to $n_c = 10^{15}$ cm^{-3} and the dashed line to $n_c = 10^{13}$ cm^{-3}.

energy, where n_c is the critical density. For H, at $T = 100\ \mu$K, we have $n_c \approx 5 \times 10^{14}$ cm^{-3} and $n_c\tilde{U} \approx 0.2\ \mu$K. This clearly shows the extreme weakness of the interaction in this system.

Qualitatively, Eq. (5) can be understood as follows. Above T_c we have, according to Eq. (3) $\nu_r \sim \alpha N^2/V_e$, where the classical effective volume V_e is defined trough $n(r) = (N/V_e)\exp(-V(r)/T)$. Below T_c we have an additional contribution associated with the condensate particles: $\nu_0 \sim \alpha N_0^2/V_0$. Explicit calculation [16] shows

$$V_0 = \frac{4}{3\sqrt{(\pi)}}(\frac{\mu - 2n_c\tilde{U}}{T_c})^{3/2}V_e \approx 2.0(\frac{n_c\tilde{U}}{T_c})^{3/5}(\frac{\Delta T}{T_c})^{3/5}V_e. \qquad (6)$$

Clearly, as a result of the weak interaction, we have $V_0 \ll V_e$. Hence, even if $N_0 \ll N$ (and accordingly $\Delta T \ll T$) the condensate can dominate the relaxation rate. In Fig. 1 the value of $\nu_r(T)/\nu_r(T_c)$ is given for H atoms for $n_c = 10^{13}$cm^{-3} ($T_c = 7\mu$K) and $n_c = 10^{15}$cm^{-3} ($T_c = 150\mu$K).

To investigate the effect of the relaxation explosion, we use a simple model for the kinetics of the cooling proces, in which we assume that the cooling proceeds sufficiently slowly to consider the system as being in thermal equilibrium. To obtain the trajectories in $N - T$ space, we start from the expressions for the internal energy of the trapped gas at temperatures above and below T_c (here we neglect $n_c\tilde{U}$):

$$U = 3NT\frac{g_4(\xi)}{g_3(\xi)} \quad ; T \geq T_c \tag{7}$$

$$U = 3NT(\frac{T}{T_c})^3\frac{g_4(1)}{g_3(1)} = 3g_4(1)\frac{T^4}{(\hbar\omega)^3} \quad ; T \leq T_c, \tag{8}$$

where ξ is the fugacity $\exp(-\mu/T)$. From this we obtain relations between the time derivatives of T, U and N:

$$\left(4\frac{g_4(\xi)g_2(\xi)}{g_3^2(\xi)} - 3\right)\frac{\dot{T}}{T} = \frac{\dot{U}}{U}\frac{g_2(\xi)g_4(\xi)}{g_3^2(\xi)} - \frac{\dot{N}}{N} \quad ; T \geq T_c. \tag{9}$$

$$\frac{\dot{T}}{T} = \frac{1}{4}\frac{\dot{U}}{U} \quad ; T \leq T_c. \tag{10}$$

If we parametrize the cooling process by an energy removal rate $R_U \equiv -\dot{U}/U$, and an accompanying rate R_N of particle removal, we can write the rate of change of particle number and internal energy as

$$\frac{\dot{N}}{N} = -R_N - \frac{\nu_r}{N}; \quad \frac{\dot{U}}{U} = -R_U + \frac{\dot{U}_r}{U}. \tag{11}$$

Here $\dot{U}_r$ is time rate of change of energy associated with relaxation. As the relaxing particles have a finite energy, the internal energy decreases with decreasing particle number. Hence, $\dot{U}_r$ is always negative. From Eq. (10) one thus finds that when $T \leq T_c$, because the energy does not explicitly depend on N, relaxation does not lead to heating. In contrast, from Eq. (9) we see that for $T > T_c$ removal of particles only leads to cooling if the energy of these particles is sufficiently high. This is the principle of evaporative cooling. In this sense, any loss of (above-condensate) particles leads to evaporative cooling for $T \leq T_c$. Note, moreover, that the ratio $\dot{T}/\dot{N}$ changes discontinuously at T_c. This behavior has the same origin as the discontinuity in the derivative of specific heat at the BEC point. A straightforward calculation shows that $\dot{U}(T) = -(9/4)T\nu_r$ for $T \gg T_c$ and $\dot{U}(T) = -(9/4)T0.8\nu_r(T_c)$ for $T < T_c$.

Using the above results we can easily obtain the trajectories in the $N - T$ plane, if we make the somewhat simplifying assumption that R_U is proportional to the elastic collision rate $R_c \equiv \frac{1}{4}\sigma v_T N/V_e$. The resulting trajectories are shown in Fig. 2 for two values of $\tilde{R}_U \equiv R_U/R_c$. Below T_c the relaxation explosion manifests itself as an attraction of the trajectories towards the BEC line. This is particularly striking when comparing with the corresponding trajectories obtained by assuming the system to obey Boltzmann statistics at any temperature and density.

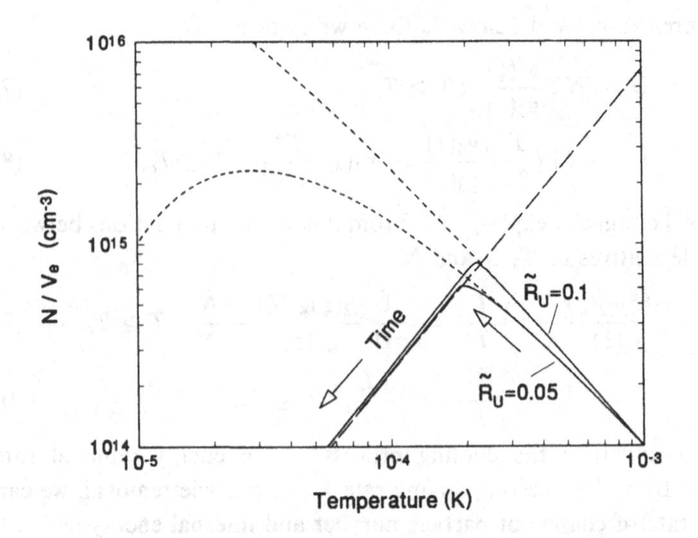

Fig. 2. Cooling trajectories for $\tilde{R}_U = 0.1$ and 0.05 (solid lines) plotted as N/V_e vs T. The long-dashed curve is the BEC line. The short-dashed curves represent the trajectories corresponding to a system obeying Boltzmann statistics.

This research was supported by the Nederlandse Organisatie voor Wetenschappelijk Onderzoek (NWO-PIONIER and NWO-07-30-002).

References

[1] H.F. Hess, G. Kochanski, J.M. Doyle, N. Masuhara, D. Kleppner, and T.J. Greytak, Phys. Rev. Lett. **59**, 672 (1987).

[2] R. van Roijen, J.J. Berkhout, S. Jaakkola, and J.T.M. Walraven, Phys. Rev. Lett. **61**, 931 (1988).

[3] J.M. Doyle, J.C. Sandberg, I.A. Yu, C.L. Cesar, D. Kleppner, and T.J. Greytak, Phys. Rev. Lett. **67**, 603 (1991).

[4] O.J. Luiten, H.G.C. Werij, I.D. Setija, M.W. Reynolds, T.W. Hijmans, and J.T.M. Walraven, Phys. Rev. Lett. **70**, 544 (1993).

[5] I.D. Setija, H.G.C. Werij, O.J.Luiten, M.W. Reynolds, T.W. Hijmans, and J.T.M. Walraven, Phys. Rev. Lett. **70**, 2257 (1993).

[6] H.F. Hess, Phys. Rev. B **34**, 3476 (1986).

[7] Yu. Kagan and G.V. Shlyapnikov, Phys. Lett. A **130**, 483 (1988).

[8] T.J. Tommila, Europhys. Lett. **57**, 314 (1986).

[9] V.V. Goldman, I.F. Silvera, and A.J. Leggett, Phys. Rev. B **24**, 2870 (1981).

[10] D.A. Huse and E.D. Siggia, J. Low. Temp. Phys. **46**, 137 (1982).

[11] V. Bagnato, D.E. Pritchard, and D. Kleppner, Phys. Rev. A **35** 4354 (1987).

[12] A. Lagendijk, I.F. Silvera, and B.J. Verhaar, Phys. Rev. B **33**, 626 (1986).

[13] H.T.C. Stoof, J.M.V.A. Koelman, and B.J. Verhaar, Phys. Rev. B **38**, 4688 (1988).

[14] Yu. Kagan, B.V. Svistunov, and G.V. Shlyapnikov, Pis'ma Zh. Eksp. Teor. Fiz. **48**, 54 (1988) [JETP. Lett. **48**, 56 (1988).

[15] H.T.C. Stoof, A.M.L. Jansen, J.M.V.A. Koelman, and B.J. Verhaar, Phys. Rev. A **39**, 3157 (1989).

[16] T.W. Hijmans, Yu. Kagan, G.V. Shlyapnikov, and J.T.M. Walraven, Phys. Rev. B **48**, 12886 (1993).

22

Quest for Kosterlitz–Thouless Transition in Two-Dimensional Atomic Hydrogen

A. Matsubara,† T. Arai, S. Hotta, J. S. Korhonen, T. Mizusaki, and A. Hirai ‡

Department of Physics
Kyoto University
Kyoto 606-01
Japan

Abstract

We have cooled atomic hydrogen adsorbed on liquid helium to very low temperatures: a surface temperature of 87 mK *was obtained with a density of* $2 \cdot 10^{12}$ atoms/cm^2. *The two-dimensional hydrogen gas was cooled on a cold spot, which was inside a large sample cell. When the cold spot is much colder than the rest of the sample cell the surface density on it is controlled by the hydrogen flux into the sample cell. We found that at least* 98% *of the recombination heat was carried away from the cold spot by the excited* H_2 *molecules. There was a heat input to the cold spot which was independent of the presence of atomic hydrogen and which prevented cooling of the adsorbed hydrogen below* 87 mK. *We attribute this heat input to thermal conduction by ripplons.*

1 Introduction

The experiments with polarized atomic hydrogen [1, 2] are usually made in sample cells whose walls are covered with a helium film. The atomic hydrogen is adsorbed on the He surface and forms a two-dimensional (2D) gas, which is expected to undergo a Kosterlitz–Thouless transition [3] to a superfluid state. The critical temperature of the transition is

$$T_{KT} = \pi \hbar^2 \sigma / 2 k_B m \ , \tag{1}$$

where m is the mass of the hydrogen atom and σ is the surface density.

If the adsorbed hydrogen is in thermal equilibrium with the bulk gas, it is difficult to control the surface density and temperature independently:

† Permanent address : Department of Physics, Osaka-City University, Sumiyoshi-ku, Osaka, 558 Japan.

‡ Deceased on 31 December, 1992.

the adsorption energy ϵ_a is about 1 K and when temperature is lowered much below 1 K the surface density tends to increase exponentially and soon approaches the saturation density $2 \cdot 10^{14}$ atoms/cm^2. This leads to rapid three-body recombination at the surface, which prevents cooling below T_{KT}.

Our method of cooling 2D-H$\downarrow$ is the following: inside a large sample cell we prepared a small cold spot. The cold spot was thermally isolated from the rest of the sample cell, and its temperature, T_{CS}, was kept much lower than that of the sample cell. At sufficiently low T_{CS} the recombination occurs mainly on the cold spot. Then the surface density is nearly independent of temperature and is determined by the input flux of hydrogen to the sample cell. We found that although the recombination occurs mainly on the cold spot, most of the recombination heat is carried away from the cold spot by the excited H_2 molecules.

2 Experimental Apparatus

We prepared doubly polarized atomic hydrogen, H$\downarrow\downarrow$, as follows. The atomic hydrogen was produced at low temperature by rf discharge. It was precooled by a baffle before it was stored in a buffer volume, which was connected to the sample cell through a flow impedance. The buffer and the sample cell were in a magnetic field of 8 T. If the flow impedance, volume of the buffer and the flux of hydrogen to the buffer are selected properly, the atomic hydrogen in the buffer volume is doubly polarized.

The sample cell was a cylinder with radius 17.5 mm and height 75 mm. Its walls were covered with superfluid ^{4}He film of saturation thickness. The cold spot was a copper disk of radius 2.2 mm. It was located at the upper part of the sample cell. The cold spot surface was partly covered with a sintered silver piece of radius 1.4 mm and estimated surface area of 0.1 m^2. The cold spot was thermally isolated from the rest of the copper sample cell by a piece of Stycast with an outer radius of 8 mm.

The discharge cell and the baffle were anchored to a heat exchanger of the dilution refrigerator at a temperature of 0.5 K. The common temperature of the buffer volume and the sample cell was fixed to a temperature T_H, between 290 and 430 mK, by a temperature controller. The cold spot was connected by a mesh of copper wires to the mixing chamber of the dilution refrigerator and its temperature was controlled at T_{CS}.

We measured two quantities: the bulk density n of the hydrogen in the sample cell and the increase in the heat input to the cold spot which was due to the introduction of hydrogen. In order to measure changes in the heat input to the cold spot we fixed the temperatures of the mixing chamber and the cold spot separately by temperature controllers. Then the change in the feedback power of the cold spot controller was equal (but opposite) to the power $\dot{Q}$ due to the introduction of hydrogen.

The bulk density n was measured by making all the hydrogen in the sample cell recombine on a bolometer. The density was calculated from the total recombination heat, which was obtained from the increase in the sample cell temperature. With this method we were able to measure only densities larger than $10^{12}\,\mathrm{cm}^{-3}$ but the absolute value of n was obtained reliably. For smaller densities n was calculated from the rise in the bolometer temperature.

3 Rate Equations for Hydrogen Densities

We assume that the atomic hydrogen in the sample cell was doubly polarized. Then the data analysis can be based on the following equations for the time evolution of the bulk density n of H↓↓ in the sample cell and the surface density σ on the cold spot:

$$\begin{aligned} \frac{dn}{dt} &= \frac{\phi}{V} - R_b^{eff} n - K_b^{eff} n^2 - L_b^{eff} n^3 - \frac{A_{cs}\bar{v}_b S_b}{4V} n + \frac{A_{cs}}{V\tau_s}\sigma, \\ \frac{d\sigma}{dt} &= \frac{\bar{v}_b S_b}{4} n - \frac{1}{\tau_s}\sigma - K_s\sigma^2 - L_s\sigma^3. \end{aligned} \qquad (2)$$

Here V is the sample cell volume' A_{cs} is the area of the cold spot; and ϕ is the flux of H↓↓ to the sample cell; K_s and L_s are the two- and three-body recombination rates on the cold spot; R_b^{eff}, K_b^{eff} and L_b^{eff} are the effective one-, two- and three-body recombination rates, which include the contribution from the recombination on the hot part of the sample cell wall and are defined through equations $R_b^{eff} n = R_s(A/V)\sigma_H$, $K_b^{eff} n^2 = K_b n^2 + K_s(A/V)\sigma_H^2$ and $L_b^{eff} n^3 = L_b n^3 + L_s(A/V)\sigma_H^3$. The sample cell area is A and $\sigma_H = n\lambda \exp(\epsilon_a/k_B T_H)$ is the surface density on the hot walls of the sample cell; λ is the thermal de Broglie wavelength. In our experiments, only the recombination constants $L_s = 1.2 \times 10^{-24}\,\mathrm{cm}^4\,\mathrm{s}^{-1}$ [4] and $R_s = 8.5\,\mathrm{cm}^{-1}$ are important. The determination of R_s is discussed in section 4. For values of K_a and L_b, see Ref. [4], and for K_s, see Ref. [5].

In Eq. (2), the thermal velocity of the atoms in the bulk gas is $\bar{v}_b = \sqrt{8k_B T_H/\pi m}$, and $\bar{v}_b n A_{cs}/4$ is the number of atoms colliding with the cold

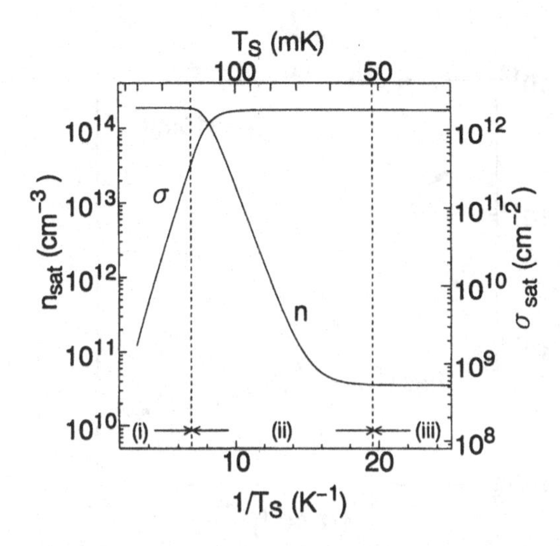

Fig. 1. The calculated saturation values of the bulk density n in the sample cell and the surface density σ on the cold spot as a function of the inverse temperature T_S^{-1} of the adsorbed hydrogen. In region (i) the cold spot does not affect the bulk density; in regions (ii) and (iii) the recombination occurs mainly at the cold spot.

spot surface in unit time. The sticking coefficient $S_b = 0.33T_H/\mathrm{K}$ [5, 6] is the probability that a surface collision leads to the adsorption of the atom. The mean residence time on the surface is $\tau_s = (4\lambda_s/\bar{v}_s S_s)\exp(\epsilon_a/k_B T_S)$, where λ, $\bar{v}$ and S must now be evaluated at the temperature of the adsorbed hydrogen, denoted by T_S. Due to the thermal resistance between the adsorbed hydrogen and the cold spot, T_S may not be the same as T_{CS}.

The saturation values of n and σ, given by Eq. (2), are plotted in Fig. 1 as a function of T_S, at constant $T_H = 310\,\mathrm{mK}$ and $\phi = 1.1 \cdot 10^{12}\,\mathrm{s}^{-1}$. There are three distinct temperature regions: (i) the high temperature, (ii) the desorption limited, and (iii) the recombination limited regions. In the high temperature region the recombination on the cold spot can be neglected. In regions (ii) and (iii) the recombination occurs mainly on the cold spot. Then σ is given by the equation $A_{cs}(K_s\sigma^2 + L_s\sigma^3) = \phi$ and is almost independent of the cold spot temperature. A desired value of σ can be obtained by selecting an appropriate hydrogen flux ϕ. In region (ii) the probability that the adsorbed atom desorbs before recombination is large, while in region (iii) nearly all the atoms adsorbed on the cold spot recombine before being desorbed.

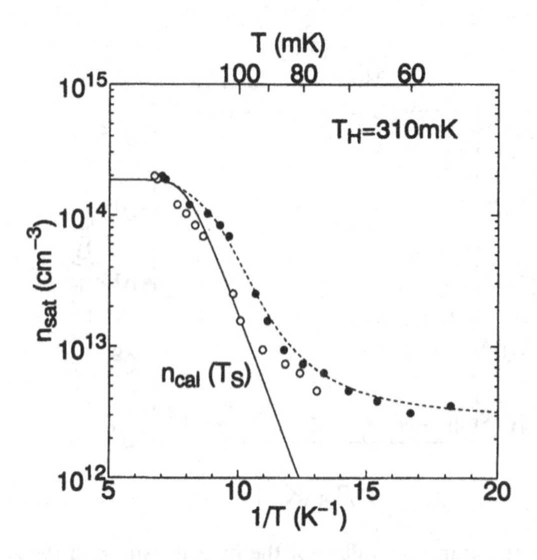

Fig. 2. The saturated bulk density of hydrogen vs. the inverse cold spot temperature T_{CS}^{-1} (filled circles) and the inverse temperature of the adsorbed hydrogen T_S^{-1} (open circles). T_S is calculated from the measured heat input to the cold spot and the thermal resistance between the ripplon and the phonon systems. The dashed line is a guide for the eyes, and the solid curve is calculated from Eq. (2).

4 Experimental Results and Data Analysis

The measurements of n were started about two hours after switching on the pulsed rf discharge. This time was needed to saturate the density of hydrogen in the buffer volume. First we determined the values of the flux ϕ and the recombination rate constant R_s. This was done at high T_S (region (i) in Fig. 1 by recording $n(t)$ after all the atomic hydrogen was removed from the cell at $t = 0$. A fit of the measured $n(t)$ to Eq. (2) gave $\phi = 1.1 \times 10^{12}$ s^{-1} and $R_s = 8.5$ s. It turned out that one-body recombination was the dominant mechanism in the bulk, because n was rather low. Next we measured the saturation value of the hydrogen density n_{sat}. The solid circles in Fig. 2 show n_{sat} as a function of the cold spot temperature T_{CS}. The solid line, denoted by $n_{cal}(T_S)$, is calculated from Eq. (2). The difference between $n_{sat}(T_{CS})$ and $n_{cal}(T_S)$ indicates that T_S was higher than T_{CS}. The observed value of n_{sat} serves as a sensitive thermometer for T_S in region (ii) in Fig. 1. The saturation value of the heat input to the cold spot, $\dot{Q}_{sat}$, is shown in Fig. 3 as a function of T_{CS}.

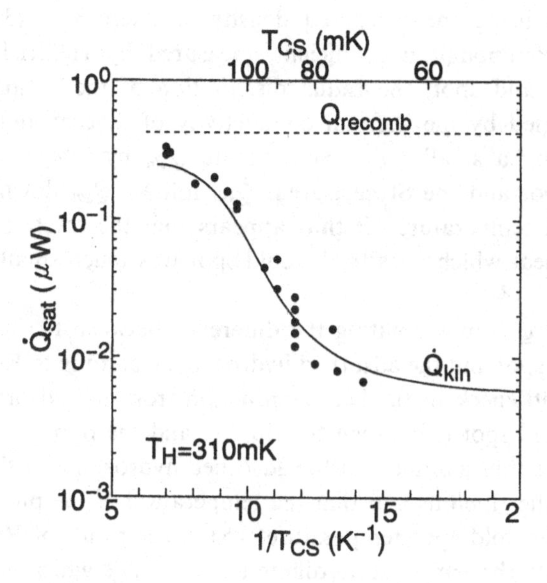

Fig. 3. The heat input to the cold spot caused by the atomic hydrogen as a function of the inverse cold spot temperature. The circles show the measured data; the solid line is the power carried as the kinetic energy of the hydrogen atoms. $\dot{Q}_{recomb}$ is the total power released in the recombinations. Only a fraction smaller than 1.4% of $\dot{Q}_{recomb}$ goes to the cold spot.

All these data were taken under conditions such that the recombination occurs mainly on the cold spot. The total recombination heat power $\dot{Q}_{recomb}$, which is calculated from the flux ϕ, is shown in Fig. 3 by the upper dashed line. Comparison with $\dot{Q}_{sat}$ reveals that less than 1.4% of the total recombination heat goes to the cold spot. This implies that the H_2 molecule, which is left in a highly excited state in the recombination, cannot penetrate into liquid helium, and the recombination heat is delivered evenly to the sample cell. A similar result was obtained by Vasilyev *et al.* [8] They measured an upper limit of 10% for the recombination energy which is deposited into helium at the recombination site. The heat input in Fig. 3 seems to vary roughly proportional to the density n_{sat} in Fig. 2. A mechanism which would show such a dependence is heat transfer from the hot bulk hydrogen gas to the cold surface. The calculated power $\dot{Q}_{kin}$, which is released when the hot hydrogen atoms thermalize after being adsorbed, is shown in Fig. 3 by the solid line. The local heat flux

$$\dot{q}_{kin} = 2k_B(T_H - T(r))n\bar{v}\alpha/4, \tag{3}$$

is calculated by using the measured density n. Here $\alpha = (3/2)S$ is the thermal accommodation coefficient, measured by Helffrich *et al.* [9] Outside the cold spot, the radial distribution of the temperature, $T(r)$, is determined by the thermal conductivity of Stycast and is not changed much by a small $\dot{q}_{kin}$. We estimate $\dot{Q}_{kin}$ by integrating $\dot{q}_{kin}$ over the cold spot and the Stycast area. $\dot{Q}_{kin}$ follows $\dot{Q}_{sat}$ down to the lowest cold spot temperature. It thus appears that the fraction of the recombination heat which goes to the cold spot was much smaller than 1.4%.

We have used $\dot{Q}_{kin}$ in estimating the difference between the temperatures of the cold spot and the adsorbed hydrogen. According to Reynolds *et al.* [7], the bottleneck in the heat conduction from the adsorbed hydrogen to the cold spot is between the ripplon and the phonon systems in the helium. We thus assume that the adsorbed hydrogen is at the same temperature as the ripplons and that the temperature of the phonons in helium equals the cold spot temperature. Using the result of Reynolds *et al.* we calculate the temperature difference $T_S - T_{CS}$ which is caused by $\dot{Q}_{kin}$ being dumped to the ripplons on the cold spot. The open circles in Fig. 2 show the measured n_{sat} as a function of the calculated T_S. Agreement with $n_{calc}(T_S)$ is good at $T_S > 90$ mK but at lower temperatures the measured $\dot{Q}_{sat}$ is too small to explain the temperature difference $T_S - T_{CS}$. This difference should be attributed to some residual heat input which does not depend on the presence of the atomic hydrogen.

The saturation of T_S can be explained by the ripplon heat conduction. Using the conduction measured by Mantz *et al.* [10, 11], we obtain a heat input of about 30 nW to the cold spot under the conditions of Figs. 2 and 3. At low temperatures the heat carried by the ripplons is thus much larger than $\dot{Q}_{sat}$ and determines the minimum T_S. The calculated minimum $T_S = 85$ mK is in good agreement with the value of 87 mK obtained from Fig. 2.

In order to clarify the effect of the ripplon heat conduction, we measured n_{sat} as a function of T_H with fixed $T_{CS} = 80$ mK. Using Eq. (2) we estimated T_S from n_{sat}; $T_S(T_H)$ is shown by the circles in Fig. 4. The lines are calculated by using the results of Reynolds *et al.* The dashed line is obtained by considering only the ripplon heat conduction, while the lower solid line results if only $\dot{Q}_{kin}$ is taken into account. Both ripplons and $\dot{Q}_{kin}$ contribute to the upper solid line. Clearly, the heat carried by ripplons becomes important for low T_H.

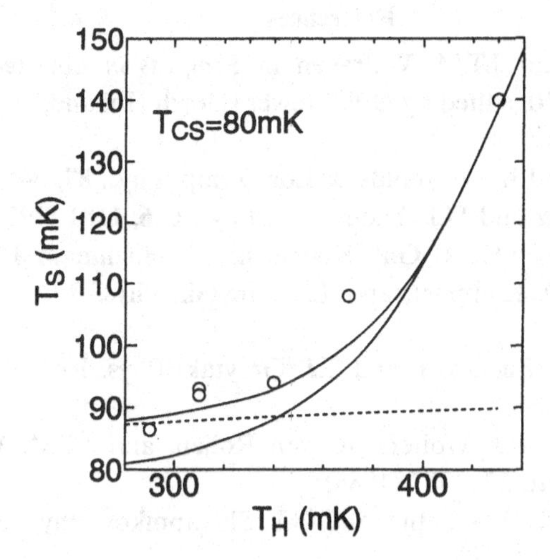

Fig. 4. Temperature of the adsorbed hydrogen T_S as a function of the sample cell temperature T_H. The temperature of the cold spot is fixed to $T_{CS} = 80$ mK. The open circles show T_S estimated from the measured n_{sat}, using Eq. (2). The lines are calculated from the thermal resistance between the ripplon and the phonon systems and the heat input to the surface. Dashed line: only ripplon heat conduction considered. Lower solid line: only $\dot{Q}_{kin}$ taken into account. Upper solid line: combined effect of $\dot{Q}_{kin}$ and ripplon heat conduction.

5 Summary

Our experiments demonstrate that two-dimensional atomic hydrogen gas can be cooled on a small cold area much below the temperature of the bulk gas. In particular, we have shown that the recombination heat was not the limiting factor of the minimum temperature of the adsorbed hydrogen: the minimum T_S was determined by the residual heat input to the ripplon system of the He-film. This residual heat flux was due to ripplon heat conduction. In future experiments we will try to cool the two-dimensional hydrogen further by using thinner ^{4}He films and ^{3}He or ^{3}He–^{4}He films.

Finally, we note that although we do not at the moment have a direct signal from the adsorbed hydrogen we believe that the KT transition can be detected even in the present set-up. This is due to a large decrease in the surface recombination rate when the KT transition takes place [12, 13]. If the measurement is done in the desorption limited temperature region (ii) in Fig. 1, the transition should be seen as a sudden increase in n when T_S is lowered below T_{KT}.

References

[1] I.F. Silvera and J.T.M. Walraven in *Progress in Low temperature Physics*, Vol. 10, edited by D.F. Brewer (North Holland, Amsterdam, 1986), p. 139.

[2] I.F. Silvera and M. Reynolds, J. Low Temp. Phys. **87**, 345 (1992).

[3] J.M. Kosterlitz and D.J. Thouless, J. Phys. C **6**, 1181 (1973).

[4] D.A. Bell, H.F. Hess, G.P. Kochanski, S. Buchman, L. Pollack, Y.M. Xiao, D. Kleppner, and T.J. Greytak, Phys. Rev. B **34**, 7670 (1986).

[5] L. Pollack, S. Buchman, and T.J. Greytak, Phys. Rev. B **45**, 2993 (1992).

[6] J.J. Berkhout, E.J. Wolters, R. van Roijen, and J.T.M. Walraven, Phys. Rev. Lett. **57**, 2387 (1986).

[7] M.W. Reynolds, I.D. Setija, and G.V. Shlyapnikov, Phys. Rev. B **46**, 575 (1992).

[8] S.A. Vasilyev, E. Tjukanov, M. Mertig, A. Ya. Katunin, and S. Jaakkola, Europhys. Lett. **24**, 223 (1993).

[9] J. Helffrich, M.P. Maley, M. Krusius, and J.C. Wheatley, Phys. Rev. B **34**, 6550 (1986).

[10] D.O. Edwards, I.B. Mantz, and V.U. Nayak, J. Phys. (Paris), Colloq. **39**, C6-300 (1978).

[11] I.B. Mantz, D.O. Edwards, and V.U. Nayak, Phys. Rev. Lett. **44**, 663 (1980).

[12] Yu. Kagan, B.V. Svistunov, and G.V. Shlyapnikov, Sov. Phys. JETP, **66**, 314 (1988).

[13] H.T.C. Stoof and M. Bijlsma, preprint (1993).

23

BEC of Biexcitons in CuCl

Masahiro Hasuo and Nobukata Nagasawa
Department of Physics
The University of Tokyo
7-3-1 Hongo, Bunkyo-ku, Tokyo, 113
Japan

Abstract

A new approach to detecting coherence associated with Bose–Einstein condensation of biexcitons at K=0 in CuCl is described. Remarkable enhancement of the phase-conjugate signal at the two-photon resonance of the biexcitons at K=0 is observed when high-density incoherent biexcitons are injected, suggesting macroscopic increase of the occupation number of the K=0 biexciton state and enhancement of the coherence. A small blue shift of the K=0 biexciton resonance is also observed, suggesting the change of the chemical potential of the biexciton due to the interaction between biexcitons.

1 Introduction

An excitonic particle such as an exciton or a biexciton (excitonic molecule) in a semiconductor crystal has a boson-like nature. If the system of excitonic particles is regarded as an ideal Bose gas, Bose–Einstein condensation (BEC) may occur at the critical density of

$$N_c = \frac{2.612}{(4\pi)^{3/2}}(\frac{2mk_B T}{\hbar^2})^{3/2}, \tag{1}$$

where m and T are the effective mass of the particle and the temperature of the system, respectively.

In most semiconductors, the biexciton, a bound complex of two excitons, is the most stable entity, with the lowest energy per electron–hole pair. The CuCl crystal is famous as such a material, the excitonic parameters of which have been well determined experimentally. For example, the effective mass of the biexciton is 5.3 times as large as the electron rest mass [1]. Thus, the corresponding N_c is estimated to be $2.2 \times 10^{17}/\mathrm{cm}^3$ at T=2 K.

A biexciton in CuCl decomposes into one photon and one exciton through a radiative process that gives characteristic luminescence called M-emission. The spectral shape of the M-emission is well known to be determined by the density of states and the distribution function of the biexciton system. When the density of biexcitons, N, is far less than N_c, the spectral shape is reproduced approximately by the Maxwell–Boltzmann distribution function [2]. With increase of N, it is expected that the spectral shape changes, reflecting the Bose statistics due to the Bose nature of the biexciton.

In fact, it has been reported that the spectral shape of M-emission is reproduced by the Bose–Einstein distribution function corresponding to μ=0 under special experimental conditions, where μ is the chemical potential of biexcitons [3]. A demonstration showing the quantum attraction between biexcitons in phase space has also been presented by the use of a pump-and-probe method on the M-emission associated with the two-photon excitation of biexcitons [4].

When the BEC occurs, it is expected that a macroscopic number of biexcitons collapses to a single quantum state of zero momentum, K=0, and forms a coherent quantum state. However, the coherence inherent in the BEC has never been demonstrated experimentally so far. Recently, we have proposed a new pump-and-probe method that aims to detect the coherence [5].

We measure the change in the intensity of the phase-conjugate signal generated from the third-order macroscopic non-linear polarization associated with the coherent biexcitons populated at K=0 under the presence of the high-density biexcitons injected incoherently. Fig. 1 shows a schematic illustration of the phase conjugation. The relevant non-linear polarization is expressed by

$$P^{(3)} = \chi^{(3)}(E_1 \cdot E_2)E_3^*, \tag{2}$$

where E_1 and E_2 are the electric fields of the counter–propagating laser beams 1 and 2 used for exciting the coherent biexcitons at K=0 [6]. Adjusting these beams' intensities, the density of the coherent biexcitons populated at K=0 is substantially lower than N_c. The electric field E_3 of the laser beam 3 is used to induce a parametric scattering process associated with the coherent biexcitons, as shown in Fig. 1. The signal light field has phase which is conjugate to that of E_3, as seen in Eq. (2). Since this phase-conjugate signal is generated from the coherent assembly of the

(a)

(b)

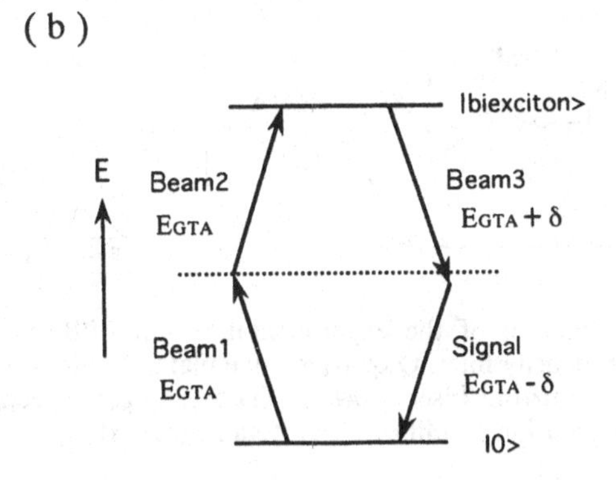

Fig. 1. Geometry (a) and energy diagram (b) for the phase conjugation.

relevant microscopic polarizations which enter into $\chi^{(3)}$, one can monitor the occupation number of the biexcitons at K=0 and its coherence. If BEC is realized, it is expected that the intensity of the phase-conjugate signal should show a remarkable increase as an indication of the onset of BEC and of the enhancement of the relevant coherence.

2 Experimental Method

Figure 2 shows a schematic diagram of the experimental set-up. Each photon energy of the counterpropagating beams 1 and 2, provided by Dye Laser A, was adjusted to be the two-photon resonance energy of the biexcitons at K=0, E_{GTA}=3.1858 eV. The spectral width of the laser light, $\gamma_{1,2}$, was adjusted to be 138 μeV for a reason mentioned below. These beams were focused in a sample. The spot size on the sample surface was about 100 μmϕ, resulting in a total intensity of 160 kW/cm^2. The phase

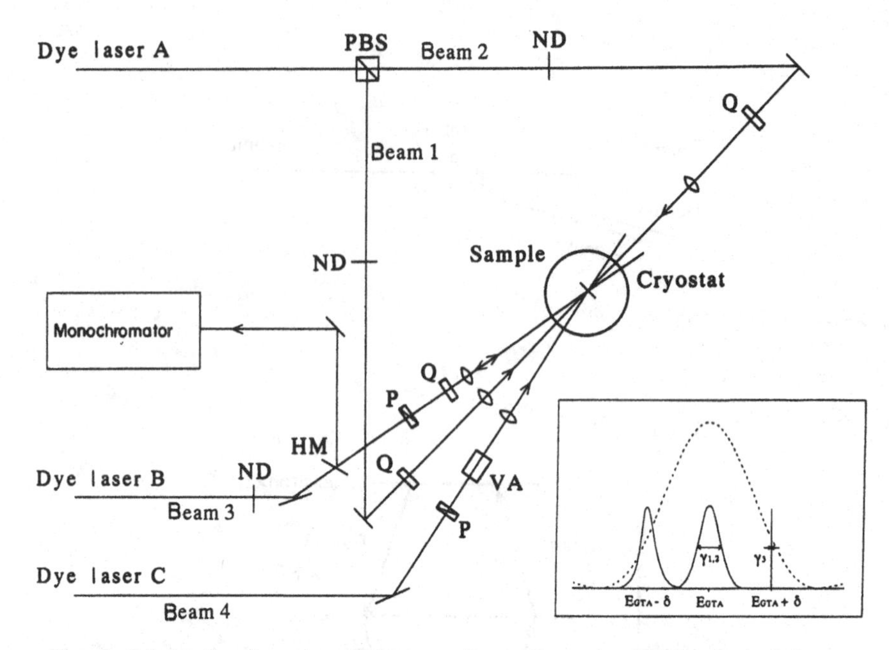

Fig. 2. Schematic diagram of the experimental set-up. PBS:polarized beam splitter, ND:neutral density filter, Q:quarter wave plate, P:polarizer, VA:variable attenuator, HM:half mirror. Inset shows a phase-conjugate spectrum expected under the present excitation conditions. For details, see text.

coherent signal from the coherent biexcitons generated by beams 1 and 2 was induced by the radiation of beam 3 from Dye Laser B. The photon energy was set at $E_{GTA}+\delta$, where δ is the detuning from the resonance energy and was 4 meV. The spectral width, γ_3, was set to be 37μeV. An inset of Fig. 2 shows a schematic illustration of a phase-conjugate signal in the spectral domain under these excitation conditions. The phase-conjugate signal appears at $E_{GTA} - \delta$, shown in the diagram. The corresponding transition is shown in Fig. 1(b). The spectral width of the phase-conjugate spectrum is determined mainly by the level width of the biexciton at K=0, γ_0, through the relevant $\chi^{(3)}$ under the condition $\gamma_{1,2} > \gamma_0 \gg \gamma_3$. The spectral shape is deformed by a phase-matching factor, f_{PM}, given by

$$f_{PM} = \frac{sin^2 \frac{\Delta KL}{2}}{\left(\frac{\Delta KL}{2}\right)^2}, \tag{3}$$

where ΔK and L are the relevant phase mismatch [7] and the sample thickness, respectively. The dashed curve shows the phase-matching factor as a function of photon energy estimated for the present experimental conditions.

The polarization of these laser beams is made circular so as to excite mainly the $K=0$ biexcitons via the relevant selection rule of the two-photon transition. The external angle between beams 1 and 3 was 10°, which corresponds to an internal angle of 3.6°. The phase-conjugate signal was analyzed with a high resolution monochromator (Jobin-Yvon THR1500, band pass 16μeV).

The fourth laser beam, from Dye Laser C, was used to generate the incoherent biexcitons through a bimolecular collision process of photo-excited excitons. We shall refer to this light as the incoherent pump light hereafter. The photon energy and the spectral width of this light were 3.2029 eV and 1.36 meV, respectively. The beam was focused on the sample surface to a spot of 200 $\mu m \phi$ in order to cover the region illuminated by beams 1, 2 and 3. The intensity of this light was adjusted continuously with a variable intensity attenuator. The maximum intensity was 2.7 MW/cm^2.

All the dye lasers (BBQ in p-dioxane) were pumped by a common excimer laser (Lambda Physik 53MSC). The pulse duration of each dye laser light was about 10 ns.

The sample was a 15 μm thick single-crystal platelet, grown from vapor phase. All the measurements were done at 2 K.

3 Results and Discussion

Figure 3 shows an example of the experimental results on the phase-conjugate spectrum. Curve (a) shows the spectrum without injection of the incoherent biexcitons. With increase in the intensity of the incoherent pump light, the intensity of the phase-conjugate signal shows an increase, as shown in curves (b)–(d). In these cases, the increase of the signal was mainly localized in the peak region of the spectrum. A small blue shift of the spectrum was also observed correlated to the increase of the signal intensity. However, this trend reversed beyond about 1 MW/cm^2. In these cases, the decrease of the signal was seen in the whole spectral region.

Figure 4 shows the peak intensity of the phase-conjugate spectrum, normalized by the spectrum without injection of the incoherent biexcitons, as a function of the intensity of the incoherent pump light. Open squares correspond to the cases shown in Fig. 3. The dotted curve

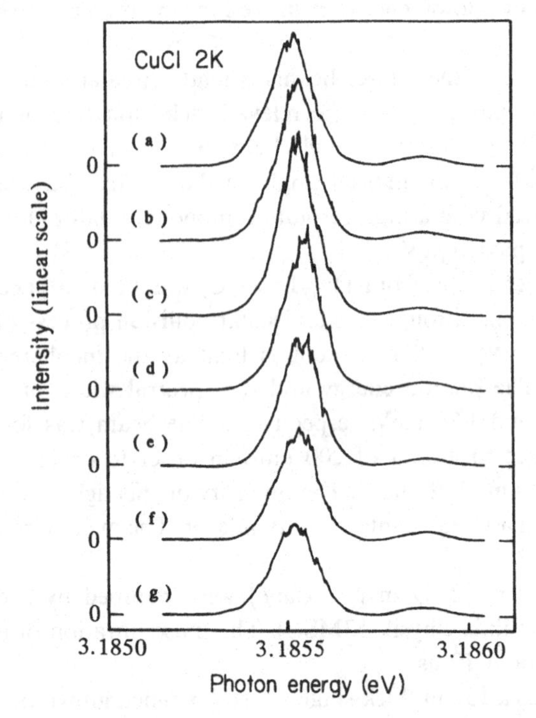

Fig. 3. Spectra of the phase-conjugate signal at 2 K with the intensities of the incoherent pump light of (a) 0 kW/cm^2, (b) 120 kW/cm^2, (c) 260 kW/cm^2, (d) 680 kW/cm^2, (e) 930 kW/cm^2, (f) 1800 kW/cm^2 and (g) 2700 kW/cm^2.

is a similar one, obtained by changing the intensity of the incoherent pump light continuously by a variable intensity attenuator. In this case, the monochromator's wavelength was fixed at the peak energy of the phase-conjugate spectrum without injection of incoherent biexcitons. The general trend was well reproduced.

The scale shown in the top of this diagram shows the density N of incoherent biexcitons that was calculated from the intensity of the incoherent pump light. This was estimated by solving the following rate equations under steady state conditions:

$$\frac{dN}{dt} = \frac{\eta}{2}n^2 - \gamma_{\mathrm{mol}}N, \tag{4}$$

$$\frac{dn}{dt} = -\eta n^2 + \gamma_{\mathrm{mol}}N - \gamma_{\mathrm{ex}}n + J, \tag{5}$$

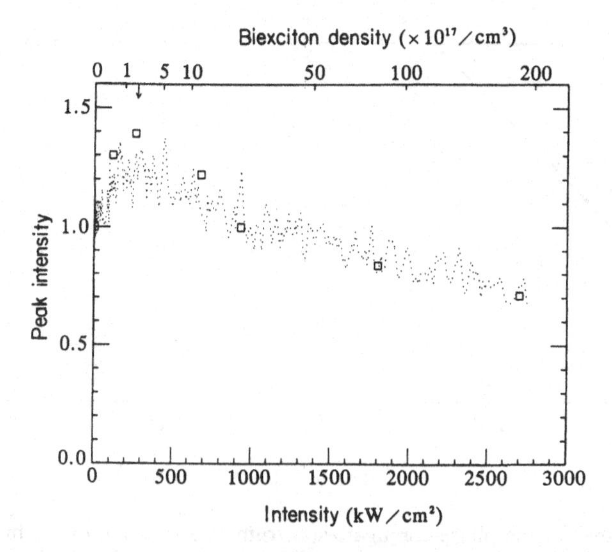

Fig. 4. The peak intensity of the phase-conjugate spectrum as a function of the intensity of the incoherent pump light. The calculated biexciton density is scaled at the top of the diagram. The arrow shows the critical density for BEC of ideal Bose gas at 2 K. For details of the estimation of the biexciton density, see text.

where n is the density of excitons. The parameters listed below were adopted for the present calculation:

η : biexciton formation rate, $6 \times 10^{-8}\,\mathrm{cm^3/s}$ [8],

γ_{mol} : radiative decay rate of a biexciton, $1/50\,(\mathrm{ps})^{-1}$ [9],

γ_{ex} : decay rate of an exciton, $1/5\,(\mathrm{ns})^{-1}$ [10].

In this analysis, the effective excitation density, J, is given by

$$J = \frac{(1-R)I}{L_{\mathrm{eff}}E}. \tag{6}$$

Here, R, I and E are the reflectivity, intensity and photon energy of the incoherent pump light, respectively; L_{eff} is the effective excitation depth and was assumed to be expressed in the form

$$L_{\mathrm{eff}} = \frac{v_g}{\gamma_{sc} + \eta n}, \tag{7}$$

where v_g and γ_{sc} are the group velocity and the scattering rate of the exciton polaritons which are converted from the incoherent pump light at the sample surface. They were adopted to be $1.5 \times 10^6\,\mathrm{cm/s}$ and $1 \times 10^9\,\mathrm{s^{-1}}$, respectively [10]. The collision rate of excitons to create the incoherent biexcitons is ηn. As seen in Fig. 4, the remarkable increase

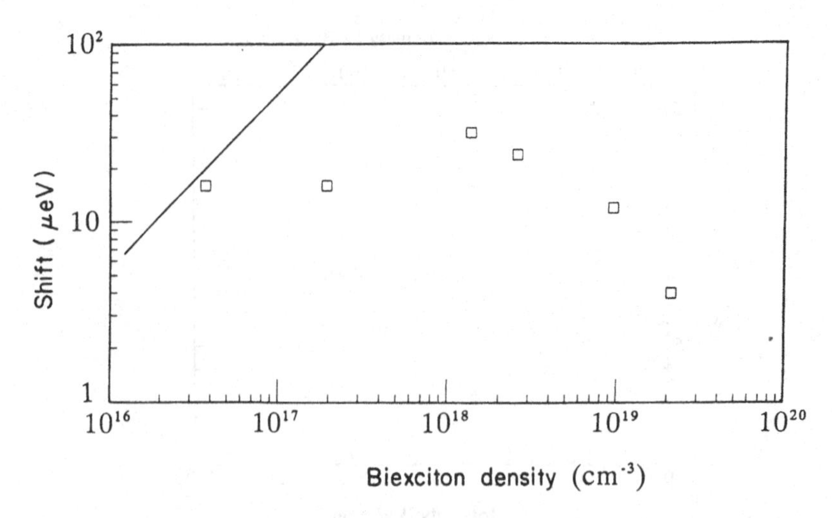

Fig. 5. Peak shift of the phase-conjugate spectrum as a function of the biexciton density. The solid line shows the change of the chemical potential of biexcitons obtained theoretically.

of the signal intensity was observed in the density region of N_c at 2 K, marked by the arrow.

Figure 5 shows the energy shift of the peak of the phase-conjugate spectrum as a function of N on a log-log scale. The change of the chemical potential due to the interaction between particles, ΔE, may be taken as

$$\Delta E = 4\pi \frac{\hbar^2 a_0}{m} N, \tag{8}$$

where a_0 is the scattering length of the particles, 2.9 nm [11, 12]. For the present experimental conditions, we take $\Delta E = 5.25 \times 10^{-16} N$ in units of μeV, shown by the solid line in the diagram. The observed shift was only reproduced quantitatively in the lowest excitation case. The origin of the deviation in the high-density region is not known.

4 Conclusion

We have presented a new approach for observing the coherence associated with the BEC of the biexciton system in CuCl. A phase-conjugate signal was used as a probe to monitor the macroscopic occupation of the biexcitons at $K=0$ and the relevant coherence under the high-density injection of the biexcitons of incoherent nature. A remarkable increase of the signal intensity at the critical density for BEC and a blue shift of the $K = 0$ biexciton resonance were observed. These results are attributed to the presence of the Bose–Einstein condensate of biexcitons.

Acknowledgements. We acknowledge the support of CNRS and JSPS through the joint collaborative program. We would like to express our sincere thanks to Dr A. Mysyrowicz for his kind support and valuable discussions at every stage of this work. We also thank Profs. T. Itoh, M. Inoue, E. Hanamura, and M. Kuwata-Gonokami and Drs A. Ivanov and T. Hatano for fruitful discussions. This work was partially supported by the Grant-in-Aid for Scientific Research from the Ministry of Education, Science and Culture.

References

[1] T. Mita, K. Sôtome and M. Ueta, J. Phys. Soc. Jpn. **48**, 496 (1980).
[2] H. Haug, and E. Hanamura, Physics Reports **33**, 209 (1977).
[3] L.L. Chase, N. Peyghambarian, G. Grynberg, and A. Mysyrowicz, Phys. Rev. Lett. **42**, 1231 (1974).
[4] N. Peyghambarian, L.L. Chase, and A. Mysyrowicz, Optics Comm. **41**, 178 (1982).
[5] M. Hasuo, N. Nagasawa, and A. Mysyrowicz, Phys. Stat. Sol.(b) **173**, 255 (1992); M. Hasuo, N. Nagasawa, T. Itoh, and A. Mysyrowicz, Phys. Rev. Lett. **70**, 1303 (1993).
[6] G. Mizutani, and N. Nagasawa, J. Phys. Soc. Jpn. **52**, 2251 (1983); L.L. Chase, M.L. Claude, D. Hulin, and A. Mysyrowicz, Phys. Rev. A **28**, 3969 (1983).
[7] Y. Masumoto, and S. Shionoya, J. Phys. Soc. Jpn. **49**, 2236 (1980).
[8] T. Hatano, doctoral thesis (The University of Tokyo, 1993).
[9] H. Akiyama, T. Kuga, M. Matsuoka, and M. Kuwata-Gonokami, Phys. Rev. B **42**, 5621 (1990).
[10] T. Ikehara, and T. Itoh, Phys. Rev. B **44**, 9283 (1991).
[11] L.V. Keldysh, and A.N. Kozlov, Sov. Phys. JETP **27**, 521 (1968).
[12] M. Inoue, and E. Hanamura, J. Phys. Soc. Jpn. **41**, 1273 (1976).

24

The Influence of Polariton Effects on BEC of Biexcitons

A. L. Ivanov and H. Haug

Institut für Theoretische Physik
J.W. Goethe-Universität
Robert-Mayer-Str. 8, D-60054 Frankfurt/Main
Germany

Abstract

The influence of the radiative decay of excitonic molecules on a possible quasiequilibrium Bose–Einstein condensation (BEC) of excitonic molecule's is examined with respect to the radiative renormalization of the excitonic molecule energy (excitonic molecule Lamb shift). For the excitonic molecule wave function, a Schrödinger equation which contains polariton effects is derived and analyzed. Both the inverse excitonic molecule radiative lifetime γ^m and the biexciton Lamb shift Δ^m depend strongly on the total excitonic molecule momentum K. The energy renormalization $\Delta^m(K)$ leads to the excitonic molecule effective mass modification and can result in a camel-back structure at $K = 0$, which opposes a BEC of excitonic molecules at $K = 0$.

1 Introduction

Observations of a quasiequilibrium Bose–Einstein condensation (BEC) of excitonic molecules (EM) have been attempted [1, 2, 3] following its theoretical prediction [4, 5] (for reviews see, e.g. [6, 7]). Recent approaches [8, 9] with high-precision techniques renewed the interest in this phenomenon. Traditionally, one tries to detect BEC of the EMs in luminescence. In the new approach the appearance of coherence in the thermal system of the EMs has been tested by means of four-wave mixing and treated as a fundamental manifestation of BEC. (See the paper by Hasuo *et al.* in this book.) Both optical methods for the BEC detection imply that the optical transition to the corresponding intermediate exciton (IE) state is dipole-active. In this case the EM state is unstable against optical decay with a "giant" oscillator strength [6]. The short radiative lifetime τ^m of the EM is a very serious obstacle for

BEC. The influence of this factor has been considered exclusively in the kinetics of the Bose condensate formation from arbitrary initial non-equilibrium distributions of EMs (see, e.g. [10]). The strong interaction with electromagnetic field also modifies the dispersion of the elementary excitations, however. A well-known example is the Lamb shift of the resonance energy of a dipole-active two-level system. Another example is the polariton spectrum of dipole-active IEs. Here we study the EM Lamb shift and show that this effect strongly influences the BEC of EMs.

2 Self-consistent Biexciton Wave Equation with Polariton Effects

Neglecting the interaction with an electromagnetic field, the complete set of EM eigenfunctions $\Psi_J(p)$ and corresponding energies Ω^m_{JK} can be obtained from the following Schrödinger equation in the momentum representation (with $\hbar = 1$):

$$\sum_{\mathbf{p}_1} \left[\left(H^x\left(\mathbf{p} + \tfrac{\mathbf{K}}{2}\right) + H^x\left(-\mathbf{p} + \tfrac{\mathbf{K}}{2}\right) \right) \delta_{\mathbf{p},\mathbf{p}_1} + W_{12}(\mathbf{p} - \mathbf{p}_1) \right] \times \Psi_J(\mathbf{p}_1) = \Omega^m_{J,\mathbf{K}} \Psi_J(\mathbf{p}). \quad (1)$$

Here, $\mathbf{p}$ and $\mathbf{K}$ are the momenta of the relative and center-of-mass motion of the EM, respectively. $H^x(\mathbf{k}) = k^2/2M$ is the kinetic energy of an IE, and $W_{12}(\mathbf{p} - \mathbf{p}_1)$ is the attractive potential between two singlet IEs with opposite internal electron spin orientation. Due to the quadratic IE dispersion, the center-of-mass motion splits off, and the wave function and energies of the relative motion are independent of $\mathbf{K}$. This equation in the underlying *e–h* picture has often been used to calculate the EM properties variationally [11, 12, 13, 14]. In order to investigate the influence of the polariton effects on the interior structure of EM we derived self-consistently [15] an EM wave equation from the complete *e–h*-photon ($e - h - \gamma$) Hamiltonian :

$$\sum_{\mathbf{p}_1} \left[\left(H^p\left(\mathbf{p} + \tfrac{\mathbf{K}}{2}\right) + H^p\left(-\mathbf{p} + \tfrac{\mathbf{K}}{2}\right) \right) \delta_{\mathbf{p},\mathbf{p}_1} + \tilde{W}_{12}(\mathbf{p}, \mathbf{p}_1, \mathbf{K}) \right] \times \tilde{\Psi}_J(\mathbf{p}_1, \mathbf{K}) = \tilde{\Omega}^m_{J,\mathbf{K}} \tilde{\Psi}_J(\mathbf{p}, \mathbf{K}), \quad (2)$$

with $H^p(\mathbf{k}) = \omega^-(\mathbf{k})$, where $\omega^\pm(\mathbf{k})$ is the dispersion of the upper (+) or lower (-) polariton branch, which are given by the roots of

$$\left(\frac{\omega^\gamma(p)}{\omega} \right)^2 = \frac{c^2 p^2}{\epsilon_0 \omega^2} = 1 + \frac{\Omega_c^2}{\omega_t^2 + \omega_t p^2/M - \omega^2}. \quad (3)$$

Here, $\omega^\gamma(p)$ is the photon frequency and ϵ_0 is the background optical dielectric constant. The polariton parameter Ω_c is defined in terms of the longitudinal–transverse ($\ell - t$) splitting $\omega_{\ell t}$ and the transverse exciton frequency ω_t:

$$\Omega_c^2 = 2\omega_{\ell t}\omega_t. \tag{4}$$

The effective interaction $\tilde{W}_{12}$ is determined by the x components of the two interacting polaritons (with $\mathbf{k}^{\pm} = \pm\mathbf{p} + \frac{\mathbf{K}}{2}$):

$$\tilde{W}_{12}(\mathbf{p}, \mathbf{p}_1, \mathbf{K}) = f(\mathbf{p}, \mathbf{K})\, W_{12}(\mathbf{p} - \mathbf{p}_1), \tag{5}$$

where

$$f(\mathbf{p}, \mathbf{K}) = \phi^-\left(\mathbf{k}^+, \omega^-\left(\mathbf{k}^+\right)\right) \phi^-\left(\mathbf{k}^-, \omega^-\left(\mathbf{k}^-\right)\right), \tag{6}$$

and

$$\phi^{\pm}(\mathbf{k}, \omega) = \frac{\omega^{\pm}(k)\,(\omega - \omega^{\gamma}(k))}{\omega[\omega^{\pm}(k) - \omega^{\mp}(k)]}. \tag{7}$$

These x components satisfy the following conditions:

$$\begin{aligned} \phi^+(\mathbf{p}, \omega^+(p)) &\geq 0; \\ \phi^-(\mathbf{p}, \omega^-(p)) &\geq 0; \\ \phi^+(\mathbf{p}, \omega^+(p)) + \phi^-(\mathbf{p}, \omega^-(p)) &= 1. \end{aligned} \tag{8}$$

In Eq. (2) only the lower polariton dispersion branch is taken into account. In the following we will consider only the ground states of the IE and EM (J=1). The linear equation, Eq. (2), contains the radiative decay of an EM state, as well as the renormalization of the EM energy (EM Lamb shift) due to the polariton effects. When an IE in the EM acquires a small momentum within the optical range (sectors A_1A_2 and B_1B_2 in Fig. 1), the EM undergoes a radiative annihilation. In other words, the linear part of the lower polariton dispersion branch corresponds to the case without a bound EM state. The radiative decay of an EM stems from the polariton dispersion of its IE rather than from the usual tunneling process out of a metastable state. In our picture the EM luminescence is a continuous evolution of the EM interior state rather than a discrete act of the EM optical conversion to γ and IE.

3 Biexciton Lamb-Shift

The non-stationary solution of Eq. (2), $\tilde{\Psi}(\mathbf{p})$, represents an outgoing spherical wave due to the radiative decay of a EM state with $\tilde{\Omega}_K^m = \Omega_K^m + \Delta^m(K) - i\gamma^m(K)$. Here, $\Omega_K^m = \Omega_{K=0}^m + K^2/2M_m$ is the unperturbed EM energy with the EM translational mass $M_m = 2M$. For small relative

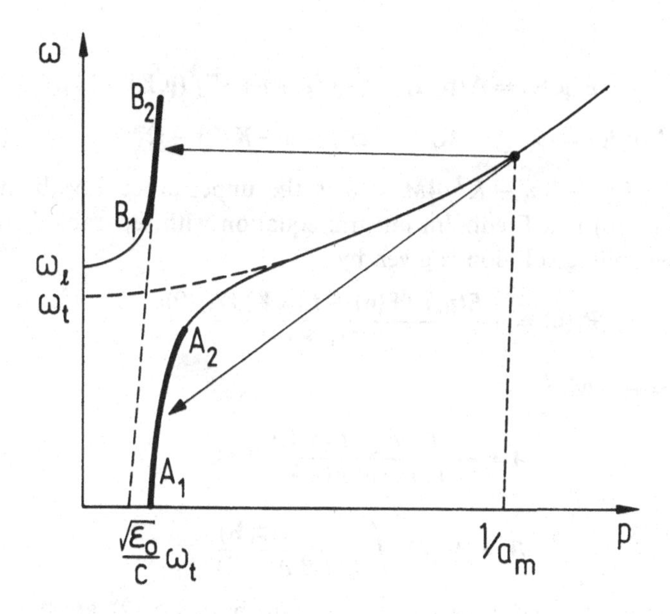

Fig. 1. Illustration of the optical decay of an EM.

momenta which belong to the optical range of the polariton spectrum, the EM wavefunction (EMWF) $\tilde{\Psi}(\mathbf{p})$ is strongly modified. This modification of the EMWF cannot be treated perturbationally. Here the EM Lamb shift $\Delta^m(K)$ and the inverse radiative lifetime $\gamma^m(K)$ are of the same order, while in atomic optics the radiative energy shifts are always substantially smaller than corresponding inverse lifetimes. In order to treat Eq. (2) we decompose the EMWF in

$$\tilde{\Psi}(\mathbf{p}) = \Psi(\mathbf{p}) + \Psi_1(\mathbf{p}, \mathbf{K}), \tag{9}$$

where $\Psi(\mathbf{p})$ is the known solution of unperturbed Eq. (1), while $\Psi_1(\mathbf{p}, \mathbf{K})$ describes the influence of the polariton effects and is large only in the optical range, that is, for $p << a_m^{-1}$, where a_m is the EM radius. For these small momenta one neglects the momentum dependence of the attractive potential $W_{12}(\mathbf{p} - \mathbf{p}_1) = W_{12}(0) =$const.

Substituting Eq. (9) in Eq. (2), one finds the following integral equation for $\Psi_1(\mathbf{p})$ in a D-dimensional system:

$$E(\mathbf{p}, \mathbf{K})\Psi(\mathbf{p}, \mathbf{K}) + \Delta(\mathbf{p}, \mathbf{K})\,\Psi_1(\mathbf{p})$$
$$+ f(\mathbf{p}, \mathbf{K})\, W_{12}(0) \int \frac{d^D p_1}{(2\pi)^D} \Psi_1(\mathbf{p}_1, \mathbf{K}) = 0, \tag{10}$$

where

$$E(\mathbf{p},\mathbf{K}) = \Delta(\mathbf{p},\mathbf{K}) - \frac{p^2}{M} f(\mathbf{p},\mathbf{K}) + \epsilon^m f(\mathbf{p},\mathbf{K})$$
$$\Delta(\mathbf{p},\mathbf{K}) = \omega^-(\mathbf{p}+\mathbf{K}/2) + \omega^-(-\mathbf{p}+\mathbf{K}/2) - \tilde{\Omega}^m_K. \tag{11}$$

Here, $\epsilon^m = \Omega^m_K - 2\omega_t - K^2/4M < 0$ is the unperturbed EM binding energy. Eq. (10) is a Fredholm integral equation with separable kernel. The corresponding solution is given by

$$\Psi_1(\mathbf{p}) = \frac{-E(\mathbf{p},\mathbf{K})\Psi(\mathbf{p}) + f(\mathbf{p},\mathbf{K})\, W_{12}(0) C}{\Delta(\mathbf{p},\mathbf{K})}, \tag{12}$$

where $C = \frac{A}{1+B}$ with

$$A = -\int \frac{d^D p}{(2\pi)^D} \frac{E(\mathbf{p},\mathbf{K})}{\Delta(\mathbf{p},\mathbf{K})} \Psi(\mathbf{p}), \tag{13}$$

$$B = W_{12}(0) \int \frac{d^D p}{(2\pi)^D} \frac{f(\mathbf{p},\mathbf{K})}{\Delta(\mathbf{p},\mathbf{K})}. \tag{14}$$

The functions $\Psi(\mathbf{p})$ and $\Psi_1(\mathbf{p},\mathbf{K})$, as well as the basic Eq. (2), are symmetric with respect to the substitution $p \to -p$. According to Eq. (12), the roots $\mathbf{p}_0 = \mathbf{p}_0(\mathbf{K})$ of the equation $\Delta(\mathbf{p},\mathbf{K}) = 0$ give rise to singularities in $\Psi_1(\mathbf{p},\mathbf{K})$. These singularities describe the two outgoing polariton waves with momenta $\mathbf{p}_0 + \mathbf{K}/2$ and $-\mathbf{p}_0 + \mathbf{K}/2$ in the decay of the EM with K. $\Delta(\mathbf{p},\mathbf{K}) = 0$ is the energy-conservation law of this decay. The graphic solution of this equation is presented in Fig. 2.

According to Eq. (2) the total EM momentum K influences the relative motion of the IEs in the EM. The polariton dispersion makes $\Psi_1(\mathbf{p},\mathbf{K})$, $\Delta^m(K)$, and $\gamma^m(K)$ K-dependent in the optical range $K \leq 2p_0(K=0)$. Formally, the solution (12) can be found for an arbitrary EM energy $\tilde{\Omega}^m_K$. In order to find the true EM energy for a given K, we change $\tilde{\Omega}^m_K$ continuously around the unperturbed EM energy Ω^m_K and find the value which minimizes the outgoing part $\int |\Psi_1(\mathbf{p},\mathbf{K})|^2 d^D p/(2\pi)^D$ and maximizes the EM radiative lifetime. Although the wave functions (12) cannot be normalized, one can compare the *relative* contributions of the outgoing parts. For simplicity we present here only numerical solutions for one-dimensional EM systems with $D = 1$. The fraction of the phase space in which polariton effects dominate over the Coulomb IE–IE interaction is of the order of $(a_m p_0(K=0))^D$. Therefore, the influence of polariton effects on the EMWF is much stronger in one dimension than in three dimensions. We use the parameters of GaAs/GaAlAs quantum-well wires [16]. For the attractive IE–IE potential $W_{12}(p - p_1)$

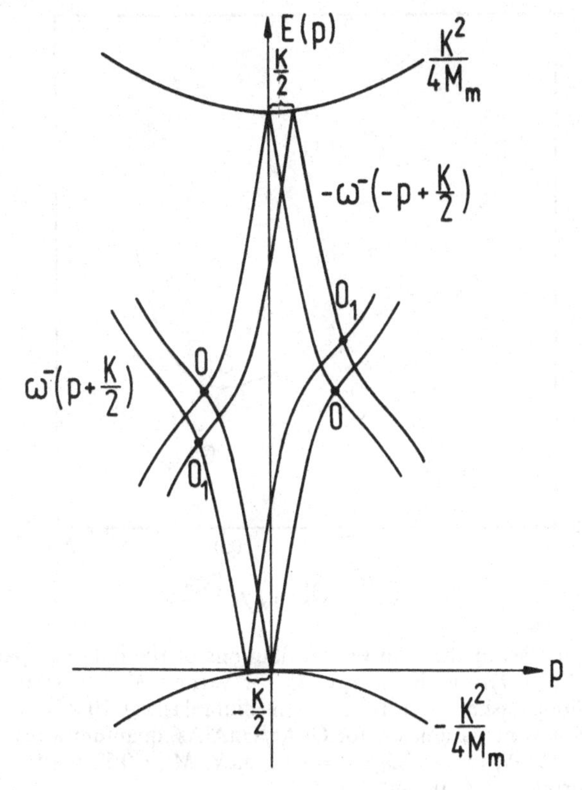

Fig. 2. Graphic solution of Eq. (12). The points O and O_1 correspond to outgoing polaritons.

we use the simple form of the deuteron model [17] in dimensionless units (normalized with the EM Rydberg and the EM radius)

$$W_{12}(p-p_1) = \frac{\beta}{(p-p_1)^2+1}. \qquad (15)$$

Eq. (1) can be treated with the trial EMWF $\Psi(p)$

$$\Psi(p) = \frac{C}{p^2+\alpha^2}. \qquad (16)$$

For $\beta = 4.0$, the parameter $\alpha = 0.5$ minimizes the energy and gives the binding energy $\epsilon^m = 0.75$. In dimensional units these parameters correspond to $\epsilon^m = -5$ meV and EM Bohr radius $a^m = 73.9$Å. The accuracy of the variational procedure is about 2.5% [17].

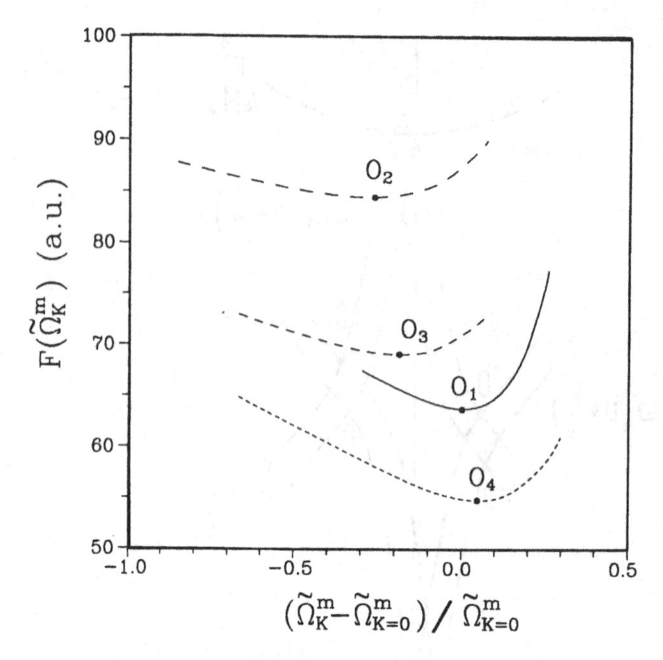

Fig. 3. Dependencies of the relative contributions of the outgoing parts of the *EMWF* (see Eq. (17)) on the energy $\tilde{\Omega}^m_K$ for various K values: 0 (full line), 3×10^5 cm^{-1} (long dashes), 15×10^5 cm^{-1} (medium dashes), 20×10^5 cm^{-1} (short dashes). The following parameters for GaAs/GaAlAs quantum wires have been used: $\omega_t = 1.515$ eV, $\epsilon_0 = 12.5, \omega_{\ell t} = 0.08$ meV, $M = 0.52 m_0, \epsilon^m = -5$ meV, polariton absorption $\gamma^x = 10$ cm^{-1}.

The set of curves

$$F = F\left(\tilde{\Omega}^m_K\right) = \int dp |\Psi_1\left(p, \tilde{\Omega}^m_K\right)|^2 / \int dp |\Psi_1\left(p, \Omega^m_K\right)|^2 \qquad (17)$$

which represent the relative contributions of the outgoing parts for a given K, is shown in Fig. 3, where the points O_i correspond to the true EM energy. The dependences $\gamma^m = \gamma^m(K)$ and $\Delta\Omega^m(K) = \hbar^2 K^2/2M_m + \Delta^m(K)$ are plotted in Fig. 4.

The EM Lamb shift $\Delta^m(K)$ gives rise to a camel-back structure with a relative maximum at $K = 0$ and a minimum at $K_0 \simeq 2p_0(K = 0)$, where $p_0(K = 0) = 2.76 \times 10^5$cm^{-1} (points O in Fig. 2). Experimentally, the EM luminescence can be observed only for EMs with $K \geq p_0(K = 0)$ both due to the strong polariton dispersion and due to the interference with the pump [2,3,7,8,9]. These EMs are strongly coupled to the electromagnetic field which results in a large EM Lamb shift, as seen in the camel-back

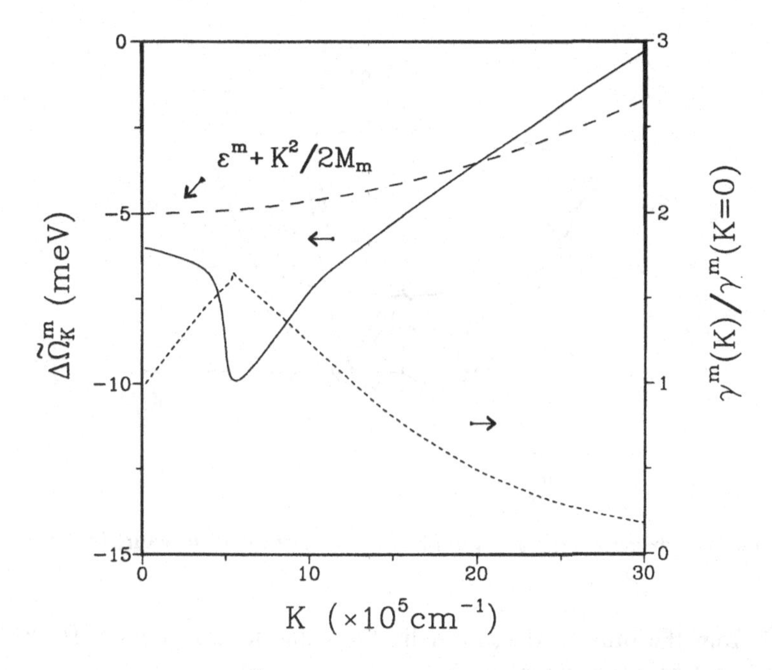

Fig. 4. Biexciton energy $\Delta\tilde{\Omega}^m_K = K^2/2M_m + \Delta^m(K)$ (full line) and inverse radiative lifetime $\gamma^m(K)$ (dashed line) versus total momentum K. All parameters are the same as in Fig. 3, except for $\gamma^x = 500$ cm^{-1}.

structure around $K = 0$. In our theory this behavior has the following explanation. Increasing the total EM momentum K from 0, one reaches the point at which the second lower branch $-\omega^-(-\mathbf{p}+\mathbf{K}/2)$ (left branch of $-\omega^-(-\mathbf{p}+\mathbf{K}/2)$ in Fig. 2) touches and crosses the first polariton branch $\omega^-(\mathbf{p}+\mathbf{K}/2)$ (right branch in Fig. 2). Thus, one receives beyond the points 0_1 additional solutions of the energy-conservation law $\Delta(\mathbf{p},\mathbf{K}) = 0$. The touching point corresponds to $v_s = \partial\Delta(p,K)/\partial p = 0$ and gives rise to a *van Hove singularity.* This singularity at $K_0 \simeq 2p_0(K=0)$ is responsible for a strong coupling of the EM with K_0 to the electromagnetic field and results in a "jump" of the EM energy to smaller values (see Fig. 4). For two dimensions/three dimensions the van Hove singularities are a continuous family of critical points in the form of a circle (sphere) with radius K_0. So the character of the van Hove singularity is independent of the dimensionality of EM. The large spatial dispersion of $\Delta^m(K)$ in the optical range $K \simeq p_0(K=0)$ allows one to discriminate the EM Lamb

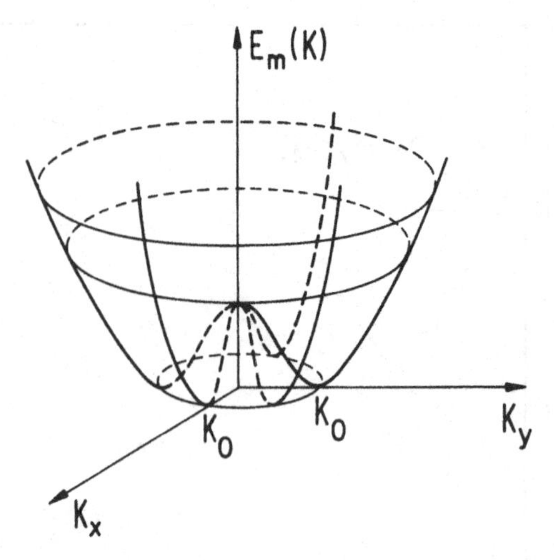

Fig. 5. Schematic representation of the "Mexican hat" structure in 2D.

shift from the other EM renormalizations due to *LO*-phonon, IE–EM and EM–EM interactions.

In CuCl, recent experiments [18]–[20] measured a radiative EM lifetime of 50 – 90 ps under relatively weak resonant excitation. This lifetime is much shorter than any other values previously reported and has been attributed to the giant oscillator strength of the IE–EM optical transition. Moreover, a strong dependence of $\gamma^m(K)$ on K has been observed [19]. Our theory gives a qualitative explanation (see Fig. 4) of these experimental results.

4 Discussion

We expect that in a three-dimensional system such as bulk CuCl a camel-back or, rather, "Mexican hat" structure will also be obtained for the renormalized energy, particularly because the heavy hole mass of this material results in a heavy EM mass of $M_m = 5.2m_0$, which favors the minimum at K_0. The magnitude of the minimum $\Delta^m(K_0)$, however, is expected to be considerably smaller than in one dimension. The appearance of this minimum will eliminate the possibility of a BEC of EMs at $K = 0$. But at the same time a condensation in the minimum at $K_0 \simeq 2p_0(K = 0)$ (see Fig. 5) is not possible, because the relation for the total number of EMs

$$N^m = \frac{V}{2\pi^2\hbar^3}\int_0^\infty \frac{K^2 dK}{\exp((\Delta\Omega_K^m - \mu)/k_B T) - 1} \tag{18}$$

yields for all finite temperatures $\mu(N^m, T) < 0$, if we scale the EM minimum energy at K_0 to be zero. Our results are consistent with the experimental data of Refs. [2, 3]. In these experiments a strong accumulation of the EMs at $K \simeq 2p_0(K = 0)$, rather than at $K = 0$, has been detected.

Acknowledgments. We appreciate valuable discussions with L.V. Keldysh, A. Mysyrowicz, N. Nagasawa, M. Hasuo and D. Snoke. This work has been supported by the Volkswagen Stiftung and by the NATO Collaborative Research Grant No. 930084.

References

[1] N. Nagasawa, T. Mita, and M. Ueta, J. Phys. Soc. Jpn. **41**, 929 (1976).

[2] L.L. Chase, N. Peyghambarian, G. Gryndberg, and A. Mysyrowicz, Phys. Rev. Lett. **42**, 1231 (1979).

[3] N. Peyghambarian, L.L. Chase, and A. Mysyrowicz, Phys. Rev. B **27**, 2325 (1983).

[4] E. Hanamura and M. Inoue, in *Proc. 11th International Conference on the Physics of Semiconductors*, Warsaw, (1972) p. 711.

[5] S.A. Moskalenko, Fiz. Tverd. Tela **4**, 276 (1962)

[6] E. Hanamura and H. Haug, Physics Reports **33**, 209 (1977).

[7] A. Mysyrowicz, J. de Physique (Supplement) **41**, C -281 (1980).

[8] M. Hasuo, N. Nagasawa, and A. Mysyrowicz, Phys. Stat. Sol. (b) **173**, 255 (1992).

[9] M. Hasuo, N. Nagasawa, T. Itoh, and A. Mysyrowicz, Phys. Rev. Lett. **70**, 1303 (1993).

[10] M. Ueta, H. Kanzaki, K. Kobayashi, Y. Toyozawa, and E. Hanamura, *Excitonic Processes in Solids* (Springer-Verlag, Berlin, 1986), Springer Series in Solid-State Sciences **60**, p. 110.

[11] O. Akimoto and E. Hanamura, Solid State Comm. **10**, 253 (1972).

[12] A.I. Bobrysheva, M.F. Miglei, and M.I. Shmiglyuk, Phys. Stat. Sol. (b) **53**, 71 (1972).

[13] W.F. Brinkman, T.M. Rice, and B. Bell, Phys. Rev. B **8**, 1570 (1973).

[14] M.I. Sheboul and W. Ekardt, Phys. Stat. Sol. (b) **73**, 165 (1976).

[15] A.L. Ivanov and H. Haug, Phys. Rev. B **48**, 1490 (1993).
[16] L. Bányai, I. Galbraith, C. Ell, and H. Haug, Phys. Rev. B **36**, 6099 (1987).
[17] S. Flügge, *Practical Quantum Mechanics* (Springer, Berlin 1974), p. 196.
[18] H. Akiyama, T. Kuga, M. Matsuoka, and M. Kuwata-Gonokami, Phys. Rev. B **42**, 5621 (1990).
[19] M. Hasuo, N. Nagasawa, and T. Itoh, Opt. Commun. **85**, 219 (1991).
[20] T. Ikehara and T. Itoh, Sol. Stat. Comm. **79**, 755 (1991).

25

Light-Induced BEC of Excitons and Biexcitons

A. I. Bobrysheva and S. A. Moskalenko

Institute of Applied Physics
Academy of Sciences of Moldova
5 Academy Street, Kishinev
Moldova

Abstract

We review the theory of coherent pairing of excitons and biexcitons in two-dimensional and three-dimensional semiconductor structures.

1 Introduction

The coherent pairing of bosons, formally analogous to Cooper pairing of electrons in superconductors, and the coexistence of particle and pair Bose–Einstein condensation (BEC) was first studied in Refs. [1–4]. BEC of excitons in semiconductors was investigated using the approximation of coherent pairing of the electrons and holes [5]. Later on, the idea of coherent pairing of bosons was applied to excitons [6–8]. In Ref. [6], it was assumed that the boson pairs consist of two electron–hole pairs instead of two excitons, and it was suggested that the new collective state is a BEC even though the two-pair bound state does not exist. It was proved independently [7, 8] that in a system of bosons and excitons with an attractive pair interaction sufficient for biexciton formation, the coherent pairing of excitons with momenta **k** and −**k** coincides with the BEC of biexcitons with zero translational momentum.

The possibility of particle and pair BEC in a system of four species of excitons with different values of spin projections of the electron and hole was studied in Ref. [9]. It is known [10, 11] that the interaction of an exciton of either type with other exciton species is repulsive on the average, but that pairs of excitons with antiparallel spins of electrons and holes interact attractively, and biexcitons can be formed. In two limiting cases, when the exciton–exciton interaction energy W is much greater or smaller than ortho–paraexciton splitting Δ_{o-p}, it was shown that the BEC of biexcitons persists without collapse due to the internal spin structure of excitons. In the case where $\Delta_{o-p} \ll W$ the biexciton is formed by all four species of excitons. When $\Delta_{o-p} \gg W$ in a system of para- and orthoexcitons, BEC of paraexcitons and biexcitons formed by two orthoexcitons with zero total spin can take place simultaneously. Taking the CuCl crystal as an example, it was proved [12] that in the presence of nonresonant polarized laser radiation, which induces the polarized excitonic BEC, coherent pairing of excitons takes place. Two induced BECs coexist: that of polarized excitons with wave vector $\mathbf{k}_0$, and the biexcitonic one with wave vector $2\mathbf{k}_0$. It was shown that the biexciton becomes anisotropic in the sense that four species of excitons Γ_{5x}, Γ_{5y}, Γ_{5z} and Γ_2 do not participate equally in biexciton formation, but have different weights. That is, the two-photon biexciton absorption becomes polarized.

In the present paper, this phenomenon is studied in a wider class of crystals with different dimensionalities $d = 2, 3$, in particular in quasi-two-dimensional semiconductor structures with quantum wells such as $GaAs/Al_xGa_{1-x}As$ and ZnSe/(Zn,Mn)Se and three-dimensional semiconductor crystals such as CuCl. Optical methods which allow the observation of the new electronic state of the crystal are of interest. Therefore biexciton hyper-Raman scattering and luminescence under the above-mentioned conditions are also studied.

2 The Exciton–Exciton Interaction

We first construct the exciton–exciton interaction Hamiltonian H_{ex-ex} for crystals which have a band structure which consists of a spin-degenerate conduction band and twofold-degenerate valence bands. This model suits the above-mentioned quantum wells (QW) which have two valence bands with a maximum at $\mathbf{k} = 0$ (Γ_6^v symmetry for heavy holes, and Γ_7^v for light holes) and the bottom of the conduction band also at $\mathbf{k} = 0$ (symmetry Γ_6^c). The CuCl bulk crystal band structure also consists of a valence band with Γ_7^v symmetry and conduction band with Γ_6^c symmetry at $\mathbf{k} = 0$. H_{ex-ex} can be constructed as follows:

$$
\begin{aligned}
H_{ex-ex} &= (2L^d)^{-1} \sum_{m=1}^{16} \sum_{\mathbf{k}_1,\mathbf{k}_2,\mathbf{k}_3,\mathbf{k}_4} v_m(\mathbf{k}_1, \mathbf{k}_2, \mathbf{k}_3, \mathbf{k}_4) \\
&\times \hat{\Psi}_m^+(J_t, M_t) \hat{\Psi}_m(J_t, M_t) \delta_{Kr}(\mathbf{k}_1 + \mathbf{k}_2, \mathbf{k}_3 + \mathbf{k}_4) \qquad (1)
\end{aligned}
$$

Here $\hat{\Psi}_m^+$ is the two-exciton creation operator characterized by the total momentum of four fermions J_t and its projection M_t; v_m is the Fourier transform of exciton–exciton interaction energy; L^d is the surface area of two-dimensional system or the volume of the three-dimensional one.

Sixteen two-exciton wave functions $\Psi_m(S, S_z)$ for a model with two spin-degenerate bands and $\Delta_{o-p} \ll W$ have been constructed by Čulik [13] in terms of four electron and hole Fermi operators. Here S, S_z mean the total spin and its projection, respectively, of four fermions. They can be easily generalized for our case by a simple interchange of the spin projections S^h of the holes by their momentum projections J_z^h. After that, four species of exciton operators characterized by the spin projection S_z^e of the electron and by J_z^h are introduced and expressed in terms of operators $a^+_{\mathbf{k}\Gamma\xi}$ for excitons belonging to a certain row ξ of the irreducible representation Γ. The final step is to represent the two-exciton operator as a linear combination of the products of two operators $a^+_{\mathbf{k}\Gamma\xi}$. In the effective-mass approximation, v_m obtained for the case of two spin-degenerate bands for three-dimensional [14] and two-dimensional [15] structures remain valid in our twofold-degenerate valence band model. Only one function Ψ_m results in an attractive interaction, implying $v \equiv v_{a,a}$. The remaining functions give rise to repulsive interaction and $v \equiv v_{s,s}$. The subscripts $a, a(s, s)$ (the first symbol refers to electrons and the second to holes) describe the symmetry of Ψ_m with respect to the permutation of creation operators of two identical fermions. For small $\mathbf{k}$ values the $v_{s,s}(\mathbf{k}_1, \mathbf{k}_2, \mathbf{k}_3, \mathbf{k}_4) \simeq -v_{a,a}(\mathbf{k}_1, \mathbf{k}_2, \mathbf{k}_3, \mathbf{k}_4)$ and $v_m(\mathbf{k}, \mathbf{k}, \mathbf{k}, \mathbf{k}) \equiv v_m(0)$. For three-dimensional systems $v_{s,s}(0) = (26/3)\pi Ry^{3D} a_{3D}^3$, but for two-dimensional ones $v_{s,s}(0) = 8\pi Ry^{2D} a_{2D}[1 - (315/2^{12})\pi^2]$.

In the case of the above-mentioned quantum wells, the four types $i = 1, 2, 3, 4$ of HH-excitons are characterized by the rows of the irreducible representation E: $E(x)$, $E(y)$ and onefold representations A_2, A_1 respectively. The types of LH-excitons are $E(x)$, $E(y)$, B_2, B_1, and for bulk CuCl $\Gamma_5(x)$, $\Gamma_5(Y)$, $\Gamma_5(z)$, Γ_2. In terms of the exciton creation operators $a^+_{\mathbf{k}\Gamma\xi}$, the H_{ex-ex} one obtains has an identical form in these three cases. It can be written as follows:

$$\begin{aligned} H_{ex-ex} &= (2L^d)^{-1} \sum_{\mathbf{k}_1, \mathbf{k}_2, \mathbf{q}} \left[\frac{1}{4} \sum_{i=1}^{4} (3v_{s,s} + v_{a,a})(iiii) \right. \\ &+ \sum_{i \neq j=1}^{4} v_{s,s}(ijji) + \frac{1}{4} \sum_{i \neq j=1}^{3} (v_{a,a} - v_{s,s})(iijj) \\ &+ \left. \frac{1}{4} \sum_{j=1}^{3} (v_{s,s} - v_{a,a})((jj44) + (44jj)) \right] . \end{aligned} \tag{2}$$

3 Coherent Pairing of Excitons Induced by Polarized Laser Radiation

The full Hamiltonian is written as

$$H = H_0 + H_{ex-ex} + H_{ex-R}, \tag{3}$$

where

$$H_0 = \sum_{i=1}^{4} E(\mathbf{k}) a^+_{i,\mathbf{k}} a_{i,\mathbf{k}}. \tag{4}$$

H_{ex-ex} has the form given by (2). We assume that laser radiation is polarized along the **x** axis and excites the exciton mode $E(\mathbf{x})$ with the wave vector $\mathbf{k}_0$; H_{ex-R} describes this exciton–electromagnetic field interaction.

The induced BEC of polarized excitons in the state $\mathbf{k}_0$ is introduced in the Hamiltonian by Bogoliubov's translation operation. The biexciton creation operator has the form

$$\hat{\Psi}^+_{2\mathbf{k}_0} = \sum_{i=1}^{4} C_i L^{-d/2} \sum_{\mathbf{q}} \Phi(\mathbf{q}) a^+_{i,\mathbf{k}_0+\mathbf{q}} a^+_{i,\mathbf{k}_0-\mathbf{q}}\,, \tag{5}$$

where the function of relative motion $\Phi(\mathbf{q})$ and the coefficients C_i obey normalization conditions. Therefore, BEC of biexcitons in the state $2\mathbf{k}_0$ is introduced by the operation of coherent pairing of excitons $(\mathbf{k}_0+\mathbf{q}, \mathbf{k}_0-\mathbf{q})$ of the same type i.

The coefficients of the (u_i, v_i) transformations are given by

$$u_i(\mathbf{q}) = \cosh(2C_i\sqrt{n^{biex}_{2\mathbf{k}_0}}\Phi(\mathbf{q})), \tag{6}$$

and

$$v_i(\mathbf{q}) = \sinh(2C_i\sqrt{n^{biex}_{2\mathbf{k}_0}}\Phi(\mathbf{q})), \tag{7}$$

where $n^{biex}_{2\mathbf{k}_0} = N^{biex}_{2\mathbf{k}_0} L^{-d}$ is the condensate biexciton density. We note that without laser radiation, the absolute values of C_i are equal.

The stability conditions of the new ground state of the system lead to the conclusion that $C_2 = C_3 = -C_4$ and allow the possibility of obtaining equations which, along with the normalization conditions, determine both C_1, C_2 as well as the condensate exciton $n^{ex}_{x,\mathbf{k}_0}$ and biexciton $n^{biex}_{2\mathbf{k}_0}$ densities. In the linear v_i approximation, we obtain:

$$C_{1,2} = (|<v>| \mp \Delta)(\sqrt{2}\sqrt{3(|<v>| + \Delta)^2 + (|<v>| - \Delta)^2})^{-1}, \tag{8}$$

where 1, 2 refer to signs minus, plus, respectively; $< v >$ is the mean potential energy of the internal motion; Δ is the biexciton detuning. It follows from Eq. (7) that for $\Delta \neq 0$, the number of excitons of each of the four types forming the biexciton are different. In this sense the biexciton is polarized. At $\Delta = 0$, fast mutual conversions of the excitons remove this inequality.

4 The Biexciton Two-Photon Absorption, Hyper-Raman Scattering and Luminescence

The probability P of a two-photon biexciton transition is obtained in the second order of perturbation theory. The biexciton wave function is given by (5). If two probe photons have the same polarization $\mathbf{e}_t$, then the transition probability depends on the orientation of $\mathbf{e}_t$ relative to the $\mathbf{x}$ axis along which the laser radiation is polarized. The relative value P/P_0 is equal to $8 \mid C_1 \mid^2$ if $\mathbf{e}_t \parallel \mathbf{x}$, and $8 \mid C_2 \mid^2$ if $\mathbf{e}_t \perp \mathbf{x}$. Here P_0 is the transition probability without pumping.

Let us consider the process of absorption of two photons with wave vectors $\mathbf{k}_0$ and energies in the spectral region $E_m(0)/2$, as a result of which a photon with wave vector $\mathbf{q}_R$ and polarization $\mathbf{e}_R$ is emitted and either a longitudinal L or transverse T exciton is created. The probability P^R of this hyper-Raman scattering process is described by third order perturbation theory. It is enough to take into account only the T-exciton and biexciton as intermediate states. It is assumed that the incident radiation propagates along the $\mathbf{z}$ axis and is polarized along $\mathbf{x}$. Then for a (001) quantum well, we obtain $P_L^R \equiv 0$, $P_T^R \sim \mid C_1 C_2 \mid^2 (\mathbf{e}_R \mathbf{y}^0)^2$, $\mathbf{q}_R \parallel \mathbf{x}^0$. The $\mathbf{x}^0$ and $\mathbf{y}^0$ are unit vectors. In the case of CuCl, we find $P_L^R \sim \mid C_1 C_2 \mid^2$, and the intensity of scattered light reaches a maximum in the direction perpendicular to $\mathbf{k}_0$. In turn, $P_T^R \sim \mid C_1 \mid^2 \left(\mid C_1 \mid^2 (\mathbf{e}_R \mathbf{x}^0)^2 + \mid C_2 \mid^2 (\mathbf{e}_R \mathbf{y}^0)^2 \right)$. The intensity of scattered light reaches a maximum in the direction parallel to $\mathbf{k}_0$.

We would expect that the biexciton polarization manifests itself in the M luminescence band in the case of two simultaneously acting pump and probe beams. The pump beam produces the biexciton polarization, but the weak unpolarized probe beam with frequency $\omega_m/2$ populates the biexciton states. In CuCl, one has $P_L^M \sim \mid C_2 \mid^2$, $P_T^M \sim \mid C_1 \mid^2 (\mathbf{e}_l \mathbf{x}^0)^2 + \mid C_2 \mid^2 (\mathbf{e}_l \mathbf{y}^0)^2$. In a (001) quantum well, $P_L^M = 0$, $P_T^M \sim \mid C_2 \mid^2 (\mathbf{e}_l \mathbf{y}^0)^2$. Here $\mathbf{e}_l$ is the luminescence polarization vector.

References

[1] J.G. Valatin and D. Batler, Nuovo Cimento **10**, 37 (1958).

[2] A. Coniglio and M. Marinaro, Nuovo Cimento **48B**, 249 (1967).

[3] A. Coniglio and F. Manciniand M. Maturi, Nuovo Cimento **63B**, 227 (1969).

[4] W.A.B. Evans and Y. Imry, Nuovo Cimento **63B**, 155 (1969).

[5] L.V. Keldysh and A.N. Kozlov, Zh. Eksp. Teor. Fiz. **54**, 978 (1968).

[6] C.A. Mavroyannis, Phys. Rev. **10**, 1741 (1974).

[7] P. Nozières and D. Saint James, J. Physique **43**, 1133 (1982).

[8] A.I. Bobrysheva and S.A. Moskalenko, Fiz. Tverd. Tela **25**, 3282 (1983).

[9] A.I. Bobrysheva, S.A. Moskalenko, and Yu.M. Shvera, Phys. Stat. Sol. (b) **147**, 711 (1988).

[10] A.I. Bobrysheva, S.A. Moskalenko, and V.I. Vybornov, Phys. Stat. Sol. (b) **76**, K51 (1976).

[11] A.I. Bobrysheva and S.A. Moskalenko, Phys. Stat. Sol. (b) **119**, 141 (1983).

[12] A.I. Bobrysheva, S.A. Moskalenko, and Hoang Ngok Cam, Zh. Eksp. Teor. Fiz. **103**, 301 (1993).

[13] F. Čulik, Czech. J. Phys. B **16**, 194 (1966).

[14] A.I. Bobrysheva, M.F. Miglei, and M.I. Shmigliuk, Phys. Stat. Sol. (b) **53**, 71 (1972).

[15] A.I. Bobrysheva, V.T. Zyukov, and S.I. Beryl, Phys. Stat. Sol. (b) **101**, (1980).

26

Evolution of a Nonequilibrium Polariton Condensate

I. V. Beloussov and Yu. M. Shvera
Institute of Applied Physics
Academy of Sciences of Moldova
5 Academy Street, Kishinev
Moldova

Abstract

We review the theory of the polariton condensate taking into account exciton–exciton interactions.

It is known [1] that coherent electromagnetic radiation resonant with an isolated exciton energy level excites in the crystal a coherent polariton wave with the wave vector $\mathbf{k}_0 \neq 0$ — the non-equilibrium polariton condensate. Different scattering processes accompanying its propagation lead to the loss of the initial coherence of the polariton wave, complete or partial depletion of the condensate, excitation of polaritons with wave vector $\mathbf{k} \neq 0$, and other phenomena.

In the present paper the effect of exciton–exciton scattering processes on the properties of a coherently excited polariton system are discussed. This scattering mechanism is of considerable interest due to recent experimental investigations [2] and many interesting results (see e.g. Refs. [3, 4]) obtained in theoretical study of dynamic and kinetic processes in a system of interacting polaritons.

According to Refs. [4, 5], exciton–exciton scattering is very important when coherent polaritons are excited in a certain spectral region in which energy and momentum conservation laws allow real processes of two-quantum excitation of polaritons from the condensate. These processes lead to the instability of the condensed state of the polariton system. The existence of this spectral region situated around the isolated exciton resonance is due to the peculiarities of the polariton dispersion relation.

In [5] the energy spectrum of non-condensate polaritons, arising as the result of decay of the coherent polariton wave, is studied. According to [5], in some regions of **k**-space the polariton modes will be damped.

The investigation reported in [5] was based on a model formally analogous to that used by N.N. Bogoliubov in Ref. [6] to study the equilibrium system of a weakly non-ideal Bose gas. In the non-equilibrium situation considered in [5], in which decay of the polariton condensate and excitation of non-condensate polaritons take place, this model is adequate for the experimental situation only at the initial stage of the condensate decay, when the number of condensate polaritons is still much greater than the total number of non-condensate polaritons. The system is essentially non-stationary, while a real energy spectrum implies a steady state of the system (see e.g. Ref. [7], p.46). Thus the results of Refs. [4, 5] for the polariton energy spectrum based on the above-mentioned model are only approximate.

Because of the essential non-stationarity of the processes in the system, the methods of non-equilibrium statistical mechanics must be used to describe it adequately. The derivation of equations that describe the kinetics of the polariton condensate decay and the excitation of quantum fluctuations have some specific features due to the degeneracy in the system. As the energy and wavevector of two non-condensate polaritons can be equal to the energy and wavevector of two condensate polaritons, respectively, there is degeneracy of two-particle states. Moreover, the presence of the condensate in the system also leads to degeneracy due to its macroscopic amplitude [8].

The correct description of the system with degeneracy requires the introduction of abnormal (or off-diagonal) distribution functions [9] $\Psi_{\mathbf{k}_0} = \langle \Phi_{\mathbf{k}_0} \rangle$ and $F_{\mathbf{k}} = \langle \Phi_{\mathbf{k}} \Phi_{2\mathbf{k}_0 - \mathbf{k}} \rangle$, together with the usual (normal) ones $N_{\mathbf{k}} = \langle \Phi^+_{\mathbf{k}} \Phi_{\mathbf{k}} \rangle$. Here $\Phi^+_{\mathbf{k}}(\Phi_{\mathbf{k}})$ are Bose operators of creation (annihilation) of a polariton on the lower branch with wave vector **k**. The appearance of abnormal averages and the coherent part of the polariton field indicates a breaking of the selection rules connected with the gauge invariance of the system [9–11]. It can take place as a result of the action of external classical sources, spontaneously, or because of non-invariant initial conditions due to the action of external sources.

An earlier attempt to obtain kinetic equations for polaritons excited in semiconductors by an external classical field has been given in Refs. [12, 13]. However, in this work the degeneracy of two-particle states was not taken into account, and the abnormal distribution functions $F_{\mathbf{k}}$ were not introduced. As a result, one expects that the equations obtained in Refs. [12, 13] will lead to unphysical singularities.

Kinetic equations describing the evolution of partially coherent polaritons which take into account the degeneracy were obtained in [14, 15] using the Keldysh method [16], presented in terms of functionals. They coincide with the equations obtained in [17] using the non-equilibrium statistical operator method [18] and do not possess unphysical singularities. One can find extended versions of Refs. [14, 15, 17] in Refs. [19–21].

According to Refs. [15, 17], the kinetics of partially coherent polaritons are described in the Born approximation by a closed set of nonlinear integro-differential equations for the coherent part of polariton field $\Psi_{\mathbf{k}_0}$ and the normal $n_{\mathbf{k}} = N_{\mathbf{k}} - \delta_{\mathbf{k},\mathbf{k}_0}|\Psi_{\mathbf{k}_0}|^2$ and abnormal $f_{\mathbf{k}} = F_{\mathbf{k}} - \Psi^2_{\mathbf{k}_0}$ distribution functions. In the absence of quantum fluctuations described by the functions $n_{\mathbf{k}}$ and $f_{\mathbf{k}}$, the equations become identities, and the equation for $\Psi_{\mathbf{k}\alpha}$ ($\alpha = 1, 2$ is the label of the polariton branch) coincides with that obtained in Ref. [22] for a interacting system of coherent excitons and photons. In the case when $\Psi_{\mathbf{k}_0} = 0$ and $f_{\mathbf{k}} = 0$, the equations obtained in [15, 17] reduce to the usual kinetic equation for the distribution function $N_{\mathbf{k}}$ (see e.g. Ref. [23]).

The right-hand sides of the equations obtained in [15, 17] include terms linear in the exciton–exciton interaction strength v, and terms proportional to v^2. Terms proportional to v correspond to the self-consistent field approximation (SCFA), which neglects the higher correlation functions. This approximation is sufficient to describe the initial stage of evolution. It is shown [15] that depletion of the coherent part $\Psi_{\mathbf{k}_0}$ and excitation of quantum fluctuations $n_{\mathbf{k}_0}$ and $f_{\mathbf{k}_0}$ in the condensate mode occurs at this stage. Thus it partially loses its coherence. There exist two regions of wavevector $\mathbf{k} \neq \mathbf{k}_0$. In the first one, describing the instability region, the number of polaritons increases monotonically; in the second one, it is a periodic function of time with an amplitude decreasing with the distance from the instability region.

As usual, the terms proportional to v^2 describe the difference between the number of processes (per unit of time) of polariton creation and annihilation in a state with wavevector $\mathbf{k}$. They only arise if non-condensate polaritons exist in the system. They describe the evolution which is slower than that described by terms proportional to v, which correspond to resonant scattering with participation of two condensate polaritons.

It should be mentioned that the SCFA takes into account the influence of the excited non-condensate polaritons on the condensate. On the other hand, it describes the fastest processes in the system. So one should expect the establishment of steady state in the isolated polariton system after the time interval $\tau \sim \hbar/vn$ (n is the condensate initial density).

Terms in the kinetic equations proportional to v^2 are responsible for its further evolution.

Using considerations similar to those used in [24], we have found the exact steady-state solution of the evolution equations given by the SCFA. Formally, it has the following nature. In the Heisenberg representation, the SCFA corresponds to the Hamiltonian [15] which has operator terms $\Phi_{\mathbf{k}}^{+}\Phi_{2\mathbf{k}_0-\mathbf{k}}^{+}$ with time-dependent coefficients $\xi = \Psi_{\mathbf{k}_0}^2 + \Delta$ (here $\Delta = \sum_{\mathbf{k}} f_{\mathbf{k}}$). At the initial stage of evolution, when $|\Psi_{\mathbf{k}_0}|^2 \gg |\Delta|$, these terms are responsible for the appearance and development of the polariton condensate instability [4, 5]. In the steady state, the two terms in ξ cancel each other, so that $\xi = 0$. Thus these terms disappear from the Hamiltonian.

We note that in [4, 5], the influence of non-condensate polaritons on the condensate was not taken into account, and the term Δ in the expression for ξ did not appear. As a result, in [4, 5] the possibility of obtaining a steady-state solution, which is the result of mutual compensation of $\Psi_{\mathbf{k}_0}^2$ and Δ, did not exist.

From the physical point of view, the results obtained here mean that while the condensate is being depleted, and polaritons are being excited in the instability region (modified by the concentration-dependent corrections), the backward processes (in which scattering of two non-condensate polaritons creates two polaritons in the condensate) become more important. It results in slowing down of the parametric instability, and finally in the establishment of a steady state. This state corresponds to a dynamic equilibrium between the condensate and non-condensate polaritons.

The steady state is characterized by the renormalized frequency of the polariton condensate $\hbar\omega_{\mathbf{k}_0} = \hbar\Omega_{\mathbf{k}_0} + \sqrt{2}vn_0$ and renormalized energies of non-condensate polaritons $E_{\mathbf{k}} = \hbar\Omega_{\mathbf{k}} + 2vn_0$ (here $\Omega_{\mathbf{k}_0}$ is the frequency of non-interacting polaritons). This result differs crucially from that obtained in [4, 5]. Distribution functions of non-condensate polaritons are localized in the regions of **k**-space, as determined by the resonant condition $E_{\mathbf{k}} + E_{2\mathbf{k}_0-\mathbf{k}} - 2\hbar\omega_{\mathbf{k}_0} = 0$.

Future investigations of this problem require numerical analysis of the set of integro-differential equations corresponding to the SCFA. Another important problem, in our opinion, is the study of polariton kinetics under the action of an external field.

We wish to thank Dr. A. Ivanov and Dr. S. Tikhodeev for helpful discussions.

References

[1] A.L. Ivanov and L.V. Keldysh, Zh. Eksp. Teor. Fiz. **84**, 404 (1983).

[2] M.V. Lebedev, Zh. Eksp. Teor. Fiz. **101**, 957 (1992).

[3] S.A. Moskalenko, *Introduction to the Theory of High Density Excitons* (Ştiinţa, Kishinev, 1983).

[4] M.I. Shmigliuk and V.N. Pitei, *Coherent Polaritons in Semiconductors* (Ştiinţa, Kishinev, 1989).

[5] V.N. Pitei, M.I. Shmigliuk, V.T. Zyukov, and M.F. Miglei, in *Bose Condensation of Polaritons in Semiconductors*, pp. 58–79 (Ştiinţa, Kishinev, 1985).

[6] N.N. Bogoliubov, J. Phys. USSR **11**, 23 (1947); see also N.N. Bogoliubov, *Selected Works*. Vol.2, pp. 242–257 (Naukova Dumka, Kiev, 1971).

[7] L.D. Landau and E.M. Lifshitz, *Quantum Mechanics* (Pergamon, Oxford, 1963).

[8] J.R. Schrieffer, *Theory of Superconductivity* (Nauka, Moscow, 1970).

[9] N.N. Bogoliubov, *Quasi-Averages in the Problems of Statistical Mechanics*. Dubna: Preprint of JINR R-1415, 1963; see also N.N. Bogoliubov, *Lectures on Quantum Statistics* (Gordon and Breach, NY, 1970), Vol. 2.

[10] N.N. Bogoliubov (Jr) and B.I. Sadovnikov, *Some Questions of Statistical Mechanics* (Vysshaya Shkola, Moscow, 1975).

[11] V.N. Popov and V.S. Yarunin, *Collective Effects in Quantum Statistics of Radiation and Matter* (Izd. LGU, Leningrad, 1985).

[12] S.A. Moskalenko, A.H. Rotaru and Yu.M. Shvera, in *Laser Optics of Condensed Matter*, J.L. Birman, H.Z. Cummins and A.A. Kaplyanskii, eds. (Plenum, NY, 1988), pp. 331–336.

[13] V.R. Misko, S.A. Moskalenko, A.H. Rotaru and Yu.M. Shvera, Phys. Stat. Sol. (b) **159**, 477 (1990).

[14] I.V. Beloussov and Yu.M. Shvera, Phys. Stat. Sol. (b) **139**, 91 (1990).

[15] I.V. Beloussov and Yu.M. Shvera, Z. Phys. B **90**, 51 (1993).

[16] L.V. Keldysh, Zh. Eksp. Teor. Fiz. **47**, 1515 (1964).

[17] I.V. Beloussov and Yu.M. Shvera, Teor. Mat. Fiz. **85**, 237 (1990).

[18] D.N. Zubarev, *Non-Equilibrium Statistical Thermodynamics* (Plenum, NY, 1974).

[19] I.V. Beloussov and Yu.M. Shvera, in *Excitons and Biexcitons in Confined Systems* (Ştiinţa, Kishinev, 1990), pp.133–176.
[20] I.V. Beloussov and Yu.M. Shvera, in *Interaction of Excitons with Laser Radiation* (Ştiinţa, Kishinev, 1991), pp.57–74.
[21] I.V. Beloussov and Yu.M. Shvera, in *Nonlinear Optical Properties of Excitons in Semiconductors of Different Dimensionalities* (Ştiinţa, Kishinev, 1991), pp.53–64.
[22] L.V. Keldysh, in *Problems of Theoretical Physics* (Nauka, Moscow, 1972), pp.433–444.
[23] A.I. Akhiezer and S.V. Peletminskii, *Methods of Statistical Physics* (Nauka, Moscow, 1977).
[24] V.L. Safonov, Physica A **188**, 675 (1992).

27

Excitonic Superfluidity in Cu_2O

E. Fortin and E. Benson

Department of Physics
University of Ottawa
Ottawa, ON K1N 6N5
Canada

A. Mysyrowicz

LOA, Ecole Polytechnique
Palaiseau
France

Abstract

Using the exciton-mediated photovoltaic effect, we examine exciton transport over large distances in Cu_2O as a function of temperature and particle density. Evidence for a phase transition at low temperatures and high densities is attributed to the onset of excitonic superfluidity.

We have performed exciton transport measurements over a range of temperatures and exciton densities in ultrapure, oriented large Cu_2O single crystals. A sketch of the experimental method is shown in Fig. 1. The crystal is illuminated on the back surface by 10 ns pulses from a frequency-doubled YAG laser ($\lambda = 532$ nm). The initial exciton density created over an absorption depth (about one micron at $\lambda = 532$ nm) can be varied by inserting calibrated neutral density filters in the laser beam, reaching values of up to $10^{19}\,\mathrm{cm}^{-3}$. The excitons which have migrated to the opposite face of the crystal are dissociated into free carriers by the high electric field near the Cu Schottky contact [1] deposited in a comb configuration together with an ohmic Au electrode, resulting in an external current. A time-resolved measurement of that current will give the velocity distribution of the excitons migrating through the crystal. This method of detection – as opposed to photoluminescence – is particularly well suited to the study of optically inactive paraexcitons in Cu_2O; moreover, since the migration time is of the order of one microsecond as compared to the lifetime of 13 μs [2] for paraexcitons, recombination processes have little influence on the measurements.

Figure 2 shows the time-resolved photocurrent at a temperature of 1.85K in a 3.56-mm-thick crystal for several illumination intensities, that is, initial exciton densities. The initial signal at $t = 0$ results from the

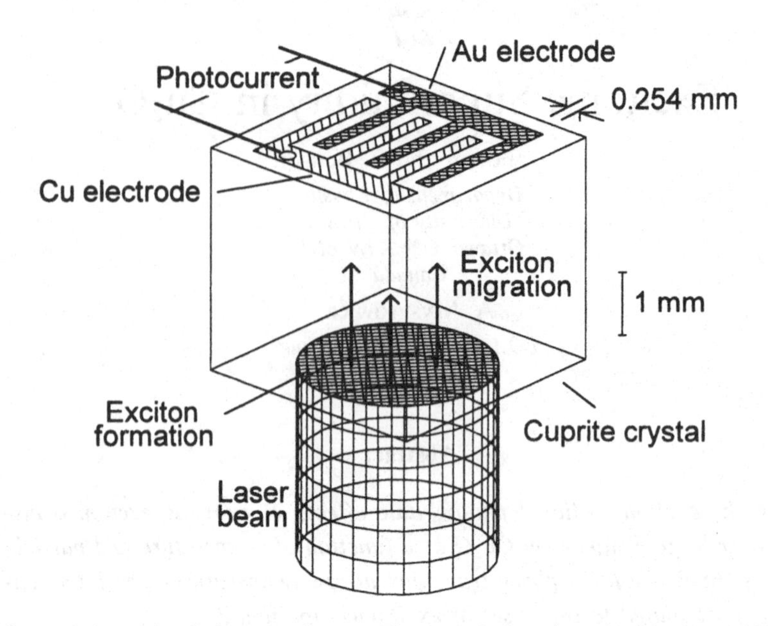

Fig. 1. Schematic of the experiment. The sample is oriented with the < 100 > direction along the edges.

direct illumination of the electrodes, and serves as a convenient measure of the exciton detector response time as well as an independent optical trigger for the 2 GHz digital oscilloscope. The delayed signal, about one μs later, results from the dissociation of the excitons having travelled across the crystal. As the illumination intensity (exciton density) is increased (bottom curve to top curve), a remarkable change occurs in the exciton transport: from a diffusive behaviour characteristic of a free gas expanding through a viscous medium, the propagation evolves, at high exciton densities, to a ballistic behaviour where the majority of particles propagate in a narrow range of speeds. The speed is found to approach asymptotically the longitudinal sound velocity in Cu_2O [3] ($v_\ell = 4.5 \times 10^5$ cm/s) [4]. Measurements on a second crystal confirm the dragfree motion over distances of nearly 1 cm. If Bose condensation is responsible for the evolution shown, the crystal temperature should strongly influence the exciton transport as well. Fig. 3 shows data for the thinner crystal at constant illumination intensity of 1.2×10^6 W/cm^2 at several temperatures. Here again the transport evolves from diffusive to ballistic as the temperature is lowered. Note the narrow range of temperatures over which this change occurs.

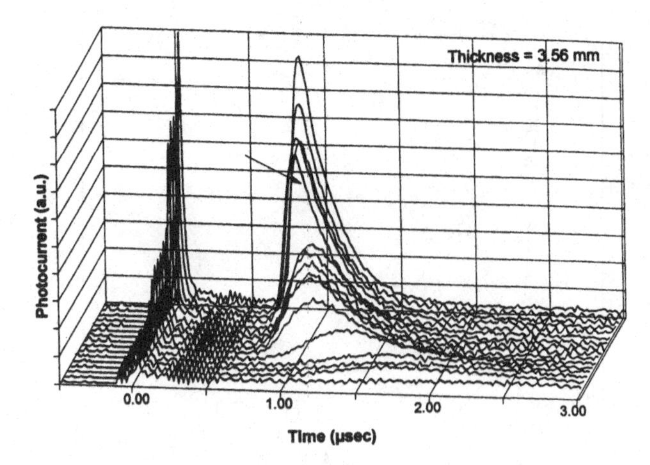

Fig. 2. Time-resolved photocurrent of sample 1 measured across a 50 Ω resistance at different illumination intensities and a constant temperature of 1.85 K. The intensity increases from bottom to top curve from 3.75×10^4 to 1.5×10^6 W/cm^2. The curve shown by the arrow is for an intensity of 6×10^5 W/cm^2.

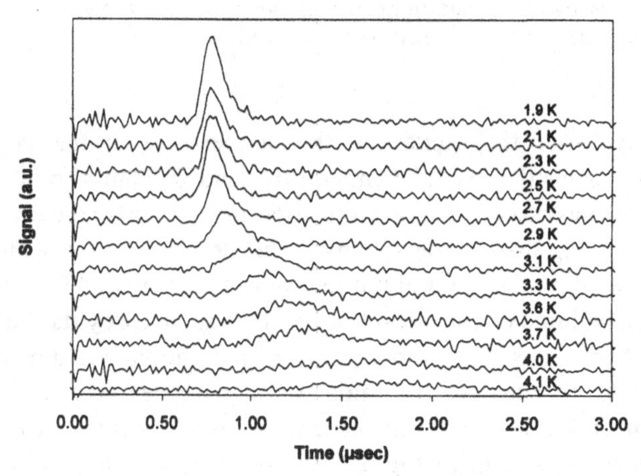

Fig. 3. Time-resolved photocurrent for sample 1 at different temperatures and constant illumination intensity of 1.2×10^6 W/cm^2.

We believe that these results are strong evidence for the onset of a superfluid excitonic phase above a critical density and below a critical temperature. Let us first compare the density of excitons inside the initial packet to the critical density for Bose–Einstein condensation as predicted by the ideal Bose model. Since the measurements are taken

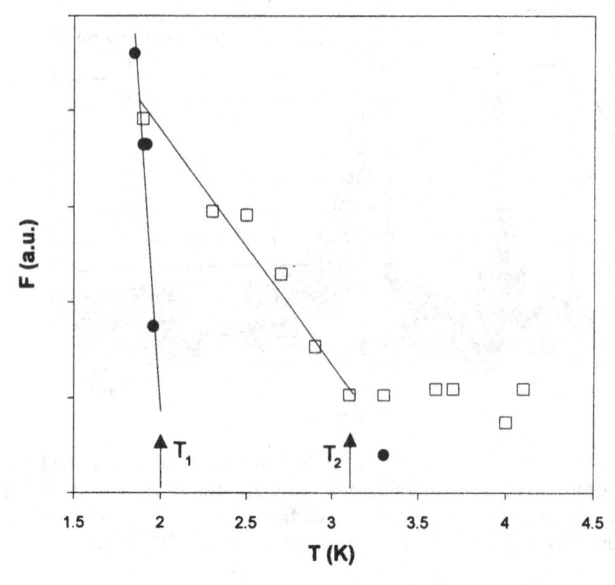

Fig. 4. Integrated photocurrent of sample 1 for a fixed time window as a function of temperature for two illumination intensities: squares, 1.2×10^6 W/cm^2; circles, 6×10^5 W/cm^2. Arrows show transition temperatures.

at long delays after the excitation, the temperature of the excitonic fluid is imposed by the crystal temperature, and the initial mixture of ortho- and para-excitons has now converted into an energetically lower population of long-lived paraexcitons. The density can be estimated from the knowledge of the initial number of excitons created by the laser pulse, and the volume of the packet at $t = 1$ μs as given by its thickness $d \simeq \Delta t v_\ell$, where Δt is the temporal width and the lateral dimensions are assumed to be equal to the initial ones (diameter = 2 mm). For an excitation of 600 kW/cm^2 (corresponding to the onset of a ballistic excitonic packet as shown in Fig. 2, fifth graph from the top), a value of $n(t = 1\mu s) \simeq 1.4 \times 10^{17}$ cm^{-3} is found, where a factor of 2 has been included to take into account reflection losses at cryostat windows and sample surface. This value is in good agreement with the calculated critical density of 6×10^{16} cm^{-3} for paraexcitons of mass $m = 2.7m_0$ at $T = 1.85$K [5].

The critical density for condensation is predicted to vary with temperature as $n_c = CT^{3/2}$. To test our results against this relationship, we plot in Fig. 4 the delayed signals of Fig. 3 integrated over a *fixed* time window (defined as the full width at half maximum at the lowest temperature of 1.9 K), as a function of temperature. A sharp transition is observed at $T = 3.1$ K, which we can interpret as the critical temperature for condensation at this particular exciton density. When the experiment is repeated at an excitation intensity that is lower by a factor of 2, a similar graph displays a transition temperature of 2K. Over a density change of a factor of 2 the theory predicts a critical temperature variation of:

$$\frac{T_2}{T_1} = 2^{2/3} = 1.58$$

This is in good agreement with the experimentally observed value of 1.55.

In conclusion, detection of exciton transport with good time and space resolution reveals the onset of an anomalous ballistic transport below a critical temperature and above a critical particle density. The conditions required for the appearance of ballistic transport are in very good agreement with those predicted for Bose–Einstein condensation in an ideal Bose gas. We interpret these results as evidence for a superfluid excitonic phase associated with Bose–Einstein condensation. For more details, see Ref. [6].

References

[1] E. Tsélépis, E. Fortin and A. Mysyrowicz, Phys. Rev. Lett. **59**, 2107 (1987).

[2] A. Mysyrowicz, D. Hulin and A. Antonetti, Phys. Rev. Lett. **43**, 1123 (1979); D. W. Snoke, A. J. Shields and M. Cardona, Phys. Rev. B **45**, 11693 (1992).

[3] A. Mysyrowicz and E. Fortin, to be published.

[4] J. Berger, J. Castaing and M. Fisher, J. de Physique **40**, 13 (1979).

[5] N. Caswell, J. S. Weiner and P. Y. Yu, Solid State Comm. **40**, 843 (1981).

[6] E. Fortin, S. Fafard and A. Mysyrowicz, Phys. Rev. Lett. **70**, 3951 (1993).

28

On the Bose–Einstein Condensation of Excitons: Finite-lifetime Composite Bosons

Sergei G. Tikhodeev
Institute of General Physics
Moscow 117333
Russia

Abstract

A brief review of experimental and theoretical work on Bose-Einstein condensation (BEC) in nonequilibrium systems is presented, with emphasis on excitons in a semiconductor. Conditions for obtaining BEC in such systems are discussed, with special attention being paid to the effects of finite lifetime, the role of spatial inhomogeneity and phonon-driven transport of excitons.

1 Introduction

After many years of study, Bose–Einstein condensation (BEC) in a dilute Bose gas has not yet been unambiguously established experimentally in any real physical system. A systematic search has been undertaken in atomic systems (atomic spin-polarized in magnetic field hydrogen [1] and more recently in cesium [2]) and in excitonic systems in solids (see e.g. [3, 4]).

Observation of BEC in an excitonic system is made easier [5, 6] because of the following reasons: (i) small particle masses (of the order of the free electron mass m_0); (ii) the possibility of reaching sufficiently high gas densities with an increase of excitation intensity; and (iii) the presence of luminescence, making it possible to determine the particle distribution function. However, excitons are finite-lifetime particles, and a number of problems arise due to this fact. The goal of the present paper is to analyze the effect of the finite lifetime on the possibility of BEC of excitons.

In the case of incoherent excitation [7], a Bose–Einstein (non-Maxwellian) distribution has been experimentally demonstrated in a sys-

Table 1. *Densities, temperatures, particle masses, degeneracy factors and Silvera–Reynolds numbers of dilute Bose gases*

	n (cm^{-3})	T (K)	m/m_0	g	δ
Atomic H[a]	$2 \cdot 10^{19}$	0.65	1836	1	0.23
Atomic H[b]	$7 \cdot 10^{13}$	10^{-4}		1	0.33
Excitons in stressed Ge[c]	$2.8 \cdot 10^{15}$	3.1	0.5	1	0.44
Orthoexcitons in Cu_2O[d,e]	$8 \cdot 10^{18}$	20	2.7	3	0.85
Paraexcitons in Cu_2O[e]	$5 \cdot 10^{18}$*	20		1	1.4

[a]E. Tjukanov *et al.*, *Physica* B **178**, 129 (1992).
[b]J. M. Doyle *et al.*, *Phys. Rev. Lett.* **67**, 603 (1991).
[c] Ref. [8].
[d] Ref. [9].
[e] Ref. [11].
* Corrected number, see Ref. [12].

tem of magnetically spin-polarized excitons in uniaxially stressed Ge [8] and of ortho- and paraexcitons in Cu_2O [9]–[12]. The best results to date are listed in Table 1, where, in addition to the data on dilute Bose-gas densities and temperatures achieved, the values of Silvera–Reynolds number [1] δ are shown,

$$\delta = \left(\frac{n}{n_c(T)}\right)^{\frac{1}{\sqrt{13}}} \left(\frac{T_c(n)}{T}\right)^{\frac{3}{2\sqrt{13}}}, \tag{1}$$

where

$$T_c(n) = \frac{3.31\hbar^2 n^{2/3}}{g^{2/3} m k_B} \tag{2}$$

is the BEC temperature of an ideal Bose gas with density n, and $n_c(T)$ is the critical density at temperature T. For comparison, the corresponding numbers are shown for atomic hydrogen and excitonic systems. The logarithm of δ is the dimensionless distance from a point (n, T) on a phase diagram to the BEC phase boundary given by (2); if $\delta > 1$, the conditions of BEC are met. (The mysterious number $13 = 2^2 + 3^2$ is the sum of the squared numerator and denominator of the exponent 2/3 in the BEC criterion (2) .) One can see from Table 1 that, up to now, the search for BEC in excitonic systems has been in the lead, and, in the case of paraexcitons in Cu_2O, the conditions for BEC in an ideal Bose gas have been met [13].

2 Experiments under Stationary Excitation

Stationary excitation was used in earlier experiments in Cu_2O [9] and in Ge [8]. The flow of thermalizing excitons should compensate the decay of condensate particles. The latter introduces some deviations of BEC conditions from the ideal Bose gas given by (2) [14, 15]. The physical reason for these modifications of the phase diagram is that during its lifetime, a newly generated exciton should have enough time to diminish its energy via exciton–exciton and exciton–phonon collisions and to reach the $\mathbf{k} = 0$ state.

To illustrate the modifications of BEC conditions, a standard kinetic equation for excitons interacting with a phonon thermostat at $T = 0$ was solved in Refs. [14, 15]. In the statistical limit $V \to \infty$, $N/V \to n = \text{const}$, this equation takes the form

$$\frac{\partial n_0}{\partial t} = n_0 \left[\int d\mathbf{q} W(\mathbf{q}, 0) f_{\mathbf{q}} - \tau^{-1} \right], \tag{3}$$

$$\begin{aligned} \frac{\partial f_{\mathbf{p}}}{\partial t} &= g_{\mathbf{p}} + (1 + f_{\mathbf{p}}) \int_{p<q} d\mathbf{q} W(\mathbf{q}, \mathbf{p}) f_{\mathbf{q}} \\ &\quad - f_{\mathbf{p}} \left[\tau^{-1} + n_0 W(\mathbf{p}, 0) + \int_{0<q<p} d\mathbf{q} W(\mathbf{q}, \mathbf{p})(1 + f_{\mathbf{q}}) \right], \end{aligned} \tag{4}$$

where τ is the lifetime and $g_{\mathbf{p}}$ is the generation rate of excitons, assumed to be a smooth function of momentum (i.e. without resonant excitation of the condensate at $\mathbf{p} = 0$). Such equations for biological systems were analyzed previously in Ref. [16, 17]. These equations can be solved analytically for some specific forms of $W(\mathbf{q}, \mathbf{p})$. It can be shown [14, 15] that even at $T = 0$, the condensate does not appear until the gas density exceeds some threshold value $n_{\text{th}} \sim V_0^{-1} \tau_{\text{phon}}/\tau$. Here $V_0^{-1} \sim (mS/\hbar)^3$ is a characteristic volume of momentum space, where the thermalization of excitons scattering with phonons slows down due to the energy-momentum conservation law, S is the velocity of sound in the semiconductor, and τ_{phon} is a large subsonic exciton–phonon scattering time.

Close to the BEC, when the occupation numbers become large, one should obviously take into account exciton–exciton collisions, which cause an increase of the excitonic flow into the $\mathbf{k} = 0$ state. However, it can be easily shown that even taking these collisions into account, we obtain a threshold density for condensate nucleation

$$n_{\rm th} = \frac{1}{4\pi^{3/4}} \frac{m}{\hbar^2 a} \sqrt{\frac{\hbar\bar{\varepsilon}}{\tau}}, \tag{5}$$

where a is the exciton scattering length and $\bar{\varepsilon}$ is the mean exciton energy. Assuming $a \sim a_{\rm ex}$ and $\bar{\varepsilon} \sim k_B T$, one can obtain $n_{\rm th} \sim 10^{13}\,{\rm cm}^{-3}$ for Ge and $10^{15}\,{\rm cm}^{-3}$ for Cu_2O. One can see that, fortunately, $n_{\rm th} \ll n$ (see Table 1) in both cases, so this mechanism cannot impede BEC in experiments [8, 9]. However, it can become significant in other excitonic systems with shorter exciton lifetimes.

3 Experiments under Pulsed Excitation. Nucleation Time

In order to achieve BEC in the case of pulsed excitation, the condensate nucleation time should be less than the lifetime of the particles. Recently the problem of the condensate nucleation time was addressed in a number of papers [18]–[20]. The situation had been dramatized by the fact pointed out for the first time in Ref. [21]: within the kinetic approach, the kinetic equation for condensate density takes the form $\frac{dn_0}{dt} \propto n_0$. (This can be seen, e.g., from (3).) Thus, if the condensate density at the initial moment is zero, the condensate nucleation time is infinite.

It was shown in Refs. [19, 20] that actually the condensate nucleation time is finite, of the order of the usual kinetic collision time $\tau_{\rm kin} \sim (nv\sigma)^{-1}$. The physical reason for this is the following [19]. In the region of small energies close to the condensate, the kinetic approach is not valid, and the system has to be described by means of a complex order parameter Ψ, obeying the Ginzburg–Landau equation:

$$i\frac{\partial\Psi}{\partial t} = -\frac{1}{2m}\Delta\Psi + \chi|\Psi|^2\Psi. \tag{6}$$

The characteristic evolution time of Ψ is the correlation time $\tau_{\rm cor} \sim \hbar/n_0\chi$. On the time scale of $\tau_{\rm cor}$, a quasicoherent state arises with suppressed density fluctuations. However, this stage of evolution is preceded by a kinetic one, which takes a time $\tau_{\rm kin} > \tau_{\rm cor}$. Thus, $\tau_{\rm kin}$ becomes a characteristic quasicondensate nucleation time.

Although the analysis of Refs. [19, 20] was devoted to atomic systems, it is also valid for excitonic systems. A consistent approach to deriving (6) for excitons was developed in Ref. [22]. One can estimate $\chi \sim 4\pi\hbar^2 a_{\rm ex}/m$. Then for excitons in Cu_2O, we obtain $\tau_{\rm cor} \sim 10^{-12}\,{\rm s}$, $\tau_{\rm kin} \sim 10^{-11}\,{\rm s}$ (at $n \sim 10^{19}\,{\rm cm}^{-3}$ and $T \sim 20\,{\rm K}$). One can see that $\tau_{\rm kin} \ll \tau$, and the excitonic system in Cu_2O has enough time to develop a quasicondensate.

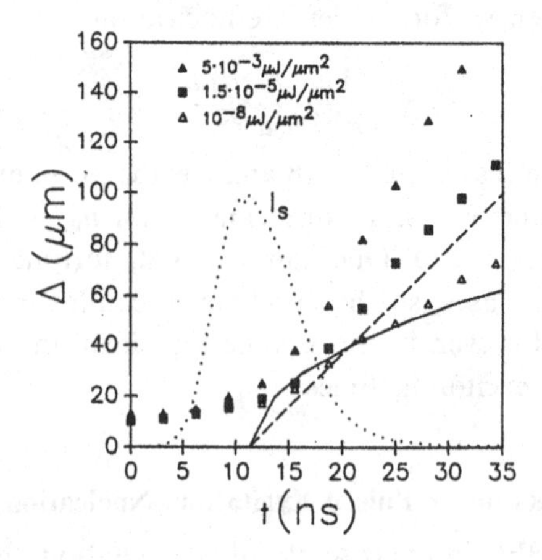

Fig. 1. Supersonic transport of orthoexcitons in Cu_2O at high power. The distance Δ is measured from the surface to the innermost half maximum of the spatial distribution, for the orthoexcitons as a function of time for several laser-power densities. The dashed line has a slope given by the velocity of sound S (from [11]).

4 Impact of the Spatial Inhomogeneity and Phonon Wind

In order to make the initial exciton density as large as possible, surface excitation was used in [10]–[12] with a photon energy well above the band gap of Cu_2O, $\hbar\omega - E_g \gg k_B T$. As a result, the excitonic system in these experiments was very inhomogeneous, and the processes of exciton thermalization and condensate nucleation were proceeding at the same time as the spreading of the exciton cloud into the sample bulk. Anomalously fast (supersonic) nondiffusive transport of excitons was observed [11] at higher excitation intensities (see Fig. 1) and was attributed to the onset of exciton superfluidity. However, this fast transport can also be explained by the joint action of diffusion and the so-called phonon wind, that is, by nonequilibrium phonon-driven transport of excitons [23].

Phonon wind, produced by ballistically propagating nonequilibrium phonons born during the thermalization of carriers, plays a very significant role in solids at low temperatures (see [24]–[26] for a review). For example, it causes fast carrier transport in experiments with electron–hole liquid in Ge [27]. Phonon wind effects are unavoidable if $\hbar\omega > E_g$. They should be important in experiments with surface excitation in Cu_2O, and a simple model of phonon-driven transport of excitons was developed in Ref. [23].

Figure 2 illustrates the dependence (calculated within this model) of

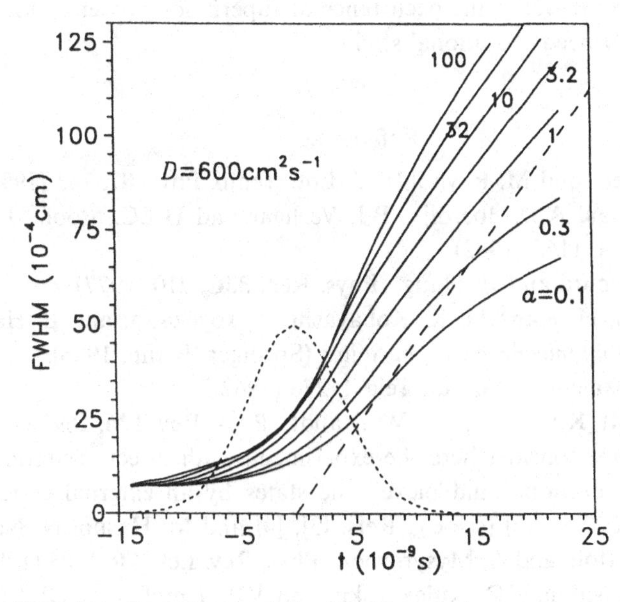

Fig. 2. The time dependence of the FWHM of the carrier concentration for different values of dimensionless excitation energy $a = E/E_s$ (Eq. (7)) in the case of surface excitation (from [23]).

the FWHM of the exciton density on time and excitation intensity, which agrees well with the experimental data. One can see in Fig. 2 that if

$$E \geq E_s = \frac{m\hbar\omega S^2 \tau_i}{\tau_r \sigma_{\text{ph}} \varepsilon_{\text{PW}}}, \tag{7}$$

a region of supersonic drift appears. (Here τ_i is the excitation pulse duration, τ_r is the exciton relaxation time, σ_{ph} is the exciton–phonon cross section and ε_{PW} is the energy (per excitation quantum) which goes into the phonon wind.) Comparing Figs. 1 and 2, one can estimate that $E_s \sim 1.5 \cdot 10^4\,\text{erg/cm}^2$. Then $\sigma_{\text{ph}}\varepsilon_{\text{PW}} \sim 2 \cdot 10^{-15}\,\text{meV cm}^2$, which is in reasonable agreement with a simple estimate of this parameter [23, 28] and also with data on phonon wind in other semiconductors [25]. Thus, the phonon wind model may explain fast supersonic transport of excitons without assuming their superfluidity. However, as was shown recently in [29], this fast transport at higher excitation intensities may also be explained by drag-free hydrodynamic flow, which may indicate that the excitonic superfluidity has indeed been observed in experiments [11]. In view of this controversy, the occurrence of superfluidity (over and above BEC) in Cu_2O needs additional study.

References

[1] I.F. Silvera and M. Reynolds, J. Low Temp. Phys. **87**, 343 (1992).
[2] E. Tiesinga, A.J. Moerdijk, B.J. Verhaar and H.T.C. Stoof, Phys. Rev. A **46**, 1167 (1992).
[3] E. Hanamura and H. Haug, Phys. Rep. **33C**, 210 (1977).
[4] M. Ueta, H. Kanzaki, K. Kobayashi, Y. Toyozawa, and E. Hanamura, *Excitonic Processes in Solids* (Springer, Berlin, 1986).
[5] S.A.Moskalenko, Fiz. Tv. Tela **4**, 276 (1962).
[6] J.M. Blatt, K.W. Boer, and W. Brandt, Phys. Rev. **126**, 1691 (1962).
[7] We do not consider here the experiments with direct excitation of coherent excitonic and biexcitonic states by an external coherent source (EM wave) (see e.g. Refs. [3], [4] and M. Hasuo, N. Nagasawa, T. Itoh, and A. Mysyrowicz, Phys. Rev. Lett. **70**, 1303 (1993)).
[8] I.V. Kukushkin, V.D. Kulakovskii, and V.B. Timofeev, JETP Lett. **34**, 36 (1981).
[9] D. Hulin, A. Mysyrowicz, and G. Benoît à la Guillaume, Phys. Rev. Lett. **45**, 1970 (1980).
[10] D.W. Snoke, J.P. Wolfe, and A. Mysyrowicz, Phys. Rev. Lett. **59**, 827 (1987).
[11] D.W. Snoke, J.P. Wolfe, and A. Mysyrowicz, Phys. Rev. B **41**, 11171 (1990).
[12] D.W. Snoke, Jia Ling Lin, and J.P. Wolfe, Phys. Rev. **43**, 1226 (1991).
[13] Further results for Cu_2O are reported in this volume. J.L. Lin and J.P. Wolfe obtain a Silvera–Reynolds number $\delta = 2.8$ (for paraexcitons in stressed Cu_2O), and E. Fortin, E. Benson, and A. Mysyrowicz report δ as large as 5.
[14] S.G. Tikhodeev, Solid State Commun. **72**, 1075 (1989).
[15] S.G. Tikhodeev, Sov. Phys. – JETP **70**, 380 (1990).
[16] H. Fröhlich, Int. J. Quantum Chem. **2**, 641 (1968).
[17] N.G. Duffield, J. Phys. A: Math. Gen. **21**, 625 (1988).
[18] D. Snoke and J.P. Wolfe, Phys. Rev. B **39**, 4030 (1989).
[19] Yu. Kagan, B.V. Svistunov, and G.V. Shlyapnikov, Zh. Eksp. Teor. Fiz. **101**, 528 (1992).
[20] H.T.C. Stoof, Phys. Rev. A **45**, 8398 (1992).
[21] E. Levich and V. Yakhot, Phys. Rev. B **15**, 243 (1977).

[22] L.V. Keldysh, in *Problems of Theoretical Physics, A Memorial Volume to Igor E. Tamm* (Nauka, Moscow, 1972), p. 433.

[23] A.E. Bulatov and S.G. Tikhodeev, Phys. Rev. B **46**, 15053 (1992).

[24] J.P. Wolfe, J. Lumin. **30**, 82 (1985).

[25] S.G. Tikhodeev, Sov. Phys. Usp. **28**, 1 (1985).

[26] L.V. Keldysh and N.N. Sibeldin, in *Nonequilibrium Phonons in Nonmetallic Crystals*, Eisenmenger and Kaplyanskii, eds. (Elsevier, Amsterdam, 1986), p.455.

[27] N.N. Sibeldin, V.B. Stopachinskii, S.G. Tikhodeev, and V.A. Tsetkov, JETP Lett. **38**, 207 (1983).

[28] V.V. Konopatskii, G.A. Kopelevich, and S.G. Tikhodeev, Zh. Eksp. Teor. Fiz., in press (1994).

[29] B. Link and G. Baym, Phys. Rev. Lett. **69**, 2959 (1992).

29

Charged Bosons in Quantum Heterostructures

L. D. Shvartsman and J. E. Golub

Racah Institute of Physics
The Hebrew University of Jerusalem
Jerusalem 91904
Israel

Abstract

We show that heterostructures may give rise to charged bosons. Bosons may be formed as a result of two quantum-well holes pairing in a repulsive potential. In the case where at least one hole has a negative effective mass, the repulsive electrostatic interaction may be converted to an effective attraction. The effect is a general one, possible in quantum layers of most cubic semiconductors. We compute the x-dependent hole-hole binding energy for $AlGaAs/In_xGa_{1-x}As/GaAs$ *quantum layers, taking account of the screening by a degenerate background gas of positive mass holes. We conclude that bound hole–hole pairs should be observable in infrared absorption experiments.*

The possibility of creating a controlled gas of bosons has stimulated years of work on excitonic systems [1–4], spin-polarized hydrogen [5], and atom trapping [6]. But the only opportunity of *charged* boson formation in solids which is widely accepted is Cooper pairing. The possibility of any other mechanism of charged boson formation in solids is a very exciting topic for investigation. Needless to say, the superfluidity of a charged boson gas in solids would be manifested as superconductivity.

In this work we discuss regular electrostatic repulsion, which, in the case where at least one of the carriers has a negative effective mass, may be converted to an effective attraction, as a mechanism of pairing. A composite boson, carrying charge $2|e|$, may be formed in such a case.

It is known that the presence of negative mass states of carriers is a general property of crystal band structure. The pairing mechanism of interest is realized in the following situations. First of all, it may be an electrostatic interaction of two carriers when one of them is near the bottom of one band and another one near the top of the next band. In such a case, the pairs are excitonic charged complexes. The other variant corresponds to the case of two carriers near the top of the band, so that both of them have a negative mass. In both situations, it is necessary to stress that we are considering an excited state of the system. As will be clear in the first case, a boson may be formed only when the reduced mass of the pair is negative and so its total mass is positive. For two negative-mass carriers, both the total mass and the reduced mass of the pair will be negative.

In order to make clear the physical nature of the pairing mechanism, let us write down a regular hydrogen-like Schrödinger equation :

$$[-(\hbar^2/2m^*)\nabla^2 + e\phi(\rho)]\psi(\rho) = E\psi(\rho) \tag{1}$$

It does not matter if it is a three- or a two-dimensional equation. In the usual case, when $m^* > 0$ is the reduced mass of the pair, and $\phi(\rho)$ is an attractive electrostatic potential, (1) describes bound states with a discrete spectrum. In the three-dimensional situation the potential well, of course, is supposed to be deep enough, and in the two-dimensional situation, as is well known, at least one confined level always exists.

It is obvious that if we put in Eq. (1) a negative reduced mass but a repulsive potential $\phi(\rho)$, we still have the same equation. This would describe a bound state of the complex not in the well but in the barrier. The kinetic energy of the internal motion has a negative value in this case.

The idea of a charged, exciton-like complex appeared as early as 1971 in connection with the observation of a reversed hydrogen-like absorption spectrum in bulk crystals of BiI_3 [7, 8]. The reversal of the usual hydrogenic spectrum for excitons was explained as a result of the existence of a bound complex of two electrons of different bands, with negative exciton reduced mass. A similar effect has been reported in bulk ZnP_2 [9]. The complex of two electrons was called a bielectron, and the complex of two holes was called a bihole.

In the present paper, we show that quantum heterostructures of cubic semiconductors may exhibit biholes, and that this characteristic is a general one, possible in nearly all quantum wells of cubic semiconductors. Specifically, we show that two holes may form a bound pair, and compute the binding energy including the effects of the two-dimensional screening. Being a quasistationary state, the pairs have a finite lifetime. We conclude that experimentally observable binding energies are realizable, and we

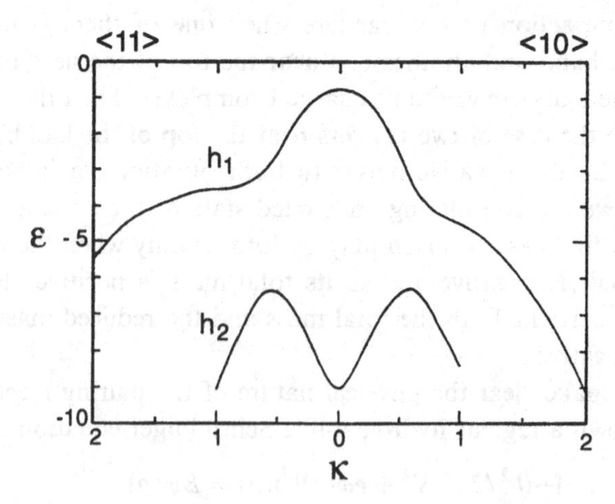

Fig. 1. Valence subband structure for a GaAs quantum layer in the approximation of an infinite potential barrier. $\kappa = kL/\pi$ is the dimensionless wavevector, where k is the usual wavevector and L is the well width. The energy ϵ is measured in units of $\hbar^2\pi^2/2m_0L^2$.

argue that the pairs are long-lived objects. A detailed calculation of the lifetime is postponed for a later publication.

The possibility of charged bosons in heterostructures follows from the form of the valence subband structure of quantum wells of most cubic semiconductors. A generic band structure is shown in Fig. 1. A general feature of the dispersion is the presence of negative mass states. These states exist in the first excited subband of two-dimensional films of almost all cubic semiconductors [10–12]. They appear as a sequence of multicomponent anisotropic structure of the Kohn-Luttinger hamiltonian. For example [10–12], in deep potential wells of some materials (e.g., GaAs, Ge, GaSb, InSb, InP, InAs), this subband is the $n = 2$ heavy hole band [10]. In others (e.g., AlSb, ZnTe, GaP), it is the $n = 1$ light hole band [10]. In shallow wells, the band ordering may be reversed. However, in all these systems, the sign of the effective mass is negative [13].

Consider the electrostatic interaction between two holes when one belongs to the ground subband (mass $m_1 > 0$), and the other to the excited subband (mass $m_2 < 0$). If the reduced mass $(m^*)^{-1} = m_1^{-1} + m_2^{-1}$ is negative, the electrostatic repulsion, as we saw above, is converted to a net attraction. It leads to the formation of a bound pair, that is, to the formation of a bihole. A necessary condition for bihole formation

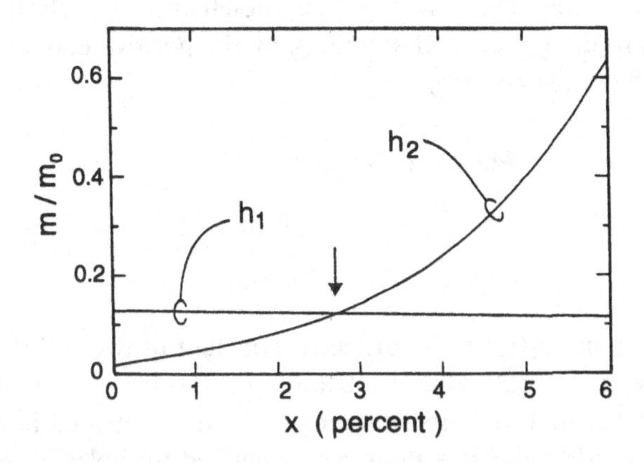

Fig. 2. Indium-mole-fraction-dependent masses of the ground and excited subbands for $L = 10$ nm. The absolute value of the h_2 band mass is shown. The arrow indicates the point at which the bihole reduced mass diverges.

is $m^* < 0$ or

$$|m_1| > |m_2|. \tag{2}$$

Eq. (2) is satisfied in infinite quantum wells of all the materials mentioned above except InSb and InAs. Eq. (2) may also be satisfied in finite wells of properly chosen width and material composition. So, the possibility of biholes is a general feature of quantum wells.

Bihole formation in quantum wells was first considered in 1983 [14]. We discuss below several important differences between experiments in bulk crystals and the present work.

We next calculate the bihole binding energy using a quantum well of AlGaAs/ $In_xGa_{1-x}As$/AlGaAs as a model system starting from a Kohn–Luttinger multicomponent formulation. In any heterostructure, changes in the doping x affect both the well-depth and the uniaxial strain. The present system has the important calculational advantage that the uniaxial strain dominates [15]. The calculation thus proceeds from a subband structure obtained assuming an infinite well, but takes into account the x-dependent uniaxial strain in the well layer. In particular, we assume the dependence of the effective masses upon the indium mole fraction x shown in Fig. 2 [15, 16]. This assumption permits us to calculate an x-dependent binding energy and then adjust x to maximize the effect within the applicability range of the model.

Now let us consider (1) in the two-dimensional case. The electrostatic potential $\phi(\rho)$ in the presence of screening by the positive mass holes of the ground subband is given by

$$\phi(\rho) = \int \frac{d\kappa}{(2\pi)^2} e^{i\kappa\cdot\rho} \phi(\kappa) \tag{3}$$

where [17–19]

$$\phi(\kappa) = 2\pi \; e/\epsilon_1(\kappa + 2/a). \tag{4}$$

Here, $\epsilon = \epsilon_2/\epsilon_1$; and $\epsilon_1(\epsilon_2)$ is the dielectric constant of the well (barrier) layers. In the numerical solutions below, we assume $\epsilon = 1$. Other cases, for example, that of a free standing film, will be treated in a later publication. $a = \epsilon_1 \hbar^2/m_1 e^2$ is a Bohr radius defined for holes in the first subband. For $\rho \gg L$, we find for $\phi(\rho)$ the asymptotic behavior

$$\phi(\rho) = (e/\epsilon_1)\{1/\rho - (\pi/a)[H_0(2\rho/a) - N_0(\rho/a)]\}, \tag{5}$$

where $H_0(N_0)$ denotes the Struve (Neumann) function of order zero. Eq. (5) expresses the rather non-coulombic force law obtained in the narrow-well limit.

The bihole binding energy now is determined from (1), where $m^* < 0$ is the reduced mass, and $\phi(\rho)$ is determined by (3) and (4). To solve (1), we use a direct variational method in k-space. We choose the two-parameter variational function

$$\psi(\rho) = [(2\lambda)^{2/\nu}/2\pi\Gamma(2/\nu)]^{1/2} \exp[-\lambda\rho^\nu], \tag{6}$$

where λ and ν are variational parameters. This function satisfies the trivial boundary conditions $\psi(0) =$ const. and $\psi(\rho \to \infty) = 0$. The prefactors are chosen for normalization. The bihole binding energy is found by maximizing (because of the reversed signs of the mass and interaction) the total energy $< T + U >_\psi$. The kinetic energy is easily found to be:

$$< T >_\psi = (\pi\hbar^2\nu/4m^*)(2\lambda)^{2/\nu}/2\pi\Gamma(2/\nu). \tag{7}$$

Calculation of the potential term is less straightforward, and will be detailed elsewhere. The result is

$$< U >_\psi = \int \rho \, d\rho \int \kappa \, d\kappa \frac{e^2}{\epsilon(\kappa + 2/a)} \, \frac{(2\lambda)^{2/\nu}}{\Gamma(2/\nu)} \, e^{-2\lambda\rho^\nu} J_0(\kappa\rho). \tag{8}$$

Combining (7) and (8), we find for the x-dependent bihole binding energy

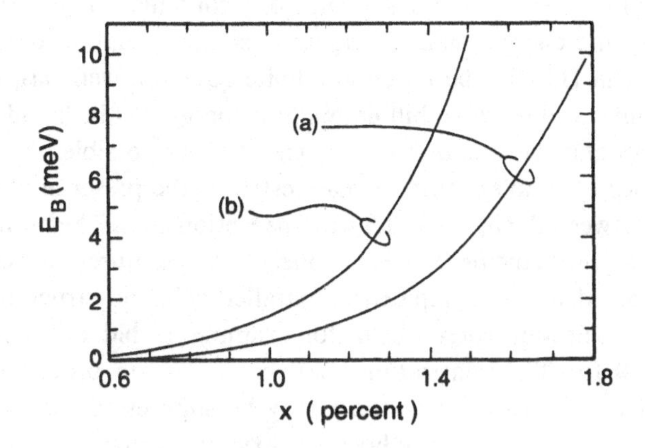

Fig. 3. Indium-mole-fraction-dependent bihole binding energy computed using (a) a single-parameter trial function, and (b) the trial function in (6).

$$\begin{aligned} E_B(x) &= \max(< T(x) + U(x) >) \\ &= \max C(\lambda, v)\Big[2\pi^2\hbar^2 v/4m^*(x) \\ &+ \int \rho d\rho\; kdk\; 2\pi e^2/\epsilon(k + 2/a(x)) \exp[-2\lambda\rho^v] J_0(k\rho)\Big], \end{aligned} \tag{9}$$

where $C(\lambda, v) = (2\lambda)^{2/v}/2\pi\Gamma(2/v)$. We have calculated this dependence numerically with the result shown in Fig. 3, which also shows the result of a single-parameter variational calculation using a hydrogenic trial function. (The one-parameter function is obtained from (6) by taking $v = 1$.) The diagram shows that the addition of as little as 1.3% indium to the GaAs quantum well-layer has the effect of increasing the bihole binding energy from a small fraction of one meV to nearly 5 meV. The reason for this increase in binding energy may be understood on simple physical grounds. According to Fig. 2, strain induced by the lattice mismatch has the effect of tuning the two hole masses to be more nearly equal and opposite. As a result, the reduced mass goes to infinity, and the binding energy diverges.

There are other means to make the hole masses nearly equal and opposite. We illustrate this point using a well of $Al_yGa_{1-y}As/GaAs$. For $y = 1$, the hole masses are dissimilar [16] with $m_1 > |m_2|$. By contrast, for $y = 0.3$, we have the opposite situation [17]: $m_1 < |m_2|$. Since the masses vary monotonically with y, there is a value of y for which $m_1 \approx -m_2$ and the bihole binding energy again diverges. A similar situation occurs in the Si/Ge system [11]. In the pure Ge infinite quantum well, $|m_2|$ is too small to permit an observable bihole binding energy. With the addition of Si, m_2 grows and the situation $m_1 \approx -m_2$ becomes possible.

The existence of biholes must be manifested as the presence of additional peaks between the h_1–h_2 and h_1–h_3 absorption lines. The h_1–h_2 and h_1–h_3 absorption lines themselves have nonzero, temperature-independent widths determined by the complex (nonparallel) subband structure [11] and the dipole selection rules which, for example, forbid h_1–h_2 transitions at $k = 0$ for TE polarization. While biholes are broadened by disorder-related inhomogeneities, these may be suppressed through the use of interrupted growth, atomic layer epitaxy, or through other technical means. The only fundamental limitation to the observation of biholes comes from temperature broadening. Since the binding energies calculated above can be as large as 5 meV ($\approx$ 60 K), the bihole line should be observable in low temperature, infrared absorption experiments. The range of validity of the present model can be understood as follows. The model is inadequate in the limit of large binding energy because of our neglect of the third (and higher lying) subbands. Thus, the model is most accurate when the bihole level is much closer to the second subband than to the third. In the opposite limit, that of very small binding energies, the model fails because of inaccuracies associated with the use of an approximate trial function, a situation common to all variational calculations.

Because dispersion curves like those displayed in Fig. 1 are nonparabolic, and bihole states lie within a continuous background of free hole–hole states, the bihole state cannot be a fully stationary state in the strict sense. The binding energy calculated above is thus the real part of a complex energy $E = E_B + i\hbar/\tau$, where τ gives the lifetime with respect to various decay processes. Let us divide the decay channels into two types: those which proceed directly, and those which require the participation of a third body. The second class of decays proceeds with the characteristic time of interaction with the third body, e.g., a phonon scattering time. These processes proceed slowly with the result that $\hbar/\tau$ is small compared to the binding energies (meV's) calculated above. The first class of (direct) decay channels is forbidden for the structure calculated here as for many structures [20, 21]. To establish this fact, we note that the complex subband dispersion prevents the spontaneous decay of a bihole to an isoenergetic hole-hole pair state. As we have checked numerically, such a decay process can not conserve both momentum and energy. As long as a third body (e.g., a phonon) is required, the bihole must be metastable.

To conclude, we have shown that bound hole–hole pairs are a generic feature of quantum wells made of most cubic semiconductors. We have calculated the binding energy of the bihole in the presence of a degenerate gas of positive mass holes and shown that, by careful choice of the quantum-well parameters, it may be made quite large. We conclude that it should be possible to realize charged bosons experimentally using heterostructures.

We gratefully acknowledge the support of the Ministry of Science and Technology (3845/1/99).

References

[1] P.L. Gourley and J.P. Wolfe, Phys. Rev. Lett. **40**, 526 (1978); J. P. Wolfe and C. D. Jeffries, in *Electron–Hole Droplets in Semiconductors*, C.D. Jeffries and L.V. Keldysh, eds. (North Holland, Amsterdam, 1983) p. 431.

[2] D. Snoke, J.L. Lin, and J.P. Wolfe, Phys. Rev. B **43**, 1226 (1991).

[3] D. Snoke and J.P. Wolfe, Phys. Rev. B **42**, 7876 (1990).

[4] D. Snoke, J.P. Wolfe, and A. Mysyrowicz, Phys. Rev. Lett. **64**, 2563 (1990).

[5] I.F. Silvera, Physica **109+110B**, 1499 (1982); T. J. Greytak and D. Kleppner, in *New Trends in Atomic Physics*, G. Grynberg and R. Stora, eds. (North Holland, Amsterdam, 1984), Vol. II.

[6] E.L. Raab, M. Prentiss, Alex Cable, Steven Chu, and D. E. Pritchard, Phys. Rev. Lett. **59**, 2631 (1987), and references therein.

[7] E.F. Gross, V.I. Perel, and R.I. Shechmamet'ev, JETP Lett. **13**, 229 (1971).

[8] E.F. Gross, N.V. Starostin, M.P. Shepilov and R. I. Shechmamet'ev, Proceedings of the USSR Academy of Sciences **37**, 885 (1973).

[9] A.V. Selkin, I.G. Stamov, N.U. Syrbu, and A.G. Umenets, JETP Lett. **35**, 57 (1982).

[10] L.D. Shvartsman, Ph.D. thesis, USSR Academy of Science Siberian Branch, Institute of Semiconductors, Novosibirsk (1984).

[11] A.V. Chaplik and L.D. Shvartsman, Poverkhnost (Surface), No. 2, 73 (1982).

[12] L.D. Shvartsman, Sol. State Commun. **46**, 787 (1983).

[13] Silicon is the only exception to this rule. See Refs. [10] and [12].

[14] A.V. Chaplik and L.D. Shvartsman, in *All Union School on Surface Physics,* (USSR Academy of Sciences, Chernogolovka, 1983), proceedings of the Third All Union School on Surface Physics held in Tashkent, October, 1983, p. 123.

[15] B. Laikhtman, R.A. Kiehl and D.J. Frank, J. Appl. Phys. **70**, 1531 (1991).

[16] O.V. Kibis and L.D. Shvartsman, Poverkhnost (Surface), No. 7, 119 (1985), in Russian.

[17] N.S. Rytova, Vestnik Moskovskogo Universiteta (Moscow University Proceedings), No. 3, 30 (1967).

[18] L.V. Keldysh, JETP Lett. **29**, 658 (1979).

[19] A.V. Chaplik and M.V. Entin, JETP **34**, 1335 (1972).

[20] G. Bastard, in *Physics and Applications of Quantum Wells and Superlattices,* E.E. Mendez and K. von Klitzing, eds. (Plenum, New York, 1987).

[21] Direct decay can always be made forbidden by slight changes in material composition, or other quantum-well parameters.

30

Evidence for Bipolaronic Bose-liquid and BEC in High-T_c Oxides

A. S. Alexandrov
Interdisciplinary Research Centre in Superconductivity
University of Cambridge
Madingley Road
Cambridge, CB3 OHE
UK

Abstract

Recent experiments on the near-infrared absorption, thermal conductivity and the critical field H_{c2} in several high-T_c oxides are interpreted as a manifestation of the Bose–Einstein condensation of small bipolarons.

1 Basic Model for High-T_c Oxides

To describe low-energy spin and charge excitations of metal oxides and doped fullerenes with bipolarons, Alexandrov and Mott [1, 2, 3, 4] have suggested that bipolarons are intersite in two possible spin states ($S = 0$ or 1), and a proportion of bipolarons are in Anderson localised states.

Our assumption is that *all electrons* are bound in small singlet or triplet bipolarons and they are responsible for the *spin* excitations. *Hole* pairs, which appear with doping, are responsible for the low-energy charge excitations of the CuO_2 plane. Above T_c, a material such as YBCO contains a non-degenerate gas of these hole bipolarons in a singlet or in a triplet state, with a slightly lower mass due to the lower binding energy.

The low-energy band-structure includes two bosonic bands (singlets and triplets), separated by the singlet–triplet exchange energy J, estimated to be of the order of a few hundred meV. The half-bandwidth w is of the same order. The bipolaron binding energy is assumed to be large ($\Delta >> T$), and therefore single polarons are irrelevant in the temperature region under consideration.

We argue that many features of spin and charge excitations in metal oxides can be described within our simple model. In this mini-review, we discuss the recent observation of the significant influence of the superconducting transition on the near-infrared absorption [5], the thermal conductivity enhancement below T_c [6] and the upper critical field, divergent at low temperature [7]. These observations can be interpreted as a manifestation of the Bose–Einstein condensation of bipolarons in copper-based oxides.

2 Evidence for Bose–Einstein Condensation from Near-Infrared Absorption

Evidence for the bipolaronic origin of the "spin" gap and for the Bose–Einstein condensation comes from the temperature dependence of the near-infrared absorption. It is well known that for frequencies much higher than the superconducting energy gap, no change with temperature in optical absorption is expected within the BCS theory. However Dewing and Salje [8] observed the effect of the superconducting phase transition on the near infrared absorption, with the characteristic frequency $\nu \simeq 0.7$ eV in $YBa_2Cu_3O_{7-\delta}$. The relative change of the integrated optical conductivity at the superconducting transition turns out to be as high as 10%.

With the sum rule for bipolaronic conductivity, one can describe the experimentally observed temperature dependence of the integrated absorption intensity $I(T)$, shown in Fig. 1 [5]:

$$\frac{I(T)}{I(0)} = i_n(T) + i_c(T), \tag{1}$$

where the non-coherent contribution, responsible mainly for the temperature dependence of the normal-state absorption, is

$$i_n(T) = 1 - \frac{3AT}{2w} \ln\left(\frac{1 - ye^{-(2w+J)/T}}{1 - ye^{-J/T}}\right) \tag{2}$$

and the coherent contribution is:

$$i_c(T) = \delta_c \frac{T^2}{T_c^2} \int_0^{2w/T} xdx \left(\frac{1}{y^{-1}e^x - 1} + \frac{3}{y^{-1}e^{x+\frac{J}{T}} - 1}\right). \tag{3}$$

The chemical potential $\mu = T \ln y$ is determined by the bipolaron density n, which is the sum of singlets and triplets :

$$\ln\left(\frac{1 - ye^{-2w/T}}{1 - y}\right) + 3\ln\left(\frac{1 - ye^{(-2w-J)/T}}{1 - ye^{-J/T}}\right) = \frac{2nw}{T}, \tag{4}$$

and $\mu = 0$ if $T < T_c$. Here n is the total number of pairs per cell and the density of states is assumed to be energy independent within bands.

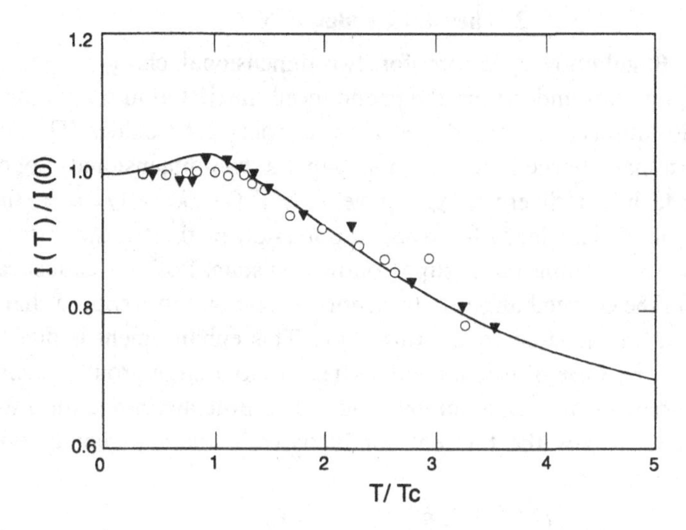

Fig. 1. Integrated optical absorption of inter-site bipolarons (solid line) and experiment [5][$YBa_2Cu_3O_7$ (triangles) and YBaCuO/Fe (circles)], $J/T_c = 4.4$, $\delta_c = 0.014$, $B = 1.6$, $w/T_c = 1.9$.

The temperature-independent constant A in (2) is proportional to the difference of the non-coherent absorption of singlet and triplet pairs. In general, singlets absorb light in a frequency window different from triplets due to an internal bonding–antibonding transition. The triplet half-bandwidth w is assumed to be the same as the singlet one, and the dimensionless constant δ_c determines the relative value of the coherent contribution (i.e. without phonon and spin-wave "shakeoff").

A remarkable change of the slope of the optical absorption below T_c (see Fig. 1) is due to the temperature-dependent coherent contribution to the sum rule. This contribution vanishes at $T = 0$ due to the Bose–Einstein condensation of singlet bipolarons. The condensed pairs on a lattice cannot absorb light coherently because of the parity and time-reversal symmetry. The pronounced decrease of the integrated absorption intensity in the normal state is explained by the temperature dependence of the singlet bipolaron density on the scale $T \sim J$. One can fit the experimental curve, Fig. 1, with the reasonable value of the singlet–triplet exchange energy $J/T_c \simeq 4.4$. The latter is close to the characteristic value of the spin pseudogap in "1–2–3" determined from neutron scattering [9], $E_{Gap}/T_c \simeq 3.5 - 5$ (depending on the doping).

3 Thermal Conductivity

Using the Bogoliubov spectrum for two-dimensional charged bosons $\epsilon(k) \sim \sqrt{k}$, one can understand the pronounced maximum in the thermal conductivity observed in two-dimensional copper-based oxides [6]. For two-dimensional charged bosons, the gapless two-dimensional Bogoliubov mode has a divergent group velocity $d\epsilon(k)/dk \sim 1/\sqrt{k}$ in the long-wave limit. This leads to strong suppression of the transport scattering rate of excitations in the superconducting state. Both the acoustical phonon and the charged impurity transport relaxation times show a sharp enhancement in the superconducting state. This enhancement is due to screening by the Bose–Einstein condensate, and to a large group velocity of the two-dimensional Bogoliubov mode. The Boltzmann equation for the excitations yields the thermal conductivity in the superconducting state [6]:

$$K_s = K_{s0}\frac{(1-t)^2}{t^2}\int_0^\infty \frac{x^4 dx}{\sinh^2 x(x^4 + \eta(1-t)^2/t^5)}, \tag{5}$$

where K_{s0} is temperature independent, $t = T/T_c$ is the reduced temperature and η is proportional to the charged impurity concentration.

With Eq. (5) one can explain the experimentally observed enhancement of the thermal conductivity below T_c, shown in Fig. 2. It should be mentioned that the commonly accepted explanation of this enhancement due to the lattice contribution to the heat transport has now been rejected [11].

For carriers with a double elementary charge, the Lorentz number should be at least four times smaller than for degenerate electrons. This fact enables one to understand [2] the near equality of the in-plane thermal conductivity above 100K in the insulating and 90K crystals of $YBa_2Cu_3O_{7-\delta}$ [12].

4 Upper Critical Field

The interacting charged Bose gas is condensed in a field lower than a certain critical value H^* because the interaction with impurities [13], or between bosons [14], broadens the Landau levels and thereby eliminates the one-dimensional singularity of the density of states. The critical field of Bose–Einstein condensation has an unusual positive curvature near T_{c0}, $H^*(T) \sim (T_{c0}-T)^{3/2}$ and diverges at $T \to 0$, where $T_{c0} \simeq 3.31n^{2/3}/\mathrm{m}$ is the critical temperature of Bose–Einstein condensation of an ideal gas in zero field.

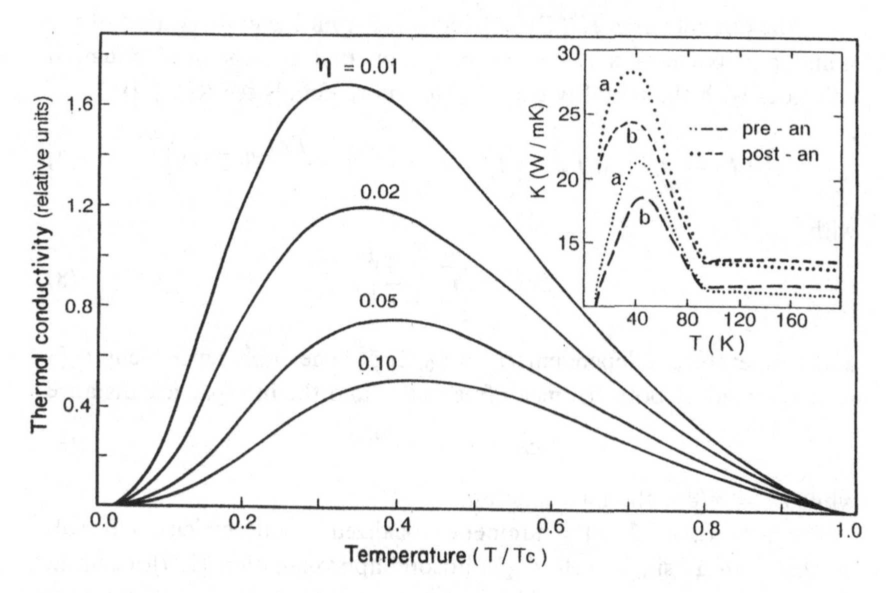

Fig. 2. Theoretical temperature dependence of the thermal conductivity of near-two-dimensional charged bosons for different charged impurity concentrations η [6]. Insert: enhancement of the in-plane thermal conductivity in the superconducting state of $YBa_2Cu_3O_{7-\delta}$ [10].

It was shown recently [7] that localization drastically changes the low-temperature behaviour of the critical field. H^* saturates with the temperature lowering at some value of the impurity concentration and at higher concentrations the re-entry effect to the normal state takes place.

H^* is determined as the field in which the first non-zero solution of the linearized stationary Ginzburg–Pitaevskii equation [15] for the macroscopic condensate wave function $\psi_0(\mathbf{r}) = \langle N|\hat{\psi}(\mathbf{r})|N+1\rangle$ ($N \to \infty$, $N/V = n = \text{const}$) appears:

$$\left(-\frac{1}{2m}(\nabla - 2ie\mathbf{A}(\mathbf{r}))^2 + U_{imp}(\mathbf{r})\right)\psi_0(\mathbf{r}) = \mu\psi_0(\mathbf{r}), \tag{6}$$

where $2e$ is the charge of a boson; $\mathbf{A}(\mathbf{r})$, $U_{imp}(\mathbf{r})$ and μ are the vector, random, and chemical potentials, respectively.

The definition of H^* in Eq. (6) is identical to that of the upper critical field H_{c2} of BCS superconductors of the second kind. Therefore H^* determines the upper critical field of bipolaronic or any "bosonic" superconductor.

The first non-trivial *delocalized* solution of Eq. (6) appears at $\mu = E_c$.

Thus the critical curve $H^*(T)$ is determined from the conservation of the number of particles n under the condition that the chemical potential coincides with the mobility edge E_c (for more details see Ref. [7]):

$$H^*(T) = H_d(T_{c0}/T)^{3/2}\left(1-(T/T_{c0})^{3/2} - \frac{Tn_L}{\gamma n}\beta(T/\gamma)\right)^{3/2}, \quad (7)$$

with

$$\beta(x) = \sum_{k=0}^{\infty}\frac{(-1)^k}{x+k}, \quad (8)$$

and temperature independent $H_d = \phi_0/2\pi\xi_0^2$. The "coherence" length ξ_0 is determined by both the mean free path l and the interparticle distance

$$\xi_0 \simeq 0.8(l/n)^{1/4}, \quad (9)$$

while $\phi_0 = \pi/e$ is the flux quantum.

The last term in (7) is the number of localized bosons, which can be calculated with a "single well–single boson" approximation [2, 7](localized *charged* bosons obey Fermi–Dirac statistics):

$$n_L(T) = \int_{-\infty}^{E_c}\frac{N_L(\epsilon)d\epsilon}{e^{(\epsilon-\mu)/T}+1}, \quad (10)$$

where the density of localized states $N_L(\epsilon)$ may be approximated in many cases by the exponential tail:

$$N_L(\epsilon) = \frac{n_L}{\gamma}\exp(\frac{\epsilon - E_c}{\gamma}), \quad (11)$$

with γ of the order of the binding energy in a single random potential well and n_L the total number of localized states per unit cell.

The localization does not change the positive "3/2" curvature of the critical magnetic field near T_c [13], as shown in Fig. 3. I believe that this curvature is a universal feature of a charged Bose-gas, which does not depend on a particular scattering mechanism or the approximations made. At low temperature ($T << \gamma$), the temperature dependence of H^* turns out to be drastically different for different impurity concentration, as seen in Fig 3. An upward curvature of H_{c2} near T_c has been observed in practically all superconducting oxides, including cubic ones.

Recently, divergent $H_{c2}(T)$ has been measured in a wide temperature range starting from the mK-level up to T_c [16]. Resistively determined H_{c2} values from $T/T_c = 0.0025$ to $T/T_c = 1$ in a $T_c = 20$K single crystal of $Tl_2Ba_2CuO_6$ follow a temperature dependence that is in good qualitative agreement with the type of curve shown in Fig. 3 for $n_L/n \simeq 1$.

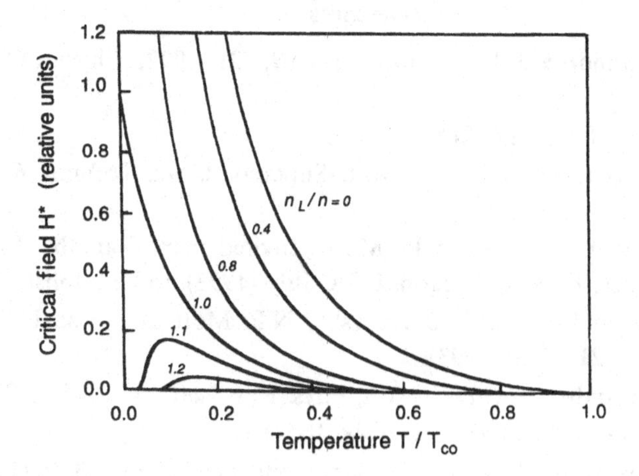

Fig. 3. Temperature dependence of the critical magnetic field of Bose–Einstein condensation [7](in units of $H_d(T_{c0} \ln 2/\gamma)^{3/2}$) for different relative number of localized states n_L/n and $\gamma/T_{c0} = 0.2$.

Osofsky *et al.* [17] also observed a divergent upward temperature dependence of the upper critical field $H_{c2}(T)$ for thin BSCO films, which was five times longer than expected for a conventional superconductor at the lowest temperature. They compared their data with Eq. (7).

5 Conclusion

Earlier evidence for small (bi)polarons in copper-based oxides comes from the photoinduced infrared absorption, measured by Heeger and by Taliany [18, 19], from the observation by Sugai [20] of both infrared and Raman-active vibration modes, and from XAFS results on the radial distribution function of apex oxygen ions [21, 22].

The experimental observations discussed above, and many other features of high-T_c copper oxides [3, 4], lead us to the conclusion that a charged Bose-liquid of small intersite bipolarons is a simple but far-reaching model of high-T_c superconductors.

Acknowledgments. I am grateful to Sir Nevill Mott, Yao Liang, Alex Bratkovskii, Ekhardt Salje and Andrew Mackenzie for helpful stimulating discussions. I appreciate the financial support from the Leverhulme Trust.

References

[1] A.S. Alexandrov, J. Low Temp. Phys. **87**, 721 (1992); Physica C **182**, 327 (1991).

[2] N.F. Mott, Physica C **205**, 191 (1993).

[3] A.S. Alexandrov and N.F. Mott, Supercond. Sci. Technol. **6**, 215 (1993).

[4] A.S. Alexandrov and N.F. Mott, Invited paper at the MOS-Conference, Eugene, Oregon, USA, July (1993), to be published.

[5] A.S. Alexandrov, A.M. Bratkovsky, N.F. Mott and E.K.H. Salje, Physica C **215**, 359 (1993).

[6] A.S. Alexandrov and N.F. Mott, Phys. Rev. Lett. **71**, 1075 (1993).

[7] A.S. Alexandrov, Phys. Rev. B **48**, 10571 (1993).

[8] H.L. Dewing and E.K.H. Salje, Supercond. Sci. Technol. **5**, 50 (1992).

[9] J. Rossat-Mignod, L.P. Regnault, P. Bourges, C. Vettier, P. Burlet and J.Y. Henry, Physica Scripta **45**, 74 (1992).

[10] J.L. Cohn *et al.*, Phys. Rev. B **45**, 13144 (1992).

[11] R.C. Yu, M.B. Salamon, J.P. Lu, and W.C. Lee, Phys. Rev. Lett. **69**, 1431 (1992).

[12] S.J. Hagen, Z.Z. Wang, and N.P. Ong, Phys. Rev. B **40**, 9389 (1989).

[13] A.S. Alexandrov, doctoral thesis, Moscow Engineering Physics Institute, Moscow (1984); A.S. Alexandrov, J. Ranninger, and S. Robaszkiewicz, Phys. Rev. B **33**, 4526 (1986).

[14] A.S. Alexandrov, D.A. Samarchenko, and S.V. Traven, Zh. Eksp. Teor. Fiz. **93**, 1007 (1987) [Sov. Phys. JETP **66**, 567 (1987)].

[15] E.M. Lifshitz and L.P. Pitaevskii, *Statistical Physics,* Part 2 (Pergamon, Oxford, 1980), p.117.

[16] A.P. Mackenzie, S.R. Julian, G.G. Lonzarich, A. Carrington, S.D. Hughes, R.S. Liu and D.C. Sinclair, Phys. Rev. Lett. **71**, 1238 (1993).

[17] M.S. Osofsky *et al.*, Phys. Rev. Lett. **71**, 2315 (1993).

[18] Y.H. Kim, C.M. Foster, A.J. Heeger, S. Cos, and G. Stucky, Phys. Rev. B **38**, 6478 (1988).

[19] C. Taliani, A.J. Pal, G. Ruani, R. Zamboni, X. Wei, and Z.V. Vardini, *Electronic Properties of High T_c Superconductors and Related Compounds,* H. Kuzmany, M. Mehring, and J. Fink, eds., Springer Series of Solid State Science **99** (Springer-Verlag, Berlin, 1990), p. 280.

[20] S. Sugai, Physica C **185-189**, 76 (1991).

[21] J. Mustre de Leon, S.D. Conradson, I. Batistic, and A.R. Bishop, Phys. Rev. Lett. **65**, 1675 (1990).
[22] J. Mustre de Leon, I. Batistic, A.R. Bishop, S.D. Conradson, and S.A. Trugman, Phys. Rev. Lett. **68**, 3236 (1992).

31

The Dynamic Structure Function of Bose Liquids in the Deep Inelastic Regime

A. Belić

International Center for Theoretical Physics
P.O.Box 586
34100 Trieste
Italy

Abstract

We study the dynamic structure function $S(k, \omega)$ of Bose liquids in the asymptotic limit $k, \omega \to \infty$ at constant $y \equiv \frac{m}{k}(\omega - k^2/2m)$, using the orthogonal correlated basis of Feynman phonon states.

1 Introduction

In the last few years there has been a growing interest in the possibilities for experimental determination of single-particle momentum distributions in non-relativistic many-body systems by means of inelastic neutron scattering at large momentum transfers. This interest was generated by the advent of pulsed-neutron sources that made possible measurements with substantially larger momentum transfers then before. For example, in the case of liquid ^{4}He a momentum transfer of 20–30 $Å^{-1}$ was achieved [1], which is much larger than the rms momentum of atoms in the liquid of ~ 1.6 $Å^{-1}$. If it is assumed that at such large momentum transfers, the potentials between the atoms in the liquid are negligible compared to the large kinetic energy of the struck atom, the deep inelastic response is completely determined by the initial single-particle momentum distribution. This assumption is known as the impulse approximation (IA) [2, 3]. At large momentum transfers k, the dynamic structure function (DSF) $S(\mathbf{k}, \omega)$ exhibits the phenomenon of y-scaling, i.e. the combination $J(y) \equiv \frac{k}{m} S(\mathbf{k}, \omega)$, called the scaling function, depends solely upon the scaling variable [2] $y = \frac{m}{k}(\omega - \frac{k^2}{2m})$ and not separately upon k and ω. The scaling function in the IA is given by:

$$J_{IA}(y) = n_c \delta(y) + \frac{1}{4\pi^2 \rho} \int_{|y|}^{\infty} dq \, q \, n(q), \tag{1}$$

where ρ is the density of the system, $n(q)$ is the single-particle momentum distribution and n_c the corresponding condensate fraction. However, the y-scaling property of the deep inelastic response is more general than the IA in the sense that it may hold even when the IA fails.

If the IA is valid, the momentum distribution can be extracted from the deep inelastic response in a model-independent way. This gives rise to the exciting possibility that the momentum distribution is a directly observable quantity. If this were true, not only would we be able to check the accuracy of our calculations and determine interparticle potentials more reliably, but also we could test fundamental physical ideas such as Bose condensation!

Unfortunately, in the most interesting cases, such as liquid ^{4}He, there is a strong repulsive short-ranged component in the interparticle potential which is not negligible compared to the recoil kinetic energy even for the large momentum transfers achieved in recent experiments. Thus the IA must be corrected to account for interaction of the recoiling particle with the surrounding medium. These corrections are called final state interaction (FSI) corrections. Clearly, it would be desirable to develop an analytic, first-principles theory of the dynamic structure function at large momentum transfers, in order to understand the FSI.

There were many attempts that start from the IA but take into account the interactions in the initial state of the target, and improve upon it by incorporating the effects of FSI [4]. This procedure becomes more and more difficult as k decreases and the FSI become more important. In contrast the response at small k and ω is easily treated by using Feynman phonon states [5]. In particular the orthogonal correlated basis (OCB) formalism [6] based on Feynman's ideas is quite successful [7] in explaining the observed $S(\mathbf{k},\omega)$ at $k \lesssim 2$ $\mathring{A}^{-1}$. The small and large k methods have different starting points; the momentum distribution $n(q)$ is the main input for the IA, while the OCB formalism uses the static structure function $S(q)$ as the main input.

Although in principle the OCB formalism can be used to study $S(\mathbf{k},\omega)$ at all k and ω, in practice the calculations [7] become technically complicated as k increases. Nevertheless we have shown [8] that, by using field-theoretical techniques in OCB perturbation theory, it is possible to calculate the $S(\mathbf{k},\omega)$ in the scaling limit, and thus extend this low k method to the $k \to \infty$ limit. In this paper, we present results of this approach as applied to the case of a Bose liquid at zero temperature.

2 OCB Calculation of $S(\mathbf{k},\omega)$ in the Scaling Limit

The DSF $S(\mathbf{k},\omega)$ can be written as

$$S(\mathbf{k},\omega) = -\frac{1}{\pi}\mathrm{Im}D(\mathbf{k},\omega), \tag{2}$$

where

$$\begin{aligned} D(\mathbf{k},\omega) &= \frac{1}{N}\langle 0|\, \rho_{\mathbf{k}}\, [\omega - H + E_0 + i\eta]^{-1}\, \rho_{\mathbf{k}}^{\dagger}\, |0\rangle \\ &= \frac{1}{N}\sum_{n=0}^{\infty}\langle 0|\, \rho_{\mathbf{k}}\, [\omega - H_0 + E_0 + i\eta]^{-1} \\ &\quad \times (H'\, [\omega - H_0 + E_0 + i\eta]^{-1})^n\, \rho_{\mathbf{k}}^{\dagger}\, |0\rangle, \end{aligned} \tag{3}$$

is the density–density response function, $\rho_{\mathbf{k}} = \sum_i e^{-i\mathbf{k}\cdot\mathbf{r}_i}$ is the Fourier component of the density, η is a positive infinitesimal, and the Hamiltonian of the system is divided into diagonal and off-diagonal parts with respect to the set of intermediate states $|i\rangle$:

$$H = H_0 + H', \tag{4}$$

$$H_{0,ij} = \delta_{ij}\langle i|\, H\, |i\rangle, \quad H'_{ij} = (1-\delta_{ij})\langle i|\, H\, |j\rangle. \tag{5}$$

If the intermediate states are taken to be $|\, i\rangle \equiv |\, \mathbf{p}\rangle_1\, |\, n\rangle_{N-1}$, where $|\,\mathbf{p}\rangle$ is a plane wave state of the struck particle, and $|\,n\rangle$ are eigenstates for $N-1$ background particles, then the expansion (3) becomes the so-called Gersh series [9]. When the interparticle potential is weak, it is an asymptotic expansion in powers of m/k around the zeroth-order term that is equal to IA. However, when the interaction contains a hard core, i.e. when the interparticle potential $v(r)$ is divergent enough at small r so that its Fourier transform $\tilde{v}(p)$ is not defined, the Gersh series fails. The reason for this failure is that the non-perturbed Hamiltonian decouples the recoiling particle from the rest of the system and does not take into account strong short-range correlations that would normally prevent the struck particle from entering the hard-core region of background particles.

Our choice for intermediate states are Feynman phonons (FP):

$$|\mathbf{p}_1,\ldots,\mathbf{p}_n) = \frac{(\rho_{\mathbf{p}_1}^{\dagger}\cdots\rho_{\mathbf{p}_n}^{\dagger})\,|0\rangle}{\langle 0|\,(\rho_{\mathbf{p}_1}\cdots\rho_{\mathbf{p}_n})(\rho_{\mathbf{p}_1}^{\dagger}\cdots\rho_{\mathbf{p}_n}^{\dagger})\,|0\rangle^{1/2}}, \tag{6}$$

where

$$(\rho_{\mathbf{p}_1}^{\dagger}\cdots\rho_{\mathbf{p}_n}^{\dagger}) \equiv \sum_{i\neq l\neq\ldots}\exp[i(\mathbf{p}_1\cdot\mathbf{r}_i + \cdots + \mathbf{p}_n\cdot\mathbf{r}_l)], \tag{7}$$

and $|\,0\rangle$ is the ground state of the system. FP states do contain the appropriate short-range correlations, and are reasonably close to the exact eigenstates of the Hamiltonian. Thus we expect H' to be small and series (3) to be rapidly convergent. However, they are not mutually orthogonal, which is the general property of correlated basis (CB) states. Such states can be used in the perturbation expansions, but require a special formalism that takes into the account their non-orthogonality, and it is not clear whether the non-orthogonality effects that are introduced when the expansion is truncated are negligible or not. In order to use normal perturbation theory, we have to orthogonalize the CB states. This can be achieved using the Löwdin transformation [10], but resulting OCB states have higher energies than CB states. Perturbative corrections move the energy down again, and both the increase in energy due to the Löwdin transform and the decrease due to perturbative corrections are larger than the net displacement [6]. In view of these difficulties, the OCB perturbation theory (OCBPT) is not the preferred tool for studies of quantum liquids. Recently, however, a new orthogonalization scheme, free of the problems mentioned above, was proposed [11]. It is expected that the convergence of OCBPT does not depend upon the specific nature of the bare interaction, i.e. if it has a hard-core or not. For these reasons, we use it to study the scaling properties of the response at large k and ω.

When the orthogonalized FP in (6) are used as intermediate states in (3), it can be shown [8] that the diagonal matrix elements $H_{0,ij}$ are of order $O((k/m)^2)$, while the off-diagonal matrix elements H'_{ij} are of order $O(k/m)$, from which the scaling property

$$\frac{k}{m}D(\mathbf{k},\omega) = D_S(y) + O(m/k) \tag{8}$$

immediately follows. The only assumption used to derive this result is that the static structure function $S(k)$ of a Bose liquid approaches unity exponentially fast as $k \to \infty$, which is true even when the interatomic interactions are hard, as for example, the Lenard–Jones 6-12 potential. Here we do not address the problem of a hard sphere gas where $(S(k)-1)$ does not go exponentially to zero at large k due to the singularity in the derivative of the pair distribution function $g(r)$ at $r = r_c$ (the hard core radius). The correction term of order $O(m/k)$ in (8) was not studied using OCBPT. It is known that it is an antisymmetric function of y, and can be estimated using sum-rule arguments [12].

An exact calculation of the scaling function $J(y)$ is difficult because of the large number of different types of vertices, representing the matrix elements H'_{ij}, in the expansion (3). However, all the diagrams based on the vertex at which the hard phonon (with momentum $\sim k$) emits or absorbs a soft phonon (with momentum $\ll k$) can be summed by the standard field-theoretical methods without further approximations. The resulting system of coupled integral equations can be solved in the closed form to obtain [8]:

$$J(y) = C \left\{ \delta(y) \quad + \quad \frac{1}{\pi} \int_0^\infty ds \cos(ys) \right. \\ \left. \times \left[\exp\left(\frac{1}{8\pi^2 \rho} \int_0^\infty dl \, \frac{\sin(sl)}{sl} f(l) \right) - 1 \right] \right\}, \quad (9)$$

where we have introduced the constant

$$C = \exp\left(-\frac{1}{8\pi^2 \rho} \int_0^\infty dl \, f(l) \right). \quad (10)$$

The function $f(l)$ depends on the static structure function $S(q)$ only, and is given by:

$$f(l) = S(l) \left[\sum_{n=0}^{\infty} \alpha_n I_n(l) \right]^2, \quad (11)$$

where the coefficients $\alpha_n = 1, -1/2, 3/8, \ldots$ are those that appear in the expansion of $(1+x)^{-1/2}$, and the functions $I_n(l)$ are defined recursively as:

$$I_0(l) \quad = \quad \frac{l(S(l)-1)}{S(l)}, \quad (12)$$

$$I_n(l) \quad = \quad \int \frac{d^3 l_1}{(2\pi)^3 \rho} \hat{\mathbf{l}}_1 \cdot \hat{\mathbf{l}} S(l_1)(S(|\mathbf{l} - \mathbf{l}_1|) - 1) I_{n-1}(l_1), \;\; n \geq 1. \quad (13)$$

Eqs. (9)–(13) represent the main results of this work.

3 Discussion and Numerical Results

In the limit $y \to 0$ the scaling function $J(y)$ is completely determined by the properties of the function $f(l)$ for small l. More precisely, from (11)–(13) it follows that

$$\lim_{l \to 0} \frac{f(l)}{l} = \lim_{l \to 0} \frac{l}{S(l)} = 2mc, \quad (14)$$

where c is the sound velocity in the liquid, and we obtain [8]

$$J(y \to 0) = C\delta(y) + \frac{1}{4\pi^2 \rho} \left[A - B|y| \right], \quad (15)$$

where the constants A and B are given by

$$\begin{aligned} A &= 4\pi\rho C \int_0^\infty ds \left[\exp\left(\frac{1}{8\pi^2\rho}\int_0^\infty dl \frac{\sin(sl)}{sl} f(l)\right) - 1\right], \\ B &= C\frac{mc}{2}. \end{aligned} \tag{16}$$

The scaling function in IA (Eq. (1)) exhibits the same small y behavior with the coefficients

$$C_{IA} = n_c, \quad A_{IA} = \int_0^\infty dq\, qn(q), \quad B_{IA} = \lim_{q\to 0} qn(q) = n_c\frac{mc}{2}, \tag{17}$$

and it follows that the ratio B/C in our theory is identical to that in IA:

$$\frac{B}{C} = \frac{B_{IA}}{C_{IA}}. \tag{18}$$

Next, we want to discuss the origin of the δ-function peak at $y = 0$. In general, the presence of the δ-function in DSF means that probe couples to a long-lived state with a given energy and momentum. The inverse lifetime of this excitation is expected to be proportional to k on the basis of the semi-classical value $\tau^{-1} \sim \rho v \sigma$, where $v \equiv k/m$ is the velocity and σ the average two-body scattering cross section. However, at $y = 0$ this argument fails, as can be seen in the second-order contribution to (3). In that order, an off-shell FP with momentum $\mathbf{k}$ and energy $\omega = k^2/2m + ky/m$ can decay into two FP of momenta $\mathbf{k} - \mathbf{l}$ and $\mathbf{l}$, thus acquiring a lifetime

$$\begin{aligned} \frac{1}{\tau} &= 2\pi \int \frac{d^3l}{(2\pi)^3\rho} \left|\langle \mathbf{k}-\mathbf{l},\mathbf{l}|\, H\, |\mathbf{k}\rangle\right|^2 \delta\left(\omega - \frac{(\mathbf{k}-\mathbf{l})^2}{2m} - \frac{l^2}{2m\, S(l)}\right) \\ &= \frac{1}{4(2\pi)^2\rho}\frac{k}{m} \int d^3l\, (\hat{\mathbf{k}}\cdot\hat{\mathbf{l}})^2 f(l)\delta(y + \hat{\mathbf{k}}\cdot\mathbf{l} + O(k^{-1})). \end{aligned} \tag{19}$$

When $y \neq 0$ we obtain $\tau \sim k^{-1}$, as expected from semi-classical arguments, but when $y = 0$ (on-shell FP) the energy-conserving δ-function forces the cosine $\hat{\mathbf{k}}\cdot\hat{\mathbf{l}}$ to be of the order k^{-1} which makes the integrand small, and gives the on-shell FP a long lifetime $\tau \sim k$.

In our calculation of the scaling function $J(y)$, we have taken into account only the processes in which one soft phonon is emitted or absorbed at a time. We can estimate the validity of this approximation by checking if $J(y)$ satisfies the sum-rules [6]. It is easy to show that the y^0 and y sum-rules:

$$\int_{-\infty}^\infty dy\, J(y) = 1, \quad \int_{-\infty}^\infty dy\, yJ(y) = 0, \tag{20}$$

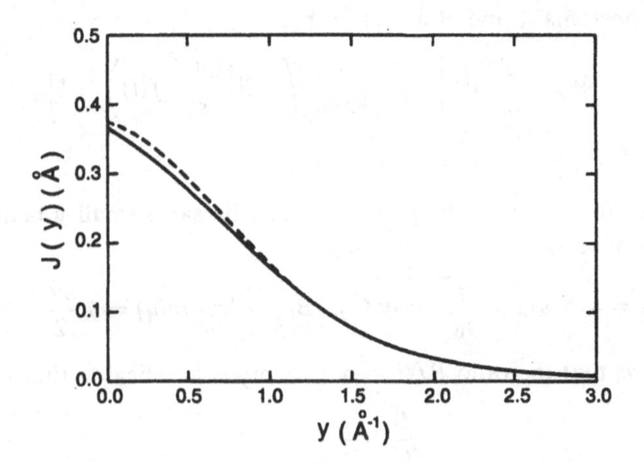

Fig. 1. The scaling functions of liquid ^{4}He at equilibrium density and zero temperature. Solid line: $J(y)$ of Eq. (9) obtained using $S(q)$ of Ref. 13; dotted line: $J_{IA}(y)$ generated from $n(q)$ of Ref. 14. The δ-function peaks at $y = 0$, with the respective strengths of 0.15 and 0.092, are not shown.

are satisfied by $J(y)$ in (9). For the y^2 (kinetic energy) sum-rule, we obtain

$$\frac{2m}{3}E_k = \int_{-\infty}^{\infty} dy\, y^2 J(y) = \frac{1}{24\pi^2\rho}\int_0^{\infty} dl\, l^2 f(l). \tag{21}$$

The degree to which this sum-rule is satisfied is a measure of importance of neglected processes.

Finally, in the uniform limit ($|g(r) - 1| \ll 1$), the scaling function $J(y)$ should be equal to the $J_{IA}(y)$. It can be shown [8] that in this limit, the function $f(l)$ becomes:

$$f_U(l) = \frac{l^2(S(l) - 1)^2}{S(l)} = 4l^2 n(l), \tag{22}$$

and that the scaling function $J(y)$ of (9)–(10) reduces to (1), as expected.

When (9)–(13) are applied to the case of liquid ^{4}He, using the measured [13] $S(q)$, the value $C = 0.15$ and the $J(y)$ shown in Fig. 1 are obtained. For comparison, $J_{IA}(y)$ generated from the variational $n(q)$ (Ref. [14]) which has the condensate fraction $n_c = 0.092$, is also shown.

Finally, using the value $E_k = 14.8$K from Ref. [14] and $S(q)$ from Ref. [13], we find that

$$\frac{3}{2mE_k}\frac{1}{24\pi^2\rho}\int_0^{\infty} dl\, l^2 f(l) = 0.97. \tag{23}$$

As was argued above, the closeness of this value to 1 suggests that the neglected terms in expansion (3) are much smaller than those taken into account.

References

[1] P.E. Sokol, in *Momentum Distributions*, R.N. Silver and P.E. Sokol, eds. (Plenum, New York, 1989).

[2] G.B. West, Phys. Rep. **18C**, 263 (1975).

[3] P.C. Hohenberg and P.M. Platzman, Phys. Rev. **152**, 198 (1966).

[4] R.N. Silver, in *Momentum Distributions*, R.N. Silver and P.E. Sokol, eds. (Plenum, New York, 1989); A.S. Rinat, Phys. Rev. B **40**, 6625 (1989).

[5] R.P. Feynman, Phys. Rev. **94**, 262 (1954).

[6] E. Feenberg, *Theory of Quantum Liquids* (Academic Press, New York, 1969).

[7] E. Manousakis and V.R. Pandharipande, Phys. Rev. B **33**, 150 (1986).

[8] A. Belić and V.R. Pandharipande, Phys. Rev. B **45**, 839 (1992).

[9] H.A. Gersch and L.J. Rodriguez, Phys. Rev. A **8**, 905 (1973).

[10] P.O. Löwdin, J. Chem. Phys. **18**, 365 (1950).

[11] S. Fantoni and V.R. Pandharipande, Phys. Rev. C **37**, 1697 (1988).

[12] A. Belić and V.R. Pandharipande, Phys. Rev. B **39**, 2696 (1989).

[13] H.N. Robkoff and R.B. Hallock, Phys. Rev. B **25**, 1572 (1982).

[14] E. Manousakis, V.R. Pandharipande, and Q.N. Usmani, Phys. Rev. B **31**, 7022 (1985); *ibid.* B **43**, 13587 (1991).

32

Possibilities for BEC of Positronium

P. M. Platzman and A. P. Mills, Jr
AT&T Bell Laboratories
600 Mountain Avenue
Murray Hill, New Jersey 07974
USA

Abstract

We review the proposal for Bose–Einstein condensation of positronium atoms. All of the ingredients necessary to achieve BEC of Ps atoms are currently available.

1 Introduction

In this volume several authors have discussed and described a variety of weakly interacting systems which might display BEC. In this short contribution we suggest that a dense gas of positronium (Ps) atoms in vacuum is a rather ideal but somewhat more exotic system that might be a very good candidate for observing a weakly interacting BEC.

Recent investigations of the interactions of positrons (e^+) and (Ps) with solids have led to extraordinary improvements in the kinds of low energy experiments we can do with the positron [1, 2]. All of the ingredients necessary to achieve BEC of Ps atoms are currently available. We envision a scenario where roughly $N \cong 10^5$ Ps atoms are trapped in a volume $V \simeq 10^{-13}$ cm^3 and allowed to cool through the Bose transition temperature of $20 - 30$K in a time of the order of nanoseconds. In the following we discuss the relevant interactions and describe how Ps BEC can be achieved.

2 Single Positronium Physics

Ps is comprised of an $e^+ - e^-$ bound in a hydrogenic orbit. Its mass, $2m_e$, is extremely light compared to H, an important ingredient for achieving reasonable Bose condensation temperatures. Its binding energy (6.8 eV) is half that of H. The ground state of Ps is a spin singlet separated from an excited triplet by an energy $\Delta E_{ST} \cong 10\text{K}$. Ps annihilates itself, i.e., it becomes high energy γ-rays [3].

The annihilation characteristics of the Ps atom are dependent on which of the two ground states it is in. The ground state singlet (s) is short lived with a lifetime $\tau_s = 1.25 \times 10^{-10}$ s. For a Ps atom at rest, the decay occurs with the emission of two 0.5 MeV γ-rays which come out precisely in opposite directions due to momentum conservation. If the Ps atom is moving with momentum p, the γ-rays come out with a small angle $\Theta \cong p/mc$ relative to each other. The nearby triplet state (t) is prohibited by selection rules from decay into the two γ-ray channel. Instead it decays by three γ-rays with a spread of energies and a much longer lifetime of $\tau_t = 1.42 \times 10^{-7}$ s. A magnetic field mixes the triplet with singlet, thus rapidly quenching the triplet.

Like hydrogen, the Ps_2 molecule exists in an overall singlet state [4]. It has a binding energy $E_M \cong 0.4$ eV. Two low energy Ps atoms scatter from one another with a cross section $\sigma \equiv 4\pi a^2$ determined by how close the bound state is to the continuum [5]. Specifically, $a_s = (2mE_M)^{-1/2} \cong 3$ Å, and $\sigma_s \cong 10^{-14}$ cm^2. On the other hand the non-singlet channel has a scattering length more like a Bohr radius, i.e., $a_t \cong 1$ Å and $\sigma_t \cong 10^{-15}$ cm^2. Of course, the experiment we are considering here would, for the first time, be sensitive to such cross sections and a measurement of them should be possible.

3 e^+ and Ps at Solid Surfaces

In order to make the case for Bose condensation we need to understand, at least qualitatively, how a mildly energetic ($\approx$ 10 keV) beam of e^+ interacts with a semi-infinite slab of some simple solid [1]. The interaction process is sketched in Fig. 1. The incident beam enters the solid and begins to make electronic excitations in the material, plasmons, inner shell ionizations, etc. The energetic e^+ reaches an energy of 10 eV or so in a short time $\tau < 10^{-13}$ s, and does so in a distance of roughly 1000 Å for a 5 keV beam [2]. The mechanism is primarily plasmon emission. At this point in its history, the e^+ begins to slow down to thermal energies by electron–hole pair excitations and phonon emission. In this regime, its motion in the solid may be described as diffusive. Calculations and experiments convincingly demonstrate that diffusion constants D_p for our superthermal e^+ are roughly 1 cm^2/s [1]. Since annihilation times in the solid τ_A are, in most materials, a fraction of a nanosecond, a typical

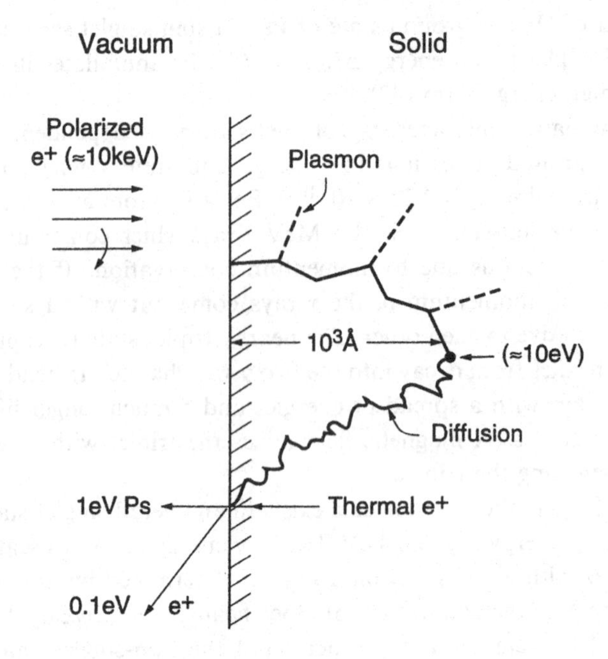

Fig. 1. Schematic of the interaction of a 10 keV beam with a typical solid such as Si.

e^+ diffuses a distance $\ell_D \cong \sqrt{D_p \cdot \tau_A} \cong 2 \times 10^3$Å. This simple estimate suggests that some large fraction (50%) of our initial 5 keV e^+ beam returns to the surface, where we must consider its energetics.

The work function (the energy difference between the lowest state inside the solid and the vacuum) for an e^+ consists of three parts [1, 2]. First is the correlation energy, which is positive, i.e., it keeps the e^+ inside the solid. It must be less then the 6.8 eV Ps binding energy. Second is the *negative* dipole part which exists because the e^- in the solid spill out into the vacuum. The value of this energy varies but is typically between 5 and 10 eV. Finally there is the negative kinetic energy part typically 2–4 eV. This part comes from the fact that the bottom of the e^+ band moves up when the e^+ is squeezed into the space between the ions. These general considerations lead one, *a priori*, to believe that the e^+ work function could be negative. In fact, simple materials [Ni, Al, W, ...] do have negative e^+ wavefunctions [1, 2].

An e^+ arriving at the surface of a negative work function material simply slides down the hill out into the vacuum. When it gets into the regime where the electron density is low enough it can, and does, pick up an electron and form neutral Ps which can also be ejected from the solid. Emission of e^+ and Ps occurs with about 50% probability. Typically, the energy of the e^+ is non-thermal, i.e., directed perpendicular to the surface and greater than 1 eV, while the Ps energy distribution is similarly non-thermal but is most often greater than 1 eV [2]. This unique surface interaction characteristic permits the generation of very bright pulsed e^+ beams (see Canter in Ref. [2]) and is essential, as we shall see, to the proposed BEC experiment.

4 Dense Positronium

The "initial" condition for the Ps condensation experiment is sketched in Fig. 2. A bunched 1 nanosecond brightness-enhanced microbeam (diameter $1\mu m$) consisting of $N \geq 10^6 e^+$ at an energy of 5 keV is incident at time $t = 0$ (1 nanosecond uncertainty) on the surface of a Si target which has, at the point of entry, a small cavity etched into it. We choose Si rather than a metal for several reasons, to be discussed shortly. The geometry we envision will typically be a cylindrical cavity of 1 μm diameter with a height of 1000 Å. These dimensions are rather easily achieved with conventional lithography techniques. The e^+, as discussed, stop in the Si at a depth of about 1000 Å. About one-quarter of the initial e^+, because of diffusion and the negative work function for Ps, will be reemitted as electron volt Ps into the cavity. The singlet Ps rapidly decays leaving us with a hot gas of bosons (triplet Ps atoms), at a density $n \equiv N/V \cong 10^{18}$ cm^{-3}.

In Ref. [6] we consider in some detail the time evolution of this confined hot gas of Ps atoms due to collisions with the wall and with other Ps atoms in the cavity. We conclude that such a gas would indeed cool to a Bose condensed state by phonon emission at the walls in a time short compared to a triplet lifetime. However, the serious difficulty which must be avoided is quenching of the triplet state by collisions of Ps atoms with the same wall and with each other.

Scattering from the walls occurs at a very rapid rate, with $\Gamma_W \cong 10^{12}$ s^{-1} for Ps atoms with a energy of 0.1 eV. If the solid wall of the cavity is metallic, i.e., there are unpaired spins, then exchange of the e^- in the triplet Ps with e^- in the metal occurs with high probability and the triplet state is quenched very rapidly. If the wall is insulating, i.e., the electron spins are paired, then exchange scattering is often energetically

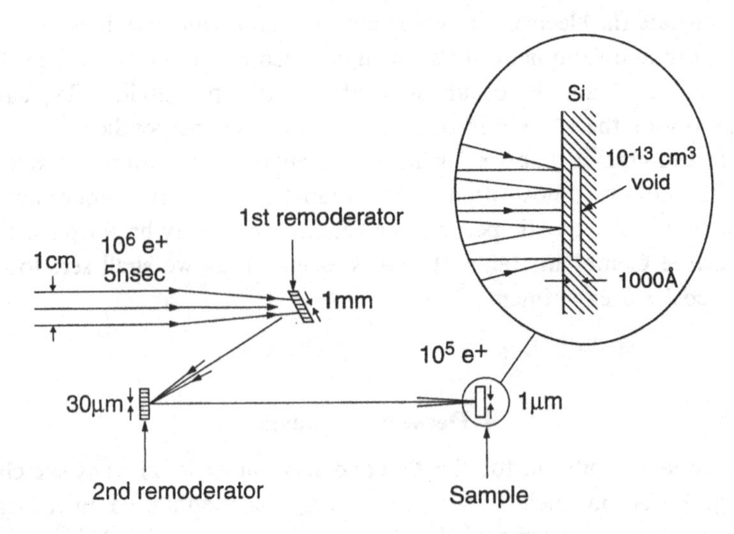

Fig. 2. Sketch of a pulsed brightness enhanced e^+ microbeam incident on a Si target with a void etched in it. The dimensions are typical dimensions for the proposed experiments.

forbidden and the only allowed process is a weak relativistic effect, which is negligible.

The scattering of triplet Ps atoms with one another is the mechanism by which the system equilibrates internally. This type of scattering can also lead to exchange scattering which will annihilate the gas before it reaches BEC. In the s wave scattering approximation the bulk triplet-triplet scattering rate characterized by an s-wave scattering length of 1Å is roughly $\Gamma_t \cong 10^{10}$ s^{-1}. It is rapid enough to equilibrate the gas as it cools.

In Ref. [6] we have shown that collisions between triplets with their spin pointing in different directions will surely lead to conversion of triplet to singlet at roughly the same rate. However, it is possible to prevent this type of annihilation by using a fully polarized e^+ beam. Fully polarized e^+ beams are commonly available since all beta-decay sources generate e^+s by a parity non-conserving process which leads to e^+ polarized with their spin in the direction of their momentum. Slowing down inside a low Z solid such as Si leaves the initially fully polarized e^+s about 50% percent polarized (see Berko in Ref. [2]).

For such a polarized beam of e^+ incident on an unpolarized e^- target, it is easy to show, based on rather general arguments [6], that triplet–

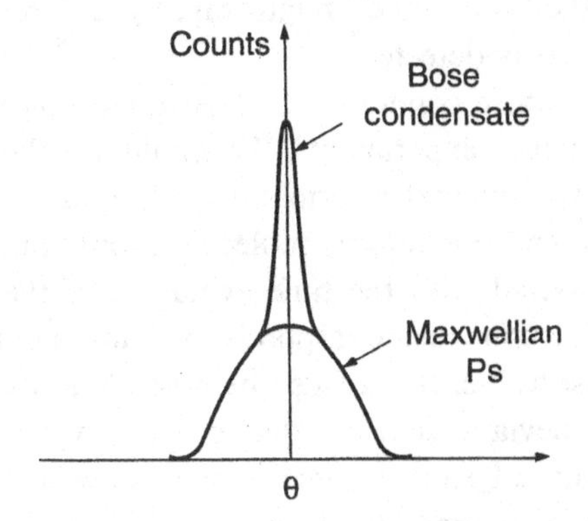

Fig. 3. Sketch of a hypothetical γ-ray angular correlation spectrum for BEC Ps. The double humped structure is characteristic of the Bose condensed fraction.

triplet collisions rapidly lead to a fully polarized Ps gas consisting of electrons and positrons with spins in the *same* direction as the initial polarized e^+ beam. This fully polarized gas decays with a long triplet lifetime and should quite happily cool through the BEC.

5 Observation of the Bose Condensate

Because of the unique annihilation characteristics of Ps, we will be able to observe the condensate directly, by converting a small but significant fraction of triplets to singlets. This can be accomplished by applying a small transverse magnetic field $\sim$ 0.1T to mix the singlet and triplet states. The small amount of singlet will annihilate into two γ-rays. The angle between the two γ-rays, as we have pointed out, directly mirrors the momentum of the Ps atom. Thus an angular correlation profile with enough resolution will indicate the presence of a zero momentum condensate. If the sample is below T_c, the angular correlation profile will resemble the angular correlation data shown in Fig. 3. The best angular resolution currently attainable is a $\Delta\Theta \cong 0.1$ milliradian. This implies that we can measure the velocity distribution function with a resolution $\Delta v \simeq c\Delta\Theta \simeq 3 \times 10^6$ cm/s, or about two times better than the velocity of a Ps atom at room temperature. With some improvement in $\Delta\Theta$, there should be unambiguous evidence of a qualitative peaking of the angular correlation profile as a function of time. If the system were to Bose condense, it might also be possible to demonstrate the existence of a superfluid by placing a small hole in the cavity containing the Ps atoms. Below T_c the "superfluid" might rapidly leak out to the vacuum, where it could be easily detected.

Observation of a Bose condensate, while interesting in itself, will not be the only goal of such experiments. We would, for the first time, begin to obtain direct experimental information about the dense Ps gas, that is, Ps–Ps collisions and possibly Ps molecule formation. Secondly, all of the problems connected with the time evolution of the condensate, for example, how such a condensate responds to changes in temperature and density, will be observable. It may also be possible to look carefully near the transition for deviations from ideal-gas behavior. Such non ideal behavior will of course become more apparent as we increase the density one to two orders of magnitude. In any event it seems likely that this exotic, but rather ideal, system will be a fascinating one to look at in the near future.

Acknowledgements. The authors would like to thank S. Berko, K. G. Lynn and K. F. Canter for numerous discussions.

References

[1] A.P. Mills, Jr, in *Positron Solid State Physics*, edited by W. Brandt and A. Dupasquier (North Holland, Amsterdam, 1983) p. 421; P.J. Schultz and K.G. Lynn, Rev. Mod. Phys. **60**, 701 (1988).

[2] *Positron Studies of Solids Surfaces and Atoms*, edited by A.P. Mills, Jr, W.S. Crane and K.F. Canter (World Scientific, Singapore, 1985).

[3] H. Bethe and E. Salpeter, *Quantum Mechanics of One and Two Electron Atoms* (Academic, New York, 1957).

[4] E.A. Hylleraas and A. Ore, Phys. Rev. **71**, 493 (1947); Y. K. Ho, Phys. Rev. A **33**, 3584 (1986). For a review see M.A. Abdel-Raouf, Fortschr. Phys. **36**, 521 (1988).

[5] T.Y. Wu and T. Ohmura, *Quantum Theory of Scattering* (Prentice Hall, New Jersey, 1962).

[6] P.M. Platzman and A.P. Mills, Jr., Phys. Rev. B **49**, 454 (1994).

33

Bose–Einstein Condensation and Spin Waves

R. Friedberg and T. D. Lee
Department of Physics
Columbia University
New York, NY 10027
USA

H. C. Ren
The Rockefeller University
New York, NY 10021
USA

The Quantum Lattice Model [1] is a natural basis for computer simulations of Bose liquids as well as for possibly realistic models of mobile bosonic excitations in a solid. It depicts particles obeying Bose statistics occupying sites of a lattice and possessing kinetic energy by virtue of a hopping term $-2\sum_{<ij>}(b_i^\dagger b_j + b_j^\dagger b_i)$ in the Hamiltonian, where $\sum_{<ij>}$ counts each nearest neighbor pair once; but it is understood that no more than one particle may occupy the same site.

It is of interest to add an attractive two-body potential at nearest neighbor separation, so that the Hamiltonian may be written

$$H = v\sum_i b_i^\dagger b_i + 2\sum_{<ij>}(b_i^\dagger - b_j^\dagger)(b_i - b_j) - \alpha\sum_{<ij>} b_i^\dagger b_j^\dagger b_i b_j. \tag{1}$$

The problem is then to study (1) in a Hilbert space *restricted* to states satisfying $b_i b_i| >= 0$ for all i. A convenient formal way to achieve this is to study the Hamiltonian

$$H_G = H + G\sum_i b_i^\dagger b_i^\dagger b_i b_i \tag{2}$$

and take the limit $G \to +\infty$ in the results [2].

In the case where $\alpha = 0$, this is the natural lattice analog of the hard-sphere boson problem, and the "infinite" potential G can be dealt with in the same way as the "infinite" potential forbidding two spheres to penetrate. One introduces the binary dressed vertex $\Gamma(\mathbf{P}, \mathbf{k}, \mathbf{q}, E)$ obtained by summing over all ladder diagrams involving two particles with initial momenta $\frac{1}{2}\mathbf{P}\pm\mathbf{k}$, final momenta $\frac{1}{2}\mathbf{P}\pm\mathbf{q}$ and external energy E. All physics can be calculated in powers of Γ, which itself remains finite as $G \to +\infty$. Particularly important is the long-wavelength scattering amplitude

$$\frac{1}{2}\gamma_0 = \frac{1}{2}\Gamma(0,0,0,0) = 4/g_0, \tag{3}$$

where g_0 is the solution at the origin of the Poisson equation with unit source. The continuum analog of $\frac{1}{2}\gamma_0$ is $4\pi a/m$, where a, m are the diameter and mass of the spheres.

If $\alpha \neq 0$ the same treatment can be applied with

$$\frac{1}{2}\gamma_0 = \frac{4-\alpha}{g_0 - \frac{1}{4}\alpha g_1}, \tag{4}$$

where g_1 is the solution of the same equation at neighboring sites; thus $\chi(g_0 - g_1) = 1$, where χ is the number of nearest neighbors. For low-momentum phenomena one replaces the α term in H by $\frac{1}{2}\gamma_0 \Sigma b_i^\dagger b_i^\dagger b_i b_i$ and omits higher ladder terms already included in Γ. The quantity γ_0 then determines the physics of the dilute system; for example the ground state energy agrees with that found by substituting $\frac{1}{2}\gamma_0$ for $4\pi a/m$ and setting $m = 4$ (in accordance with the hopping term of (1)) in Lee and Yang's formula [3], but at higher powers of density, the lattice departs from the continuum.

An insight into the numerator $4 - \alpha$ in (4) is gained by considering the anisotropic spin-$\frac{1}{2}$ Heisenberg model

$$H_{Heis} = \sum_{\langle ij\rangle} (-\boldsymbol{\sigma}_i \cdot \boldsymbol{\sigma}_j + A\sigma_i^z \sigma_j^z) - h\Sigma_i \sigma_i^z. \tag{5}$$

If we define

$$c_i = \frac{1}{2}(\sigma_i^x + i\sigma_i^y),\ c_i^\dagger = \frac{1}{2}(\sigma_i^x - i\sigma_i^y) \tag{6}$$

so that

$$[c_i, c_j^\dagger] = \sigma_i^z \delta_{ij} = (1 - 2c_j^\dagger c_i)\delta_{ij}, \tag{7}$$

then (6) becomes

$$\begin{aligned} H_{Heis} &= \sum_{\langle ij\rangle} [-2(c_i^\dagger c_j + c_j^\dagger c_i) - (1-A)(1-2c_i^\dagger c_i)(1-2c_j^\dagger c_j)] \\ &- h\sum_i (-c_i^\dagger c_i). \end{aligned} \tag{8}$$

Now if we could only replace the c_i by boson operators b_i (ignoring the term $2c_j^\dagger c_i$ in (7)) we see that H_{Heis} would be the same, apart from a constant term, as H in (1), with the identifications

$$\alpha = 4(1 - A), \tag{9}$$

$$v + 2\chi = h + 2\chi(1 - A). \tag{10}$$

Thus the isotropic Heisenberg model ($A = 0$) corresponds to the case $\alpha = 4$ or $\gamma_0 = 0$ in (4).

Indeed, Dyson [4] has shown that the isotropic Heisenberg model can be made to correspond with a bosonic system provided that in H we replace the α term with a certain *nonhermitian* two-body scattering term whose form is such that it vanishes at zero momentum. In this way Dyson could treat the dynamics of spin waves.

We have found it simpler to treat (5) by relating it to (2). That is, spin excitations are just like bosons with same-site exclusion, and we back this up with a theorem [2] showing that those eigenvalues of H_G that remain finite as $G \to +\infty$ do tend toward eigenvalues of $H_{Heis} + const$, and that the respective eigenvectors of H_G tend toward permitted states (no two particles on a site) corresponding to the respective eigenvectors of H_{Heis}.

Armed with this theorem, we can treat the anisotropic Heisenberg model by converting it to (2), thus gaining the advantage of diagrammatic expansion of a Hermitian interaction, while retaining Dyson's insight that when $A = 0$, the low-momentum spin waves "slide past" one another.

The physical equivalence of a bosonic to a fermionic system raises an apparent paradox: the commutator (7) is always different from the bosonic commutator no matter what the value of G, and yet this commutator is the zero-time value of the fourier transform of the propagator which reflects the physics of the system. This contradiction is resolved by noting that the bosonic propagator for any fixed energy converges as $G \to +\infty$ to the fermionic propagator, but that this convergence is not uniform in the energy. Thus the fourier transform converges to agreement for any value of time other than zero. The one anomalous point is that of measure zero and does not affect the physics.

Bose condensation in (2) can take place for $\alpha < 4$ ($\gamma_0 > 0$); the corresponding (4) with $A > 0$ has "x–y-like" anisotropy, and the condensation involves breaking the symmetry about the z-axis. For $A < 0$ this symmetry is not broken, and the corresponding (2) with $\alpha > 4$ and $\gamma_0 < 0$ has a two-body attraction strong enough to bind, so that the system will separate, "Ising-like," into a packed liquid and a gas, with no Bose condensation.

Bose condensation in a dense system corresponds to the anisotropic Heisenberg model with $A > 0$ and $|h|$ not too large; then there is a large spontaneous symmetry breaking represented by the transverse components of $< \boldsymbol{\sigma}_i >$. Matsubara and Matsuda [1] studied the dense Bose condensate by exploiting this connection and subjecting (4) to a rotation:

$$H_{Heis} \rightarrow H_{Heis}^{\theta} = e^{\frac{1}{2}i\theta\Sigma\sigma_j^y} H_{Heis} e^{-\frac{1}{2}i\theta\Sigma\sigma_j^y}. \tag{11}$$

The parameter θ is chosen so that H_{Heis}^{θ} leads to a zero value of $< \sigma_i^x >$. Then, by keeping only those terms of H_{Heis}^{θ} quadratic in c_j and $c_j^{\dagger}$, they obtained an approximate spectrum of excitations in the presence of the condensate. This spectrum is phonon-like at low momentum; the phonons are the Goldstone bosons associated with the breaking of azimuthal symmetry by setting $\theta \neq 0$.

We have made (2) the starting point of a systematic expansion by using the connection between (2) and (4) a second time in reverse; that is, we replace H_{Heis}^{θ} by a new bosonic Hamiltonian $\overline{H}^{\theta}$ obtained by replacing c_i by a new boson operator $\overline{b}_i$ and adding a term $G\sum_i \overline{b}_i^{\dagger}\overline{b}_i^{\dagger}\overline{b}_i\overline{b}_i$ where $G \rightarrow +\infty$ [2]. Because of the θ-rotation, $< \overline{b}_i >= 0$; θ, not $< \overline{b} >$, is the order parameter for Bose condensation. The new Hamiltonian $\overline{H}^{\theta}$ contains terms in $\overline{b}_i^{\dagger}\overline{b}_j^{\dagger}$ and $\overline{b}_i\overline{b}_j$ so that the ground state has nonzero values of $< \overline{b}_i^{\dagger}\overline{b}_i >$ and $< \overline{b}_i^{\dagger}\overline{b}_j^{\dagger} >$, etc. These and other quantities (speed of sound, spectrum, ground state energy) can be calculated from diagrams as a power series in A. For furthur details, see Ref. [2].

References

[1] T. Matsubara and H. Matsuda, Prog. Theor. Phys. **16,** 569 (1956).

[2] R. Friedberg, T.D. Lee and H.C. Ren, Ann. Phys. (N. Y.) **228**, 52 (1993).

[3] T.D. Lee and C.N. Yang, Phys. Rev. **105,** 1119 (1957); **113,** 1165 (1959).

[4] F. Dyson, Phys. Rev. **102,** 1217, 1230 (1956).

34

Universal Behaviour within the Nozières–Schmitt-Rink Theory

F. Pistolesi and G. C. Strinati
Scuola Normale Superiore
I-56126 Pisa
Italy

Abstract

We show that the natural variable to follow the crossover from Cooper-pair-based superconductivity to Bose–Einstein condensation within the model of Nozières and Schmitt-Rink is the product $k_F\xi$*, where* k_F *is the Fermi wave vector and* ξ *is the coherence length for two-electron correlation. In terms of this product, the results of the model do not depend on the detailed form of the (separable) pairing potential, and the crossover turns out to be restricted to the universal region* $\pi^{-1} \lesssim k_F\xi \lesssim 2\pi$*. Experimental estimates indicate that* $k_F\xi \approx 10$ $(> 2\pi)$ *for high-*T_c *superconductors.*

Evolution from weak to strong coupling superconductivity has been considered by Nozières and Schmitt-Rink [1] (hereafter referred to as NSR) following the pioneering work by Leggett [2]. After the discovery of high-T_c superconductivity, the interest in this problem has grown, and many papers on this subject have appeared [3]. In the present work, we show that working within the simplified treatment by NSR, it is already possible to isolate the essential features of the crossover.

Central to the work of NSR and Leggett is the argument [4] that the BCS wave function has the Bose–Einstein condensation (BEC) built in as a limiting case. (See the review by Randeria in this volume.) NSR study the evolution from BCS to BEC through the increase of the coupling strength associated to an effective fermionic attractive potential, and conclude that the evolution is "smooth." Although this result is appealing from a theoretical point of view, it does not allow for a direct comparison with the experimental data since the coupling strength of the effective fermionic attraction is not a quantity that is easily obtained from experiments. Thus it is not possible to know whether a given physical system is either in the BCS limit or in the BEC limit, or in a genuine intermediate regime.

It would clearly be useful to have a more accessible variable to track. The new variable should satisfy the following criterion: the evolution from BCS to BEC should be *universal*, that is, independent of the details of the interaction potential and of the single-particle density of states. In the following, we propose to identify the product $k_F\xi$ as the desired parameter.

We briefly review the NSR model. We introduce the model fermionic Hamiltonian

$$\begin{aligned} H &= \sum_{\mathbf{k},\sigma} \epsilon_{\mathbf{k}} a^{\dagger}_{\mathbf{k},\sigma} a_{\mathbf{k},\sigma} \\ &+ \sum_{\mathbf{k},\mathbf{k}',\mathbf{q}} V_{\mathbf{k},\mathbf{k}'} a^{\dagger}_{\mathbf{k}+\mathbf{q}/2,\uparrow} a^{\dagger}_{-\mathbf{k}+\mathbf{q}/2,\downarrow} a_{-\mathbf{k}'+\mathbf{q}/2,\downarrow} a_{\mathbf{k}'+\mathbf{q}/2,\uparrow}\,, \end{aligned} \tag{1}$$

where $a_{\mathbf{k},\sigma}$ is the destruction operator for fermions with wave vector $\mathbf{k}$ and spin σ, $\epsilon_{\mathbf{k}}$ is a single-particle dispersion relation, and $V_{\mathbf{k},\mathbf{k}'}$ is an "effective" fermionic attraction. In its simplest form, $\epsilon_{\mathbf{k}} = \mathbf{k}^2/2m^* - \mu$, where m^* is an effective particle mass and μ is the chemical potential.

The variational calculation with the ground state BCS wave function

$$|\Phi\rangle = \prod_{\mathbf{k}} \left(u_{\mathbf{k}} + v_{\mathbf{k}} a^{\dagger}_{\mathbf{k},\uparrow} a^{\dagger}_{-\mathbf{k},\downarrow} \right) |0\rangle \tag{2}$$

leads to the two *coupled* equations

$$2\epsilon_{\mathbf{k}}\varphi_{\mathbf{k}} + (1 - 2v_{\mathbf{k}}{}^2) \sum_{\mathbf{k}'} V_{\mathbf{k},\mathbf{k}'}\varphi_{\mathbf{k}'} = 0, \tag{3}$$

$$n = \frac{1}{\Omega} \sum_{\mathbf{k}} \left(1 - \frac{\epsilon_{\mathbf{k}}}{E_{\mathbf{k}}} \right), \tag{4}$$

where n is the particle density, Ω is the quantization volume, and $\varphi_{\mathbf{k}} = 2u_{\mathbf{k}}v_{\mathbf{k}} = \Delta_{\mathbf{k}}/E_{\mathbf{k}}$ with $\Delta_{\mathbf{k}} = -\sum_{\mathbf{k}'} V_{\mathbf{k},\mathbf{k}'} u_{\mathbf{k}'} v_{\mathbf{k}'}$ and $E_{\mathbf{k}} = \sqrt{\epsilon_{\mathbf{k}}^2 + \Delta_{\mathbf{k}}^2}$ [5]. Provided that $v_{\mathbf{k}}{}^2 \ll 1$ for *all* $\mathbf{k}$, (3) reduces to the Schrödinger equation for the relative motion of two particles with equal mass m^*, interacting via $V_{\mathbf{k},\mathbf{k}'}$, and with eigenvalue 2μ. In this limit, complete bosonization of bound-electron pairs is achieved.

Solution of (3) and (4) is simplified with the NSR choice of a separable pairing potential:

$$V_{\mathbf{k},\mathbf{k}'} = V w_{\mathbf{k}} w_{\mathbf{k}'}, \qquad w_{\mathbf{k}} = \frac{1}{\sqrt{1+(\mathbf{k}/k_0)^2}}, \tag{5}$$

where $V(<0)$ and k_0 are parameters specifying the interaction. With this potential, $\Delta_{\mathbf{k}} = \Delta_0 w_{\mathbf{k}}$ and the two-body eigenvalue problem has (in three dimensions) the eigenvalue $2\mu = -\epsilon_0 = -(k_0^2/m^*)(G-1)^2$ for $G > 1$, where $G = -V\Omega m^* k_0/4\pi > 0$, while the bound-state radius has the asymptotic behavior $r_0 \sim k_0^{-1}G^{-1/2}$ for $G \to \infty$ [6]. This latter result should be contrasted with that of NSR, that $r_0 \to k_0^{-1}$, which led NSR to conclude that bosonization can be achieved only in the "dilute" limit. In agreement with NSR, we find that the solutions (Δ_0, μ) of (3) and (4) are smooth functions of G *for a given* k_0, although (Δ_0, μ) both depend strongly on k_0 for given G.

One way of introducing a more physical variable in place of G stems from the observation that, for strong coupling, the $(T{=}0)$ coherence length ξ is expected to be quite short. This length ξ can be calculated from the pair correlation function (with opposite spins)

$$g(\mathbf{r}) = \frac{1}{n^2}\left|\langle\Phi|\Psi_\uparrow^\dagger(\mathbf{r})\Psi_\downarrow^\dagger(0)|\Phi\rangle\right|^2, \tag{6}$$

where $\Psi_\sigma(\mathbf{r})$ is the fermion field operator, using the expression

$$\xi^2 = \frac{\int d\mathbf{r} g(\mathbf{r})\mathbf{r}^2}{\int d\mathbf{r} g(\mathbf{r})} = \frac{\sum_{\mathbf{k}} |\nabla_{\mathbf{k}}\varphi_{\mathbf{k}}|^2}{\sum_{\mathbf{k}} |\varphi_{\mathbf{k}}|^2}. \tag{7}$$

Our definition (7) recovers the Pippard expression $\xi_0 = (d\epsilon_{\mathbf{k}}/dk)_{k_F}/\pi\Delta_{k_F}$ in the weak-coupling limit ($\xi \to \xi_0\pi/2\sqrt{2} = 1.11\xi_0$ when $G \ll 1$) and the bound-state radius r_0 in the strong-coupling limit. The behavior of ξ versus G turns out to be "smooth", although it strongly depends on the value of the parameter k_0 of the interaction. We attempt to eliminate the coupling constant G by replacing it by a dimensionless parameter containing ξ. Since k_F^{-1} is the only other independent physical length scale in the problem, we replace the original pair of variables (G, k_0) by the alternative pair $(k_F\xi, k_0)$ and study the crossover from BCS to BEC as a function of $k_F\xi$ for given k_0.

In Fig. 1 the chemical potential μ is plotted versus $k_F\xi$ for different values of the reduced density n/k_0^3 ($= 10^\alpha$ with $\alpha = -5, -4, \ldots, +4$). Positive values of μ have been normalized by the Fermi energy ϵ_F (equal to $k_F^2/2m^*$, by our definition), while negative values of μ have been normalized by half of the eigenvalue ϵ_0 of the two-body problem.

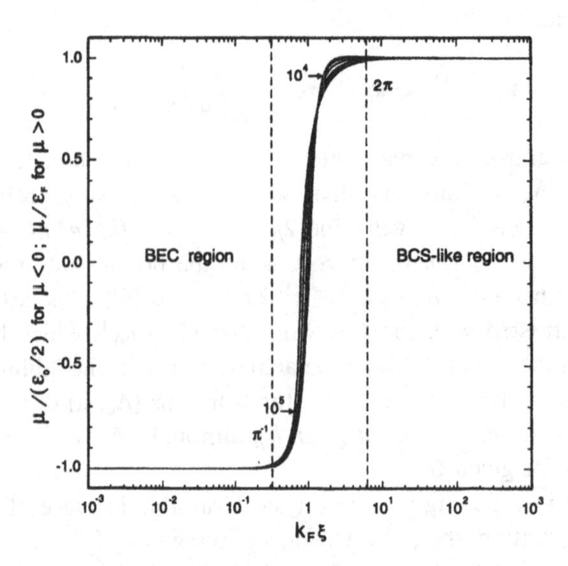

Fig. 1. Chemical potential μ versus $k_F\xi$ (at zero temperature). For the normalization of μ and the meaning of the different curves see the text. The two limiting curves corresponding to the values 10^{-5} and 10^4 of n/k_0^3 are marked with arrows.

The two curves shown for the reported extreme values of n/k_0^3 may be regarded as limiting curves for all practical purposes. The remarkable feature of Fig. 1 is that, when expressed in terms of $k_F\xi$, the behavior of the chemical potential is *universal* (i.e., independent of the parameter k_0 of the interaction potential), but possibly for an "intermediate" range $\pi^{-1} \lesssim k_F\xi \lesssim 2\pi$ [7]. This universal behavior of μ versus $k_F\xi$ suggests that $k_F\xi$ is the *appropriate variable* to follow the evolution from BCS to BEC.

Note also from Fig. 1 that μ is pinned to (about) the normal-state value ϵ_F when $k_F\xi \gtrsim 2\pi$, and that μ drops *rather abruptly* from ϵ_F at $k_F\xi \simeq 2\pi$. Fig. 1 thus shows that, when the coherence length ξ equals the Fermi wavelength $\lambda_F = 2\pi/k_F$, the system becomes unstable against bosonization and the Fermi surface disappears. We expect that the instability of the Cooper-pair-based superconductivity when $k_F\xi \simeq 2\pi$ should persist beyond the limits of validity of the procedure we have adopted to establish it. The stability criterion $k_F\xi \gtrsim 2\pi$ could then be regarded as the analog of the Ioffe–Regel criterion for transport in disordered systems.

One of us (F.P.) gratefully acknowledges partial research support from Europa Metalli-LMI S.p.A.

References

[1] P. Nozières and S. Schmitt-Rink, J. Low. Temp. Phys. **59**, 195 (1985).

[2] A.J. Leggett, in *Modern Trends in the Theory of Condensed Matter,* A. Pekalski and J. Przstawa, eds. (Springer-Verlag, Berlin, 1980), p. 13.

[3] S. Schmitt-Rink, C.M. Varma, and A.E. Ruckenstein, Phys. Rev. Lett. **63**, 445 (1989); R. Friedberg and T.D. Lee, Phys. Rev. B **40**, 6745 (1989); M. Randeria, J. Duan, and L. Shieh, Phys. Rev. Lett. **62**, 981 (1989) and Phys. Rev. B **41**, 327 (1990); L. Belkhir and M. Randeria, Phys. Rev. B **45**, 5087 (1992); M. Randeria, N. Trivedi, A. Moreo and R.T. Scalettar, Phys. Rev. Lett. **69**, 2001 (1992); R. Côté and A. Griffin, Phys. Rev. B **48**, 10404 (1993).

[4] Cf., e.g., J.R. Schrieffer, *Theory of Superconductivity* (Benjamin, New York, 1964), Chapter 2.

[5] We have eliminated the Hartree–Fock-like terms by setting $V_{\mathbf{k},\mathbf{k}} = 0$ for the diagonal components. This choice will not invalidate our results, since these terms turn out to be irrelevant in the parameter region of interest.

[6] NSR (Ref. [1]) state that $r_0 \sim k_0^{-1}$ for $G \rightarrow \infty$. Since bosonization can be achieved only when $r_0^3 \ll n^{-1}$, NSR are able to follow the evolution from BCS to BEC as a function of G *only* in the "dilute limit" $n/k_0^3 \ll 1$ for the reduced (three-dimensional) density, that is, for given density n only when $k_0 \gg k_F$. This limitation prevented NSR from connecting the two limits (BCS and BEC) irrespective of k_0. Our conclusion that $r_0 \sim k_0^{-1} G^{-1/2}$, on the other hand, enables us to satisfy the bosonization condition $(n/k_0^3)/G^{3/2} \ll 1$ even in the "dense limit" $n/k_0^3 \gg 1$, provided G is large enough.

[7] We have verified that the universal behavior of Fig. 1 for $k_F\xi \gtrsim 2\pi$ and $k_F\xi \lesssim \pi^{-1}$ is *independent* of the choice of the single-particle dispersion relation $\epsilon_{\mathbf{k}}$ and of the shape of the interaction potential $w_{\mathbf{k}}$ (barring pathological cases).

35

Bound States and Superfluidity in Strongly Coupled Fermion Systems

G. Röpke
Fachbereich Physik
Universität Rostock
Universitätsplatz 3
D 18051 Rostock
Germany

Abstract

The two-particle spectrum in a dense Fermion system is treated using a thermodynamic Green function approach. A self-consistent description of possible bound states and a superfluid condensate in a correlated medium is given.

A detailed understanding of superfluidity and superconductivity in correlated Fermion systems, especially the transition from the Cooper-paired state (weak-coupling limit) to the Bose condensed state of tightly bound pairs of Fermions (strong coupling limit) [1], is of great interest for very different physical systems. The problem of a unified treatment of Bose–Einstein condensation (BEC) and the Bardeen–Cooper–Schrieffer (BCS) phase arises not only in describing the electron structure of strongly correlated electron superfluids such as superconductors [2], the electron–hole system in semiconductors [3], spin-polarized hydrogen and liquid ^{3}He [4], but also in the theory of nuclear matter [5] and quark–gluon systems [6].

Recently, there have been several new approaches to this stimulating problem. A Monte-Carlo simulation of a finite He system at zero temperature has been performed in Ref. [7]. The microscopic theory of strongly coupled quantum fluids has been treated within the Jastrow approximation (cf. Ref. [8]) to obtain the ground state and low-lying excited states of strongly correlated boson quantum fluids. The crossover from weak to strong coupling superconductivity has been considered using a functional integral representation [2] (see also Ref. [7]). The critical temperature for the onset of a superfluid state has been obtained within the Matsubara Green function approach neglecting the interaction between pairs [9] (see also Ref. [10] for pairing in two dimensions). A finite temperature Green function approach describing both two-particle (Brueckner ladders) and pairing correlations has been considered in Ref. [11], but a self-consistent solution of the gap equation has not been obtained so far (see Zimmermann [3]).

Clearly one would like an improvement of the Hartree–Fock–Gorkov mean-field description. As is well known, the Gorkov equation corresponds to a Hartree–Fock approximation in the normal state. A thermodynamic Green's function approach has been developed for the normal state which is able to describe the formation of bound states and their dissolution at higher densities (Mott effect, see Ref. [9]). We will give an approach which includes both correlations and bound state formation in the superfluid state. In this way, the equation of state is obtained for the normal state and the superfluid state in a consistent way. Furthermore, a unified description of BEC and BCS can be given, with a generalized Gorkov equation.

The fermion system is described by the Hamiltonian

$$H = \sum_{1} E(1) a_1^+ a_1 + \frac{1}{2} \sum_{121'2'} V(121'2') a_1^+ a_2^+ a_{2'} a_{1'} \tag{1}$$

with $1 = \{\vec{p}_1 \sigma_1\}$ denoting both momentum and internal quantum numbers such as spin. Due to the interaction $V(p\alpha, k\beta; p'\alpha', k'\beta')$, which is assumed to be proportional to $\delta_{\alpha\alpha'} \delta_{\beta\beta'} \delta_{p+k,p'+k'}$, correlations and bound states can arise. We restrict ourselves in this paper to two-particle correlations. We note, however, that the generalization to higher order correlations is straightforward.

In general, the many-particle system described by (1) should be treated within a nonequilibrium approach. To derive kinetic equations for the reduced density matrices, we follow the scheme cited in Ref. [12] based on the Zubarev approach [13] to the nonequilibrium statistical operator. In Ref. [12], a Boltzmann-type equation is derived in the normal state for the single-particle diagonal distribution function

$$\langle a_2^+ a_1 \rangle^t = n(12;t) = \delta_{12} n(1,t). \tag{2}$$

Here, we also include the time evolution of the pair amplitude $\langle a_2 a_1 \rangle^t = F(12;t)$ and the two-particle distribution function

$$\langle a_1^+ a_2^+ a_{2'} a_{1'} \rangle^t = n(1'1,t) n(2'2,t)_{\text{ex}} + c(12,1'2',t), \tag{3}$$

where the index *ex* denotes antisymmetrization. For simplicity, we restrict ourselves to the special case of a homogeneous system (see Eq. (2)).

Furthermore, we consider only singlet pairing with total momentum $2q$, and thus

$$F(12,t) = F(p,t)e^{-i\alpha_p(t)}\chi(12)\delta_{p_1+p_2,2q}, \quad (4)$$

with $\chi(12)$ denoting the spin function, and $\vec{p}$ the relative momentum.

We introduce new operators b, b^+ which are defined by the time-dependent Bogoliubov–Valatin transformation

$$\begin{aligned} a_{q+p\uparrow} &= u_p b_{q+p\uparrow} + v_p b^+_{q-p\downarrow}, \\ a_{q-p\downarrow} &= u_p b_{q-p\downarrow} - v_p b^+_{q+p\uparrow}, \end{aligned} \quad (5)$$

where

$$\begin{matrix} 2|u_p|^2 \\ 2|v_p|^2 \end{matrix} \equiv 1 \pm (1+\vartheta_p^2)^{-1/2} \quad (6)$$

with

$$\vartheta_p = \frac{\sqrt{2}|F(12)|}{1-n(1)-n(2)}.$$

The anomalous (or off-diagonal) mean values $\langle b_2 b_1 \rangle^t$ vanish.

Expressing $\hat{H} = H - \sum_1 \mu_1 a_1^+ a_1$ in the b-representation, we have

$$\hat{H} = h_0 + \sum_1 h(1) b_1^+ b_1 + \frac{1}{2} \sum_{121'2'} W(12,1'2') b_1^+ b_2^+ b_{2'} b_{1'} - (\mathrm{MF}) + H^{\mathrm{off}}, \quad (7)$$

where the mean-field (MF) contributions are contained in $h(1)$. Additional "off-diagonal" contributions involving b, b^+ such as $b_2 b_1$ are collected in H^{off}. In the b-representation, the modified interaction is (for example)

$$\begin{aligned} & W(q+p_1\uparrow, q-p_2\downarrow; q+p_1'\uparrow, q-p_2'\downarrow) \\ & = V(q+p_1\uparrow, q-p_2\downarrow; q+p_1'\uparrow, q-p_2'\downarrow) u^*_{p_1} u^*_{p_2} u_{p_2'} u_{p_1'} \\ & + V(q+p_2'\uparrow, q-p_2\downarrow; q+p_1'\uparrow, q-p_1\downarrow) v_{p_1} u^*_{p_2} v^*_{p_2'} u_{p_1'} \\ & + V(q+p_1\uparrow, q+p_2'\uparrow; q+p_2\uparrow, q+p_1'\uparrow)_{\mathrm{ex}} u^*_{p_1} v_{p_2} v^*_{p_2'} u_{p_1'} \\ & + (p\uparrow \leftrightarrow -p\downarrow, u \leftrightarrow v^*). \end{aligned} \quad (8)$$

Similar expressions are obtained for other spin combinations. In terms of these new operators, in place of the original distribution function given in (2)–(4), we can use the mean values $< b_1^+ b_1 >^t, < b_1^+ b_2^+ b_{2'} b_{1'} >^t$ to describe the system. We will not derive the kinetic equations for the corresponding quantities $N(1,t)$, $\vartheta(12,t)$, $C(12,1'2',t)$ (see Ref. [12]), but will only discuss these expressions for thermal equilibrium. A systematic perturbative treatment can be given of the statistical operator

$Z_0^{-1}\exp[-\beta\hat{H}]$ with the help of the usual Matsubara Green function technique. As pointed out above, we assume that only single-particle and two-particle states are of relevance. In the cluster-Hartree–Fock approximation (i.e., interactions involving the single-particle and two-particle states in the medium are taken in the Born approximation), we obtain the following diagrammatic representation for the single-particle and two-particle Green functions (compare Ref. [14] for the normal state case):

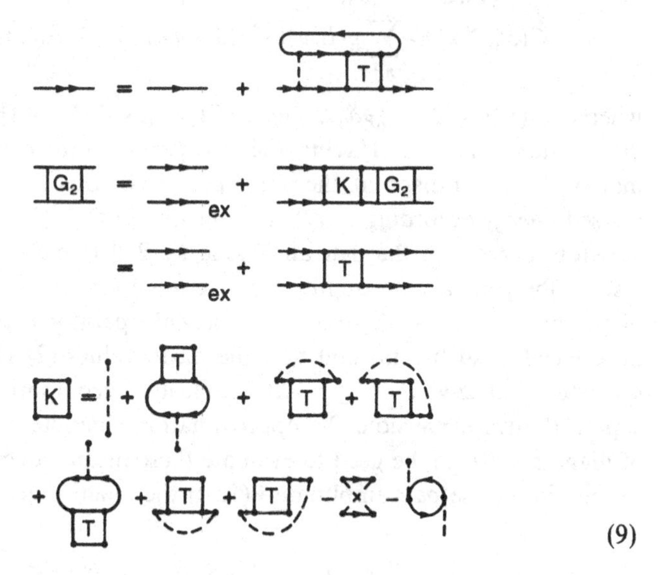

(9)

In (9), a line with an arrow signifies the mean-field single-particle Green function, while the broken line denotes the effective interaction W, such as given by (8). One can show that it is consistent to neglect terms arising from H^{off} if we work in the Born approximation.

The solution of the Bethe–Goldstone equation (9) can be given via a spectral representation of the single-frequency, two-particle Green function, where the two-particle states $\varphi_{nP}(12)$ and eigenvalues E_{nP} follow from the Schrödinger-like wave equation

$$[h(1)+h(2)-E_{nP}]\varphi_{nP}(12) \qquad (10)$$
$$+\sum_{1'2'}\left[W(12,1'2')(1-N(1)-N(2))+\Delta H_2(12,1'2')\right]\varphi_{nP}(1'2')=0.$$

In addition to the mean-field contributions in $h(1)$ and the Pauli blocking terms, additional "in-medium" effects are described by ΔH_2. For simplicity, we consider only two-particle correlations in the singlet spin channel, so that $\varphi_{nP}(12) = \varphi_n(p)\chi(12)\delta_{p_1+p_2,P}$. Evaluating the diagrams in (9), we find in the low-density limit

$$\Delta H_2(12, 1'2') = \sum_{343'4'} W(1234, 1'2'3'4')C(3'4', 34), \tag{11}$$

with

$$\begin{aligned} W(1234, 1'2'3'4') &= W(13, 1'3')_{\rm ex}\delta_{4'2}\delta_{42'} \\ &+ W(14, 3'2')\delta_{4'2}\delta_{31'} - W(14, 3'4')\delta_{2'2}\delta_{31'} + (1 \leftrightarrow 2, 3 \leftrightarrow 4), \\ C(34, 3'4') &= \sum_{nP} g(E_{nP})[\varphi^*_{nP}(3'4')\varphi_{nP}(34) - \delta_{nP}(34)\delta_{nP}(3'4')], \end{aligned} \tag{12}$$

where $\delta_{nP}(12) = \delta_{p_1+p_2,P}\delta_{n,p}$, and $g(E) = [\exp(\beta E) - 1]^{-1}$ is the Bose distribution function. Having the two-particle Green function (or T-matrix T_2) at our disposal, the single-particle density can be obtained via the self-energy according to (9), as discussed in Ref. [9].

Before evaluating the pair amplitude $F(12, t)$ (see Eq. (4)), we briefly discuss the general nonequilibrium case (compare Ref. [12] for the case of the normal state). A relevant statistical operator $\exp(-S(t))$ can be introduced in such a fashion that the mean values (2)–(4) are correctly reproduced at any value of t. After a cluster decomposition of $S(t)$ and a perturbative expansion, the approximation characterized by the class of diagrams (9) can be used to evaluate these mean values. In particular, we obtain for the pair amplitude $F(12, t)$ the kinetic equation

$$\begin{aligned} \frac{d}{dt}F(12, t) &= \frac{i}{\hbar}\epsilon \int_{-\infty}^{0} dt' e^{\epsilon t'} e^{i(\mu_1+\mu_2)t'/\hbar} \\ &\times \mathrm{Tr}\{e^{-S(t'+t)}e^{-i\hat{H}t'/\hbar}[H, a_2a_1]e^{i\hat{H}t/\hbar}\}. \end{aligned} \tag{13}$$

To find the stationary solution $F(12, t) \sim \exp(-i\omega t)$, we can perform a perturbative expansion with respect to $H' \equiv \hat{H} - S/\beta$. In zeroth order, the t' integral can be immediately performed and yields a nonvanishing result in the limit $\epsilon \to 0$ only if the condition $\hbar\omega = \mu_1 + \mu_2$ is fulfilled. We find

$$\begin{aligned} &[E^{HF}(1) + E^{HF}(2) - \mu_1 - \mu_2]F(12, t) \\ &+ \sum_{1'2'} V(12, 1'2')[1 - n(1) - n(2)]F(1'2', t) \\ &- \sum_{1'33'} [Y(13, 1'3')C(13, 1'3') - \tilde{Y}(13, 1'3')C^*(13, 1'3')] \\ &+ (1 \leftrightarrow 2) = 0, \end{aligned} \tag{14}$$

where we have defined

$$\begin{aligned}
&Y(q+k\uparrow,q-p\downarrow;q+k'\uparrow,q-p'\downarrow)\\
&=V(q+k\uparrow,q-p\downarrow;q+k'\uparrow,q-p'\downarrow)v_k u_p^* u_{p'} u_{k'}\\
&+V(q+k\uparrow,q-k'\downarrow;q+p\uparrow,q-p'\downarrow)v_k v_p u_{p'} v_{k'}^*\\
&+V(q+k\uparrow,q+p'\uparrow;q+p\uparrow,q+k'\uparrow)_{\text{ex}} v_k v_p v_{p'}^* u_{k'}.
\end{aligned} \tag{15}$$

$\tilde{Y}(13,1'3')$ follows from $Y(13,1'3')$, with $k\uparrow\leftrightarrow -k\downarrow$, $p\uparrow\leftrightarrow -p\downarrow$, and $u\leftrightarrow v$.

In this manner, together with the treatment of the two-particle propagator (9), a self-consistent description of the stationary state for a superfluid system with two-particle correlations is obtained from (10) and (14). At a given temperature T, the remaining parameters μ_i and $\vec{q}$ are fixed by the particle and current densities. In the trivial case $C=0$, the Gorkov equation is obtained from (14), and the BCS theory results in the weak-coupling limit. In the strong-coupling limit, we find Bose condensation of bound states. In the limit of low temperatures and low densities, the gap equation (14) only has solutions if $\mu_1+\mu_2$ is equal to the bound state energy.

The main theoretical advance given here is to take into account the influence of "in-medium" correlations and bound states, as described by the gap equation (14) and the modification of the two-particle properties (10) in the superfluid state. In contrast to the Gorkov approach, which corresponds to a mean-field theory, in the normal state ($F(12,t)\equiv 0$) the equation of state of a correlated many-particle system [9] is obtained where the two-particle states are described according to (10).

To illustrate the modification of the two-particle spectrum in a simple exploratory calculation, we consider the one-dimensional case, where only a short-range δ-like interaction is assumed to act between opposite spins:

$$V_{\alpha\beta}(p_1p_2,p_1'p_2') = -2\hbar(\varepsilon/m)^{1/2}L_0^{-1}\delta_{\alpha,-\beta}\delta_{p_1+p_2,p_1'+p_2'}. \tag{16}$$

The parameter ε is the binding energy of the isolated two-particle system. The two-particle energies are shown in Fig. 1 as a function of the total momentum P at given values of $T, N/L_0, q$. The shifts of bound state energies $E^{\text{b}}(P)$ and the continuum edge of scattering states $E^{\text{sc}}(P)=2E_1(P/2)$ are determined in this case by a Hartree term, which is constant, and the Pauli blocking term. The Pauli blocking term, which is large for two-particle states with total momentum P near $2q$, leads to a deformation of the two-particle spectrum. It should be noted out that the

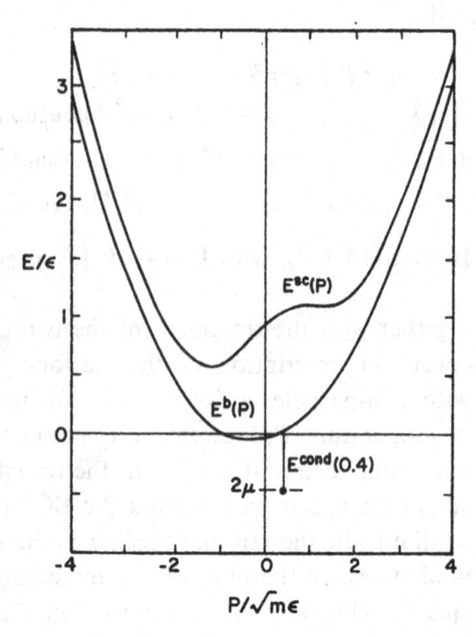

Fig. 1. Two-particle energy spectrum in a strongly correlated superfluid system as a function of the total momentum P: $E^{\mathrm{b}}(P)$ is the energy of the bound state; $E^{\mathrm{sc}}(P)$ is the energy of the continuum edge of scattering states; $E^{\mathrm{cond}}(2q) = 2\mu$ is the energy of the superfluid condensate. Parameter values: temperature $\beta^{-1} = 0.1\varepsilon$, density $N/\Omega_0 = 0.5(m\varepsilon)^{1/2}/\hbar$, $q = 0.2(m\varepsilon)^{1/2}/\hbar$; ε is the binding energy of the isolated two-particle system.

two-particle energies lie well above the energy of the Bose condensate at $E^{\mathrm{cond}}(2q) = 2\mu$. Furthermore, the two-particle binding energy (difference between the continuum edge of scattering states and the bound state energy) depends on the total momentum and decreases with increasing density. The Mott effect arises when the bound state energies merge with the continuum of scattering states. This illustrates that our approach to strongly coupled superfluidity corresponds, in the normal phase, not only to a Hartree–Fock approximation, but also to a self-consistent description of two-particle correlations [9]. In particular, our present approach includes the formation of bound states in a correlated medium and their disappearance (or unbinding) at high densities. The Pauli blocking is responsible not only for the dissolution of the bound states, but it also arises as the essential mechanism to establish the superfluid condensate in the strong-coupling limit.

An interesting result of our approach is the description of the transition

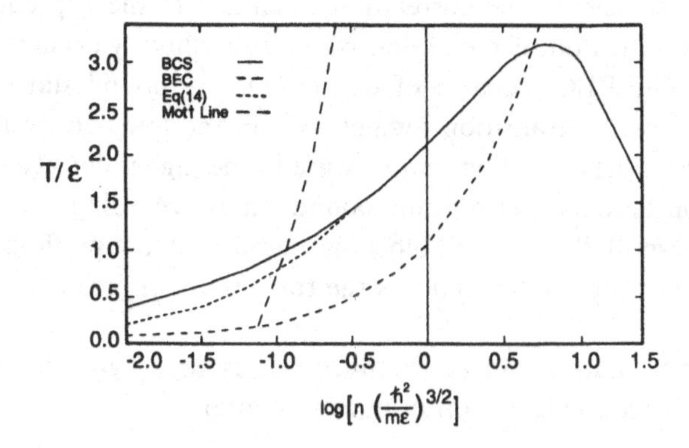

Fig. 2. Phase transition region to the superfluid state for a Fermionic system with separable interaction (17), $\gamma = 6.25$. The solution of the gap equation (14) including correlations in the medium is compared with the solution of the Gorkov equation for vanishing gap (BCS) and the Bose–Einstein condensation of noninteracting bound states (BEC). Bound states are "blocked out" for densities above the Mott line.

from the normal state to the superfluid state. We present a calculation for a simple separable interaction of the Yamaguchi type

$$V_{\alpha\beta}(\mathbf{p}_1\mathbf{p}_2, \mathbf{p}_1'\mathbf{p}_2') = -8\pi\gamma(1+\gamma)^2\Omega_0^{-1}(\hbar^2/m\varepsilon)^{3/2}\varepsilon \\ \times g(\mathbf{p}_1 - \mathbf{p}_2)\, g(\mathbf{p}_1' - \mathbf{p}_2')\, \delta_{\alpha,-\beta}\, \delta_{p_1+p_2, p_1'+p_2'}, \tag{17}$$

where $g(\mathbf{p}) = (\hbar^2 p^2/m\varepsilon + \gamma^2)^{-1}$. For large values of the parameter γ, the Hartree energy shifts are small compared with the Fock terms. In Fig. 2, the transition temperature from the normal state to the superfluid state is shown for the Gorkov equation (BCS, no correlation in the medium), for non-interacting bosons (BEC) and for the solution of the gap equation (14), including self-consistent correlations in the medium. A value $\gamma = 6.25$ was choosen corresponding to the interaction in the triplet (deuteron) channel in nuclear matter. The Mott line shows the densities above which the bound states are "blocked out". A homogeneous state has been considered, neglecting first-order phase transitions (see Ref. [9]).

The Gorkov equation overestimates the transition temperature in the low density region, where bound states are of importance. Because few bound states are excited, the medium is described in terms of free-quasiparticle states which are relatively weakly populated. In the high density region, the bound states are shifted into the continuum of scattering states and are blocked out (Mott effect). Here, the usual BCS picture is applicable. In general, the correlations included in the gap equation (14) leads to a reduction of the region where superfluidity occurs.

In contrast, the BEC scenario of non-interacting bound states leads to small values of the transition temperature in the low density region, but overestimates the transition temperature in the higher density region. The interaction between the bosonic bound states according to the gap equation (14), especially the Pauli blocking, stabilizes the superfluid phase in the low density region, but reduces the transition temperature at high densities.

The present formalism can be extended to describe higher order correlations, such as α-condensation in nuclear matter.

References

[1] P. Nozières and S. Schmitt-Rink, J. Low Temp. Phys. **59**, 159 (1985); M. Randeria, this volume.

[2] M. Drechsler and W. Zwerger, Ann. Physik **1**, 15 (1992); A.J. Leggett, in *Modern Trends in the Theory of Condensed Matter*, A. Pekalski and J. Przystawa, eds. (Springer, Berlin, 1980). M. Randeria, J.M. Duan, and L.Y. Shieh, Phys. Rev. Lett. **62**, 981 (1989); Phys. Rev. B **41**, 327 (1990); see also M. Randeria, this volume.

[3] H. Haug and S. Schmitt-Rink, Progress in Quantum Electronics **9**, 3 (1984); R. Zimmermann, *Many-Particle Theory of Highly Excited Semiconductors* (Teubner, Leipzig, 1987).

[4] I.F. Silvera, J. Low Temp. Phys. **89**, 287 (1992); D. Vollhardt and P. Wölfle, *The Superfluid Phases of Helium-3* (Taylor and Francis, London, 1990).

[5] R.K. Su, S.D. Yang and T.T.S. Kuo, Phys. Rev. C **35**, 1539 (1987); J.M.C. Chen, J.W. Clark, E. Krotscheck, and R.A. Smith, Nucl. Phys. A **451**, 509 (1986); T. Alm, G. Röpke, and M. Schmidt, Z. Phys. A **337**, 355 (1990).

[6] W. Weise and U. Vogl, Progr. Part. Nucl. Phys. **27**, 195 (1991); S. Klevanski, Rev. Mod. Phys. **64**, 649 (1992).

[7] S.A. Chin and E. Krotschek, Phys. Rev. Lett. **65**, 2658 (1990); Phys. Rev. B **33**, 3158 (1986); see also D.M. Ceperley and E.L. Pollock, Phys. Rev. Lett. **56**, 351 (1986); Phys. Rev. B **36**, 8343 (1987).

[8] C.E. Campbell and B.E. Clements, in *Elementary Excitations in Quantum Fluids*, K. Ohbayashi and M. Watanabe, eds. (Springer, Berlin, 1989).

[9] M. Schmidt, G. Röpke, and H. Schulz, Ann. Phys. (N.Y.) **202**, 57 (1990).

[10] S. Schmitt-Rink, C.M. Varma, and A.E. Ruckenstein, Phys. Rev. Lett. **63**, 445 (1989).

[11] W.H. Dickhoff, Phys. Lett. B **210**, 15 (1988); W.H. Voderfecht, C.C. Gearhart, W.H. Dickhoff, A. Polls and A. Ramos, Phys. Lett. **253**, 1 (1991).

[12] G. Röpke and H. Schulz, Nucl. Phys. A **477**, 472 (1988).

[13] D.N. Zubarev, *Nonequilibrium Statistical Thermodynamics* (Plenum, New York, 1974).

[14] G. Röpke, T. Seifert. H. Stolz, and R. Zimmermann, Phys. Stat. Sol. (b) **100**, 215 (1980).

36

Onset of Superfluidity in Nuclear Matter

A. Hellmich, G. Röpke, A. Schnell, and H. Stein
Fachbereich Physik and MPG-AG "Theoretische Vielteilchenphysik"
Universität Rostock
PF 999, D 18051 Rostock
Germany

Abstract

Within the Gorkov approach, the onset of superfluidity in nuclear matter is considered in the spin-singlet (S=0) *and triplet* (S=1) *channel. In the triplet channel, a transition from Bose–Einstein condensation (BEC) of deuterons at low densities to a BCS state at high densities is obtained. It is shown that correlations and bound state formation in the medium should be included to improve the Gorkov approach. The drastic change in the composition of the system due to the Mott effect is investigated.*

1 Introduction

Nuclear matter is an interesting example of strongly coupled fermions showing a transition to a superfluid state at densities $n \lesssim n_0 = 0.17 \text{ fm}^{-3}$ and temperatures of the order $T \sim 1$ MeV. We consider symmetric nuclear matter containing equal ratios of protons and neutrons. The nucleon–nucleon interaction is taken in the form of a separable potential [1]

$$V_\alpha^{LL'}(k,k') = \sum_{i,j=1}^{n} v_{\alpha_i}^L(k)\lambda_{\alpha ij}^{LL'} v_{\alpha_j}^{L'}(k'), \tag{1}$$

where $v_{\alpha_i}^L(k)$ denotes the form factor in the channel $\alpha = {}^{2S+1}L_J$ for angular momentum L, $\lambda_{\alpha ij}^{LL'}$ are coupling constants .

Calculations are performed for a rank $n = 1$ potential of the Yamaguchi type describing scattering states and a bound state (deuteron) in the S-channel. Nuclear matter is an example of a system showing Bose–Einstein condensation (BEC) of bound states (deuterons) at low densities and Cooper pairing (BCS) at higher densities. (See also the article by Randeria in this volume.) Therefore, we expect a smooth transition from strong coupling in the BEC regime to weak coupling in the BCS regime. We apply the formalism given by Nozières and Schmitt-Rink [2] (see also [3]) to nuclear matter.

In Section 2 the critical temperature is derived using two alternative approaches, the equation of motion for thermodynamic Green functions and the T-matrix approach. For the gap and the critical temperature model, calculations using the Yamaguchi potential are given in Section 3. The equation of state and the composition of nuclear matter, the Mott effect and the onset of a superfluid phase are discussed in Section 4. For convenience we put $\hbar$, c, k_B =1.

2 Determination of the Critical Temperature for Superfluidity in Fermionic Systems

2.1 Gap Equation from BCS Theory

The gap equation can be derived by the real-time finite temperature Green function method (see [4]). Starting with a Hamiltonian $\tilde{H} = H - \mu N$,

$$\tilde{H} = \sum_i (\frac{k_i^2}{2m} - \mu) a_i^+ a_i + \frac{1}{2} \sum_{i,j,l,m} \langle ij \mid V \mid lm \rangle a_i^+ a_j^+ a_m a_l \,, \tag{2}$$

the single-particle retarded Green function $G_{12}(t-t') = \langle\langle a_1(t), a_2^+(t') \rangle\rangle$ and the off-diagonal Green function $F_{12}(t-t') = \langle\langle a_1^+(t), a_2^+(t') \rangle\rangle$ are introduced. Gorkov-type coupled equations for G and F are obtained from the equations of motion after decoupling in the abnormal pair cut-off approximation. From the solution for F_{12}, a temperature dependent gap equation results:

$$\Delta(k) = -\frac{1}{2} \sum_{k'} V(k,k') \frac{\Delta(k')}{\sqrt{\varepsilon_{k'}^2 + \Delta(k')^2}} \tanh\left(\frac{1}{2T} \sqrt{\varepsilon_{k'}^2 + \Delta(k')^2} \right), \tag{3}$$

with $\varepsilon_k = (k^2/2m - \mu) + g \sum_{k'} \langle k,k' \mid V \mid k,k' \rangle n_{k'} - \sum_{k'} \langle k,k' \mid V \mid k',k \rangle n_{k'}$, $n_k = \langle a_{k,\alpha}^+ a_{k,\alpha} \rangle$ and $g = (2s+1)(2\tau+1)$ denoting single-particle spin and isospin degeneracies.

2.2 T-matrix Approach

Bound and scattering states in nuclear matter are obtained from a non-relativistic Bethe–Goldstone equation $T = V + VG_2^0 T$ for the T-matrix.

The two-fermion propagator has the form

$$G_2^0(k,K,z) = \frac{\int \left[1 - f_1(\frac{\mathbf{K}}{2} + \mathbf{k}) - f_2(\frac{\mathbf{K}}{2} - \mathbf{k})\right] \frac{d\Omega}{4\pi}}{z - \left[\frac{K^2}{4m} + \frac{k^2}{m} + \Delta\varepsilon_1(k,K) + \Delta\varepsilon_2(k,K)\right]}, \tag{4}$$

where $\mathbf{k}$ is the relative and $\mathbf{K}$ the total momentum and $\Delta\varepsilon(k,K)$ is the angle averaged quasiparticle energy shift due to the interactions with the medium. In this approximation, the effects of the medium enter solely through the phase-space occupation factors $f(\mathbf{k},\mathbf{K})$ and the quasiparticle energy shift $\Delta\varepsilon(k,K)$.

With a separable interaction (1), the T-matrix equation reduces to an algebraic one (we consider uncoupled channels $L = L'$) [3]:

$$T_\alpha(kk'K,z) = \sum_{ijk} v_{\alpha i}(k)[1 - J_\alpha(K,\mu,T,z)]_{ij}^{-1} \lambda_{\alpha jk} v_{\alpha k}(k'), \tag{5}$$

where

$$J_\alpha(K,\mu,T,z)_{ij} = 4\pi \int \frac{dkk^2}{(2\pi)^3} \sum_n \lambda_{\alpha in} v_{\alpha n}(k) v_{\alpha j}(k) G_2^0(k,K,z). \tag{6}$$

The critical temperature T_c for the onset of pairing is given by the Thouless criterion that a pole at $z = \mu_1 + \mu_2$ arises for the two-particle T-matrix at $K = 0$. For the T-matrix, this pole condition reads

$$\det[1 - \mathrm{Re} J_\alpha(K = 0, T = T_c, z = \mu_1 + \mu_2)]_{ij} = 0. \tag{7}$$

If the pole corresponds to a two-body bound state, the imaginary part of J_α vanishes. Furthermore, at this particular energy $z = \mu_1 + \mu_2$, the imaginary part of J_α vanishes when it lies in the continuum of scattering states, because the Pauli blocking factor $(1 - f_1 - f_2)$ contained in (4) vanishes in this case. Thus, the pole condition (7) for the onset of superfluidity holds for positive as well as negative values of $\mu_1 + \mu_2$. It coincides with the solution of the BCS theory for vanishing gap. This is most easily demonstrated for the $\alpha={}^1S_0$ channel (see Eq. (11) below), but it holds also in other channels as well as in isospin asymmetric nuclear matter [5, 6].

3 Model Calculation

3.1 The Separable Yamaguchi Nucleon–Nucleon Interaction

In the following calculation, we use the separable Yamaguchi potential of rank $n = 1$ [7]:

$$\langle \mathbf{k} \mid V \mid \mathbf{k}' \rangle = -\lambda_\alpha v(\mathbf{k}) v(\mathbf{k}'), \tag{8}$$

with the form factor

$$v(k) = \frac{1}{k^2 + \beta^2}. \tag{9}$$

The coupling of the $^3S_1$ state to the $^3D_1$ state is neglected. The parameters β, λ_s (singlet channel) and λ_t (triplet channel) are fitted to the empirical nucleon–nucleon scattering phase shifts and the deuteron binding energy

$$E_b^0 = -2.22 \text{ MeV}.$$

With the inverse potential range β=1.4488 fm^{-1}, empirical values for the potential parameters λ_α are

$$\lambda_t = (2\pi)^3 \frac{(\sqrt{-mE_b^0} + \beta)^2}{m\pi^2} \beta = 4263 \text{ MeV fm}^{-1},$$

$$\lambda_s = 2994 \text{ MeV fm}^{-1},$$

where $m = 938.9$ MeV denotes the nucleon mass.

This simple rank 1 potential has been choosen to illustrate the many-particle effects in nuclear matter. In order to obtain the effects of a repulsive core, more realistic nucleon–nucleon interaction should be considered. Starting from a general BCS theory for pairing in arbitrary channels [8] this is done in [6] for the Graz-II potential which also considers the coupling of the $^3S_1$ - $^3D_1$ channel. The fact that different realistic potentials yield very similar results has been discussed in Refs. [9, 10].

3.2 Solution of the Gap Equation

For the separable potential (8), the solution of the gap equation (3) reads

$$\Delta(k) = C(\mu, T) \frac{1}{k^2 + \beta^2}, \tag{10}$$

where C contains the summation over k'. Fig. 1 shows the shape of the gap which is proportional to the form factor of the potential. At the critical

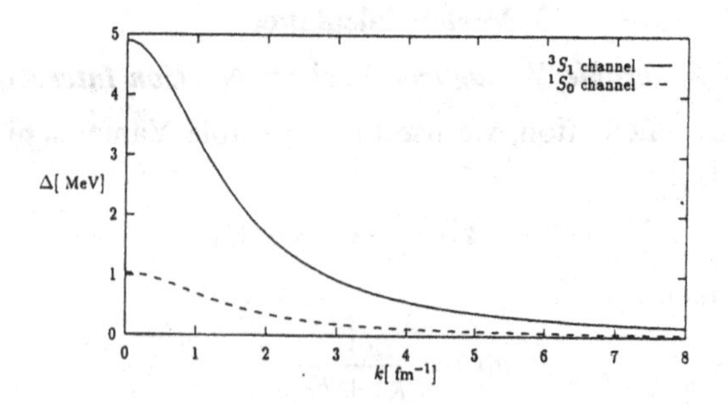

Fig. 1. Gap function $\Delta(k)$ for triplet and singlet channel at $\mu_{\text{eff}} = 2$ MeV and $T = 0.3$ MeV.

temperature T_c, the gap vanishes, which corresponds to $C(\mu, T) = 0$. To determine T_c, we find from (3):

$$1 = \frac{1}{2}\frac{\lambda_\alpha}{2\pi^2}\int_0^{+\infty} dk' \frac{k'^2}{(k'^2+\beta^2)^2\varepsilon_{k'}} \tanh\left(\frac{1}{2T_c}\varepsilon_{k'}\right), \tag{11}$$

where

$$\varepsilon_k = \frac{k^2}{2m} + \Delta\varepsilon^{HF}(k) - \mu\,. \tag{12}$$

We have introduced the k-dependent Hartree–Fock (HF) shift of quasi-particle energy

$$\Delta\varepsilon^{HF}(k) = -3\lambda_\alpha \frac{1}{(2\pi)^3}\int d^3\mathbf{k}' v\left(\frac{\mathbf{k}-\mathbf{k}'}{2}\right) v\left(\frac{\mathbf{k}-\mathbf{k}'}{2}\right) f(k')\,, \tag{13}$$

where $f(k)$ is the Fermi distribution function which contains the shift self-consistently.

We introduce an effective chemical potential $\mu_{\text{eff}} = \mu - \Delta\varepsilon(0)$ so that the continuum of scattering states starts at $\mu_{\text{eff}} = 0$ for $k = 0$. Thus (12) reads

$$\varepsilon_k = \frac{k^2}{2m} + \Delta\varepsilon^{HF}(k) - \Delta\varepsilon^{HF}(0) - \mu_{\text{eff}}\,.$$

As shown in Fig. 2, the critical temperature in the $^3S_1$ channel is larger than that in the $^1S_0$ channel due to the stronger attraction. In both channels, the maximum of the critical temperature reduces remarkably if the Hartree–Fock shift is considered self-consistently (see also [6]).

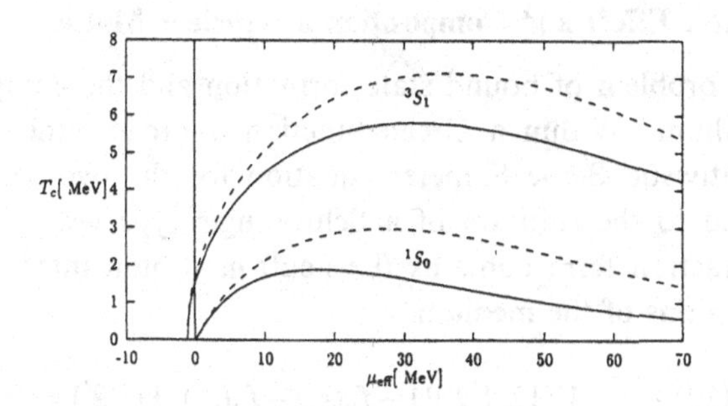

Fig. 2. Critical temperature in the $^3S_1$ and $^1S_0$ channel without (dashed) and with (bold) selfconsistent HF shifts.

3.3 *On the Relation between Bose Condensation and Cooper Pairing*

In the case of $^1S_0$ pairing, the condition for the critical temperature (7) yields non trivial solutions only for $\mu_{\text{eff}} > 0$. For $^3S_1$ pairing, the superfluid region shown in Fig. 2 extends down to negative values until $\mu_{\text{eff}} = E_b^0/2$ at $T_c = 0$. This limit represents the formation and Bose condensation of deuterons.

In order to understand the difference between Bose condensation of deuteron-like bound states and the pairing phenomenon described by BCS theory, we discuss the two cases $\mu_{\text{eff}} < 0$ and $\mu_{\text{eff}} > 0$ separately. A pole in the T-matrix at $E_b = 2\mu_{\text{eff}} < 0$ defines the critical temperature T_c for Bose condensation of a bound-state with in-medium binding energy E_b. In the low-density limit, T_c tends to 0 and μ_{eff} to $E_b^0/2$. In this limit, the phase-space occupation and the self-energy effects in (4) can be neglected and nuclear matter is an ideal gas of structureless bosons (strong coupling limit).

For $\mu_{\text{eff}} > 0$, we find the formation of Cooper pairs according to BCS theory, which leads to the familiar expression for the critical temperature in the weak-coupling limit. Of special interest is the region around $\mu_{\text{eff}} = 0$, where the in-medium bound states with $K = 0$ disappear due to the Mott effect [3]. Here the pole of the T-matrix, which defines the critical temperature, moves continuously from negative to positive energy values.

The smooth transition between the two limiting cases of Cooper pairing and Bose condensation has already been discussed in a more general context [2, 11, 12].

4 Mott Effect and Composition of Nuclear Matter

We discuss the problem of bound state formation and the composition in a dense medium. Within a Green function approach, the solution of the non-relativistic Bethe–Salpeter equation for the two-particle T-matrix is related to the solution of a Schrödinger-type wave equation where the interaction term contains the neutron–proton interaction as well as contributions of the medium,

$$(\varepsilon_1+\varepsilon_2-E_b)\Psi(12)+\sum_{1'2'} V(12,1'2')[1-f_n(\varepsilon_1)-f_p(\varepsilon_2)]\Psi(1'2')=0. \quad (14)$$

The solution gives a shift of the binding energies and a modification of the wave function. Effects of the medium enter via the HF-shift (13) in $\varepsilon_{1,2}$ and the Pauli blocking factor $[1-f_n(\mathbf{K}/2+\mathbf{k})-f_p(\mathbf{K}/2-\mathbf{k})]$, which accounts for the phase-space occupation ($\mathbf{k}$: relative momentum). Due to the Pauli blocking, the deuteron binding energy E_b becomes dependent on temperature, chemical potential and total momentum $\mathbf{K}$ and increases with increasing density until the bound state disappears (Mott effect) at the so-called Mott density. Increasing either the temperature or total momentum shifts the Mott density to higher values. For each density higher then the Mott density at $K=0$, bound states exist only for K above a minimum total momentum K_{mott}. The question of effective two-body scattering in a medium in a T-matrix formalism has been treated in Refs. [13, 14].

We consider the composition of the nuclear matter system to obtain an equation of state for the normal phase which is applicable until the critical temperature given by (11) is reached. The importance of correlated pairs in this temperature and density region is investigated by splitting the total density in a contribution due to free nucleons, a deuteron density and a contribution due to scattering states:

$$n_{\text{tot}} = n_{\text{free}}(\mu, T) + 2n_{\text{corr}}(\mu, T)$$

with

$$n_{\text{corr}}(\mu, T) = n_{\text{deut}}(\mu, T) + n_{\text{scat}}(\mu, T).$$

The different contributions are derived from a Green function approach [3] so that a generalized Beth–Uhlenbeck formula is obtained, where

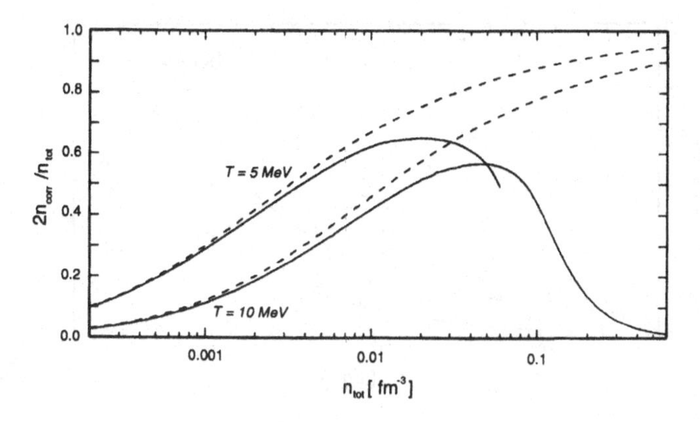

Fig. 3. Correlated pair abundancies showing dependence on total density $n_{\rm tot}$ using the classical (dashed) and generalized (bold) Beth–Uhlenbeck formula at two temperatures.

$$
\begin{aligned}
n_{\rm free} &= 4\sum_k f(\varepsilon_k),\\
n_{\rm deut} &= 3\sum_{K>K_{\rm mott}} g(E_{\rm cont}+E_b(K,\mu,T)),\\
n_{\rm scat} &= -3\sum_{K>K_{\rm mott}} g(E_{\rm cont}) \\
&\quad -3\sum_K\int_0^\infty \frac{dE}{\pi}\left[\frac{d}{dE}g(E_{\rm cont}+E)\right]\sum_\alpha\left(\delta_\alpha-\frac{1}{2}\sin(2\delta_\alpha)\right),
\end{aligned}
\qquad (15)
$$

with a degeneracy factor 3 for the triplet and singlet channels, f is the Fermi, g is the Bose distribution function, $E_{\rm cont}=K^2/4m+2\Delta\varepsilon(K,k=0)$ is the continuum edge and $\delta_\alpha(E,K,\mu,T)$ are the generalized scattering phase shifts in the $^3S_1$ and $^1S_0$ channels. Binding energies E_b are calculated according to (14). Here the quasiparticle shifts are considered only in a rigid shift approximation (effective mass $m^*=m$), which represents the same level of approximation as in Ref. [2] or [3].

As expected from the discussion of the binding energy the Mott effect already leads, for lower temperatures, to a strong suppression of the correlated pair density at low densities. Fig. 3 compares this with a classical Beth–Uhlenbeck calculation (nucleons and deuterons considered as ideal Boltzmann particles), where the contribution of correlated density is monotonically increasing with total density. Because the Green's

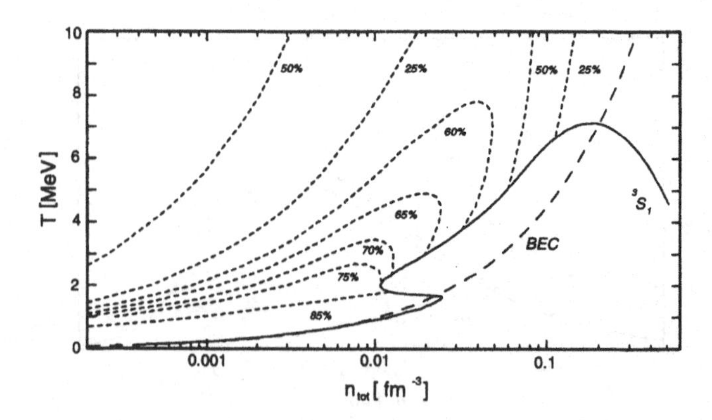

Fig. 4. Temperature–density plane of nuclear matter showing lines of equal concentration of correlated pairs $2n_{\mathrm{corr}}/n_{\mathrm{tot}}$ (dashed). For comparison the critical temperatures in the ${}^3S_1$ channel (bold) and the Bose–Einstein condensation curve (long dashed) are plotted.

function treatment was performed in the normal state, the composition for $T = 5$ MeV is shown only for densities $n_{\mathrm{tot}} < 0.06\ \mathrm{fm}^{-3}$.

In Fig. 4 lines of equal concentration of correlated pair density are plotted in the temperature–density plane outside the superfluid region according to (15). Due to the break up of deuterons (Mott effect) and medium-dependent scattering, the isolines of composition have a maximum. The critical temperature in the ${}^3S_1$ channel from (11) is plotted versus total density (compare with the upper curve in Fig. 2). It shows clearly the transition from the Bose–Einstein condensation of ideal bosons (strong-coupling limit) in the low-density, low-temperature region to the BCS pairing (weak-coupling limit) in a fermionic medium at higher densities. The negative slope of the critical line in a small range of densities around 0.02 fm^{-3} reflects the lack of a unique solution for $\mu(n_{\mathrm{tot}})$. In that region bound states disappear (the Mott effect), which may lead to a phase instability.

5 Conclusion

It has been shown that the solutions of both the Gorkov equation and the T-matrix equation coincide at the critical temperature for the onset of superfluidity. In the low-density limit the latter is identical with the Schrödinger equation for bound states so that we obtain the correct description in the low-density limit (BEC) as well as in the high-density limit (BCS). However, the Gorkov gap equation is not correct in the entire temperature–density plane because it uses a HF type decoupling procedure for the higher Green functions, equivalent to the simple ladder summation with account taken of HF single particle Green functions in the T-matrix approach. This approximation means all correlations in the medium are neglected. However, in a system such as nuclear matter, we have strongly interacting fermions which are correlated. In particular, they can form bound states (deuterons). Therefore, a HF type approach is not always justified.

Thus, we use a generalized Beth–Uhlenbeck formula as an equation of state for the normal phase to calculate the composition of the system. It has been found that bound states are of major importance in the region of low temperatures and low densities. With increasing density, bound states disappear due to the Pauli blocking (Mott effect) so that at high densities compared with the Mott density, the approximation of uncorrelated quasiparticles is possible.

To consider bound states in the medium, the Gorkov equation needs to be improved. A possible approach with Green functions could be a cluster Hartree–Fock calculation as performed in [14]. The correlated medium enters via effective phase-space occupation and cluster-Hartree–Fock shifts. In the region considered, a liquid–gas phase transition occurs, as shown in [14], which is important for the investigation of phase instabilities. For more realistic calculations of nuclear matter, the Yamaguchi potential has to be replaced by a more sophisticated nucleon–nucleon interaction. Further investigations should also include larger clusters such as tritons and α-particles.

The authors thank T. Alm for his valuable suggestions and support of this work.

References

[1] J. Haidenbauer and W. Plessas, Phys. Rev. C **30**, 1822 (1984).
[2] P. Nozières and S. Schmitt-Rink, J. Low Temp. Phys. **59**, 195 (1985).
[3] M. Schmidt, G. Röpke, and H. Schulz, Ann. Phys. **202**, 57 (1990).
[4] R.K. Su, S.D. Yang, and T.T.S. Kuo, Phys. Rev. C **35**, 1539 (1987).
[5] T. Alm, GSI-93-07 Report (dissertation), (1993).
[6] T. Alm, B. Friman, G. Röpke, and H. Schulz, Nucl. Phys. A **551**, 45 (1993).
[7] Y. Yamaguchi, Phys. Rev. **95**, 1628 (1954).
[8] R. Tamagaki, Prog. Theor. Phys. **44**, 905 (1970).

[9] M. Baldo, J. Cugnon, A. Lejeune, and U. Lombardo, Nucl. Phys. A **515**, 409 (1990).
[10] M. Baldo, I. Bombaci, and U. Lombardo, Phys. Lett. B **283**, 8 (1992).
[11] M. Randeria, J.M. Duan, and L.Y. Shieh, Phys. Rev. Lett. **62**, 981 (1989).
[12] M. Randeria, this volume.
[13] R.F. Bishop, H.B. Ghassib, and M.R. Strayer, Phys. Rev. A **13**, 1570 (1976).
[14] G. Röpke, M. Schmidt, L. Münchow, and H. Schulz, Nucl. Phys. A **379**, 536 (1982) and A **399**, 587 (1983).

Appendix. BEC 93 Participant List

International Workshop on Bose–Einstein Condensation
(Levico Terme, 31 May – 4 June, 1993)

Claudio Albanese, ETH, Zurich, Switzerland
A.S. Alexandrov, University of Cambridge, UK
Gordon Baym, University of Illinois at Urbana-Champaign, USA
Aleksandar Belić, ICTP, Italy
David Brink, Università di Trento, Italy
G.E. Brown, SUNY at Stony Brook, USA
Christoph Bruder, Universität Karlsruhe, Germany
Herbert Capellmann, RWTH Aachen, Germany
Yvan Castin, Laboratoire de Physique de l'ENS, France
Eddie Cheng, Université Paris-Sud, Orsay, France
Maria Luisa Chiofalo, Scuola Normale Superiore, Pisa, Italy
Eric Cornell, University of Colorado, USA
Franco Dalfovo, Università di Trento, Italy
Nguyen Dinh Dang, Università di Catania, Italy
E. Fortin, University of Ottawa, Canada
Frank Y. Fradin, Argonne National Laboratory, USA
Richard Friedberg, Columbia University, USA
Stefano Giorgini, Università di Trento, Italy
S. Giovanazzi, Università di Trento, Italy
Tom Greytak, MIT, USA
Allan Griffin, University of Toronto, Canada
Mashahiro Hasuo, University of Tokyo, Japan
H. Haug, Universität Frankfurt am Main, Germany
Anke Hellmich, Universität Rostock, Germany
Kristian Helmerson, NIST (Gaithersburg), USA
T.W. Hijmans, University of Amsterdam, The Netherlands
Kerson Huang, MIT, USA
F. Iachello, Yale University, USA
A. Ivanov, Universität Frankfurt am Main, Germany
Simo Jaakkola, University of Turku, Finland
Y.M. Kagan, Kurchatov Institute, Russia
G. Kavoulakis, University of Illinois at Urbana-Champaign, USA
L.V. Keldysh, Lebedev Physical Institute, Russia

Franck Laloë, Laboratoire de Physique de l'ENS, France
Andrea Lastri, Università di Trento, Italy
Anthony Leggett, University of Illinois at Urbana-Champaign, USA
R. Leonardi, Università di Trento, Italy
Jia Ling Lin, University of Illinois at Urbana-Champaign, USA
Bennett Link, Los Alamos National Laboratory USA
Akira Matsubara, Kyoto University, Japan
S.A. Moskalenko, Academy of Sciences of Moldova, Moldova
A. Mysyrowicz, Ecole Polytechnique, France
Nobukata Nagasawa, University of Tokyo, Japan
Fabio Pistolesi, Scuola Normale Superiore, Pisa, Italy
Ludovic Pricavpenko, Universite Paris-Sud, Orsay, France
Mohit Randeria, Argonne National Laboratory, USA
Julius Ranninger, CNRS-CRTBT, Grenoble, France
John D. Reppy, Cornell University, USA
Meritt Reynolds, University of Amsterdam, The Netherlands
Mannque Rho, Saclay, France
Gerd Röpke, Universität Rostock, Germany
G.V. Shlyapnikov, Kurchatov Institute, Russia
L.D. Shvartsman, The Hebrew University of Jerusalem, Israel
Yu.M. Shvera, Academy of Sciences of Moldova, Moldova
Isaac Silvera, Harvard University, USA
David Snoke, The Aerospace Corporation, Los Angeles, USA
Paul E. Sokol, Penn. State University, USA
Holger Stein, Universität Rostock, Germany
H.T.C. Stoof, Eindhoven University of Technology, The Netherlands
Sandro Stringari, Università di Trento, Italy
Eric Svensson, AECL Research, Canada
E. Tiesinga, Eindhoven University of Technology, The Netherlands
Sergei Tikhodeev, General Physics Institute, Russia
Jacques Treiner, Université Paris-Sud, Orsay, France
D. van der Marel, University of Groningen, The Netherlands
Silvio A. Vitiello, Università degli Studi di Milano, Italy
Jook T.M. Walraven, University of Amsterdam, The Netherlands
Tilo Wettig, S.U.N.Y. at Stony Brook, USA
Jim Wolfe, University of Illinois at Urbana-Champaign, USA

Index

abnormal distribution function, 514–515
AlSb, 534
Anderson localization, 26
Anderson mode, 380, 388
anomalous Green's functions, 264, 407
anomalous self energy, 266, 271
antivortices, 35
atomic cesium, 166, 175, 177, 179, 188, 190, 195, 226, 227, 465–470
  Cs–Cs interaction, 467
  hyperfine splitting, 466–470
  s-wave scattering length, 469
atomic fountain, 226
attractive Fermi gas, 359–367, 373–376, 569

BCS theory, 3, 5, 24, 126, 243, 249, 261, 334, 355–388, 394–395, 409, 422, 453, 585, 586
  and BEC, *see* BEC to BCS crossover
BEC
  compressibility, 40, 89
  deformed, 430, 434
  direct evidence for ("smoking gun"), 4, 6, 52, 53, 68, 81, 320, 444
  ferromagnetism analogy, 127, 229–231, 234, 452–455, 461
  in a potential minimum, 6, 15, 147–153, 168, 203, 223–224, 311, 320–325, 472–474
  in boson–fermion mixture, 396, 401–407, 434–436
  in one dimension, 87–90
  in random potential, 10, 26, 42–48, 93
  in two dimensions, 15, 34, 46, 87–90, 163, 203, 270, 411
  infrared divergence, 55, 61, 62, 64, 76, 88–90, 117–118, 127, 270, 294
  of atomic gas, 100, 188–190, 197, 227, 465–470
  of biexcitons, *see* biexcitons, condensation of
  of bipolarons, *see* bipolarons, condensation of
  of deuterons, 584
  of excitons, *see* excitons, condensation of
  of finite systems, 9, 418–422, 436, 574
  of ^{4}He, *see* ^{4}He, superfluid
  of hydrogen, *see* spin-polarized hydrogen, condensation of
  of ideal gas, 15, 17, 31, 39, 41, 53–55, 89, 100, 101, 117, 168, 203–204, 234–236, 294, 298, 487
  of kaons, 3, 438–450
  of nucleons, 429–432
  of pairs, 122–126, 335, 363–367, 507
  of pions, 3
  of polaritons, *see* polaritons, condensation of
  of positronium, *see* positronium, condensation of
  of quarks, 3, 574
  of system with finite lifetime, 7, 202, 228, 241, 298, 336, 524–530
  of weakly interacting gas, *see* weakly interacting Bose gas
  order parameter, 33, 34, 205, 232–236, 241, 453, 455, 568
  sum rules, 86–97, 543, 553, 556
  timescale for onset of, 7, 9, 149, 202–224, 226–243, 298, 310, 336, 497, 527
  to BCS crossover, 5, 125–126, 334, 355–388, 569–572, 574–582, 585, 589
  to laser crossover, 9, 197, 255–257
Belyaev method, 103
Beth–Uhlenbeck formula, 99, 116, 365, 590, 591, 593
biexcitons, 7, 202, 247, 249–252, 262–277, 287, 326, 332–334, 358, 487, 496–505, 507–511

biexciton–biexciton interaction, 335–336, 504
binding energy, 333
condensation of, 265, 337, 340–346, 487–496, 505, 507, 510
creation, 333–334, 345
Lamb shift, 497–504
luminescence, 337–344, 488, 511
momentum distribution, 338
s-wave scattering length, 249, 270, 494
single-photon absorption, 333
two-photon absorption, 265, 333–334, 339, 489, 496, 508, 511
two-photon recombination, 342
biholes, 532–539
BiI_3, 533
binary collision approximation, 101
bipolarons, 22, 358, 395–414, 541–547
condensation of, 395, 397, 400, 407–413, 542–547
bismuth films, 28
bismuth oxide superconductors, 397
black holes, 444–450
Bogoliubov approximation, 21, 44, 101, 335, 379
Bogoliubov dispersion law, 92, 209, 220, 243, 374, 544
Bogoliubov inequality, 87, 89, 91
Bogoliubov sum rule, 87
Bogoliubov transformation, 40, 466, 510, 576
Bohr solution of hydrogen atom, 282, 331–332
Boltzmann equation, 202, 204, 206–208, 227, 228, 236, 239, 277, 544, 575
Bose gas, *see* weakly interacting Bose gas, hard sphere Bose gas
Bose glass, 26, 27, 43, 46
Bose liquid, 18, 22, 58, 160, 265, 358, 551–557
condensate fraction, 57
driven, 264
hydrodynamic equations, 221
phonons, 89, 552, 554, 555
Bose narrowing, 6, 293
Bose–Einstein distribution, 204, 292, 293, 302, 305, 314, 340, 347, 488, 524
bosonization, 9, 418–436, 570, 571
broken gauge symmetry, 1–3, 7, 21, 32–35, 52, 93, 234, 236, 241, 368, 372, 376, 377, 387, 439, 452–462, 567
BSCO, 397, 412, 547

cesium, *see* atomic cesium
charge density wave, 252, 361
charge transfer gap, 397–399
charged bosons, 377, 380–381, 395, 411, 532, 544–547
chiral symmetry, 3, 439
coherence length, 18, 357, 371, 395, 546
coherent pairing, 507, 510–511
coherent potential approximation, 367
collisions, 204, 208, 214, 215, 228
atoms, 103, 190, 465–470
biexcitons, 249
excitons, 295, 296, 300, 308, 312, 318, 513
^4He, 66
positronium, 561–564
spin-polarized hydrogen, 134–136, 139–140, 143, 145, 147, 148, 161, 227, 228
composite bosons, 1, 4, 24–25, 247, 253, 283, 355, 356, 358, 371, 387, 423–425, 465, 532
condensate
amplitude fluctuations, 20, 205, 210, 243
in momentum distribution, 55, 58, 61, 68, 100, 206, 220, 234, 294, 302
phase coherence, 21, 22, 25, 32–35
phase fluctuations, 8, 34, 35, 46, 203, 205, 217, 220, 243
wavefunction, 32, 33, 205, 234, 456
Cooper pairs, 5, 249, 258, 334, 355–358, 370, 373, 375, 387, 388, 394, 395, 421, 453, 455–458, 507, 532, 572, 585
in disordered alloys, 27
correlation length, 163, 205
CP violation, 226
critical exponents, 4, 46, 48, 79
Cu_2O, 6, 21, 253, 281–326, 346–353, 439, 442, 519–523, 525, 526
band structure constants, 283
exciton lifetime, 286, 288, 346–347
excitonic relaxation processes, 300
optical phonons, 286
CuCl, 7, 336–346, 487–495, 504, 508, 511
cuprate superconductors, 358, 382, 397, 414, 542, 547
CuO_2 planes, 397–401, 413, 541

DeBroglie wavelength, 53, 105, 137, 145, 190, 198, 235, 293, 466
dilute Bose gas, *see* weakly interacting Bose gas
Doppler cooling, 176–177, 179

electron–hole droplets, *see* electron–hole liquid
electron–hole exchange, 262, 286, 335, 346
electron–hole insulator, 254
electron–hole liquid, 247, 249–252, 254, 287, 335, 358, 528
electron–hole plasma, 249

electron–phonon coupling, 394–395
electronic phase transition, 252
Eliashberg theory, 395
entropy loss, 312, 442
exchange effects, 17, 103–110, 119–122, 425, 435
excitonic crystal, 249
excitonic gas, 247–255, 263, 281–284, 288, 295–325, 438
excitonic insulator, 249, 250, 257
excitonic molecule, *see* biexcitons
excitons, 7, 10, 18, 21, 24, 25, 82, 100, 202, 246–277, 281–326, 330–353, 355, 426, 439, 442, 487, 519–530
  Auger process, 296, 300, 308, 311, 315, 325
  binding energy (excitonic Rydberg), 248, 282–283, 285, 331
  Bohr radius, 248, 282, 331
  condensation of, 6, 39, 100, 203, 254, 302–305, 307, 312, 316–320, 332, 334–353, 357, 377, 507, 510, 520–523
  creation, 253, 265, 284, 295, 296, 332
  diffusion, 289–291, 296, 320, 332, 349, 520, 528
  dumbbell, 8
  exciton–exciton interaction, 263–264, 269, 296, 297, 312, 320, 326, 335, 501, 507–510, 513–516, 526
  exciton–phonon interaction, 288–291, 298–300, 305, 310, 320, 336, 352, 526, 529
  in a stress field, 291, 313–325
  in two dimensions, 8, 270
  luminescence, 284–325, 347–349
  momentum distribution, 286, 298, 301–320
  ortho, 21, 286–325, 332, 346–349, 507
  para, 21, 286–325, 332, 346, 349, 507, 508, 519, 525
  phonon-assisted spin relaxation, 288, 300, 302, 314
  photovoltaic detection, 349–352, 519
  polarized, 508, 510
  "quantum saturation", 21, 223, 307–312, 314–315, 325, 349
  recombination, 9, 252, 253, 284–289, 298
  s-wave scattering length, 297, 312
  single-photon absorption, 331–332
  superfluidity of, 7, 39, 203, 252–253, 305, 311, 334–335, 349–353, 520–523, 528, 529
  two-body spin relaxation, 21, 300, 308
  virtual, 262

Fermi gas, *see* attractive Fermi gas
Fermi liquid, 247, 249, 259, 358–367, 404
Fermi–Dirac distribution, 254, 261
  of localized bosons, 8, 546
four-wave mixing, 265, 273–274, 496
  phase conjugation, *see* optical phase conjugation
fractional quantum Hall effect, 90
fullerenes, 397, 414, 541

GaAs, 500, 502, 508, 534, 535, 537
Galitsky method, 103
GaP, 534
GaSb, 534
Ge, 250, 253, 293, 323, 525, 526, 528, 534, 538
Gell-Mann–Levy model, 442
Ginzburg–Landau theory, 33–35, 442, 527
  time-dependent, 238, 368–372
Goldstone bosons, 568
gravitational atomic cavity, 194–197
Gross–Pitaevsky equation, *see* nonlinear Schrödinger equation
gun, smoking, *see* BEC, direct evidence for

hard core Bose gas, *see* hard sphere Bose gas
hard sphere Bose gas, 23, 38–42, 44, 56, 64, 100, 102, 117–118, 284, 297, 361, 379, 553, 565
Hartree–Fock approximation, 17–19, 103, 121, 237, 577, 580, 593
healing length, 148, 149
^{3}He, 18, 485
  superfluid, 5, 355, 357, 453, 455, 457, 574
^{4}He, 6, 35–38, 42, 51–82, 92, 160, 162, 168, 248, 254, 286, 355, 478, 550, 551
  gas, 250
  in a porous medium, 10, 27, 43, 46
  in two dimensions, 8
  lambda transition, 42
  momentum distribution, 4, 53, 58–82, 286, 551
    correlated basis function calculation, 60
    Green's function Monte Carlo calculation, 60–62, 64, 74, 76, 77, 80
    path integral Monte Carlo calculation, 60, 62, 70, 73, 77, 79, 81
    shadow wave function calculation, 60
    variational calculation, 60, 61, 64
  phase diagram, 56
  solid, 23, 56, 80
  superfluid, 1, 9, 135, 142, 161, 221, 229, 355, 459
    condensate fraction, 62, 64, 68–81, 101
    maxons, 96
    phonons, 35, 36, 60, 61, 63, 64, 96
    rotons, 63, 96

Higgs boson, 3
Hohenberg–Mermin–Wagner theorem, 87–90
holons, 35
Hubbard model, 409, 413
  attractive, 360–361, 367, 373, 377, 379, 381, 387
  repulsive, 358, 381
hydrogen, *see* spin-polarized hydrogen
  molecules, 142, 332–333, 335
  under high pressure, 359
hyper-Raman scattering, 511

InAs, 534
InP, 534
InSb, 534
interacting boson model, 418, 426, 427
Ising model, 23, 567

Josephson effect, 21
Josephson junction, 454, 455, 458

kaons, 438–450
  kaon–nucleon interaction, 439–440
  kaonic atoms, 439
Kosterlitz–Thouless transition, 4, 8, 18, 35, 38, 163, 203, 368
  of ^{4}He, 8
  of excitons, 8
  of spin-polarized hydrogen, 8, 168–170, 478–485

Lévy flights, 194
lambda transition, *see* ^{4}He, lambda transition
Landau–Ginzburg theory, *see* Ginzburg–Landau theory
laser, 4, 9, 255
  and BEC, *see* BEC to laser crossover
  excitonic, 255–257
  neutrino, 9
laser cooling of atoms, 100, 166, 173–198, 202, 466
  Doppler cooling, 154
  recoil limit, 190
  Sisyphus effect, 179, 182–186, 189
  velocity-selective trapping, 190–194
laser generation of excitons, *see* excitons, creation
linear response theory, 36, 38
lithium, 227
local density approximation, 152
long-range phase coherence, *see* off-diagonal long-range order
LSCO, 397

magneto-optical trap, 166, 177–179, 188–190
magnetoexcitons, 8
maser, 132
Maxwell–Boltzmann distribution, 54, 288, 289, 298, 299, 302, 314, 338, 347, 488
mean-field approximation, 23, 34, 100, 101, 121, 259, 260, 269, 271, 419, 575
Mott transition, 26, 27, 575, 581, 590, 592

neutrinos, 441, 447
neutron scattering, 4, 53, 62–81, 550–557
  final-state effects, 66–68, 71, 73, 74
  impulse approximation, 63, 64, 66, 550–552, 555, 556
neutron stars, 439, 444
nonequilibrium Green's functions
  electron–hole, 257–262
  exciton, 264–269
nonequilibrium phase transition, 255, 514
nonlinear optics, 255, 257, 262, 263, 265, 277
nonlinear Schrödinger equation, 34, 103, 205, 207, 242, 371, 545
nuclear ferromagnetism, 19
nuclear matter, 418, 574, 582, 584–593
  equation of state, 359, 585
nuclear spin waves, 104, 132
nucleons, 3, 418–436
  nucleon–nucleon interaction, 584, 587
nucleosynthesis, 448–449

off-diagonal long-range order, 1, 4, 8, 33, 57, 59, 206, 220–223, 356, 368, 387
one-body density matrix, 32, 57, 60, 94, 100, 104, 109, 122
Onsager–Feynman quantization, 33
optical lattices, 180, 186–187
optical molasses, 176, 179, 185–188, 194
optical phase conjugation, 344–346, 488–495

Pauli exclusion, 253, 254, 292, 379, 387, 425, 579–582, 590
phase locking, *see* condensate, phase coherence
phase transition
  first order, 32, 41, 42
  second order, 34, 41, 79, 229–232, 443
phase-space filling, 262
phonon wind, 528–530
phonons, 220, 222
  in Bose liquid, *see* Bose liquid, phonons
  in crystal lattice, *see* electron–phonon coupling, exciton–phonon interaction
  in superfluid helium, *see* ^{4}He, superfluid, phonons

in weakly interacting Bose gas, *see* weakly interacting Bose gas, phonons
plasmons, 377, 381, 388, 559
polariton bottleneck, 256
polaritons, 255–262, 268–269, 336, 338, 342, 497–504, 513–516
condensation of, 9, 257, 513–516
dispersion law, 255, 339
polarons, 395, 398
positronium, 7, 283, 285, 287, 300, 331
condensation of, 558–564
molecule, 333, 559, 564
momentum distribution, 563
s-wave scattering length, 562
superfluidity of, 564
two-body spin relaxation, 562
two-photon recombination, 563
"precondensate", 211
pseudogap, 407, 408, 410, 413, 414, 543
pseudopotential method, 101

QCD theory, 439
quantum heterostructures, 8, 533
quantum lattice model, 565
quantum Monte Carlo method, 381, 382
quantum tunneling, 454
quantum well, 8, 10, 270, 500, 508, 509, 533
quantum wire, 502
quasicondensate, 8, 163, 168, 203, 205, 208–211, 218, 220, 223, 243, 320, 527
quasiequilibrium, 213, 253, 267, 295, 300, 336
"quasipairs", 126
quasiparticles, 35, 36, 38, 40, 41, 119, 122, 242, 378, 591

Rabi oscillations, 257, 275, 277
radiation pressure, 175
random phase approximation, 204–206, 377, 378, 380, 381, 383

s–d boson model, *see* interacting boson model
s-wave scattering length, 39, 126, 211, 237, 360, 362, 371
atomic cesium, *see* atomic cesium, s-wave scattering length
biexcitons see – biexcitons, s-wave scattering length, 249
excitons, *see* excitons, s-wave scattering length
positronium, *see* positronium, s-wave scattering length
spin-polarized hydrogen, *see* spin-polarized hydrogen, s-wave scattering length
self-consistent field approximation, 266, 515, 516
self-phase modulation, 265
shell model, 419, 426
Si, 250, 253, 293, 323
SiGe, 538
Silvera–Reynolds number, 525
single-particle density matrix, *see* one-body density matrix
smoking gun, *see* BEC, direct evidence for
solitons, 353
spin density wave, 252
spin gaps, 358, 382–387, 414, 542
spin waves, 565–568
spin-polarized hydrogen, 7, 18, 19, 100, 131–157, 160–171, 202, 203, 226–228, 241, 295, 442, 453, 525, 574
compression, 136
condensation of, 39, 100, 131–156, 160–161, 163–167, 227, 228, 231–232, 236, 241
evaporative cooling, 138–147, 154, 163, 165, 227, 236, 475
hyperfine splitting, 133, 161, 227
in two dimensions, 162–163, 168–170, 478–485
magnetic trapping of, 134–157, 161, 227, 472–475
microwave trapping of, 163–167
relaxation explosion, 151, 164, 165, 472–475
s-wave scattering length, 145, 148–149, 227
superfluidity of, 39, 132, 160
three-body recombination, 132, 134, 137, 161, 165, 169, 210–211, 472, 479–484
two-body spin relaxation, 135, 140, 143, 147, 162, 165, 300
spontaneous symmetry breaking, *see* broken gauge symmetry
stimulated emission, 6
stimulated scattering, 6, 273, 292, 344, 352
strangeness condensation, 443
strongly interacting Bose gas, 160
superconductivity, 3, 5, 24, 25, 249, 355, 356, 361, 374, 394, 403, 421, 426, 453, 532, 574
collective excitations, 377–381
high T_c, 5, 35, 357–358, 382, 386–387, 394–414, 462, 541–547, 569
and bipolarons, *see* bipolarons, condensation of
in disordered system, 10, 358
Schafroth, 395, 396
type II, 10
superfluid density, *see* two-fluid model, superfluid density

superfluid gas, 128
superfluidity, 3–5, 8, 21, 23–25, 27, 33, 35–38, 42–48, 51, 59, 87, 101, 127, 160, 163, 206, 207, 217, 220, 223, 355, 356, 387, 407, 453, 455–462, 574–582, 584–593
supernova 1987A, 440, 449, 450
supersolids, 23
symmetry breaking, *see* broken gauge symmetry

thermodynamic limit, 9, 16, 117–118, 228, 235, 243, 452, 457, 526
topological long-range order, 206, 217–220, 222
two-body density matrix, 94–97, 101
two-body spin relaxation
  of excitons, *see* excitons, two-body spin relaxation
  of hydrogen, *see* spin-polarized hydrogen, two-body spin relaxation
  of positronium, *see* positronium, two-body spin relaxation
two-fluid model, 4, 36–38, 221–222
  superfluid density, 41, 43, 45–47, 81, 368
two-photon absorption
  by biexcitons, *see* biexcitons, two-photon absorption
  by hydrogen, 154–157
two-photon recombination
  biexcitons, *see* biexcitons, two-photon recombination
  excitons, 285
  positronium, 285, 559

ultrafast optics, 298
"universal phase standard", 455, 459
Ursell operators, 102–103, 110–128

vortices, 21, 34, 35, 206, 217, 219, 220, 222, 223

weakly interacting Bose gas, 5–7, 82, 91–93, 95, 100–128, 148–149, 160, 163, 168, 203, 204, 226, 284, 291–294, 325, 336, 371, 372, 380, 438, 524, 526
  blue shift, 320, 494
  equation of state, 110–118
  far from equilibrium, 203–224, 227–243, 514
  phonons, 206, 218, 220–223, 388
  speed of sound, 149, 218, 221, 379
weakly interacting Fermi gas, 357, 360
Wigner crystal, 380

YBCO, 382, 386, 387, 397, 399–401, 412, 542–545

ZnP_2, 533
ZnSe, 508
ZnTe, 534

Printed in the United States
By Bookmasters